Advances in Very High Energy Astrophysics

The Science Program of the Third Generation IACTs for Exploring Cosmic Gamma Rays

Advances in
Very High Energy Astrophysics

The Science Program of the Third Generation IACTs for Exploring Cosmic Gamma Rays

editors

Reshmi Mukherjee
Columbia University, USA

Roberta Zanin
Max Planck Institut für Kernphysik, Germany

World Scientific

NEW JERSEY • LONDON • SINGAPORE • BEIJING • SHANGHAI • HONG KONG • TAIPEI • CHENNAI • TOKYO

Published by

World Scientific Publishing Co. Pte. Ltd.
5 Toh Tuck Link, Singapore 596224
USA office: 27 Warren Street, Suite 401-402, Hackensack, NJ 07601
UK office: 57 Shelton Street, Covent Garden, London WC2H 9HE

Library of Congress Control Number: 2023932607

British Library Cataloguing-in-Publication Data
A catalogue record for this book is available from the British Library.

ADVANCES IN VERY HIGH ENERGY ASTROPHYSICS
The Science Program of the Third Generation IACTs for Exploring Cosmic Gamma Rays

ISBN 978-981-327-571-3 (hardcover)
ISBN 978-981-327-572-0 (ebook for institutions)
ISBN 978-981-327-573-7 (ebook for individuals)

For any available supplementary material, please visit
https://www.worldscientific.com/worldscibooks/10.1142/11141#t=suppl

Desk Editor: Carmen Teo Bin Jie

Typeset by Stallion Press
Email: enquiries@stallionpress.com

https://doi.org/10.1142/9789813275720_fmatter

Prologue

The idea behind this book was to compile a set of topics on the advances in very high energy gamma-ray astronomy in the last two decades, in a way that would be pedagogical and educational and could function as a textbook. While there are many review papers available, that are comprehensive and topical, we felt there wasn't really a simple text book that could be used by a student or early-career scholar in the field of high energy gamma-ray astronomy, who is just venturing out into this exciting area of research. Results from the third-generation of imaging atmospheric Cherenkov telescopes have propelled the field in new directions and set the stage for advanced next-generation instrumentation such as the upcoming Cherenkov Telescope Array Observatory (CTAO). With this book, we aim to convey the excitement of the wide variety of science available at energies greater than tens of GeV to a new generation of researchers. We hope this book would be appropriate at the senior undergraduate level, or as a textbook for graduate students.

We are grateful to the authors of the chapters for their patience during the entire editing process. Despite our best efforts, the book faced delays due to the impact of the Covid-19 pandemic. We thank the authors for keeping the material appropriately updated. We appreciate the support from the World Scientific editors for their constant help. We acknowledge the help of Julia Coccaro, who as a student at Barnard College, worked painstakingly as a copyeditor for the book. Finally, we would like to express our gratitude to Prof. Felix Aharonian, who has encouraged us, as well as inspired us in this adventure.

We are fortunate to have had the chance to work with the excellent set of authors who have contributed to this volume. Any errors or omissions are entirely the responsibility of the two editors. We hope you enjoy this book.

Prof. Reshmi Mukherjee
Barnard College, Columbia University
New York, USA

Dr. Roberta Zanin
CTA Observatory
Bologna, Italy

https://doi.org/10.1142/9789813275720_fmatter

Contents

Part 1

Advances in Very-High-Energy Astrophysics

https://doi.org/10.1142/9789813275720_0001

Chapter 1

Introduction to Gamma-Ray Astronomy

David Hanna*,‡ and Reshmi Mukherjee†,§

*McGill University
Montreal, Quebec, Canada
†Barnard College, Columbia University
New York, USA
‡hanna@physics.mcgill.ca
§rm34@columbia.edu

1.1. Introduction

Gamma-ray astronomy concerns photons that are found at the short-wavelength end of the electromagnetic spectrum. As is well known, astronomy began in ancient times and was limited until comparatively recently to observations made in the visible spectrum, albeit with increasing sensitivity and precision. Around the middle of the last century advanced technologies made it possible to use other wavelengths to open new windows into the Universe. The range of wavelengths brought into service, infrared and radio on one side of the visible region and ultraviolet and X-ray on the other, led to important discoveries of new astrophysical objects and processes, as well as improved understanding of known phenomena. The latest energy band to be explored is that of gamma rays, the topic of this chapter.

Gamma rays are more energetic than X-rays; the line dividing the two regions is fuzzy, but 1 MeV is a not-unreasonable choice for the boundary. There is general agreement that there is a division at about 10 GeV between High-Energy (HE) and Very-High-Energy (VHE) gamma rays, but further divisions at even higher energies are less codified. The HE/VHE divide is where ground-based detectors take over from space-borne telescopes.

Astronomy with gamma rays provides a way to examine the HE non-thermal Universe. In addition to discovering and characterizing emitters such as pulsars, supernova remnants and active galactic nuclei (AGNs), and providing observations needed for understanding their emission mechanisms, gamma-ray astronomy can be used for investigations into more exotic phenomena. Examples of this include gamma-ray production from annihilation of dark-matter particles (WIMPs) in dwarf galaxies, and final-stage evaporation radiation from primordial black holes (PBHs). Measurements of photon arrival times as a function of energy from distant AGNs can be used to look for the breakdown of Lorentz invariance. VHE gamma rays traversing from large distances through intergalactic space indirectly probe diffuse radiation fields. Gamma-ray bursts (GRBs), less of a mystery now than when first discovered, continue to be of interest as records in energy and brightness are broken from time to time.

The original and still-relevant motivation for gamma-ray astronomy is the search for the origin of cosmic rays. Cosmic rays are mostly protons and helium nuclei, together with trace amounts of heavier elements, that bombard the Earth from all directions and with energies that extend up to 10^{20} eV following a power-law spectrum with an index of about -2.7. Collectively, they contribute a significant fraction of the energy density of the interstellar medium (ISM) but, just over a century after their discovery, their origin remains a mystery. This is because, being electrically charged, their trajectories are deviated by Galactic and intergalactic magnetic fields so they do not "point back" to their birthplaces. Gamma rays are unaffected by magnetic fields and could be used as tracers since it seems reasonable that they should be produced in the same HE processes and places as the cosmic rays. For energies up to a PeV (10^{15} eV) or so — or what is known as the "knee" in the cosmic-ray spectrum — there is strong circumstantial evidence that cosmic rays are produced in supernova explosions where shock acceleration can occur. At higher energies, however, things are not so clear. The term "PeVatron" is being increasingly used to describe cosmic-ray accelerators which can accelerate particles to PeV energies.

Gamma-ray astronomy deals with HE photons, some of which could come from the decay of neutral π-mesons (pions) produced by protons that have been accelerated to HE. It is natural to expect production of charged pions in the same processes and these would result in a comparable flux of neutrinos that could be detected by sufficiently sensitive detectors. The IceCube collaboration has recently presented evidence of such a flux (Aartsen *et al.*, 2020a). Additionally, detections of gravitational waves have been reported by Ligo, Virgo and KAGRA collaborations (Abbott *et al.*, 2016c). Although the link to gamma rays is less certain, these detections motivate follow-up observations by gamma-ray telescopes. The idea of combining data from non-electromagnetic probes with more traditional measurements is exciting; it appears that we are entering an era of multi-messenger astronomy.

1.2. Space-Borne Detectors

In the following section, we will outline some of the historical developments made in the field with a focus on the detector technologies used in the different energy ranges.

1.2.1. *Technologies*

X-ray and gamma-ray astronomy are recent arrivals as subdisciplines due to the fact that the atmosphere is opaque to such radiation; it was necessary to wait for the advent of the space age in order to make use of orbiting detectors. Moreover, HE photons are not amenable to the optics traditionally used in astronomy so new instrumentation, with hard performance limits imposed by the physics of detection, needed to be developed.

For example, soft X-rays can be focused using Wolter-type grazing-incidence mirrors, such as those found in the Chandra X-ray telescope (Schwartz, 2004), but to image hard X-rays or soft gamma rays requires a detector based on coded-aperture masks, as used in the burst alert telescope (BAT) on the Neil Gehrels Swift observatory (Burrows *et al.*, 2005).

For photons with energies in the MeV range, useful for nuclear astrophysics studies, the instrument of choice is a telescope based on the phenomenon of Compton scattering. COMPTEL onboard the Compton Gamma-Ray Observatory (CGRO, NASA 1991–2000) is the only large-scale telescope of this type deployed so far and has limited precision and sensitivity, thus this energy range remains relatively under explored (Schoenfelder *et al.*, 1993).

At energies above a few MeV, e^+e^- pair production is the dominant interaction for photons with matter, so electron-pair telescopes have been used in almost all missions covering this energy range. The basic design incorporates several layers of heavy metal plates interleaved with tracking detectors, followed by a calorimeter; this assembly is surrounded by a plastic-scintillator anti-coincidence shield to veto charged particles. Gamma rays convert to pairs in the metal plates; the kinematics of the reaction ensures that the directions of the two outgoing particles are very close to that of the incident photon. Their directions are measured in the tracker, which was made from spark chambers in earlier telescopes although now silicon strips are the preferred solution, and their energies are determined using the calorimeter, usually made from a heavy scintillator like NaI(Tl) or CsI(Tl) with good energy resolution and stopping power.

The upper limit in energy reach for space-borne telescopes is set primarily by the finite size of the instrument and the steeply falling energy spectrum of the sources of interest, but in some cases may be due to leakage from the calorimeter. In a calorimeter, the incident particle generates a shower of secondary particles whose energies are converted to a measurable signal, usually scintillation light. The thickness of the calorimeter needs to increase, typically logarithmically, to deal with increasing incident energy, so there is a limit to what can be reliably measured once the thickness has been set.

1.2.2. *History*

The first space-borne gamma-ray astronomy detector was a scintillator-based instrument aboard the OSO-3 satellite (Kraushaar *et al.*, 1973) launched in March 1967. It had an energy threshold of 50 MeV and limited directional capability, but was able to detect 621 photons during its 16-month lifetime. The first pair-telescopes were SAS-2 (NASA 1972-73) (Fichtel *et al.*, 1975) and COS-B (ESA 1975-82) (Bignami *et al.*, 1975), both using spark-chamber technology. Only COS-B had a dedicated calorimeter; SAS-2 used the degree of multiple-scattering of the electron tracks to provide an energy estimate. They provided a sky survey and COS-B was able to produce a catalog of 25 sources (Swanenburg *et al.*, 1981).

These pioneering instruments motivated the building and deployment of the CGRO, one of NASA's "great observatories" (Gehrels and Shrader, 1996).[a] CGRO was deployed from the space shuttle Atlantis in 1991 and was de-orbited in 2000 following the failure of a stabilizing gyroscope. The spacecraft carried four instruments: COMPTEL (Schoenfelder, 1991),

[a]The program also included the Hubble Space Telescope, the Chandra X-ray Telescope, and the Spitzer Infrared Telescope.

a Compton-effect telescope mentioned earlier, the Burst and Transient Source Experiment (BATSE) (Brock *et al.*, 1992), the Oriented Scintillation Spectrometry Experiment (OSSE) (Johnson *et al.*, 1993), and the Energetic Gamma-Ray Experiment Telescope (EGRET) (Kanbach *et al.*, 1989).

EGRET was a pair-telescope with an effective area of 0.15 m^2 and an energy range from 30 MeV to 30 GeV. It comprised a stack of spark chambers and a calorimeter made from sodium iodide crystals. Observations made with EGRET resulted in a source catalog with 271 entries, including five pulsars and multiple extra-Galactic sources of the AGN type. It showed that the GeV sky was well populated and provided the impetus for further investigations at these energies.

1.2.3. *Currently operating instruments*

Soon after the success of EGRET had been established, a proposal for a similar but more powerful telescope was developed for NASA based on silicon-strip detectors that had been developed for experiments in particle physics. Among the many advantages offered by this technology was the elimination of the need for spark-chamber gas, a key factor limiting mission lifetimes. Other features included a wide field of view (2.4 sr — 20% of the sky) which ensured that the telescope would see the entire sky every three orbits while observing in its default $\pm$ 50° rocking mode. The device was launched in 2008, together with a companion instrument for detecting gamma-ray bursts. On reaching orbit, the combination was given the name Fermi Gamma-Ray Space Telescope (Fermi) with the two subsystems called the Large Area Telescope (LAT) and the Gamma-Ray Burst Monitor (GBM) (Michelson *et al.*, 2010).

Fermi-LAT continues to operate today and is regarded as an unqualified success, having detected and measured over 5000 sources such as pulsars, binary systems, star-forming regions, globular clusters, galaxies, starburst galaxies, supernova remnants, pulsar wind nebulae and novae (Abdollahi *et al.*, 2020). Approximately 3100 of the sources are AGNs of the blazar class, a significant increase from the numbers detected by *CGRO*-EGRET.

A smaller device using very similar technology, Astro-rivelatore Gamma a Immagini LEggero (AGILE), was launched a year before Fermi by the Italian Space Agency in 2007 (Barbiellini *et al.*, 2001). The main detector components of AGILE are similar to those of *Fermi*-LAT, consisting of the following: a tracker made of silicon for detecting tracks of gamma-ray interactions in the material; an X-ray detector called the Super-AGILE;

an anti-coincidence system for eliminating the charged-particle background; and a calorimeter made of CsI(Tl). In 2019, the AGILE Collaboration published the second AGILE catalog of gamma-ray sources (2AGL Catalog) detected by the AGILE gamma-ray imaging detector (GRID) in the energy range 100 MeV to 10 GeV. This catalog consists of data from the first 2.3 years of operations between 2007 and 2009 taken in "pointing mode", and includes 175 high-confidence sources (Bulgarelli *et al.*, 2019). The advantage of having two gamma-ray satellites in orbit at the same time is the synergy and complementarity of the missions, allowing observations of the gamma-ray sky in survey mode in addition to pointed observations.

1.3. Gamma-ray Astronomy with Cherenkov Telescopes

At energies greater than about 10 GeV, space-borne telescopes are not large enough to cope with the steeply falling energy spectra of astrophysical sources. Fortunately, it is at about this energy that ground-based detectors are able to take over, using the atmosphere itself as part of the detector.

Incident gamma rays interact near the top of the atmosphere, giving rise to extensive air showers made up of electrons, positrons and lower-energy gammas. These highly collimated cascades of particles propagate through the atmosphere following the trajectory of the original gamma rays such that if they are suitably measured, the energies and directions of those gamma rays can be reconstructed and astronomy is enabled. For energies in the 10 GeV to 100 TeV range, the particles in the shower do not penetrate to ground-level detectors, but the Cherenkov light generated by their passage through the air does, and is detected by Imaging Atmospheric Cherenkov Telescopes (IACTs). At higher energies, the shower particles can reach arrays of counters on the ground and be directly measured; this will be covered in the next section.

In both cases, the "footprint" of the Cherenkov light pool or the spread of shower particles is large ($>10^5$ m^2), and it is this that constitutes the effective area of the detector, since one need only sample a small fraction of shower products to make a measurement.

Unfortunately, charged cosmic rays — far more abundant than astrophysical gamma rays (typically by a factor of 10^4) — also generate air showers, a serious source of background. The showers are different in particle content and morphology; initiated by nucleons, they are hadronic showers while gamma-initiated showers are electromagnetic. Background rejection techniques exploit the differences.

1.3.1. History

The Cherenkov effect was discovered experimentally in the early 1930s and was understood theoretically shortly thereafter. If a particle passes through a medium with refractive index n, with speed greater than c/n, it will give rise to the emission of photons, mostly blue and near-ultra-violet, in a cone about its direction of motion. This happens with the particles in an air shower and, since the refractive index of air is so close to 1.0, the Cherenkov photons and the particles that generate them travel at almost the same speeds. This results in all of the light arriving at a detector in a pulse with a rise-time of order one nanosecond. The short pulse is momentarily much brighter than the night-sky background of light from stars and can easily be detected using a fast photo-multiplier tube (PMT). A verification of this scenario was made in 1953 (Galbraith and Jelley, 1953), but its successful use in gamma-ray astronomy was not achieved until 1989 with the detection of a signal from the Crab Nebula.

The years preceding the Crab detection were a time of heroic but unsuccessful efforts to achieve the necessary sensitivity, involving a variety of techniques to suppress the background from cosmic rays. The solution to the problem turned out to be imaging. It was shown that if one installed a "camera" comprising a cluster of PMTs in the focal plane of a large reflector, Cherenkov light from air showers would form images that could be recorded using the PMT pixels. Images from electromagnetic showers generated by gamma rays were smooth ovals, but those from hadronic showers were irregular, and a moments analysis could be used to distinguish them (Hillas, 1985a). This technique enabled the Whipple collaboration, working with a 10-m reflector and a 37-pixel camera (Fig. 1.1) to publish a 9-sigma signal from the Crab Nebula in 1989 (Weekes *et al.*, 1989). The energy threshold was approximately 1 TeV.

It was subsequently shown, during the following decade, that sensitivity could be improved with the use of an array of telescopes; the HEGRA array of five 8.5 m^2 reflectors, each with a 271-pixel camera, achieved the same sensitivity as the much larger Whipple telescope (Daum *et al.*, 1997). A fine-grained camera could also help; comparable performance was demonstrated with the single CAT telescope, which had a mirror area of about 18 m^2 and a camera with 546 PMTs (Barrau *et al.*, 1998). A key advantage of using an array of telescopes is the suppression of background triggers due to muons. Muons result from the decay of charged pions in hadronic showers and, having a long lifetime and limited energy loss, penetrate to ground level.

Fig. 1.1. The 10-m imaging atmospheric Cherenkov telescope of the Whipple Observatory that pioneered the field of TeV astronomy (Catanese and Weekes, 1999).

Their associated Cherenkov light can trigger an IACT if its threshold is low enough. Requiring simultaneous triggers from two or more widely spaced telescopes eliminates this background.

During this time, the wave-front sampling technique was also explored. Using arrays of mirrors, several groups were able to detect the Crab Nebula by fitting the arrival times of the Cherenkov light to a model of the shower.

The ASGAT (Goret *et al.*, 1993) collaboration built a detector based on seven 7-m mirrors, each with a cluster of seven PMTs at its focus, and using infrastructure at the Thémis solar power plant in the French Pyrenees. They obtained a 5.7σ signal at energies above 600 GeV. The THEMISTOCLE (Baillon *et al.*, 1994) collaboration, working at the same site with a detector based on 18 0.8-m mirrors, each with a single PMT, obtained a 5.8σ signal but with a threshold of 3 TeV.

In an attempt to lower the threshold energy by increasing the total mirror area, several groups built detectors that utilized many large heliostats in a central-tower solar-power installation (Smith, 2006).

STACEE (Hanna *et al.*, 2002) in New Mexico and CELESTE (Paré *et al.*, 2002) at Thémis were the most successful and achieved thresholds below 100 GeV on the Crab Nebula.

The wave-front sampling technique was shown to work, but it lacked the sensitivity of the imaging technique and was ultimately abandoned.

1.3.2. *Current detectors*

The success of the imaging technique in ground-based astronomy and the growing catalog of HE sources emerging from the EGRET telescope encouraged the VHE community to plan for arrays of IACTs incorporating mirrors with diameters of order 10 m and cameras with order 10^3 pixels. These numbers were driven by the requirements of increased sensitivity and lower energy thresholds. The proposals resulted in the construction of the present generation of instruments: The High Energy Stereoscopic System (H.E.S.S), major atmospheric gamma-ray imaging Cherenkov (MAGIC), and very-energetic radiation imaging telescope array system (VERITAS). A fourth array of this class, CANGAROO-III (Enomoto *et al.*, 2002), was completed in 2004 in Woomera, Australia, by a Japanese–Australian collaboration, but it is no longer being operated.

1.3.2.1. *High-energy spectroscopic system*

The H.E.S.S is an array of five Cherenkov telescopes located at an altitude of 100 m in the Khomas highlands of Namibia. Four of the five are identical and are installed at the corners of a 120-m square. Each has a 12-m tessellated mirror with an area of 107m^2 and a 13-m focal length, directing light onto a 960-pixel camera with a field of view (FoV) diameter of 5°. The array has an energy threshold of 0.1 TeV and can make a 5σ detection of a source with 0.7% of the Crab Nebula flux with 50 h of data (Aharonian *et al.*, 2006d).

This array began operations in 2004 and has been the most prolific of the present generation of IACTs. In addition to the excellence of the instrument and the collaboration, this success is due to its location in the Southern Hemisphere where sources near the Galactic center transit close to Zenith.

In 2012, an additional larger telescope called H.E.S.S. II was added at the center of the array to enable observations at lower energies (Bolmont *et al.*, 2014). Its mirror is parabolic and is 32.6 m $\times$ 24.3 m (equivalent diameter of 28 m) with a focal length of 36 m; its effective area is 614 m^2.

Fig. 1.2. A photograph of the H.E.S.S. telescopes currently operating in Namibia (Courtesy: The H.E.S.S. Collaboration).

The associated camera has 2048 pixels and a 3.2° FoV. Figure 1.2 shows a photograph of the H.E.S.S. telescopes.

1.3.2.2. *MAGIC*

Major Atmospheric Gamma-Ray Imaging Cherenkov (MAGIC) telescopes[b] are a pair of IACTs located 85 m apart on La Palma in the Canary Islands at an altitude of 2225 m (Fig. 1.3). Each telescope has a parabolic mirror with diameter of 17 m and effective area of 236 m^2 and a camera with 576 pixels and 3.5° FoV (Lorenz and Martinez, 2005).

The first telescope began observations in 2004 and two-telescope observations have been possible since 2009. The addition of the second telescope has eliminated much of the muon background and the stereo views improve the energy and angular resolution of the array. The energy

[b]They are now officially called the MAGIC Florian Goebel Telescopes to honor the project manager of MAGIC II who died in an accident while its construction was being completed.

Fig. 1.3. A view of the MAGIC telescopes located in the Canary Islands (Courtesy: The MAGIC Collaboration).

threshold is currently about 30 GeV and the sensitivity is comparable to H.E.S.S. and VERITAS.

One of the design features of MAGIC is its lightweight construction; the mirror support systems are made from carbon-fiber composites, allowing the enormous structures to be skewed to observe any point in the sky in less than 22 s. This allows MAGIC to respond quickly to transient alerts such as those from GRB detections by space-borne instruments.

1.3.2.3. *VERITAS*

The Very-Energetic Radiation Imaging Telescope Array System (VERITAS) is an array of four telescopes located at an altitude 1275 m near Tucson, AZ (Holder *et al.*, 2008). Each telescope has a 12-m tessellated mirror with effective area of 106 m^2 and a 499-pixel camera with 3.5° FoV. For historical reasons, the telescope's locations form an irregular trapezoid with average inter-telescope spacing of about 100 m (Fig. 1.4). Construction of the array was completed in 2007 and an upgrade was made in 2012. During the upgrade, all the camera PMTs were replaced with models having higher quantum efficiency and one of the telescopes was relocated to make

Fig. 1.4. Photograph of the VERITAS telescopes located at the Fred Lawrence Whipple Observatory in Arizona (Courtesy: The VERITAS Collaboration).

a more symmetric array. These changes, and improved trigger electronics, resulted in a lower energy threshold and better resolution. VERITAS has an energy threshold and flux sensitivity comparable to H.E.S.S.

1.3.3. *Future detector: CTAO*

Present-generation detectors continue to be operated and new results continue to be produced. The number of astrophysical sources detected by these IACTs now exceeds 250 and includes sources from 10 different source classes, namely AGNs (BL Lac objects, flat-spectrum radio quasars (FSRQs), FRI galaxies), binaries, pulsars, pulsar wind nebulae, SuperNova Remnants (SNRs), TeV haloes, GRBs, massive star clusters, novae, and sources with unknown classifications (see Chapter 2). Studies requiring longer observations and follow-ups of transient phenomena detected by other instruments are routinely conducted. It was recognized shortly after the commissioning of these detectors, however, that further progress at the same rate would require a larger detector with more collecting area and more telescopes. A worldwide consortium, comprising groups already active in the field along with new colleagues previously engaged in other activities, is now building the next-generation instrument, the Cherenkov telescope array observatory (CTAO). More than 1500 people from 25 countries are engaged in this effort.

To cover the entire sky at all energies, CTAO will use three classes of telescopes at two sites, one at La Palma and the other at Paranal in Chile. For observations at the lowest possible energy threshold, the northern site will have four Large-sized telescopes (LSTs), which resemble the MAGIC telescopes but have mirrors that are 23 m in diameter. These are designed to cover the energy range from 20 GeV to 3 TeV. Medium-sized telescopes

(MSTs), 14 at the southern site and 9 at the northern site, cover the energy range from 80 GeV to 50 TeV. The MSTs are similar to the H.E.S.S. and VERITAS telescopes, although a variant based on a design by Schwarzschild and Couder (SC) is also under investigation. For the highest energies, above 1 TeV, where fluxes are low but Cherenkov light levels are high, a sparse array of 37 Small-sized telescopes (SSTs) will be built at the southern site. The SSTs will employ SC optics using a primary reflector 4.3 m in diameter.[c]

The Cherenkov telescope array (CTA) will be an order of magnitude more sensitive than present-generation detectors and will extend energy coverage considerably in both directions. Considerable research and development has gone into the project over the last decade and many options have been studied. Hard decisions have been made to arrive at the final design and construction has begun. When exactly it will be completed is uncertain and subject to funding levels, but it is clear that CTA will begin delivering data while partially complete, as did the detectors that came before.

1.4. Gamma-Ray Astronomy with Particle Detectors

The Cherenkov light from air showers propagates through the atmosphere with little attenuation, but most of the particles in the showers are absorbed before reaching the ground. This means that IACTs can work with energy thresholds defined less by the altitude of the instruments and more by the number and size of the mirrors. By contrast, gamma-ray detectors that use the particles from the shower directly need to be built at high altitudes and typically operate with higher energy thresholds. Despite these disadvantages, direct detection has been pursued in parallel with IACTs and has proven to be a powerful and complementary technique.

Particle-detector arrays do not require clear, dark skies to operate and therefore have a much larger duty factor. They operate in drift-scan mode, viewing a large amount of the overhead sky, typically 2 sr or more, and are therefore unbiased and good for surveys. They can detect strong transients without an external alert and are more sensitive than IACTs for extended sources, where angular resolution is not as important a factor.

[c]As CTAO is currently under development, the numbers and types of telescopes are subject to change. The latest configurations may be found at the CTAO website: https://www.cta-observatory.org/.

A detector typically comprises an array of counters designed to produce a signal in response to the passage of one or more charged particles. To first approximation, the particles from an air shower arrive at the array in the form of a thin disk perpendicular to the trajectory of the originating gamma ray. By fitting the signal times as a function of the counter positions, the direction of incidence can be determined and by fitting the signal amplitudes to a model of shower-particle distribution, the position of impact of the shower core and the energy of the incident gamma ray can be estimated. With more and larger counters, located at higher altitudes, more shower particles are detected and the quality of these measurements improves.

The energy threshold is a key factor in the design and location of the array. To achieve a low threshold, it is important to have a dense array since low-energy showers have fewer particles in them and are more compact than HE showers. In the case of showers initiated by gamma rays, the particles are all electromagnetic: electrons, positrons and low-energy gammas with approximately three gammas for every electron or positron. Hence, the detectors need to be sensitive to gammas; this can be achieved by including a lead-sheet to convert the gammas into electron–positron pairs in the design.

The angular resolution of the array depends on the number and spacing of the counters and on accounting for the shape of the shower front, which is not strictly planar, but slightly conical. This means that for the best resolution, it is important to use only events where the shower core is contained within the confines of the array, a constraint that reduces the effective area of the detector.

As always, background from showers initiated by cosmic rays is a problem. The primary tool for rejecting this is the identification of highly penetrating particles such as muons and, at higher altitudes, charged hadrons. The standard procedure is to install counters under a layer of shielding and veto events that have signals from any of them.

1.4.1. *History*

Air shower detector arrays have existed since the discovery of extensive air showers in the 1930s. Early on they were used to study the nature of the showers themselves and the cosmic rays that generate them. During the 1980s, claims from some air-shower groups of gamma-ray signals from astrophysical sources such as Cygnus X-3 and Hercules X-1 were published, but statistical significances were marginal. Confirmation from other groups and reproducibility by the claimants were problematic. It was clear that a large-scale detector was needed to resolve the situation.

The Chicago Air Shower Array and Michigan Muon Array (CASA-MIA) constitute such a detector. Installed at an altitude of 1450 m in Dugway, Utah, it consisted of 1089 scintillation counters, each 1.5 m^2 in area and covered with a 6-mm layer of lead, installed on a square grid with 15 m spacing (Borione *et al.*, 1993). An additional 1024 scintillation counters, each 1.9 m by 1.3 m, were buried 3 m deep in 16 groups of 64 counters. These were used to identify showers that contained muons. The CASA-MIA group made interesting and useful measurements of interest to the cosmic-ray community, but detected no sources of astrophysical gamma rays.

The first group to detect an astrophysical gamma-ray source was the Tibet AS_γ collaboration. Using a detector located at 4300 m altitude near Yangbajing, they detected the Crab Nebula at a significance level of 5.5σ in an energy range above 3 TeV (Amenomori *et al.*, 2015). Their detector was a two-part array; the first part consisted of 185 0.5 m^2 scintillation counters installed on a square grid with 15 m spacing, covering an area of 36,900 m^2. This was supplemented with a higher density array made from 109 counters deployed on a 7.5 m grid and covering an area of 5175 m^2. Lead converters, 5 mm thick, were used on all counters.

In 2003, the Milagro collaboration reported a 6.4σ detection of the Crab nebula using a novel detector based on a water reservoir at an altitude of 2630 m near Los Alamos, NM. The reservoir was 80 m by 60 m in area and 8 m deep. It was instrumented with a layer of 450 8" PMTs installed at a depth of 1.4 m and a deeper layer, at 6 m, of 273 PMTs. The PMTs were on grids with 2.8 m square cells. They were estimated to be sensitive to gamma rays with energies greater than 100 GeV, with a median energy of approximately 4 TeV. Using the bottom layer of PMTs to measure the penetrating component of the air showers, the group was able to reject 92% of the showers due to cosmic rays but retain 50% of the gamma-initiated showers. The use of water tanks to detect Cherenkov radiation at ground-level from particle array detectors has proved to be successful (see results from the Milagro experiment, Abdo *et al.* 2009d).

Another particle array detector that was able to see a signal from the Crab Nebula was the ARGO-YBJ detector, located at an altitude of 4300 m near Yangbajing in Tibet. It consisted of a 74 × 78 m^2 "carpet" made from Resistive Plate Chambers (RCPs), parallel-plate gas-discharge devices that can detect single minimum-ionizing particles and have ns-level timing resolution; they are thus similar to plastic scintillators. The carpet area was 92% active and surrounded by a partially ($\sim$ 20%) instrumented area up to the detector boundary, 100 m $\times$ 110 m. The total active area was approximately 6600 m^2.

Operating from 2007 November to 2013 February, with a duty cycle of 86%, the collaboration obtained 1.12 $\times$ 10^4 h of Crab Nebula observations and were able to report a 21σ detection. The detection method relied on the angular resolution provided by fitting the shower front to the RCP timing data and using cuts on event morphology allowed by the fine-grained, high-density measurements enabled by the RCPs. The final signal was obtained by subtracting off-source flux measured using neighboring right-ascension bins.

1.4.2. *Current detectors*

There are presently two leading air-shower detectors active in gamma-ray astronomy: high-altitude water Cherenkov (HAWC) and large high-altitude air-shower observatory (LHAASO). HAWC is based on water-Cherenkov detectors and has been running for several years while LHAASO is a more classical particle array detector using a variety of elements. LHAASO has only recently been completed, but has already produced impressive results with data acquired while the array was under construction.

1.4.2.1. *High-altitude water Cherenkov*

The High-Altitude Water Cherenkov (HAWC) gamma-ray observatory employs large tanks of pure water at high altitude to detect the particles in air showers initiated by HE gamma rays. It is located at an altitude of 4100 m on the slopes of the Sierra Negro volcano in the Mexican state of Puebla and is operated by a collaboration that is an expanded version of the Milagro group, including institutions mainly from the US and Mexico. It has an effective area of 22,000 m^2 and covers the energy range from 0.1 to 100 TeV. Construction was completed in early 2015.

The detector itself comprises 300 closely packed tanks of water, each of which is 7.3 m in diameter and 4.5 m high and is instrumented with four PMTs: three 8-inch PMTs from the decommissioned Milagro detector and one 10-inch high-quantum-efficiency PMT. The PMTs are installed at the bottom of the tanks, looking up. Their signals, produced by the Cherenkov light from particles moving through the water, are used to determine the energy and direction of incoming showers. The high density of the water tanks is similar to the continuous pond used in Milagro, but their optical isolation is an improvement, as is the increased effective area of array. As with wide-field arrays, HAWC can survey approximately two-thirds of the sky and has a nearly 100% duty cycle (Albert *et al.*, 2020b).

Rejection of hadronic background is achieved by cuts on the morphology events in the tank–gamma-ray showers are smoother than those due to cosmic rays — and is better than 99% for energies above 3 TeV. Angular resolution improves with energy, since more tanks are hit by more particles, and goes below 0.1° above 10 TeV.

At energies above 10 TeV, the air-shower footprint is larger than the HAWC detector and the shower core is not well reconstructed due to degeneracies between core position and shower energy. To mitigate this, an array of outrigger detectors has been added over an area four times the size of the main detector. It is built from 345 water tanks, each 1.55 m in diameter and 1.65 m high, instrumented with a single 8" PMT at the bottom. These tanks are separated from one another by 12-18 m. Figure 1.5 shows a schema of the HAWC layout. The Installation took place between September 2017 and September 2018.

1.4.2.2. *LHAASO*

The Large High-Altitude Air-Shower Observatory (LHAASO) is located at an altitude of 4400 m on Mt. Haizi in Daocheng County, Sichuan Province, China. Its main component is KM2A, an array of 5195 scintillation counters, each with an area of 1 m^2 and 20 mm thickness, read out with wavelength-shifting fibers coupled to a single PMT, and covered with a 5-mm lead-sheet converter. The counters are located on a hexagonal grid with 15 m spacing, resulting in an array area of just under 1 km^2.

1188 muon detectors, on a 30-m grid, are buried under a 2.5-m layer of rock and soil. They are bladders, made from white reflective material, with an area of 36 m^2 and thickness of 1.2 m and filled with pure water. Cherenkov light is detected by a single 8″ PMT.

A large water-Cherenkov detector is located at the center of LHAASO. It has a total surface area of 78,000 m^2 and is 4.4 m deep. Black plastic curtains divide the counter into 5 × 5 m^2 square cells, each with a single upward-looking 8″ PMT at the bottom.

The final component of the LHAASO installation is a set of eight wide-field-of-view Cherenkov telescopes (WFCTA), but this is a pointed instrument and can only be used when skies are clear and dark. It is used

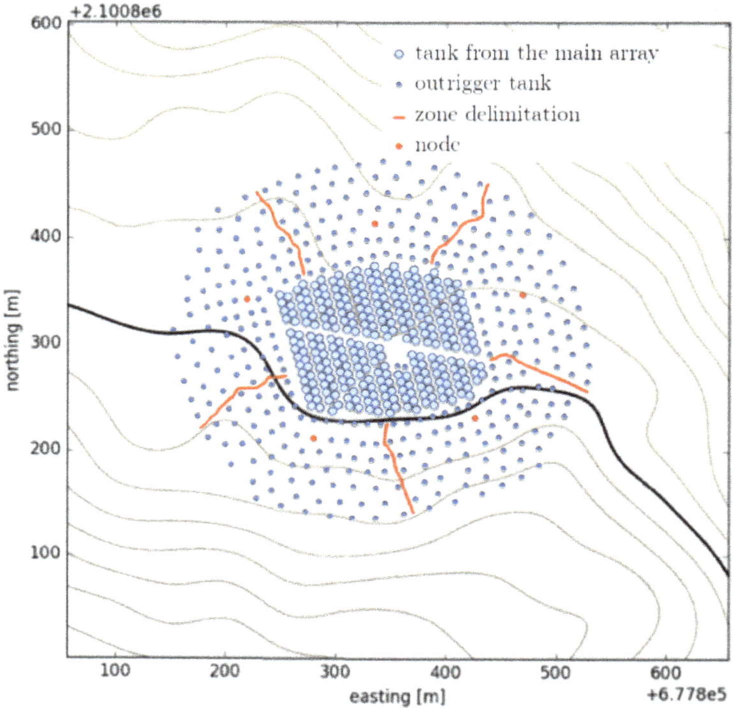

Fig. 1.5. Schema of the layout of the HAWC detector showing the positions of the water tanks and the array of outrigger detectors around the main detector (Jardin-Blicq and Joshi, 2017).

for other purposes. A schema showing the layout of the various LHAASO detectors is shown in Fig. 1.6.

1.4.3. *Future detectors*

Using data acquired over one year with a partially completed detector, the LHAASO collaboration detected 12 sources at energies above 100 TeV and up to 1.4 PeV with a statistical significance greater than seven standard deviations (Cao *et al.*, 2021). These sources have been associated with pulsar wind nebulae, supernova remnants and star-forming regions, but their nature remains unresolved.

HAWC has also been a great success; in 2020, the collaboration released its third catalog of sources which included 65 sources detected with

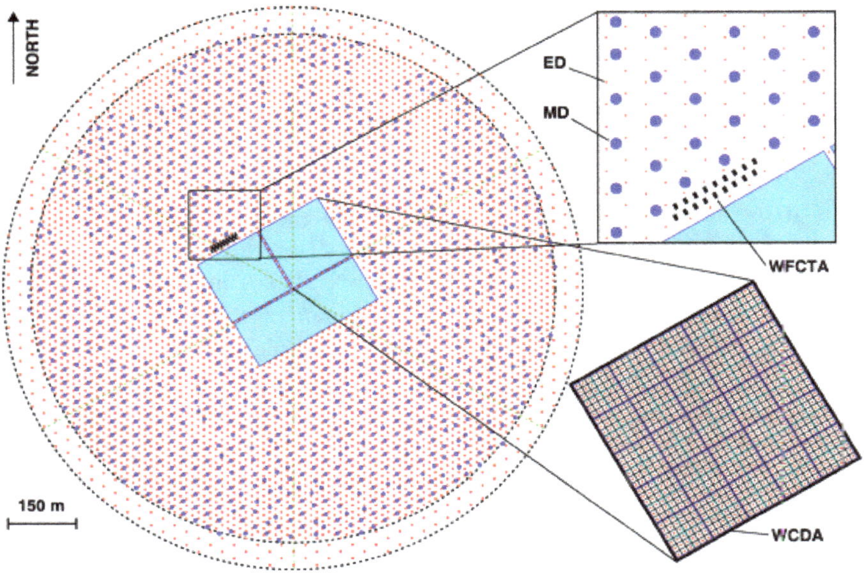

Fig. 1.6. Schema of the layout of the LHAASO detectors located in Sichuan Province, China (Cao *et al.*, 2021).

significance greater than 5σ (Albert *et al.*, 2020a). These two detectors underline the importance of altitude and area, as well as hadronic-shower rejection using hardware, and analysis techniques. The community has come to the same point as the IACT groups did when they found the key ingredients to build detectors that worked well from the start.

The next step in gamma-ray astronomy at TeV energies is to construct an array that can see the southern skies. This is the goal of the Southern Wide-Field Gamma-Ray Observatory (SWGO) collaboration, which is presently in the process of making the science case, as well as deciding on design details and working toward a funding proposal (Albert *et al.*, 2019).

1.5. Gamma-Ray Bursts

This discussion of gamma-ray astronomy has concentrated on the instruments and techniques developed for observing sources that are relatively stable, emitting a steady flux of gamma rays, with transients of various intensities and duration occurring from time to time. An important phenomenon has been left out until now; Gamma-Ray Bursts (GRBs).

GRBs are intense flashes of gamma rays emitted by sources in distant galaxies that usually last for several seconds and are followed by longer afterglows as seen in lower-energy radiation ranging from X-rays to radio waves. They were first detected in the late 1960s by the Vela satellites operated by the US military as part of a program to verify compliance with bans on nuclear testing (Klebesadel *et al.*, 1973).

GRBs remained mysterious for many years, largely due to the lack of detectors observing them and the fact that counterparts at longer wavelengths had not been identified. Statistics were improved when the BATSE detector was deployed in 1991 (Fishman *et al.*, 1994). The arrival directions of GRB photons were seen to be isotropic, indicating that their sources were extragalactic, and the timing structures of the bursts seemed to fall into two broad distributions: "short" (average length about 0.3 s) and "long" (average length about 30 s). A breakthrough occurred when the BeppoSAX X-ray satellite (Boella *et al.*, 1997) was able to provide coordinates of burst afterglows that were precise enough for follow-up by optical telescopes. This enabled red-shift measurements and established that GRB sources were at cosmological distances.

Current GRB research relies on alerts coming primarily from two instruments: the Fermi GBM (Michelson *et al.*, 2010) and the burst alert telescope on the Neil Gehrels Swift observatory (Burrows *et al.*, 2005). The IACT detectors have been responding to alerts for most of their lifetime but are handicapped by the fact that they are pointed instruments with limited fields of view and can only be operated under clear, dark skies. Also, given that VHE gamma-rays are absorbed by pair-production reactions with extragalactic background light (EBL) (Dwek and Krennrich, 2013), an energy-dependent attenuation limits the distance out to which IACTs can see. Despite this, both MAGIC (Acciari *et al.*, 2019a; Fukami *et al.*, 2022) and H.E.S.S. (Abdalla *et al.*, 2019a, 2021b) have successfully detected signals from GRBs.

The air-shower array detectors like HAWC and LHAASO that do not need dark skies are sensitive to GRBs in their fields of view, but they lack short-term sensitivity and suffer from the EBL-imposed horizon for HE photons, so expectations of their detecting GRBs have not been high. It is therefore very interesting to note that LHAASO has recently detected emission from the extraordinarily bright GRB 221009A (Huang *et al.*, 2022).

https://doi.org/10.1142/9789813275720_0002

Chapter 2

Very-High-Energy Gamma-Ray Emitters

Ruo-Yu Liu[*,§], Amanda Weinstein[†,¶] and Roberta Zanin[‡,||]

[*] *Nanjing University, Nanjing, P.R. China*
[†] *Iowa State University, Iowa, USA*
[‡] *Max-Planck-Institut fur Kernphysik, Heidelberg, Germany*
[§] *ryliu@nju.edu.cn*
[¶] *amandajw@iastate.edu*
[||] *roberta.zanin@mpi-hd.mpg.de*

2.1. Gamma-Ray Production Mechanisms

Gamma rays are produced in astrophysical sources by a variety of processes such as Bremsstrahlung, synchrotron radiation, Inverse Compton (IC) radiation and hadronic processes resulting from the interactions of relativistic hadrons with matter. Gamma-ray production and absorption mechanisms are described at length in earlier textbooks on Very-High-Energy (VHE) cosmic gamma radiation (see, for example, Aharonian, 2004). Roughly speaking, radiation is produced as a result of interactions of particles with matter, interactions with photon fields, interactions with magnetic fields and in relativistic electron–photon cascades.

Interactions with matter include processes such as electron–Bremsstrahlung and pair production, electron–positron annihilation, and gamma rays produced by relativistic protons, such as in neutral pion decays and in nuclear gamma-ray line emission.

Interactions with photon fields include IC scattering, photon–photon pair production, and interactions of hadrons with radiation fields. Finally, gamma rays may be produced by interactions of charged particles with magnetic fields, such as in the case of synchrotron radiation, pair production and synchrotron radiation of protons.

Here, we briefly describe three mechanisms that are most relevant for high-energy gamma-ray astrophysical sources. Other gamma-ray production mechanisms, such as matter–antimatter annihilation, dark matter annihilation and nuclear transformation, are not discussed in this chapter, but may be found elsewhere in this book.

2.1.1. *Synchrotron radiation*

Accelerated charges emit radiation. For non-relativistic charges, the power of this radiation is given by the Larmor formula. If an electron is ultra-relativistic and accelerated by a magnetic field, the radiation it emits is termed synchrotron radiation (also known as magneto-Bremsstrahlung). Examples of synchrotron radiation are commonly seen in astronomical objects, where relativistic electrons spiral through magnetic fields. One example is the Crab Nebula and the associated Crab pulsar, where the gamma-ray production is attributed to synchrotron emission from energetic electrons in the strong magnetic field around the pulsar. Synchrotron emission is characterized by a non-thermal power-law spectrum and polarized emission, as can be seen in the Crab Nebula, where polarized emission is detected at energies from 0.1 to 1.0 MeV (Dean *et al.*, 2008). The synchrotron radiation of electrons in general, however, does not produce VHE gamma-ray radiation due to the synchrotron burn-off limit. To be more specific in this issue, let us assume electrons are accelerated and radiate in the same region with a magnetic field strength B. We may estimate the maximum accelerated electron energy by equating the acceleration timescale and the synchrotron cooling timescale of electrons. Considering the Fermi-type acceleration, the acceleration timescale in the most efficient (but unrealistic) limit is $t_{\rm acc} = E_e/eBc$, while the synchrotron cooling timescale is $t_{\rm acc} = 6\pi m_e^2 c^4/\sigma_T B^2 E_e$. The maximum electron Lorentz factor (energy) is given by

$$\gamma_{e,\max} = \left(\frac{6\pi e}{\sigma_T B}\right)^{1/2}. \tag{2.1}$$

The characteristic synchrotron photon energy radiated by electrons of this energy reads as

$$E_{\rm syn,max} = \frac{3\gamma_{e,\max}^2 eB}{4\pi m_e c} \approx 200\,\mathrm{MeV}, \tag{2.2}$$

which is independent of the magnetic field. There are nevertheless two possibilities to boost the synchrotron energy to the VHE regime. First, the radiation region has an extremely high bulk Lorentz factor of approximately 1000 or above. Second, the magnetic field in the acceleration region is weaker than that in the radiation region. It could be due to the transport of electrons or a sudden increase of the magnetic field in the acceleration region caused by a certain mechanism.

2.1.2. *Inverse Compton scattering*

IC scattering entails the energy transfer to low-energy ambient photons from ultra-relativistic electrons, thereby upscattering the low-energy photons. The term "inverse" is used to denote the fact that it is opposite to the standard Compton scattering process that results in a decrease in the energy of the photon. The energy of the upscattered photon E_1 is related with the incident low-energy photon E_0 via $E_1 \sim \gamma_e^2 E_0$. IC scattering is an efficient mechanism for the generation of gamma-ray photons. Relativistic electrons with $\gamma \sim 100$–1000 are commonly found in astrophysical situations. For a value of $\gamma_e \sim 1000$, an optical photon could be upscattered to MeV energies. This information is commonly used in modeling the spectral energy distribution of gamma-ray Active Galactic Nuclei. As for VHE emission, it generally arises from electrons of $\gamma_e \sim 10^7$–10^8 (or 10 TeV–100 TeV) upscattering infrared photons or cosmic microwave background (CMB) photons.

When the energy of the incident photon in the rest frame of electron is higher than the rest energy of electrons $m_e c^2$, a quantum mechanical effect, namely the Klein–Nishina (KN) effect, will work and result in the reduction of the cross section of the IC process. In the KN limit, the electron will transfer almost all its kinetic energy to the incident photon, leading to $E_1 \sim \gamma_e m_e c^2$.

2.1.3. *Gamma-ray production by relativistic protons*

Gamma rays can also be produced in interactions of high-energy protons with matter or radiation; such processes are called "hadronic" processes, as opposed to "leptonic" mechanisms where electrons are involved (as discussed above). We give a brief overview here of the two most commonly modeled hadronic interactions responsible for high-energy gamma-ray production.

2.1.3.1. π^0-*decay gamma rays*

High-energy protons (or nuclei) can produce gamma rays via photohadronic interactions ($p\gamma$ interactions) with radiation field or hadronuclear interactions (pp collision) with matter. The primary channel for photohadronic interaction is the $\Delta^+(1232)$ resonance (Mücke *et al.*, 2000).

$$p + \gamma \rightarrow \Delta^+ \rightarrow \begin{cases} p + \pi^0 \\ n + \pi^+, \end{cases} \tag{2.3}$$

at $E_\gamma^r = 0.34\,\text{GeV}$ (where E_γ^r is the photon energy in the rest frame of the proton) with a branching ratio of 2:1 for π^0 production to π^+ production. However, when considering the contribution of the direction of the pion production channel, which is dominated by π^+ production, the overall branching ratio is 1:1 for π^0 production to π^+ production. Above $E_\gamma^r \approx 1\,\text{GeV}$, the multi-pion production channel dominates and the amounts of π^0, π^+ and π^- are roughly equal in products. As for the hadronuclear interactions, the amounts of the three types of pion are also roughly equal. In each photohadronic or hadronuclear interaction, the proton loses on average 20% of its kinetic energy into the produced pion.

The decay of neutral pions leads to the production of gamma rays in a very short time ($\sim 10^{-17}$ ns) by the following process:

$$\pi^0 \rightarrow \gamma\gamma.$$

The decay of charged pions, on the other hand, leads to the production of neutrinos, which may be observed by neutrino observatories operating on Earth (see Chapter 9 for a full discussion):

$$\pi^+ \rightarrow \mu^+\nu_\mu \rightarrow e^+\nu_e\bar{\nu}_\mu\nu_\mu,$$
$$\pi^- \rightarrow \mu^-\bar{\nu}_\mu \rightarrow e^-\bar{\nu}_e\nu_\mu\bar{\nu}_\mu.$$

Indeed, in hadronic processes, the production of neutrinos and gamma rays are closely linked, and the detection of neutrinos and gamma rays simultaneously from the same source point to regions where cosmic rays are accelerated. Discovering the nature of the highest-energy cosmic rays is one of the outstanding goals of current multi-messenger astronomy. The energy spectra of gamma rays, electrons and neutrinos produced in proton–proton interactions at VHE has been described by Kelner *et al.* (2006) in great detail, to which we refer the reader. If the environment where these interactions are taking place has an ambient photon field with low photon density, pp interactions will dominate over $p\gamma$ interactions. The cross section of pp interactions has an energy threshold of approximately 1.2 GeV.

It should be noted that if the hadronic process takes place in a radiation-rich environment leading to a high IC radiation efficiency of electrons, the IC radiation of secondary electron–positron pairs produced in $\pi^{+/-}$ decays could be non-negligible. In particular, if the proton spectrum is harder than E^{-2}, the energy output from the interactions is largely at high energies, and the VHE gamma-ray emission may come mostly from IC scattering of secondary pairs.

2.1.3.2. *Proton synchrotron radiation*

Another mechanism that is sometimes considered for gamma-ray emission in some astrophysical objects such as blazars is proton synchrotron radiation. Like electrons, relativistic protons can radiate in the magnetic field when they gyrate. Additionally, due to a much higher mass, protons do not radiate as efficiently as electrons. The synchrotron cooling timescale of a proton is a factor of $(m_p/m_e)^4 \sim 10^{10}$ times longer than that of an electron of the same energy; therefore, the synchrotron burn-off limit is generally negligible for protons. On the other hand, for the protons of energy E_p, the characteristic synchrotron frequency is given by

$$E_{p,\mathrm{syn}} = 100 \left(\frac{E_p}{10^{19}\mathrm{eV}}\right)^2 \left(\frac{B}{100\mathrm{G}}\right) \mathrm{GeV}. \tag{2.4}$$

An extremely high proton energy and strong magnetic field are required, then, to produce synchrotron photons in the VHE band. Considering the Doppler boost from relativistic bulk motion of the radiation region would alleviate the harsh requirement of this mechanism to a certain extent, but in general we expect uncommon conditions to produce VHE gamma-ray emission via the proton synchrotron radiation.

2.2. Very-High-Energy Gamma-Ray Emitters

High-energy cosmic accelerators can simplistically be thought to have two types of power sources: electromagnetic or gravitational. An example of the former is a highly magnetized rotating object, such as a pulsar. The latter includes the core-collapse of massive stars, and accretion onto compact objects, such as Active Galactic Nuclei (AGNs), Black Holes (BHs) and Neutron Stars (NSs). The latest 2022 catalog of known VHE sources (Wakely and Horan, 2007) lists more than 250 sources over ten broad categories both in and outside of the Milky Way Galaxy (see Chapter 4). The source list includes about 90 extragalactic sources that are active galaxies as well as three starburst galaxies. In addition, a fraction of sources at gamma-ray energies are "unidentified" with no known counterparts at other wavebands, and remain a mystery — perhaps to be solved by the next generation of instruments. Here, we give a quick overview on the different classes of gamma-ray emitters, discussed further in subsequent chapters.

2.2.1. *Relics of supernovae*

Since 185 A.D., astronomers have been recording stars that show a brief, intense period of brightening and subsequent fading. Historically, these types of events have been observed in visible light, but in recent years these events have been observed across much of the electromagnetic spectrum and — in one memorable case — using another messenger: neutrinos (Schaeffer *et al.*, 1987). There are multiple ways of classifying supernovae. The first type of supernova taxonomy classifies them according to absorption features that appear in the electromagnetic spectrum of the source (Type I and II, with the various a, b and c designations). The second, more fundamental classification, which we use here, divides supernovae according to the nature of the star that resulted in the supernova (progenitor star) and the process that produced them.

Core-collapse supernovae result from the collapse of a star that was born with a mass at least eight times greater than the mass of the sun (this birth mass roughly ensures that the star's mass at the end of its life will be at least 1.4 solar masses, i.e. the Chandrasekhar limit). These stars exist in a state of constant tension between the inward gravitational pull of the inner layers of the star on its outer layers, and radiation pressure produced by nuclear fusion in the star's core. This radiation pressure counteracts the star's tendency to collapse in on itself due to gravitational forces.

Over the course of its life, a massive star evolves through various stages of nuclear burning, synthesizing successively heavier elements from the burning products of the prior stage, and ultimately producing a degenerate iron or oxygen-neon-magnesium core. Prior to collapse, the star can be pictured as composed of layers (from outermost to inmost) formed of unprocessed hydrogen-rich material, helium, carbon, neon, magnesium, oxygen, silicon-group elements and the central iron-rich core. Because no energy can be gained from the nuclear fusion of iron, this is the natural endpoint of the processes that fueled the star and resisted gravitational collapse. The core of the star begins to collapse in on itself and form what is called a "proto–neutron star", and an initial burst of neutrinos is then formed via the fusion of protons and electrons ("neutronization").

If the core mass is greater than three solar masses or so, the collapse process continues, thus creating a black hole. This process, while interesting, is outside the scope of this text — otherwise, once the core reaches a certain density (about 2.4×10^{14} g/cc), strong repulsion (also called neutron degeneracy pressure) abruptly halts further collapse. The core then rebounds slightly and creates a shock wave when it meets the infalling gas.

One of the curious things about this process is that in most simulations of this process to date, the shock wave stalls some tens of milliseconds later (at roughly the point when the amount of matter swept up by the shock wave equals that carried by the shock itself). At this point, one of two things can happen. In the first case, the explosion fails and matter continues to accrete onto the nascent neutron star, eventually reaching the three solar-mass minimum and triggering the formation of a black hole. In the other case, the shock re-ignites a few hundred milliseconds or so after the core bounce, possibly due to convective heating by the trapped neutrinos. It is this scenario that produces a core-collapse supernova. In this circumstance, the proto-neutron star goes on to form a neutron star.

It is worth noting here that the energy released in the supernova burst (10^{51} erg) constitutes a small fraction of the gravitational binding energy of the core ($\sim 10^{53}$ erg). The overwhelming majority is released in the form of neutrinos.

In addition to the neutron star itself, core-collapse supernovae are known to leave behind three other types of relic phenomena that are bright in VHE gamma rays: SuperNova Remnants (SNRs), gamma-ray (TeV) halos and Pulsar Wind Nebulae (PWNe). These phenomena are distinguished by their power sources, the mechanism that determines their boundary and extent, and their dominant particle acceleration and transport processes. While SNRs, halos and PWNe co-evolve, they do so on different timescales and do not have the same degree of longevity.

Thermonuclear supernovae are thought to occur when a low-mass white dwarf star and a red giant companion star orbit one another. Matter from the companion star accretes onto the white dwarf until it exceeds the Chandrasekhar limit. At this point, a runaway thermonuclear fusion reaction occurs that results in the observed supernova explosion; in this case, the only relic produced is the expanding shock wave.

2.2.1.1. *Supernova remnants*

The supernova explosion injects both kinetic energy (10^{51} erg) and fragments of the destroyed progenitor star ("ejecta") into its environment in the form of an expanding, supersonic blast wave. The ejecta has an initial velocity that scales with the ratio of the supernova explosion energy and the ejecta mass, $(2E_{SN}/M_{\rm ej})^{\frac{1}{2}}$. For thermonuclear supernovae, this is of the order $10^4\,{\rm kms}^{-1}$, while for core-collapse supernovae it is of the order $5000\,{\rm kms}^{-1}$ (Reynolds, 2008). The expansion of this blast wave and its interaction with the surrounding InterStellar Medium (ISM) can persist for tens of thousands of years. This phenomenon is referred to as a "supernova remnant". While historically, the term has been used to refer to more than one type of supernova relic, these relics are distinct astrophysical phenomena and we will use the term SNR here to refer only to the long-term, persistent blast wave.

The forward shock created by the collision of supersonic ejecta with the ISM is an effective accelerator of charged particles to high energies. These particles are generally a mix of the type of particles that physicists refer to as leptons (primarily electrons) and those known as hadrons (protons and ionized nuclei of heavier elements). The dominant acceleration mechanism is thought to be "diffusive shock acceleration". Energy loss by these particles is due to the interaction with both stellar and interstellar material as well as electromagnetic fields, resulting in visible non-thermal radiation composed of some mix of radio, X-rays and gamma rays. Because the mechanisms by which highly energetic (fast-moving) particles produce non-thermal radiation (and particularly gamma rays) differ between leptons and hadrons — and are also sensitive to the energies of the particles involved — the "broadband" spectrum of electromagnetic radiation from an SNR can carry critical information about the nature of the particles being accelerated.

Although precise nomenclature and numbers of phases referred to in the literature vary, most divide the evolution of the supernova remnants into four phases:

(1) The "ejecta-dominated phase", sometimes referred to as the "free-expansion phase". The mass of the supernova ejecta (M_{ej}) dominates the matter content of the shock. The shock velocity remains relatively stable throughout this period. As the shock expands, it sweeps up material from the ISM, until the mass of swept-up material (M_{sw}) becomes comparable to that of the supernova ejecta (Reynolds, 2008;

Vink, 2012). At this point, the added mass of the interstellar material begins to significantly influence the shock expansion and the SNR moves into the next phase of its evolution. Shock speeds in the free-expansion phase are of the order of 10^3 km/s for thermonuclear and 10^4 km/s for core-collapse SNRs (Reynolds, 2008). The free-expansion phase typically lasts for a period of 100–200 years, approximately when the mass of the material swept up by the shock wave becomes greater than the mass of the ejected material (Vink, 2012). As the forward shock expands into the ISM, the abrupt deceleration of the ejecta when the forward shock slows due to encountering the ISM leads to a reverse shock moving toward the SNR center, reheating the ejecta. This two-shock structure can persist anywhere from hundreds to thousands of years (Reynolds, 2008).

(2) The "Sedov-Taylor phase", sometimes referred to as the "adiabatic-expansion" phase. This is reached as the mass of the ISM swept up by the shock wave increases and $M_{sw} > M_{ej}$, but for which radiative losses are not energetically important. This phase typically lasts for a period of order 10,000 and 20,000 years (Vink, 2012).
(3) The "pressure-driven" phase, sometimes referred to as the "radiative" or "snowplow phase". Here, radiative cooling has become energetically important (Vink, 2012) and energy is radiated much more efficiently than the thermal X-rays and synchrotron emission produced earlier in the process.
(4) The "merging phase", also known as the "dispersion phase". The shock velocity becomes equal to the turbulent velocity of the ambient ISM and the temperature behind the shock likewise cools to that of the ISM. In this phase, the SNR fades as it becomes increasingly difficult to distinguish from its surroundings. Ultimately, the SNR merges with the ISM and disappears (Vink, 2012).

The typical SNR age and duration associated with each phase of evolution is heuristic. In reality, the relationship between the "dynamical" age (i.e. its phase of evolution) and the chronological age of the SNR is variable and heavily influenced by the density and distribution of interstellar material surrounding the remnant, as well as that overtaken by/within the supernova shock. A shock expanding into a region of higher density will accumulate material more rapidly and slow more quickly than one expanding in a region of lower density (Reynolds, 2008). Thus, a dense environment has the effect of prematurely aging the SNR as well as retarding its expansion, while a rarefied environment (such as that of an SNR expanding into a bubble blown by the progenitor winds) may result in an SNR that evolves more slowly. The SNR may also age unevenly; large-scale inhomogeneities in the SNR environment (e.g. a dense molecular cloud on one side of an SNR and a less-dense ISM on the other) can result in a remnant where different regions of the shock are in different phases of evolution. In this case, some parts of the remnant will expand further than others in the same period of time, creating an asymmetric morphology.

The first two phases are where the acceleration of electrons and ions to high energies are most likely to take place.

Conversely, recent work suggests that dense clumps of material engulfed by the supernova shock can harden the overall gamma-ray spectrum of the SNR (Celli *et al.*, 2019). The argument is roughly as follows: hadronic gamma-ray emission arises in the SNR when protons and heavier nuclei interact with material to produce neutral pions. The intensity of this emission is proportional to the total number of these interactions, which is determined both by the number of protons and the density of the material. Hadrons with higher energies can also penetrate these clumps more easily–therefore, the gamma-ray emission from a large number of dense clumps is likely to be both brighter and to have a harder spectrum than that arising from interactions with the more diffuse material between the clumps.

Composite supernova remnants are a subclass of SNRs for which we observe not only the shell but also the PWN. A prototype for this kind of source is G21.5-0.9, shown in Fig. 2.1.

2.2.1.2. *Gamma-ray pulsars*

Pulsars are rapidly rotating NSs which form highly magnetized magnetospheres, usually described as magnetic dipoles. The magnetosphere is filled with charged particles extracted from the stellar surface by the induced electric fields and traveling along the magnetic field lines. These charged particles are believed to be mainly electrons and positrons, although protons and ions can be present (Amato and Arons, 2006).

The original magnetospheric model assumes that these charged particles satisfy the ideal magneto hydro dynamic (MHD) — $E\dot{B} = 0$ — and Free-Force (FF) — $\rho E + J \times B = 0$ — conditions (Goldreich and Julian, 1969). In this view, the magnetic field lines cannot remain closed beyond the light cylinder (LC), where the angular velocity of the NS is equal to the speed of light. The LC defines the pulsar co-rotating magnetosphere beyond which

Fig. 2.1. G21.5-0.9. Image obtained combining 150 h of archived *Chandra* data. The central bright PWN is surrounded by a distinctive shell of hot gas. At the center, its pulsar is visible. Credits to NASA/CXC/U. Manitoba/H. Matheson and S. Safi-Harb.

the plasma flows along the open magnetic field lines. Particles escaping the pulsar magnetosphere along asymptotically monopole field lines (Michel, 1973) form the "pulsar wind region", as shown in Fig. 2.2.

To be able to explain the particle acceleration that is responsible for the production of the detected pulsed emission, theoreticians postulated the existence of regions where the ideal MHD and FF conditions cannot be fulfilled because the electric field is not totally screened by the plasma. These regions were invoked as possible sites of particle accelerations and generally referred to as "gaps" or low-density plasma regions. In the last decade, however, a more detailed description of these accelerating regions was put forward; it involves the definition of "current sheets", i.e. current-carrying surfaces where particles can be accelerated via relativistic reconnection (Cerutti *et al.*, 2015). In particular, the existence of an

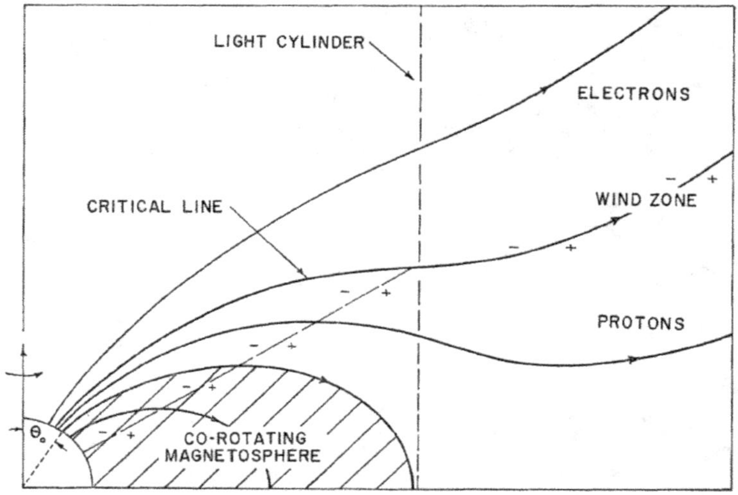

Fig. 2.2. Schema showing the co-rotating magnetosphere and the pulsar wind zone. The NS is at the lower left. Taken from Goldreich and Julian (1969).

equatorial current sheet was originally proposed by Contopoulos to ensure the closure of the large-scale currents flowing from the star to infinity and back (Contopoulos *et al.*, 1999), but only recent particle-in-cell (PIC) simulations of plasma-filled magnetospheres have provided the evidence for its existence, as well as its possible extension down to the NS surface, separating the open and closed field lines (this names them "separatrix") (Spitkovsky, 2006; Kalapotharakos and Contopoulos, 2009; Tchekhovskoy *et al.*, 2013; Cerutti *et al.*, 2015). The equatorial current sheet separates the field lines that originate from the two pulsar magnetic poles. When the rotation and magnetic axes are not aligned (oblique pulsars), this current sheet develops corrugation whose amplitude increases linearly with the distance from the star. The equatorial plane is then divided into stripes of magnetic fields of opposite polarities, hence the name "stripped wind" (Coroniti, 1990; Bogovalov, 1999; Kirk *et al.*, 2002).

In this new paradigm of the pulsar magnetosphere, two classes of pulsars are defined depending on the particle density inside the magnetosphere: the high-multiplicity pulsars where pair production takes place within the magnetosphere, that are the young pulsars with high stellar magnetic fields, and the low-density pulsars where particles are supplied only from the NS surface and are identified with the old pulsars (Contopoulos, 2016).

For the first class of pulsars, the FF solution is a good approximation of their magnetosphere characterized by the presence of the "separatrices" and the equatorial current sheet. They are weakly dissipative (Coroniti, 1990; Bogovalov, 1999; Kirk *et al.*, 2002; Cerutti *et al.*, 2015; Contopoulos, 2016). The low-density pulsars exhibit a charge-separated magnetosphere with no separatrices since the equatorial sheet is electrostatically supported. They show high dissipation ($> 40\%$) beyond the LC.

Pulsars are observed from radio frequencies up to gamma rays. Radio emission is believed to be produced at low altitudes, close to the stellar magnetic poles, by a coherent process — either maser amplification or the coordinated motion of the group of charges. At higher frequencies, pulsed emission is incoherent, most likely synchrotron radiation produced by relativistic electron–positron pairs. At high energies, pulsed emission was long attributed to synchro-curvature by particles accelerated in magnetospheric "gaps". According to the location of the gap inside the magnetosphere, gamma-ray pulsar models were classified into "polar caps", if close to the magnetic poles (Harding *et al.*, 1978), "outer gaps" if at high altitudes (Cheng *et al.*, 1986a), or "slot gaps" if along the last open field lines (Muslimov and Harding, 2003). In particular, such a synchro-curvature emission has a maximum energy limited by radiation losses or magnetic and $\gamma - \gamma$ pair absorption. The recent advances in high-energy pulsar physics due to the discovery of more than 230 gamma-ray pulsars by *Fermi*-LAT (Abdo *et al.*, 2013; Limyansky, 2019) suggest that pulsed gamma rays are produced at high altitudes favoring the outer gap models. This conclusion is mainly drawn from two observational facts:

- The spectral cutoff at a few GeV is exponential, and often even sub-exponential for young pulsars. The smoother spectral break is interpreted as the overlapping of emission beams coming from different regions and/or with different cutoff energies.
- The majority of the gamma-ray pulsars present a double-peaked pulsar profile, which in low-altitude emission models would require a specific geometric configuration (both viewing and inclination angles close to 90°), too special to meet the observed numbers (Romani and Watters, 2010).

In the context of the new paradigm of the pulsar magnetosphere, however, the high-energy emission could entirely or partially be of synchrotron origin produced in the equatorial current sheet beyond the LCs (Petri, 2012; Mochol and Petri, 2015; Cerutti *et al.*, 2016).

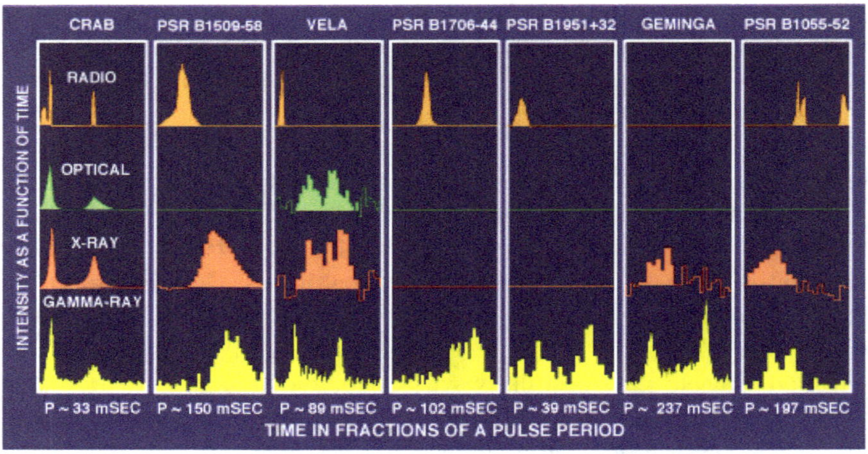

Fig. 2.3. Pulse shapes of the first seven pulsars detected in gamma rays by *CGRO*-EGRET. Credit: D.J. Thompson (NASA/GSFC).

The first pulsars detected at gamma-ray energies are the Crab and Vela pulsars. *CGRO*-EGRET was the first to carry out a systematic study of gamma-ray pulsars. Figure 2.3 shows the broadband pulse profiles of the seven pulsars detected by *CGRO*-EGRET and corresponds to the brightness of each pulsar as the neutron star makes one complete rotation. It was the last generation of gamma-ray satellites, *Fermi*-LAT, that carried out a real revolution in terms of pulsar physics, not only in terms of the number of detected sources, but also producing the first population studies. *Fermi*-LAT, for instance, showed that one-third of the detected gamma-ray pulsars are radio quiet, i.e. with no counterpart at radio frequencies, proving that Geminga, the first pulsar discovered in gamma rays (Bignami *et al.*, 1983), was simply the brightest representative of a new (sub)class of sources.

Pulsars rotate at very stable speeds and are thus often considered to be the most precise clocks. Still, they slow down as they emit radiation and lose their rotational energy. The loss rate of rotational energy, dubbed as "spin-down luminosity", is expressible in terms of the pulse period P:

$$\dot{E} = I\Omega\dot{\Omega} = -\frac{4\pi I\dot{P}^2}{P^3}, \tag{2.5}$$

where I is the NS moment of inertia and $\Omega = 2\pi/P$ is the spin (angular) velocity of the pulsar. On the other hand, the surface magnetic field of the pulsar is usually approximated as a magnetic dipole, and as a consequence, the spindown luminosity is equal to the luminosity of the magnetic dipole radiation, which reads as

$$\dot{E} = -\frac{B^2 R^6 \Omega^4 \sin^2 \alpha}{3c^3}, \tag{2.6}$$

where B is the magnetic field strength at the magnetic pole, R is the radius of the pulsar and α is the inclination angle between the rotational axis and the magnetic axis. Comparing Eqs. (2.5) and (2.6), one can find

$$\dot{\Omega} = -\frac{B^2 R^6 \Omega^3 \sin^2 \alpha}{6Ic^3}, \tag{2.7}$$

and the spindown evolution of the pulsar can be derived as

$$\Omega = \Omega_0 \left(1 + \frac{B^2 R^6 \Omega_0^2 \sin^2 \alpha t}{3c^3}\right)^{-1/2}, \tag{2.8}$$

leading to

$$\dot{E} = \dot{E}_0 \left(1 + \frac{B^2 R^6 \Omega_0^2 \sin^2 \alpha t}{3c^3}\right)^{-2} \tag{2.9}$$

where Ω_0 is the initial spin velocity of the pulsar and $\dot{E}_0 = B^2 R^6 \Omega_0^4 \sin^2 \alpha / 3c^3$ is the initial loss rate of the rotational energy. Apparently, we may define a critical timescale $\tau_0 = 3c^3 / B^2 R^6 \Omega_0^2 \sin^2 \alpha$, before which the spindown luminosity almost keeps constant and after which the spindown luminosity declines quickly. Given P and $\dot{P}$ of the pulsar measured at the current time, we may estimate the age of the pulsar to be

$$T = \frac{P}{2\dot{P}} \left[1 - \left(\frac{P_0}{P}\right)^2\right]. \tag{2.10}$$

With a further assumption that the initial pulsar period $P_0 = 2\pi/\Omega_0$ is much shorter than its current period P, the pulsar's age, τ, can be estimated as $\tau = P/2\dot{P}$.

In pulsar evolutionary history, the NSs that live in a binary system while becoming old can increase their spinning velocities. Their current speed results from hundreds of millions of years spent accreting the mass (and the relative angular momentum) transferred by a low-mass companion star through an accretion disk. Whereas accretion makes them bright sources of X-rays, as Low-Mass X-Ray Binaries (LMXBs), they switch on as "recycled" radio/gamma-ray MilliSecond Pulsars (MSPs) when the mass transfer stops. At this stage, the rotation of the pulsar magnetic field becomes the prime source of energy and powers a strong pulsar wind which might even evaporate the companion star.

An important unexpected discovery by *Fermi*-LAT was the establishment of MSPs as gamma-ray-emitting sources (Abdo *et al.*, 2009b). This is mainly because such old, recycled pulsars were not supposed to be ideal gamma-ray emitters, as their surface magnetic fields are 10^4 times weaker than those of young NSs. Despite this prediction, nowadays we know that nearly half of the known gamma-ray pulsars are MSPs (Abdo *et al.*, 2013).

2.2.1.3. *Pulsar wind nebulae*

When the magnetic axis and rotation axis of the pulsar are not aligned, the pulsar can produce a wind that exits the pulsar magnetosphere with a magnetization given by

$$\sigma = \frac{B^2}{4\pi m n \Gamma_w c^2}, \tag{2.11}$$

where B is the magnetic field, n and m are the number density and mass of the particles, respectively, and Γ_w is the bulk Lorentz factor of the wind (Slane, 2017). This quantity can be interpreted as the ratio between the Poynting flux and particle energy flux and is typically around 10^4 at the point where the wind leaves the pulsar magnetosphere. Farther out, the energy density of the wind converts from being Poynting-dominated to particle-dominated. The structure of the wind is governed by the geometry of the pulsar, with collimated jets along the rotation axis and an axisymmetric, toroidal structure around the pulsar in the plane orthogonal to the jets.

Electron/positron pairs in the wind are accelerated at the termination shock formed when the pulsar wind encounters the slowly expanding ambient medium (either the SN ejecta at the early stage or the ISM at the late stage). The radius of the termination shock is determined by the equilibrium between the ram pressure of the wind and the total pressure at the interior of the nebula (Gaensler and Slane, 2006). Either the Fermi-type acceleration or the magnetic reconnection may account for the efficient acceleration of pairs in the termination shock. A pulsar wind nebula (PWN) is the bubble of these energetic electron/positron pairs producing nonthermal radiation over a broad range of frequencies, from the radio band up to the multi-TeV or even the PeV γ-ray band. GeV–PeV gamma-ray emission of a PWN mainly arises from the IC scattering of accelerated pairs against both the interstellar radiation field and the synchrotron radiation emitting the pairs themselves. PWNe are essentially powered by the rotational energy of pulsars, so we may generally expect that those young energetic pulsars have bright PWNe.

Similar to the evolution of SNRs, the dynamic evolution of PWNe can also be generally divided into some distinct phases (Reynolds and Chevalier, 1984; van der Swaluw *et al.*, 2001). In the early stages, the shocked wind bubble expands supersonically into the cold SN ejecta and sweeps the ejecta material into a shell. As the pulsar wind is continuously injected into the bubble, the expansion of the bubble is accelerated and the radius of the PWN $R_{\rm PWN}$ evolves with time as $t^{6/5}$. The acceleration phase stops at $t = \tau_0$, when the pulsar's spindown luminosity starts a quick decline and the energy input into the bubble thus becomes unimportant. This leads to the beginning of the second phase, in which the bubble no longer overtakes the ejecta material and will coast at a constant velocity. The third phase is related to the interaction with the SNR. The SNR reverse shock propagating back into the ejecta will eventually crush the PWN in a time $t_{\rm RS} = 6(M_{\rm ej}/8M_\odot)^{5/6}(E/10^{51}{\rm erg})^{-1/2}(n_{\rm ISM}/1{\rm cm}^{-3})^{-1/3}$ kyr, compress it, and lead to an increase of its pressure and a subsequent sudden expansion of the PWN. The system will reverberate several times, resulting in oscillation of the PWN on a timescale of a thousand years, which cannot be described analytically. During this process, the interior of the SNR also relaxes toward a Sedov solution. After the passage of the reverse shock and the dissipation of the reverberations, the PWN will re-expand subsonically, in pressure equilibrium with the interior of the SNR that has been heated by the reverse shock, as $R_{\rm PWN} \propto t^{3/10}$. In short, the temporal evolution of the radius of the PWN can be described as

$$R_{\rm PWN} \propto \begin{cases} t^{6/5}, & t < \tau_0 \\ t, & \tau_0 \le t < t_{\rm RS} \\ t^{3/10}m, & t_{\rm RS} \le t, \end{cases} \tag{2.12}$$

where we neglect the reverberation phase. Note that for some PWN-SNR systems, $t_{\rm RS} < \tau_0$ can happen. The ejecta is heated at $t > t_{\rm RS}$ so that the bubble expands subsonically. With the energy injection from the pulsar wind, the radius of the PWN evolves with time as $R_{\rm PWN} \propto t^{11/15}$:

$$R_{\rm PWN} \propto \begin{cases} t^{6/5}, & t < t_{\rm RS} \\ t^{11/15}, & t_{\rm RS} \le t < \tau_0 \\ t^{3/10}, & \tau_0 \le t. \end{cases} \tag{2.13}$$

The re-expansion phase is not the end of the evolution of a PWN. Most pulsars will get a natal kick at birth due to the asymmetric SN explosion, resulting in proper motions of pulsars at typical velocities of a few hundred km/s. In the early stages, the distance traveled by the pulsar from its birth place is not important. At a certain stage of the re-expansion phase, however, the cumulative offset of the pulsar's position due to proper motion may be large so that the pulsar may escape the original shocked wind bubble, leaving behind a "relic" PWN and forming a new one around its current position. Furthermore, at a time $t_{\rm cross} \simeq 45\,(E_{\rm SN}/10^{51}{\rm erg})^{1/3}(n_{\rm ISM}/1\,{\rm cm}^{-3})^{-1/3}(v_{\rm p}/400\,{\rm kms}^{-1})^{-5/3}\,{\rm kyr}$, the pulsar may even escape the SNR and enter the ISM. At this late stage, a distinct type of new VHE gamma-ray source, coined as pulsar halo, may be formed, which will be introduced later.

While the IC process is extensively discussed as the origin of VHE gamma-ray emission from PWNe, the hadronic origin via the *pp* interactions is also possible. Protons can in principle be accelerated at either the magnetospheres of pulsars or the termination shocks of PWNe, and are injected into PWNe. The matter of the SN ejecta can serve as the target for the *pp* collisions. There are hints of hadronic components in spectra of PWNe, but no concrete evidence has been found. Sensitive high-energy neutrino observations in the future would be a smoking gun for the proton acceleration and radiation in PWNe. Note that most of the MHD models for PWNe are based on the assumption that the lost rotational energies of pulsars are converted into the energies of pairs and magnetic fields. The presence of a non-negligible proton component in PWNe would challenge our current understanding of PWNe.

2.2.1.4. *TeV gamma-ray halos*

At the late stage of ISM evolution, the pulsar may escape the related SNR and travel in the ISM supersonically. The ram pressure exerted upon the PWN by the incoming ISM confines it to the direction opposite that of the pulsar's proper motion, forming a cometary bow shock with one or more extended pulsar "tails" trailing behind. Pairs accelerated in either the PWN or the pulsar magnetosphere itself (Giacinti *et al.*, 2020b; Slane, 2017) can escape and are transported away from the PWN, resulting in an enhanced electron/positron population density around the PWN relative to the average one in the ISM. The IC radiation of the pairs produces an extended gamma-ray halo around the pulsar. Contrary to PWNe, pulsar halos do not

Fig. 2.4. A false-color composite of the famous Crab pulsar and surrounding nebulae. Infrared data from the *Spitzer* telescope (purple) and optical data from HST (red, yellow) data show the large-scale nebula and filamentary structure, while the high-resolution X-ray image from *Chandra* (blue) shows the characteristic torus and jets. Credit: X-ray: NASA/CXC/SAO/F.Seward. Optical: NASA/ESA/ASU/J.Hester & A.Loll. Infrared: NASA/JPL-Caltech/Univ. Minn./R.Gehrz.

have clearly delineated boundaries. Figure 2.5 illustrates the relationship between and relative sizes of an SNR, PWN and halo that all arise from the same core-collapse supernova. A PWN can have a size of the order of 0.1–1 pc, and in some rare cases, up to 10 pc. An SNR could extend from 1 to 100 pc, although this depends on the evolutionary stage and environment of the SNR and on the properties of the progenitor.

Recently, extended VHE emissions surrounding two nearby middle-aged pulsars, namely the Geminga pulsar and the Monogem pulsar, emerged as a TeV gamma-ray source class distinct from gamma-ray bright PWNe, which are also coined as "TeV halos". The scale of the halo is significantly larger than the typical size at the gamma-ray or X-ray band.

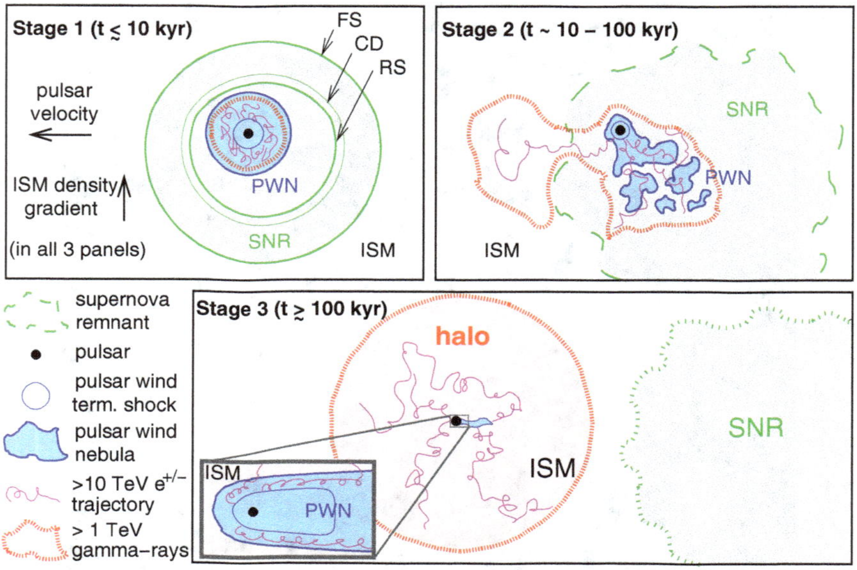

Fig. 2.5. Conceptual diagram showing the three-stage evolution of a PWN and pulsar halo. Reproduced from Giacinti *et al.* (2020a).

For example, the scale of the TeV gamma-ray halo of the Geminga pulsar is approximately 20–30 pc as opposed to < 1 pc for the corresponding X-ray bow-shock nebula. The measured surface brightness profile supports a diffusion-dominated particle transport around the PWN. When compared to theoretical predictions, however, this scale is far below the expected 500 pc with the average diffusion coefficient in ISM inferred from the secondary-to-primary cosmic ray measurement. The diffusion coefficient obtained from the halos of Geminga and Monogem is approximately $4.5 \times 10^{27}\mathrm{cm}^2\mathrm{s}^{-1}$ at an energy of 100 TeV, a factor of 300 lower than the average one (Abeysekara *et al.*, 2017a). The origin of the slow diffusion is unclear yet. It has been suggested that self-generated turbulence or a specific magnetic field geometry may account for the small diffusion coefficient.

Strictly speaking, a halo can form around any source that accelerates electrons and permits them to escape before cooling, but as of this date, the only established TeV gamma-ray halos are detected around mature pulsars. The lower-energy counterparts of pulsar halos have not been firmly detected.

2.2.2. *Gamma-ray binaries*

In addition to TeV gamma-ray halos, the most recent generation of gamma-ray observatories also established "gamma-ray binaries" as a new class of gamma-ray emitters. These binaries are believed to comprise a fast-rotating pulsar and a massive star. The interaction between the stellar wind and the pulsar wind produces a relativistic shock that accelerates particles, which then produces gamma-ray emission by the usual mechanisms up to several TeV (Porth *et al.*, 2017; Dubus, 2013, 2015).

Figure 2.6 shows two conceptual schemas of a gamma-ray binary. In the case of a pulsar wind scenario, the interaction of the two winds generally occurs at 10^4 light cylinder radii (a fraction of an AU) from the pulsar. This is much closer to the pulsar than the corresponding interaction of the pulsar wind with the ISM for PWNe. The emission varies strongly depending on where the pulsar is along the orbit.

A few VHE gamma-ray binaries have been discovered so far: PSR B1259-53, LSI + 61° 303, 1FGL J018.6-5856, HESS J1832-093, HESS J0632 + 057, LMC P3, and PSR J2032 + 4127/MT91 213. In most cases, the nature of the compact object has not been confirmed. For these binaries, an alternative interpretation — in which the gamma-ray emission powered by accretion onto the compact object rather than the pulsar rotation —

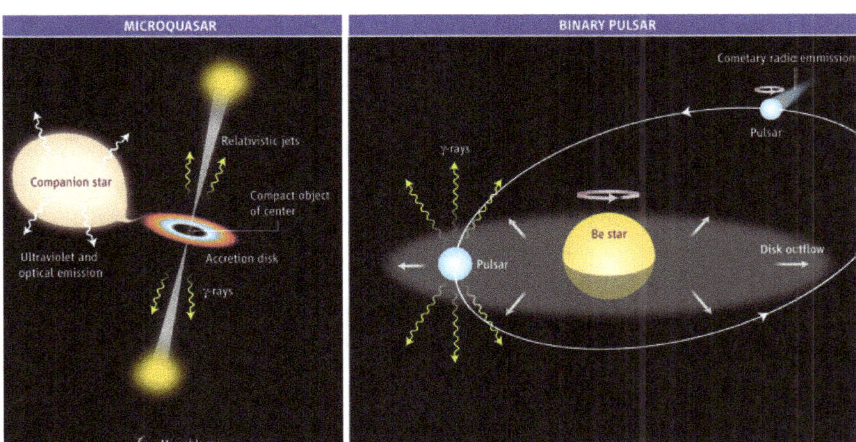

Fig. 2.6. Two conceptual views of a gamma-ray binary. *Left*: microquasar with a stellar-mass black hole accreting material from a massive companion star. *Right*: orbit of a pulsar around a Be star, the interaction of the pulsar wind with the stellar wind, and the photon field and circumstellar disk of the Be star (Mirabel, 2006).

remains viable; at this point, however, the pulsar wind scenario is generally favored. HAWC has claimed gamma-ray emission from SS 433, a binary system where the compact object has a high probability of being a black hole.

"Colliding wind binaries", a system of two massive stars (as opposed to a massive star and pulsar), are known in a few cases to emit gamma rays. Each of the stars in the system may drive a powerful, supersonic stellar wind, and charged particles are believed to be accelerated in the shock of each side of the contact discontinuity formed at the collision of the two stellar winds. H.E.S.S. has detected VHE emission from the direction of η Carinae, which is a binary system composed of a luminous blue variable primary star of $\sim 100\,M_\odot$ and an O- or B-type companion of $\sim 30\,M_\odot$, shortly before and after the minimum in the X-ray lightcurve close to the periastron. The radiation mechanism of the VHE emission from the binary is not determined yet, and a more precise measurement on the spectrum and lightcurve by the next-generation VHE gamma-ray detector would facilitate our understanding on the nature of colliding wind binaries.

2.2.3. *Gamma-ray bursts*

Gamma-ray bursts (GRBs) are intensely violent explosions observed in distant galaxies that emit X-ray and gamma-ray photons. GRBs are episodic in nature and unpredictable, but are the most luminous sources of gamma rays when a GRB event happens. The intense, highly variable emission of GRBs lasts a short period of time, typically ranging from ~ 0.1 s to ~ 100 s, called "prompt emission". GRBs are classified as "short" (lasting less than 2 s) or "long" (lasting longer than 2 s and up to several hours) depending on the duration of the bursts. Long GRBs, which constitute the majority of GRB events, are most likely produced at the collapse of massive stars during a supernova explosion, or when a high-mass star implodes to form a neutron star or black hole. Compared to stars that ended with the supernova explosions, progenitor stars of GRBs require additional specific conditions, such as extremely rapid rotation and low metallicity. Short GRBs are likely produced in the merger of neutron star binaries (D'Avanzo, 2015) or the neutron star–black hole merger. A significant contribution to GRB studies came from the BATSE instrument (Meegan *et al.*, 1996) onboard CGRO that detected 2704 GRBs and discovered the bimodal distribution of GRB distributions (Kouveliotou *et al.*, 1993). In recent years, the GRB monitor (Meegan *et al.*, 2009) and the Large Area Telescope onboard the Fermi Observatory have together studied GRBs over a broad energy range spanning from several keV to 100 GeV. The spectra of GRB prompt emission can be usually described by the Band function (Band *et al.*, 1993) which is a smoothly connected, broken power-law function, or by the "Comptonized" model — a power-law function with a high-energy cutoff. The spectrum typically peaks around the (0.1–1) MeV gamma-ray band. There is an extra spectral component extending to the GeV regime with a power-law behavior occurring in some GRBs, but so far no VHE gamma-ray emission has been reported during the prompt emission phase of GRBs. Despite the difficulty in capturing the prompt emission of GRBs by TeV gamma-ray instruments, the strong absorption of these energetic photons by the intense keV–MeV emission of GRBs is probably the intrinsic reason for the absence of VHE emission from the prompt emission phase of GRBs.

Following the prompt emission phase, an "afterglow" emission may be detected across the electromagnetic spectrum at gamma-ray, X-ray, ultraviolet, optical, infrared, microwave and radio bands. The afterglow emission arises from the emission of electrons swept by and accelerated in the shock formed when the GRB jet propagates in the circumburst medium, which could be the ISM, stellar wind or progenitor star. The picture is similar to the case of the SNR, but the shock is driven by a narrow jet instead of a quasi-spheric ejecta, and the shock is ultrarelativistic instead of nonrelativistic. Figure 2.7 shows a cartoon illustrating the standard model for afterglow emission in a GRB. The gamma rays are produced in relativistic shocks, as explained by a single-zone model, with the blast wave moving into the wind of the progenitor star or the ISM. Energetic electrons downstream a relativistic shock emit broadband radiation via the synchrotron radiation in a shock-generated magnetic field (Granot and Sari, 2002) and via the synchrotron self-Compton scatterings. The IC component predicted by the standard afterglow model was not concretely detected until 2019, when the VHE emission from the afterglow phase of GRB 190114C was announced by the MAGIC Collaboration, making GRB 190114C the first-ever GRB detected by a ground-based Imaging Atmospheric Cherenkov Telescope (IACT) (Acciari *et al.*, 2019b). Since then, several other GRB have been detected by ground-based IACTs, opening up a new era in high-energy studies of GRBs (Derishev and Piran, 2021). In particular, GRB 221009A was detected at energies above 10 TeV by LHAASO within 2000 s after the trigger of burst, challenging the standard picture of the GRB model.

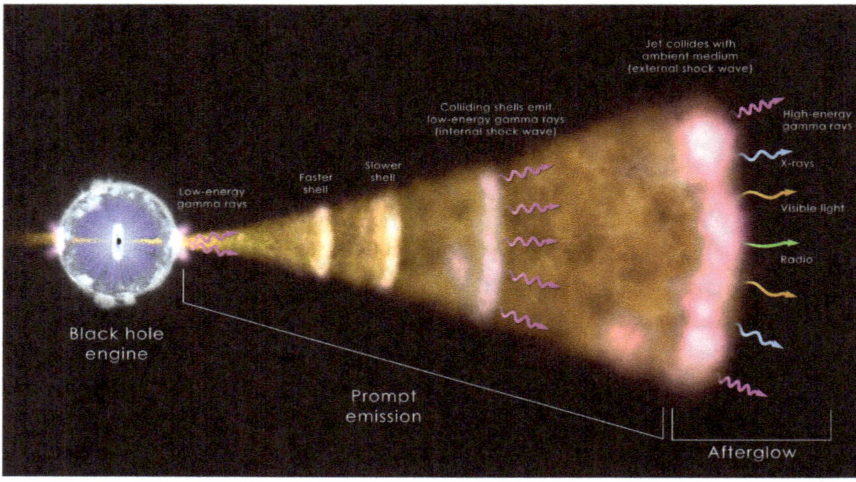

Fig. 2.7. Cartoon illustrating a popular model for afterglow emission from a GRB. Credit: NASA.

2.2.4. *Active galactic nuclei*

Almost all galaxies host in their center Supermassive Black Holes (SBHs) of approximately 10^6 to 10^9 solar masses. When SBHs accrete surrounding matter, they give rise to active galactic nuclei (AGNs). During this accreting phase, the galaxy nuclei become so bright that they can outshine their whole galaxy host. The gravitational energy is released in the form of radiation by the accretion disk in its surrounding and the out-flowing jets emitted perpendicularly to the plane of the disk.

The thin and hot accretion disk can be surrounded by a hot thermal corona and two regions filled with clouds of gas: a Narrow Line Region (NLR) and a Broad-Line Region (BLR). It is generally accepted that thermal and broad-line emission are related, and their weakness points toward an accretion occurring in an Advection Dominated Accretion Flow (ADAF) regime. The latter is, in fact, radiatively inefficient and characterized by a dramatic decrease in ionizing UV photons that are unable to photo-ionize the clouds of gas responsible for the broad lines. An obscuring torus of dust can embed the disk and its corona. All these main features of a jetted AGN are illustrated in the sketch in Fig. 2.8.

The term AGN applies to several additional classifications. A simplified — hence very common — classification is the one based on

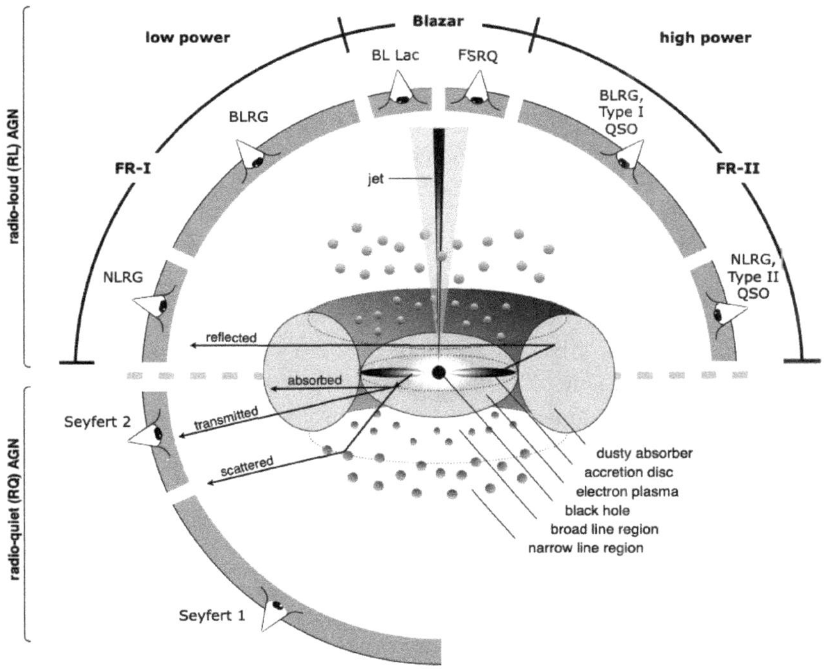

Fig. 2.8. Schematic representation of an AGN showing its main features: the central SBH, accretion disk, an obscured torus, the BLR, and the NLR. The sketch also shows the unified classification scheme of the AGN zoo. Figure taken from Beckmann and Shrader (2012).

only two observational parameters, namely, the radio loudness and the width of emission lines (Urry and Padovani, 1995; Dermer and Giebels, 2016). The radio loudness identifies the source power at radio frequencies in comparison to the optical one. When plotting radio luminosity versus optical luminosity for complete samples of optically selected sources, there are two populations; the "radio-quiet" one (that should be better renamed "radio-weak") having, for the same optical power, a radio power about 3–4 orders of magnitude smaller than the radio-loud one.

Despite the fact that 90% of the known AGNs are radio-quiet, classified into "Seyfert galaxies" and "quasi-stellar objects (QSOs)" (also dubbed as "quasars"), the most interesting population for gamma-ray astronomers is the one of the radio-loud — almost all AGNs seen at high energies are of this kind. In the radio-loud population, another important observational parameter is the angle of the jet to the line of sight. Based on the

latter, radio-loud AGNs are classified as "blazars" with jets closed to out line of sight, and "radio galaxies" which are the misaligned counterparts of blazars. Each of these two categories is further split depending on its broad-line emission: blazars include "BL Lacs" displaying weak/no broad-line emission, and "Flat-Spectrum Radio Quasars (FSRQs)" with the equivalent width of optical lines ≥5Å. Similarly, "radio galaxies" are classified into the broad-line radio galaxy (BLRG) and narrow line radio galaxy (NLRG) depending on which is the dominant emission-line region. In addition, "radio galaxies" have a further independent categorization, the Fanaroff–Riley (FR) one, based on the radio luminosity in relation to the distance from the SBH: the "FR-I" or "FR class-I" whose luminosity decreases with increasing distance from the central object, and the "FR-II" or "FR class-II" exhibiting increasing luminosity in the lobes, i.e. where the jets interact with the ISM.
Such a unified scheme of the AGN zoo is illustrated in Fig. 2.8 based on Beckmann and Shrader (2012).

The majority of AGNs detected in the gamma-ray band are blazars. Blazars are characterized by emissions that include high radio and optical polarization, strong variability at all wavelengths and non-thermal continuum spectra across several decades in energy. Gamma-ray observations of blazars have demonstrated two interesting characteristics. First, in the majority of the blazars, the gamma-ray emission carries a considerable — if not dominant — fraction of the bolometric power in the spectral energy distributions of these objects. Second, blazars are characterized by short-timescale variability, often of the order of days to months. This was first seen in gamma-ray blazars observed in the GeV band by EGRET (Mukherjee *et al.*, 1997; Mukherjee, 2001), and subsequently by Fermi-LAT (Ajello *et al.*, 2020). For most detected blazars, the typical isotropic-equivalent gamma-ray luminosity ranges in 10^{43}–10^{48} erg/s with a few reaching 10^{49} erg/s. Such high luminosities even exceed the Eddington luminosity of their SBH, implying strongly beamed emission. In fact, measurements with the very-long baseline interferometer (VLBI) suggest that a high fraction of gamma-ray blazars are superluminal, suggesting that (a) gamma-ray blazars have blobs of particles ejected relativistically along a jet and (b) the jet is inclined at small angles with respect to the line of sight. The isotropic-equivalent luminosity is amplified by a factor of δ_D^4 with respect to the luminosity in the blob's comoving frame, where δ_D is the Doppler factor of the jet.

The SED of non-thermal continuous spectra of gamma-ray blazars shows a double-hump structure: one at lower energies and one at higher energies in the X-ray to gamma-ray energies. This effect is visible in Fig. 2.10 showing the SEDs of a sample of blazars detected by *Fermi*-LAT (Ghisellini *et al.*, 2017). Depending on the peaking frequency of the low-energy hump, BL Lacs are classified as

- **High-frequency BL Lacs (HBLs)**: The low-energy hump peaks at frequencies $> 10^{15}$ Hz;
- **Intermediate BL Lacs (IBLs)**: The low-energy hump peaks at frequencies $> 10^{14} - 10^{15}$ Hz;
- **Low Bl Lacs (LBLs)**: The low-energy hump peaks at frequencies $< 10^{14}$ Hz.

whereas the low-energy hump of FSRQs generally peak at low frequency, similar to LBLs. The two-peak behavior of the blazar SED can be explained by assuming that the same population of electrons is responsible for both peaks, with the low-energy emission being of synchrotron nature and the high-energy one being due to IC scattering off low-energy photons. Nevertheless, hadronic processes may also be responsible for VHE emission in blazars.

In some blazars, very-rapid variability of gamma-ray emission at minute timescales has been discovered, implying a quite compact radiation region of $\sim$ 10 Schwarzschild radii for a 10^8 solar mass black hole. This is because the size of the radiation region is limited by $R < \delta_D c \delta t$ with δt being the timescale of the variability, otherwise the variability would have been smoothed by the difference in the arrival time of the emission from different parts of the radiation region. On the other hand, observation of minute-scale VHE gamma-ray variability from blazars would require an extremely high Doppler factor of the radiation region in order to make the compact radiation region transparent to the TeV photons, which is, however, inconsistent with the velocity inferred from the superluminal motion observed in the radio band. The tension might imply a more complex topology of the radiation region, or other origins of the VHE emission such as the jet-star/cloud interaction.

VHE emission has been also detected from some nearby radio galaxies. Figure 2.9 shows the radio and optical jet of the radio galaxy Messier 87 (M87) at a distance of 18 Megaparsec (Mpc), which is powered by an SBH

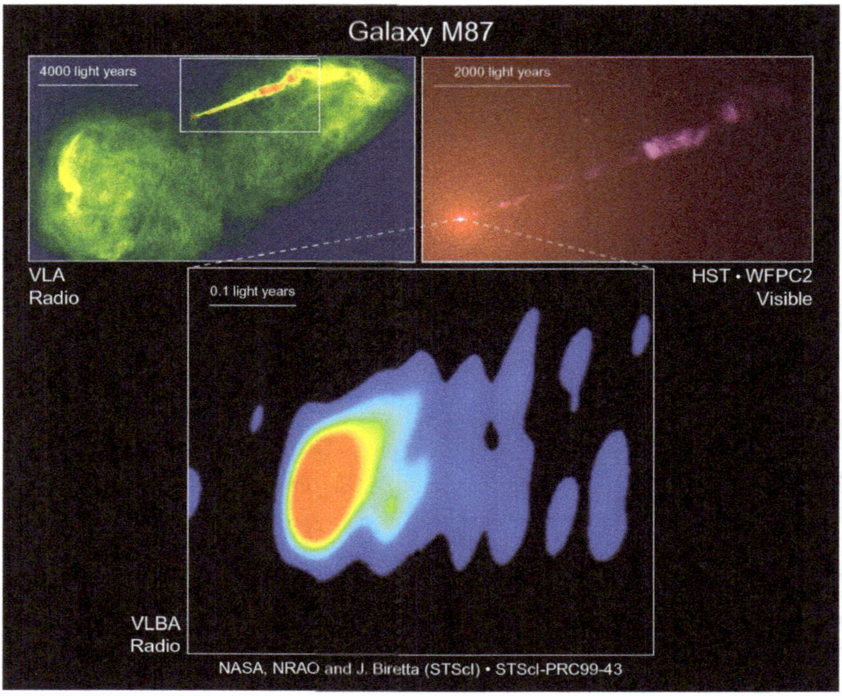

Fig. 2.9. A close-up view of the jet of the radio galaxy Messier 87 (M87) shown at three different scales. *Top left*: radio image taken with the very large array (VLA) radio telescope in February 1989. The jet is thought to be constituted of relativistic subatomic particles. The jet of M87 is believed to be powered by an SBH of ~ 3 billion solar masses. *Top right*: optical image of M87 taken with the NASA Hubble Space Telescope's Wide-Field Planetary Camera 2 in February 1998. The relativistic jet of M87 is seen across thousands of light years. *Bottom image*: X-ray jet of M87 measured with *Chandra*. (Credit: NASA, National Radio Astronomy Observatory/National Science Foundation, John Biretta (STScI/JHU), and Associated Universities, Inc.)

of ~ 3 billion solar masses (Doeleman *et al.*, 2012). The jet is presumed to contain relativistic particles that are responsible for the multi-wavelength emission from M87. This radio galaxy has been detected at TeV energies by the IACTs H.E.S.S., MAGIC and VERITAS, although the jet has not been resolved at high energies. In another nearby radio galaxy, Centaurus A, which is located 3.8 Mpc away, H.E.S.S. has resolved its kiloparsec (kpc)-scale jet and showed extended VHE emission along the jet (Abdalla *et al.*, 2020c). It demonstrates the presence of ultra-relativistic electrons of at least 50 TeV along the jet via certain "distributed" acceleration mechanisms.

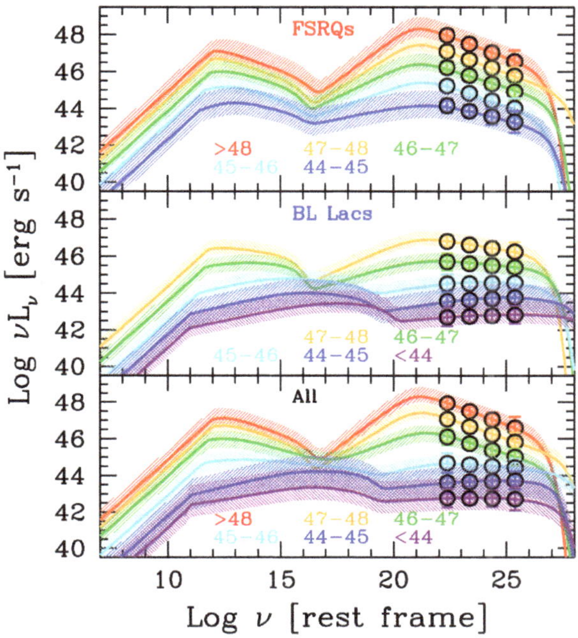

Fig. 2.10. SEDs of a sample of AGNs detected by *Fermi*-LAT. The low-energy peak of the SEDs is typically in the IR-UV or X-ray energies, depending on whether the blazar is an FSRQ or BL Lac object. The high-energy peak of the SEDs is in the gamma-ray energy band. Figure taken from Ghisellini *et al.* (2017).

2.2.5. *Starburst galaxies*

Starburst galaxies are gas-rich galaxies undergoing an exceptionally intense phase of star formation, with a star-formation rate up to $\gtrsim 1000$ times greater than that in a normal galaxy. At such high rates, the supply of gas and dust within the galaxy would be exhausted within 10^7–10^8 years, implying that these bursts of star formation are rather recent and will end in a not-too-distant future. In this view, starburst galaxies are not a separate class of galaxies, but rather an evolutionary phase in the life of a star-forming galaxy, which is probably triggered by gravitational interactions between galaxies, ranging from a close passage to a complete merger.

The starburst activity brings an enhanced formation rate of massive stars and supernova explosions, both of which are suggested to be promising particle accelerators, and thus may lead to a corresponding enhancement of a cosmic-ray production rate by orders of magnitude compared with our Galaxy. The starburst activity is not uniformly spread across the galaxy, but rather localized in a small region, often around the nucleus. The densities of the gas and radiation fields are very high in the starburst region, providing a favorable environment for the *pp* collision of protons and IC scatterings of electrons. Several starburst galaxies, as well as ordinary star-forming galaxies in the local Universe, have been detected in the GeV gamma-ray band. The gamma-ray emissions of star-forming galaxies are usually suggested as the pionic emissions from cosmic-ray protons. Interestingly, it is found that there exists a quasi-linear scaling relationship between the GeV gamma-ray luminosity and the total infrared luminosity of star-forming galaxies in the logarithmic space (Ackermann *et al.*, 2012b). It might be interpreted as the interplay between the cosmic-ray production rate and the *pp* collision efficiency in star-forming galaxies.

Among all starburst galaxies with GeV gamma-ray detection, only M82 and NGC 253, two close-by galaxies both located about 3–4 Mpc away, have been detected above 1 TeV. The TeV emission could be a natural extension of the GeV component from pionic emission of protons but produced by protons at higher energies. With plausible parameters, the starburst galaxies can be very efficient hadronic GeV gamma-ray emitters or even calorimeters; the protons lose all the energies before escaping the starburst region. As the energy goes higher, the escape timescale may be shorter than the *pp* collision timescale, meaning the galaxies are not calorimeters beyond a certain energy, and hence a steepening may appear in the gamma-ray spectrum under this interpretation, although the value of the critical energy depends on some properties of the galaxies and is not very certain. An alternative scenario is ascribing the TeV emission to PWNs and pulsar halos in the galaxies since the enhanced formation rate of massive stars also leads to an exceptional birth rate of pulsars. In this scenario, the VHE emission arises from the IC processes of leptons powered by pulsars while those at GeV may still come from the pionic emission of cosmic ray hadrons. A spectral hardening, therefore, may be expected in this scenario. An accurate measurement of the VHE spectrum is helpful to differentiate the two scenarios, which would facilitate our understanding of the origin of the diffuse Galactic gamma-ray emission from our Galaxy.

We are talking about starbursting regions of massive galaxies, such the central region of M82 or 30 Doradus in the Large Magellanic Cloud (LMC), at odds with the global starburst galaxies. The latter are indicated by an exceptionally high, specific star-formation rate (or star-formation rate per unit stellar mass in the galaxy). They are rather rare, but very interesting because they are tracers of earlier epochs of the universe, when they were abundant and bright. These are mainly "dwarf galaxies" and "IR galaxies".

https://doi.org/10.1142/9789813275720_0003

Chapter 3

The Third Generation of Imaging Atmospheric Cherenkov Telescopes: H.E.S.S., MAGIC and VERITAS

Julian Sitarek*,‡ and Gianluca Giavitto†,§

*University of Lodz
Lodz, Poland
†Deutsches Elektronen-Synchrotron
Zeuthen, Germany
‡jsitarek@uni.lodz.pl
§gianluca.giavitto@desy.de

3.1. Third Generation of IACTs

In the first decade of the 21st century, three major installations of the third generation of IACTs began operating: the High-Energy Stereoscopic System (H.E.S.S.), the Major Atmospheric Gamma-Ray Imaging Cherenkov (MAGIC) and the Very-Energetic Radiation Imaging Telescope Array System (VERITAS) arrays. All three have been highly successful and are still taking data at the time of writing.

Other contemporary IACTs are not described in detail in this book, but are worth mentioning nonetheless. The third phase of the Collaboration of Australia and Nippon (Japan) for a Gamma-Ray Observatory in the Outback (CANGAROO) project was an array of four 10-m telescopes operated between 1999 and 2011 in Australia by a Japanese–Australian collaboration (Enomoto *et al.*, 2002). Although it was a third-generation IACT array, it stopped operating early. The First G-APD Cherenkov Telescope (FACT) is a 3.5 m-diameter single telescope operating at the MAGIC site since 2011, and whose camera was the first to use Geiger-mode avalanche photodiodes (G-APDs) as pixels. This paved the way for astrophysicists to consider solid-state photo-detector technology for the

H.E.S.S.	MAGIC	VERITAS
Location		
Namibia Khomas Highland Farm Goellschau 23.3° S 16.5° E 1800 m a.s.l.	Spain Canary Islands La Palma, ORM[a] 28.8° N 17.8° W 2225 m a.s.l.	U.S.A. Arizona FLWO[b] 31.75° N 110.95° W 1268 m a.s.l.
First Light		
2002	2004	2007
Number of IACTs in final configuration		
5	2	4

[a]Obs. el Roque de los Muchachos. [b]F. L. Whipple observatory.

next-generation CTAO (Anderhub *et al.*, 2013). In India, the 9.5 m-diameter TeV Atmospheric Cherenkov Telescope with Imaging Camera (TACTIC) has been in operation since 2001 (Koul *et al.*, 2007) and the installation of its 21-m diameter successor, the Major Atmospheric Cherenkov Experiment Telescope (MACE) (Borwankar *et al.*, 2016), was finished in October 2020 (Bhatt *et al.*, 2021).

Due to its location in the Southern Hemisphere, H.E.S.S. mostly covers the southern sky, while MAGIC and VERITAS observe northern sources, with an overlap in the region between $\sim -30°$ and $\sim +30°$ of declination. The three experiments use the principle of imaging atmospheric Cherenkov technique and share a number of common features. Each of them is a system of 2–5 telescopes with apertures ranging from 12 to 28 m, working together in the stereoscopic mode. The large dimensions of the telescopes require the use of tessellated mirror segments on the dish and Alt-Az mounts. All cameras use Photo-Multiplier Tubes (PMTs) as pixels.

The comparison of sensitivity (i.e. the lowest-detectable flux in a given time) obtained for various Cherenkov telescopes of the previous, current, and upcoming generation is shown in Fig. 3.1. The advent of the third generation of IACTs made it possible to detect sources an order of magnitude lower in flux. Moreover, these technologies rendered observable the energy range of around 100 GeV, which was previously inaccessible to the Cherenkov telescopes. A similar advancement is expected from the next generation of Cherenkov telescopes: the Cherenkov Telescope Array (CTA).

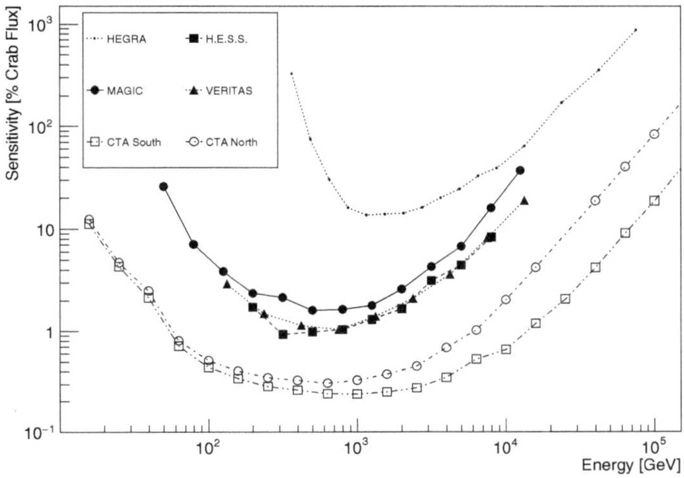

Fig. 3.1. Comparison of sensitivity of various IACT instruments obtained in 50 h of observations. Second generation: HEGRA (small markers, Aharonian *et al.* 2004b). Third generation: H.E.S.S. (filled squares, Holler *et al.*, 2015), MAGIC (filled circles, Aleksić *et al.*, 2016b) and VERITAS (filled triangles, Park *et al.*, 2015) The planned next generation of IACTs: CTAO-North (empty circles) and South (empty squares), https://www.cta-observatory.org/science/cta-performance/ (version prod3b-v1). All sensitivities have been recalculated to 5σ definition and 5 bins per decade in energy.

Over the last 30 years, the number of known sources has grown from a single source in 1989 (the standard candle of VHE gamma-ray astronomy: the Crab Nebula, Weekes *et al.*, 1989) to 200+ in 2018. The speed at which the VHE gamma-ray field has been growing is often expressed in terms of the "Kifune" plot, showing the integral number of detected sources until a given date (see the left panel of Fig. 3.2). Over a long timescale, the number of known sources has grown exponentially. Over a short timescale, however, one can see the effect of the new instruments, including a rapid growth a few years after the third generation of IACTs was constructed. The right panel of Fig. 3.2 depicts the distribution of the flux of the VHE gamma-ray sources expressed in units of the flux of the Crab Nebula.[a]

In the following subsections, we briefly describe the H.E.S.S., MAGIC and VERITAS telescopes.

[a]Note that many of those sources show variable emission, in such cases a "characteristic" flux is plotted.

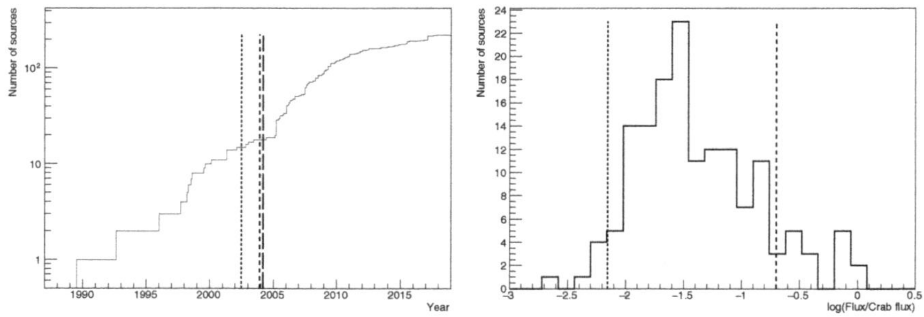

Fig. 3.2. *Left*: Total number of sources detected up to a given date. Dotted, short-dashed and long-dashed vertical curves mark the start of the operations of the third generation of IACTs: H.E.S.S., MAGIC and VERITAS, respectively. *Right*: Distribution of flux of the VHE gamma-ray sources. Vertical dashed and dotted lines show the typical sensitivities of the second and third generation of IACTs. The plots have been produced with data obtained from TeVCAT (Wakely and Horan, 2007).

3.1.1. *The high-energy stereoscopic system*

H.E.S.S.[b] (Hofmann, 2001) is the only operational IACT array presently located in the Southern Hemisphere — more precisely, in the Khomas highlands of Namibia (23.3° N, 16.5° E) — at about a 2-h drive from the capital Windhoek, at an altitude of 1800 m above sea level. From this location, H.E.S.S. is able to observe the inner parts of the galaxy with optimal sensitivity. This proved to be a crucial advantage with respect to the other arrays in the Northern Hemisphere: so far, H.E.S.S. has detected the majority of the known TeV gamma-ray sources, most of them Galactic. It is also the only IACT array that has performed a survey of the Galactic plane (Abdalla *et al.*, 2018j).

The H.E.S.S. array commenced operations in mid-2002, when a first 12-m diameter telescope was built. Three more 12-m telescopes came online during 2003, and by December 2003, the commissioning of phase I of the array was completed. These telescopes were named CT1-4, following a tradition established by their predecessor, HEGRA. H.E.S.S. took data in this configuration until 2012, when phase II started with the construction and commissioning of CT5, a new, 28-m diameter telescope located at the center of the old square 12-m array. The inclusion of a much bigger telescope lowered the threshold of the system from 100 GeV down to 20 GeV (Abdalla *et al.*, 2018d) and enhanced its overall sensitivity. In order to improve the

[b]https:///https://www.mpi-hd.mpg.de/hfm/HESS.

Fig. 3.3. H.E.S.S.-II array: the larger CT5 telescope is in the center, surrounded by the smaller CT1-4. Image courtesy of the H.E.S.S. Collaboration.

performance of the array and increase data-taking efficiency, the cameras of CT1-4 were upgraded in 2015–2017; the CT5 camera, in 2019. Both of these camera upgrades fielded prototypes of the technologies that will be used for the upcoming fourth-generation IACT array, the CTA.

3.1.2. *Major Atmospheric Gamma-Ray Imaging Cherenkov*

MAGIC[c] started scientific operations as a single telescope, MAGIC-I (M1), in 2004 (Baixeras *et al.*, 2004). In 2009, the construction and commissioning of the second telescope, MAGIC-II (M2) — an improved copy of MAGIC-I — was completed (Aleksić *et al.*, 2012d), and from that time on the telescopes have operated in stereoscopic mode, recording for analysis only events that simultaneously triggered both telescopes. The two telescopes are separated by 85 m and are located in the Observatorio El Roque de los Muchachos (28.7° N, 17.9° W) on the Canary Island La Palma, Spain (see Fig. 3.4) at an altitude of 2200 m above sea level. In 2018, the first prototype of the Large-Size Telescope (LST) of the next-generation CTA was built at close distance to the MAGIC telescopes; stereo observations with all three are now possible (Ohtani *et al.*, 2021).

MAGIC has been designed with an aim to provide a low-energy threshold in order to fill the gap between the satellite experiments observing in the GeV range and the previous generation of IACTs observing above a few hundred GeV (Barrio *et al.*, 1998). To this end, it has been equipped with a large, 17-m diameter mirror, the largest among all of the IACTs operating (until the construction of H.E.S.S. CT5 in 2012) and with high-quantum-efficiency (~30%) PMTs. On the other hand, the gain (the number of secondary electrons obtained from a single observed ph.e.) of the PMTs used in MAGIC has been intentionally kept low. This allows them

[c]https://magic.mpp.mpg.de/.

Fig. 3.4. MAGIC telescopes: MAGIC-I on the left and MAGIC-II on the right.

to also operate during moonlight conditions, doubling the duty cycle of the telescopes at the price of a larger energy threshold for observations taken during Moon time (Ahnen *et al.*, 2017d).

The MAGIC Telescopes have undergone a number of upgrades in their history (see Fig. 3.5). In 2007, a new readout system increasing the sampling speed from 300 MHz to 2 GHz was installed, allowing for a more efficient usage of the pulse timing information in the analysis (Aliu *et al.*, 2009a). The most notable effect on the performance was the construction of the M2 telescope (Aleksić *et al.*, 2012d) in 2009, which improved the sensitivity by approximately a factor of two. In 2011 and 2012, a series of upgrades of the key systems of MAGIC were performed, the most important of which was the replacement of the readout system of both telescopes (Sitarek *et al.*, 2013). This lowered the electronic noise and improved the low-energy performance of the telescopes. At the same time, the camera of the M1 telescope was upgraded to a more finely pixelized one, a 1039 pixel camera identical to the M2 camera. This upgrade also extended the trigger region of M1, improving the off-axis performance of the array (Aleksić *et al.*, 2016a).

3.1.3. *The Very Energetic Radiation Imaging Telescope Array System*

VERITAS is a four 12-m IACT array located at the Fred Lawrence Whipple Observatory at the base of Mt. Hopkins in Arizona, USA. It is the lowest

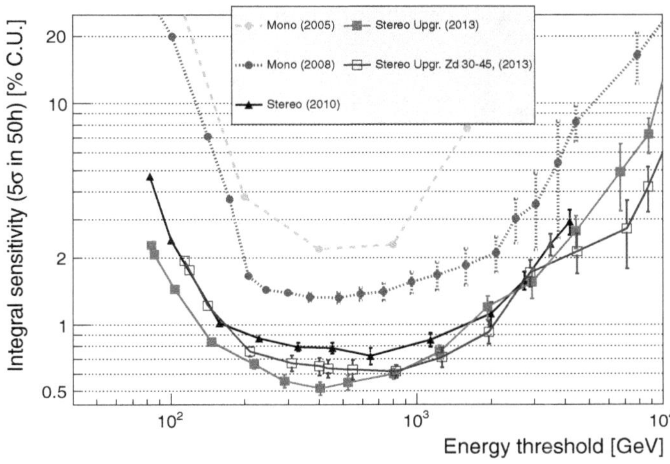

Fig. 3.5. Evolution of the integral sensitivity (the weakest source that produces a five-times-larger excess of gamma rays than the fluctuation of background) of the MAGIC telescopes. Gray circles: sensitivity of the MAGIC-I single telescope with the 300 MHz (light gray, long dashed) and 2 GHz readouts (dark gray, short dashed). Black triangles: stereo before the 2011–2012 upgrade. Squares: stereo after the 2011–2012 upgrade zenith angle below 30° (filled), 30° –45° (empty). Image reproduced from (Aleksić *et al.*, 2016b).

IACT array in operation, with an altitude of 1280 m above sea level. Even though its first prototype began taking data in 2003, four-telescope array operations started only in 2007, a delay mainly due to disputes over the site location. Despite being the youngest of the third-generation IACTs, it has perhaps the most ancient roots, being the direct successor of the very first IACT, Whipple, and a brainchild of Trevor C. Weekes, one of the pioneers of the field of VHE gamma-ray astronomy.

The VERITAS array underwent a reshaping in 2009; the first telescope, T1, which was located too close (35 m) to T4, was relocated further out. The wider spacing eliminated an important source of background due to the closest telescope pair triggering on local muons, thereby improving the sensitivity of the array significantly. The array trigger was also improved thanks to new electronics. In the summer of 2012, all 2000 PMTs of the four cameras were replaced with high-quantum-efficiency cameras. This allowed the cameras to detect twice as much light as before, resulting in a lowering of the threshold by 30% — from 100 to 70 GeV. More recently, VERITAS has showcased the ability of observation during moonlit nights (up to the full Moon) by means of a reduction of the PMT HV and UV filters placed in front of the camera. This capability improves the duty cycle of VERITAS by about 15%.

3.2. Technology

The third generation of IACT arrays has been in operation for more than a decade and has been very successful, increasing the number of known VHE sources by an order of magnitude, as shown in Fig. 3.2. Much of this success was due to the improvement in the telescope technology; this section reviews some of these developments.

3.2.1. *Structures*

One the most significant improvements of the third-generation IACTs, with respect to the previous generation, is the size and number of the telescopes themselves. Each of the three IACT arrays employs a different design solution to overcome the challenging technological hurdle of building affordable high-precision mounts for telescopes with $\sim$15 m-diameter reflectors.

<table>
<tr><td rowspan="2"></td><td colspan="2">H.E.S.S.</td><td rowspan="2">MAGIC</td><td rowspan="2">VERITAS</td></tr>
<tr><td>CT1–4</td><td>CT5</td></tr>
<tr><td colspan="5">Mount</td></tr>
<tr><td>Type</td><td colspan="3">Circular rail and fork</td><td>Antenna</td></tr>
<tr><td>Structure</td><td colspan="3">Space frame truss</td><td>Pillar</td></tr>
<tr><td>Material</td><td colspan="4">Steel</td></tr>
<tr><td colspan="5">Dish</td></tr>
<tr><td>Structure</td><td colspan="4">Space frame truss</td></tr>
<tr><td>Material</td><td colspan="2">Steel</td><td>Carbon-fiber</td><td>Steel</td></tr>
<tr><td colspan="5">Camera support</td></tr>
<tr><td>Structure</td><td colspan="2">Quadrupod</td><td>Arch</td><td>Quadrupod</td></tr>
<tr><td>Material</td><td colspan="2">Steel</td><td>Aluminium</td><td>Steel</td></tr>
<tr><td colspan="5">Azimuth positioning system</td></tr>
<tr><td>Type</td><td colspan="3">Circular rail and 6 bogies</td><td>Commercial</td></tr>
<tr><td>Drive</td><td>2 friction drives</td><td>4 motors</td><td>2 motors</td><td>Positioner</td></tr>
<tr><td>Speed</td><td>1.7 deg/s</td><td>3.3 deg/s</td><td>7 deg/s</td><td>1 deg/s</td></tr>
<tr><td colspan="5">Elevation positioning system</td></tr>
<tr><td>Type</td><td colspan="3">Circular rail and toothed ring</td><td>Commercial</td></tr>
<tr><td>Drive</td><td>1 friction drive</td><td>2 motor</td><td>1 motor</td><td>Positioner</td></tr>
<tr><td>Speed</td><td>1.7 deg/s</td><td>1.7 deg/s</td><td>7 deg/s</td><td>1 deg/s</td></tr>
</table>

The twin ~17 m telescopes of MAGIC were designed for fast re-pointing (7°/s), which is especially important in the observations of fast phenomena such as Gamma-Ray Bursts (GRBs). Their telescope structures are the most lightweight of the three third-generation IACT arrays, relative to the reflector size. A base frame rotates around the azimuth axis and supports the Optical Support System (OSS), or the "dish", which rotates around the elevation axis. The MAGIC base undercarriage is a steel space frame which moves on six bogeys on a circular rail. To reduce weight, the dish is a lightweight space-frame structure made of aluminium and carbon-fiber-reinforced plastic tubes, assembled with a proprietary rod-and-knot system. Its altitude movement is implemented with a circular rack and pinion mechanism. The total weight moving in Azimuth direction is 64 tons, and in zenith, 20 tons. There is one altitude and two azimuth motors. The camera is supported by a single tubular arch made by aluminium attached to the top and bottom of the reflector dish, secured by pre-tensed steel cables attached to the sides.

The structure of the H.E.S.S. telescopes is also based on the circular rail-and-bogies design, but the base, pillars and dish are made entirely of more conventional steel space frames. The structure of the smaller telescopes weighs around 60 tons, that of the larger telescope, around 580 tons, but its four azimuth and two altitude motors are so powerful that it can be re-pointed by 180° in approximately 50 s (3.3°/s in azimuth, 1.7°/s in altitude). The motor design of the 12-m telescopes, a friction drive, is twice as slow (1.7°/s). The cameras are supported by a tubular steel quadrupod attached at the vertices of the reflector dish.

The VERITAS telescopes use instead a commercial, single-pillar antenna positioner to suspend a trussed steel-tube dish and thus both the azimuth and altitude drive motors are located at the junction between the pillar and dish. Their maximum slew speed is 1°/s. The cameras are supported by tubular steel quadrupods attached to the reflector dish about two-thirds of the way from its center. This design is an improvement over the 50-year-old 10 m Whipple telescope (Fazio *et al.*, 1968).

All telescopes use "starguider" CCD cameras located at the center of the dish, whose images are used to build a structure bending-and-pointing model, which is then applied to correct the data offline. This procedure provides greater accuracy than using the information from the shaft encoders alone. To provide starguiding, the Field of View (FoV) of the starguider camera contains the Cherenkov camera and a portion of the sky next to it.

3.2.2. *Mirrors*

	H.E.S.S.		MAGIC	VERITAS
	CT1–4	CT5		
Reflector optics				
Shape	Hexagonal	Rectangular	Octagonal	Circular
	D.-C.[a]	Parabolic		D.-C.[a]
Diameter	12.5 m[b]	28 m[b]	17 m[b]	12 m
f	15 m	36 m	17.5 m	12 m
FoV	5°	3.2°	3.5°	3.5°
Mirror facets				
Shape	Round	Hexagonal	Square	Hexagonal
Size	60 cm	90 cm	49.5 or 99 cm	60.9 cm
Bulk	Glass		Al or mix	Glass
Area	108 m^2	614 m^2	236 m^2	110 m^2
Quantity	382	875	974 or 246	350
Coating	Quartz			Anodized Al
Alignment	2 actuators			3 screws

[a]Davies-Cotton design; [b]Circular equivalent; f: Focal length.

The mirror surfaces of the third-generation IACTs are all made of tessellated mirror facets, as was the case for the previous generations. The most relevant improvements of the third-generation IACTs are in the mirror fabrication and coating techniques, ensuring long-lasting high reflectivity, as well as in the systems of mirror facet alignment, providing better optical characteristics.

The shape of the facets is square for MAGIC, circular for the H.E.S.S. 12-m telescopes and hexagonal for the VERITAS and H.E.S.S. 28-m telescopes. Circular facets are easier and cheaper to build, but hexagonal and square facets allow complete edge-to-edge tiling of the reflecting surface. The structure of the mirrors has several layers: in the case of MAGIC-I, the frame is a lightweight aluminium honeycomb onto which the reflective surface — a 5-mm layer of diamond-ground and polished AlMgSi alloy — is glued. The mirror surface is coated with a thin layer of quartz for protection against the elements. MAGIC-II sports a mix of 104 aluminized, cold, slumped glass mirrors in the outer part of the dish and 142 aluminium honeycomb mirrors in the inner part of the dish. The mirrors of H.E.S.S. and VERITAS are instead all made of aluminized slumped glass (with a protective quartz coating for the former and an anodized aluminium layer for the latter); a cheaper, heavier design. The reflectivity of the mirrors is between 80% and 90% in the blue region of the spectrum (between 280 and 450 nm), degrading, however, by about 3–6% per year due to weathering. To address this problem, H.E.S.S. and MAGIC have gone through mirror replacement campaigns every 5–7 years of operation, while VERITAS has an on-site re-aluminizing and re-anodizing facility that constantly refurbishes the mirrors.

The mirror facets of the VERITAS and H.E.S.S. CT1–4 telescopes are arranged approximating a spherical surface whose radius of curvature is half that of the facets, an arrangement known as the Davies–Cotton design (Davies and Cotton, 1957) originally developed for solar furnaces. The reflector surfaces of the MAGIC and H.E.S.S. CT5 telescopes instead approach a paraboloid using spherical mirror facets with different focal lengths depending on their distance to the center of the dish. The parabolic mirror surface is isochronous, therefore, the time spread of the reflected photon mirrors arriving at the camera is minimized. This, combined with the fast response of the camera trigger electronics, allows for the exploitation of the very-short duration of Cherenkov light flashes to minimize spurious events due to the light from the night sky.

Both reflector designs suffer from optical aberrations (i.e. the images become less focused) with increasing angular distance from the optical axis, with the Davies–Cotton design being about 20% better than the parabolic design beyond two degrees off-axis. To describe these aberrations, it is customary to refer to the Root Mean Square (RMS) width of the Point Spread Function (PSF), which is the response of an optical system to a point source of light at infinity. The designs employed for the third-generation IACT reflectors achieve typical PSF RMS widths of the order of 0.05°on-axis and 0.1°at two degrees off-axis.

The large structures of the MAGIC and H.E.S.S. CT5 telescopes deform appreciably under their weight, changing the optical properties of their reflectors as the telescope moves. To keep the optical PSF of the segmented reflector optimal during observation, two newly introduced techniques are employed: for MAGIC, the position of each mirror facet is dynamically adjusted using two actuators (Biland *et al.*, 2007); for the stiffer H.E.S.S. CT5 structure, less complex corrections are needed, so the camera is simply moved along the optical axis. The H.E.S.S. mirrors are also equipped with mechanical actuators, but these are only used every couple of years or after mirror exchange campaigns to re-optimize the reflector optical PSF. The

VERITAS mirrors are supported by a system of three screws and have to be re-adjusted by hand after each refurbishing. These mirror PSF adjustment methods all employ a CCD camera mounted at the center of the dish, similar to the starguider camera, but with a FoV centered on and enclosing the Cherenkov camera.

3.2.3. *Cameras*

The cameras are critical elements of IACTs, since they are the detectors that transform Cherenkov light from atmospheric particle cascades into digitally processed signals. The main components of an IACT camera are the photo-detector, readout and trigger systems, all supported by mechanics and environmental control.

Photo-multiplier tubes				
Telescopes	H.E.S.S.			
	CT1–4		CT5	
	2002–2015	2015–	2012–2019	2019–
Make	Photonis		Photonis	Hamamatsu
Model	XP 2960		XP 29600	R12992
Count	960		2048	1758
Diameter	29 mm		29 mm	38 mm
FoV	0.16°		0.067°	0.09°
Cathode	Bi-alkali		Bi-alkali	“Super”
Peak Q.E.	25%		29%	35%
Dynodes	8		8	7/8 mix
Gain	$\sim 2 \times 10^6$		$\sim 2 \times 10^6$	$\sim 2 \times 10^4$
Telescopes	MAGIC		VERITAS	
	M1	M2 & M1	T1-4	
	2004–2011	2009–/2012–	2007–2012	2012–
Make	Electron Tubes	Hamamatsu	Photonis	Hamamatsu
Model	ET 9116/9117	R10408 Mod	XP 2970	R10560 Mod
Count	577(397+180)	1039	499	
Diameter	25.4/38.1 mm	25.4 mm	29 mm	
FoV	0.1°/0.2°	0.1°	0.15°	
Cathode	Bi-alkali	“Super”	Bi-alkali	“Super”
Peak Q.E.	30%	35%	25%	34%
Dynodes	6		10	8
Gain	$\sim 3.5 \times 10^4$		$\sim 2 \times 10^6$	

In the H.E.S.S. cameras, all subsystems are all located within the camera body at the focal plane of the reflector. The cameras are cooled by air and parked inside a protective shelter at ground-level, except for the newest CT5 camera, FlashCam, which is water-cooled and doesn't require a shelter. In the VERITAS cameras, signal detection and amplification happens at the focal plane, and its digitization and trigger generation is done by electronics located in a container adjacent to the telescope. The MAGIC cameras convert the photo-detector signal into light using vertical-cavity surface-emitting lasers (VCSELs), then sent through light fibers to the "counting house" located between the two telescopes, where the readout and trigger systems of the entire array are located. VERITAS and MAGIC cameras are both waterproof and parked horizontally on top of a camera access tower.

3.2.3.1. *Photo-detectors*

The photo-detection technology employed in third-generation IACT cameras is based on classic vacuum-tube Photo-Multiplier Tubes (PMTs). Their low-cost per sensitive surface area, robustness, and recent performance improvements have kept them as the main choice, ahead of alternative solid-state or hybrid detectors.

Currently, the cameras of the third-generation IACTs (except the H.E.S.S.-I cameras) all employ PMTs using "super" bi-alkali cathodes, capable of peak quantum efficiency of around 35% in the blue part of the spectrum (Mirzoian *et al.*, 2006). Since the entrance window of a PMT is typically round (flat or domed), IACT cameras employ light concentrators in the form of modified "Winston cones" (Winston, 1970). These non-imaging optical elements are paraboloidal funnels with hexagonal entrances and a highly reflecting interior surface, typically achieved by coating or aluminized foil. Their entrances are arranged in a honeycomb pattern to achieve very-high-filling factors and are shaped to restrict the maximum light entrance angle to the angular size of the dish to avoid stray light from the ground ("albedo protection"). In the case of MAGIC, they were also optimized to increase the probability of photon-cathode double-crossing for better PMT quantum efficiency. The cameras of MAGIC also employ a Plexiglass window in front of the cones for improved environmental control. The VERITAS and MAGIC camera entrance windows can be covered by specially designed UV-pass filters for observations with very-bright moonlight conditions (Moon phase greater than 70%).

3.2.3.2. *Readout*

The readout system is responsible for the amplification and digitization of the electrical signals produced by PMTs. This amplification is achieved by means of discrete, very-low-noise, analog, cross-impedance operational amplifiers. The amplification stage usually also splits the signal into two or three branches of a different gain factor; these branches are then fed to the digitizer and trigger systems. The analog bandwidth of the amplifiers of the third generation of IACTs is between ~80 (original H.E.S.S.-I cameras) and ~700 MHz (MAGIC cameras after upgrade), and their linearity is around 1% or better.

Readout				
Telescopes	H.E.S.S. CT1–4 2002–2015	H.E.S.S. CT1–4 2015–	H.E.S.S. CT5 2012–2019	H.E.S.S. CT5 2019–
Ch./pixel	2		2	1
Data type	Integrated charge & Peak time			Waveform
Exposure	16 ns			128 ns
Sampling	1 GS/s		1 GS/s	250 MS/s
Resolution	8-bit	12-bit	12-bit	12-bit
Technology	SCA		SCA	FADC
Rate	250 Hz	500 Hz	3.6 kHz	3.2 kHz

Telescopes	MAGIC M1 2004–2007	MAGIC M1–2 2007–/2016–	VERITAS T1–4
Ch./pixel	1		1
Data type	Waveform		Waveform
Exposure	30 ns		48 ns
Sampling	300 MS/s	2/1.64 GS/s	500 MS/s
Resolution	8-bit	12/14-bit	8-bit
Technology	FADC	SCA†	FADC
Rate	350 Hz	700 Hz	400 Hz

†: The M1 camera used 12-bit 2GS/s FADC from 2007 to 2011.

For the signal digitization, two different technologies are currently used: Flash Analog-to-Digital Converters (FADCs) and Switched Capacitor Arrays (SCAs). The MAGIC-I camera up to 2011, the VERITAS cameras and the H.E.S.S. CT5 FlashCam uses the former; the H.E.S.S. CT1–4 and the current MAGIC-II cameras, the latter. The inner workings of these electronic devices are profoundly different. FADCs directly digitize the input electrical signal by means of a voltage ladder and several comparators in parallel. SCAs, by contrast, are analog memory buffers that store the electrical signals by charging an array of capacitors in sample-and-hold mode that are cyclically switched to the input signal at a fast rate; the stored charge is later digitized by regular ADCs. The digitizing rate of FADCs and the storing rate of SCAs employed in the third-generation IACTs is of the order of ~ 1 GHz. FADCs are readily available commercially, relatively easy to use and have no dead-time; but their power consumption, channel density and cost are limiting factors. To overcome this, the MAGIC-I readout system in use until 2011 multiplexed 16 pixels onto a single FADC (300 MS/s, up to 2007 then 2 GS/s) requiring optic fiber multiplexers and delay lines. VERITAS instead uses a custom-built, 8-bit 500 MS/s FADC design (Holder *et al.*, 2006). The FlashCam of the H.E.S.S. CT5 employs 12-bit 250 MS/s FADCs, hence its name.

SCAs are much cheaper than FADCs, can reach higher sampling rates and consume a tenth of the power ($\sim$10 mW per channel); they cannot sample signals continuously, however, and therefore suffer from readout dead-time. H.E.S.S. was the first IACT array to employ them — the ARS chips (Druillole *et al.*, 2001) were installed in 2002 in the H.E.S.S.-I cameras. The design of these chips evolved quickly and subsequent generations were used in H.E.S.S.-II (SAM Delagnes *et al.* 2006 for the CT5 telescope from 2012 until 2019 and NECTAr Naumann *et al.* 2013 for the CT1-4 cameras after their upgrade in 2015). The same happened to the Domino Ring Sampler (DRS) chip family used in MAGIC from 2009 on; the DRS-2 (Ritt, 2004) was used in M2 between 2009 and 2011, followed by DRS-4 (Ritt, 2008) later on for both telescopes.

The experience from the third-generation IACTs showed that both FADC- and SCA-based readout systems fulfill the needs of the instruments well — therefore, both technologies will be used for different camera types in the next-generation IACTs.

3.2.3.3. *Trigger*

All third-generation IACT cameras are self-triggered detectors, that is, the camera readout is initiated by a signal produced by electronic logic built to distinguish fast, localized flashes of light over a continuous stochastic background. This is usually a pipelined logic, beginning with a fast discrimination at the pixel level (level zero, L0), then a logic singling out local patterns at camera level (level one, L1 and sometimes two, L2), and ending with a simple 2-coincidence between telescopes at the array level (level three, L3). The implementation of these trigger levels varies considerably between the three existing systems.

At the pixel-level trigger, the front-end amplification stage provides a dedicated analog trigger signal per pixel, from which the L0 signal, i.e. the digital signal carrying the information about the presence of a pulse above a certain threshold, is generated. In H.E.S.S., this is accomplished by means of a fast comparator (Funk *et al.*, 2004) whose advantage is its fast switching time; in MAGIC with a software-programmable discriminator (Paoletti *et al.*, 2007) whose advantage is its versatility; and in VERITAS with a constant fraction discriminator (Hall *et al.*, 2003), whose advantage is that it keeps the timing information of the pulse. The individual pixel threshold for the L0 trigger is usually set to 4–5 ph.e. for moonless observations — MAGIC and VERITAS, however, regulate it dynamically based on the trigger rates, with the goal of homogenizing the camera trigger response in the presence of background light (i.e. stars or stray light).

The camera-level trigger implementations are even more diverse. The H.E.S.S. CT1-4 cameras use a multiplicity trigger within overlapping 64-pixel sectors similar to that of the second-generation IACT experiment, CAT (Barrau, 1997). A camera trigger is issued whenever three or more L0 signals are simultaneously active in a time window of 1.3 ns in any of the 38 contiguous 64-pixel sectors mapping the camera. The very-short time window is achieved by the use of comparators and the careful sorting of PMTs according to their transit times. The older H.E.S.S. CT5 camera also used a similar scheme — however, an additional trigger was implemented (Tluczykont and Fontaine, 2005) with the goal of optimizing the data flow by a fast image analysis of the binary amplitude information available at the camera trigger level. The newest CT5 camera, FlashCam, has no L0 and instead triggers when the sum of the signals inside any 9-pixel group is above a certain threshold. To avoid trigger instabilities due to the light of the night sky, individual pixels inside the patch may be excluded from the sum if they are brighter than a certain level. The MAGIC cameras instead implement a compact next-neighbor scheme in which a camera trigger is issued whenever the L1 signals of a 3-pixel compact next-neighbor group are active. Since MAGIC L1 signals can be delayed and stretched, an optimization algorithm is needed to find the optimal delays, allowing the reduction of the L1 gate to 5.5 ns (Aleksić *et al.*, 2016b). An alternative camera trigger logic has been successfully implemented in MAGIC and was used in 2008 to detect the Crab pulsar for the first time (Aliu *et al.*,

2008): the "sumtrigger" (Rissi *et al.*, 2009). With this logic, the camera is triggered whenever the sum of the analog signals of pixels within a camera-contiguous sub-region ("patch") is above a certain threshold, thereby completely foregoing the L1. To eliminate spurious triggers due to PMT afterpulsing, the signals are clipped before being summed. This sumtrigger system has since been improved and implemented for both telescopes (Dazzi *et al.*, 2021). Another trigger sub-layer, also exploiting fast image analysis techniques, was implemented in MAGIC-I in 2003 (Paoletti *et al.*, 2007), but has since been decommissioned. The VERITAS camera trigger system implements a pattern logic where coincidences of three neighbor pixels (compact and in a row) trigger the camera. The coincidence window is 5 ns (Zitzer *et al.*, 2013).

An exception to all of the above is the trigger system of the newest camera of the H.E.S.S. CT5 telescope, FlashCam. It implements a fully digital trigger directly from the FADC readout, foregoing dedicated electronics and using the programmable logic elements of 96 on-board FPGAs instead.

The array trigger (L3) is implemented as for all three third-generation IACTs as a coincidence of two or more camera triggers, except for the H.E.S.S.-II array, where triggers from CT5 alone are also accepted as array triggers. The coincidence between the L2 triggers happens after the signals are delayed according to the telescope pointing position to account for the travel time of the Cherenkov light from different telescopes. The coincidence windows are 80 ns for H.E.S.S. (Funk *et al.*, 2004), 50 ns for VERITAS (Weinstein, 2007) and 100 ns for MAGIC (Paoletti *et al.*, 2008).

3.2.4. *Auxiliary devices*

All third-generation IACT arrays employ auxiliary devices for a variety of purposes. The most important one is calibration of the light throughput of the camera, usually achieved by a device that illuminates the telescope with uniform ns-long pulses of blue light. The main purpose of this device, called a "flatfielding unit", "calibration system" or "LED flasher", is to check the response of the camera to a light pulse of constant intensity. This calibration can be relative (i.e. a measure of the light detection and conversion efficiency of the system) or absolute (i.e. a measure of the conversion factor, or "gain" between incoming photons and the recorded digital signal). The "flatfielding unit" needs to provide a homogeneous light intensity across the whole camera — thus, it is usually placed at or near the optical axis of the telescope, at the center of the reflector or on the quadrupod. The light pulses are emitted by commercial blue LEDs and diffused via a holographic or opal diffuser; MAGIC uses the light of a blue laser diffused by an integrating sphere. The intensity can be varied by means of filter wheels or by flashing several LEDs simultaneously. Lasers can deliver faster, brighter, more consistent and monochromatic light pulses, while LEDs are cheaper, easier to operate and maintain, and have light intensity that is easier to regulate and make homogeneous. The two technologies are therefore interchangeable and, in fact, VERITAS went from using a laser to an LED-based device and MAGIC went the opposite way, while the 28-m H.E.S.S.-II telescope has both types of calibration systems installed.

Another important auxiliary system is the absolute timing system, which is usually implemented using a GPS-marshalled rubidium oscillator, either commercial or custom-built, operated in conjunction with the array or camera trigger. The timestamps from this system are generated at each trigger and sent to the data acquisition system where they are appended to each recorded event in the data stream. To cross-check the validity of these timestamps, third-generation IACTs use a specially modified "central pixel" system that digitizes at some kHz the DC-coupled signal from one of the PMTs of the cameras. This system is then activated when pointing the telescope at the Crab pulsar (the brightest optical pulsar); the data are then timestamped and the pulsar phases are calculated with suitable pulsar ephermerides from radio observatories. If it is possible to retrieve the expected optical pulsar light curve from this data, then the validity of the absolute timing system is confirmed.

Other auxiliary devices of IACTs are mainly used to determine the transmission properties of the atmosphere; these include weather stations, pyrometers, all-sky cameras, Light Detection And Rangings (LIDARs), dust-concentration sensors and even small battery-operated optocopters illuminating the array.

3.3. Progress in Analysis Methods

Cherenkov telescopes observe gamma rays in an indirect way by recording the images of Cherenkov light induced by the shower propagation through the atmosphere. The image information for all of the telescopes is combined to extract three pieces of information from each event: its parent particle, the arrival direction of the primary particle and the energy of the particle.

To achieve this, the signal in each pixel of the image is first extracted from the night sky background noise and calibrated to provide information about the arrival time of the signal and the total charge gathered in the pixel.

In most of the analysis methods, a pre-selection of pixels forming the actual image (and not induced solely by night sky background and electronic noise) is required — also referred to as image cleaning, often a two-stage process (see, e.g. tail cuts image cleaning in Punch *et al.*, 1991). The image is then composed of "core" pixels with a signal above a given threshold and "boundary" pixels with a signal above a lower threshold, but neighboring a core pixel. A similar procedure has also been used for the analysis of the data of H.E.S.S. telescopes (Aharonian *et al.*, 2006d). It is also being used now for the analysis of VERITAS telescope data; the thresholds are not the same for all of the pixels, however, and depend on the noise registered in a given pixel (Punch *et al.*, 1991; Daniel *et al.*, 2007). The timing information of the pulses is used in the two-pass method (Holder *et al.*, 2005) to improve the reconstruction of individual pulses. In the first pass, pulses are extracted in a fixed broad window, then parametrized, and the time gradient along the main axis is computed. That gradient is then used to derive the optimal extraction windows for the pulses, lowering the effect of the NSB. The analysis of the data from MAGIC telescopes also exploits additional knowledge of the arrival time of the signals. Since only signals occurring close in time in neighboring pixels are accepted, one can lower the cleaning thresholds without increasing the fraction of "fake" events from NSB accidental events (Aliu *et al.*, 2009a). In addition, a slightly different algorithm is used where compact groups of two, three or four neighbors are searched with a signal close in time and above a given threshold (Shayduk *et al.*, 2005; Aleksić *et al.*, 2016b).

To develop and apply the analysis methods, the response of the telescopes to gamma-ray (and sometimes also background) events needs to be known. The IACT technique is rather complex; the information registered by the telescopes depends on the development of extensive air shower (EAS) in the atmosphere, absorption of the Cherenkov light in it and the response of the telescope itself. Even more importantly, contrary to particle detectors in accelerator experiments, the environment is not fully controlled and the response of the detectors cannot be calibrated with a test beam. Therefore, Monte Carlo (MC) simulations are essential in IACT analysis (Hillas, 1982). They consist of a full simulation from a particle entering the Earth's atmosphere until the signal recorded with the IACTs's camera is stored. First, individual EASs are simulated in the atmosphere, for which task the CORSIKA program is commonly used (Heck *et al.*, 1998). The Cherenkov photons are then passed through the chain of procedures simulating the reflection from the mirror, conversion of photons into ph.e.s in the camera, trigger logic, etc. Such simulations are being done with dedicated programs developed for each telescope system, such as sim_telarray for H.E.S.S. (Bernlöhr, 2008). The MC simulations are needed mainly for two purposes: first, they are used to train the analysis methods such as gamma/hadron separation or energy estimation, and second, the full analysis chain (including event reconstruction and gamma-ray selection cuts) is applied to samples of MC simulations to calculate the energy-dependent collection area of the telescopes for a given analysis. The collection area is then used to convert the rates of gamma-ray events into fluxes expressed in absolute units. Finally, MC simulations are also crucial to evaluate the performance of the instrument (see Section 3.4) and to develop new analysis methods.

In order to fully exploit the technological novelties of the third generation of IACTs, the hardware development has to be matched by simultaneous progress in analysis methods. Figure 3.6 shows the comparison of the detection of the standard candle of VHE gamma-ray astronomy, the Crab Nebula, with an early IACT, Whipple, and the third-generation IACT, MAGIC. In the case of the first Whipple observations with a 37-pixel camera, despite 98% rejection of background events, the signal was just a small excess on top of a huge background. In the case of HEGRA, the background rejection is already much stronger with a very clear signal. In the case of the third-generation IACTs such as MAGIC, a very strong signal on top of a heavily reduced background is visible.

Despite 30 years of progress in analysis methods of Cherenkov telescopes, most of them are still based on the analytical parametrization of images as ellipses proposed by Hillas (1985a). Even though various, more sophisticated methods were already attempted in the second generation of IACTs (see, e.g. Reynolds and Fegan 1995), they were mostly not proven superior and hence did not become standard. At the start of the third generation of IACTs, simple box cuts were still commonly used, allowing ranges of individual image parameters and possibly taking into account simple correlations of those parameters (Daum *et al.*, 1997). The advent of H.E.S.S., MAGIC and VERITAS stimulated work on more advanced analysis methods. Correlations between different parameters can be easily exploited by using, e.g. artificial neutral networks (see, e.g. Dumora *et al.*

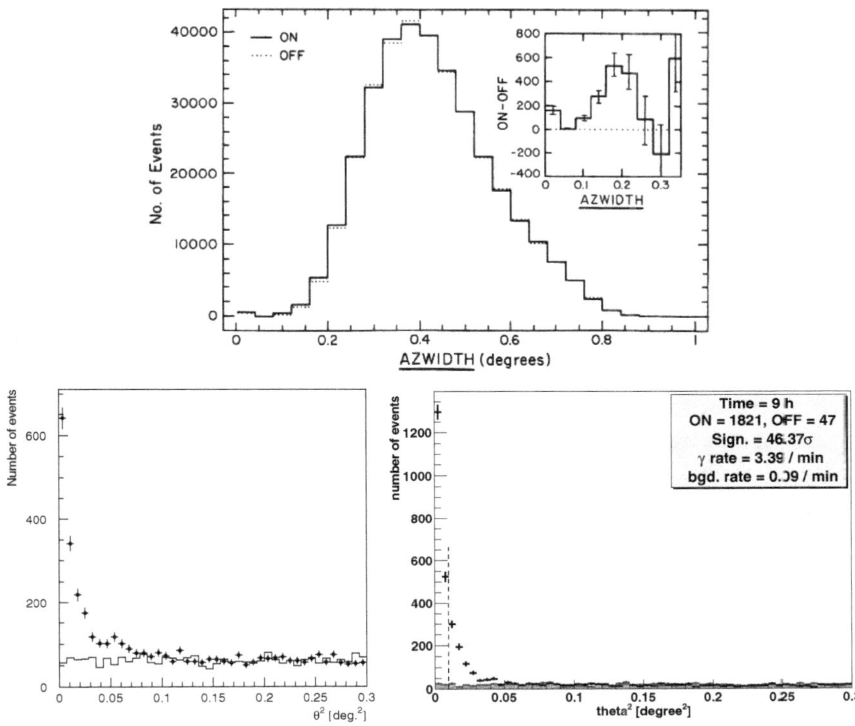

Fig. 3.6. Comparison of the source-detection plots of Crab Nebula with Whipple (*top panel*), HEGRA (*bottom left panel*) and MAGIC telescopes (*bottom right panel*). *Top panel*: the distribution of the Azwidth parameter (width of the shower calculated perpendicularly to the direction of the source). *Bottom panels*: the θ^2 parameter, which is the squared angular distance between the nominal and event-wise reconstructed source position. In both panels, the gamma-ray emission appears as excess at small values of Azwidth or θ^2. Top panel is reproduced from (Weekes *et al.*, 1989), bottom left from (Aharonian *et al.*, 2000), bottom right from (Aleksić *et al.*, 2012d).

2001), random forest (see, e.g. Albert *et al.* 2008a) or boosted decision trees (see Ohm *et al.* 2009; Becherini *et al.* 2011; Krause *et al.* 2017). Finally, the so-called model-analysis (see, e.g. de Naurois and Rolland 2009) approach does not use the simple elliptical parametrization of images — instead, the images are fit with an average template shape corresponding to different physical parameters of the showers. Currently a possibility to use the "deep learning algorithms" (convolutional neural networks) is being investigated for Cherenkov telescopes (see, e.g. Nieto *et al.* 2017). It should be noted, however, that the more complicated the analysis method, the more prone it is to systematic uncertainties caused by small differences between the real data and the MC simulations describing the response of the telescopes. The advanced analysis methods allow us to get closer to the physical limits of IACT performance (see Maier and Knapp, 2007; Sobczyńska, 2009a,b; Sitarek *et al.*, 2018).

3.3.1. *Gamma/hadron separation with box cuts*

Even for a bright source, before the event selection, there are three orders of magnitude more background events than gamma rays. Strong background suppression can be achieved by exploiting the image shape information (the hadronic images are normally broader and more irregular, with additional separated parts of the image, i.e. islands) or by applying cuts in the reconstructed arrival direction of the shower. The images can be easily parametrized in an analytical way (Hillas, 1985a; Fegan, 1997) as two-dimensional Gaussian distributions. The central moments calculated along the main axes of this distribution are called the "width" and "length" of the shower (see Fig. 3.7). The width and length parameters are often represented as half-widths of the main axis of an ellipse.

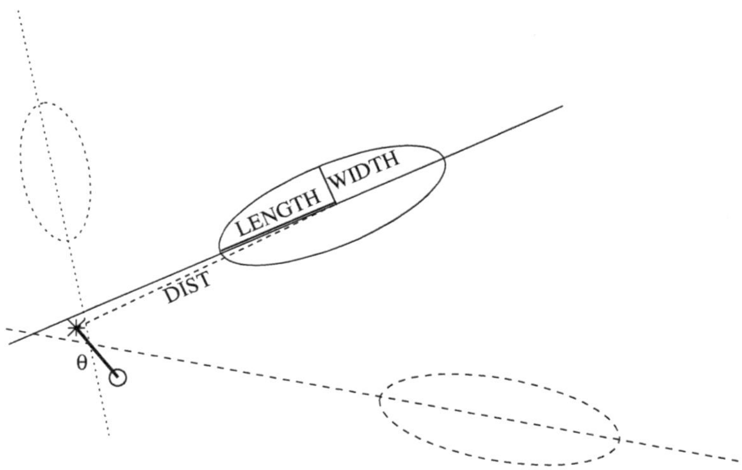

Fig. 3.7. Parametrization of Cherenkov images. The image (solid line ellipse) is parametrized as an ellipse, with width and length being the central moments of the 2D Gaussian distribution. The estimated direction of the primary gamma ray (star) is evaluated from the crossing point of the main axes of overlaid images from all of the telescopes (dotted and dashed lines). The angular distance between the true (circle) and estimated source position (thick line) is called θ.

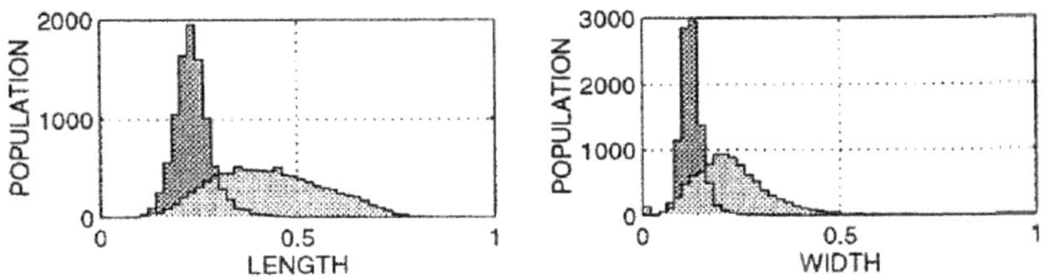

Fig. 3.8. Distribution of the image parameters length (*left*) and width (*right*) for simulated gamma-ray (dark gray) and real data hadronic background (light gray). Image reproduced from Fegan (1997).

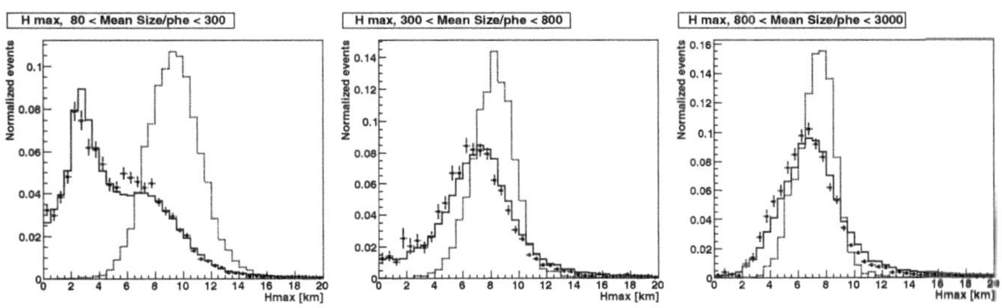

Fig. 3.9. Distribution of $H_{\max}$ in different ranges of the size parameter for MC gamma rays (dotted line), MC protons (full circles) and real MAGIC telescope background data (solid line). Image reproduced from Aleksić *et al.* (2012d).

By applying cuts in the image width and length parameters, a strong suppression of the background can be achieved (see, Fig. 3.8). The values of the cuts have to be chosen to provide optimal gamma/hadron separation (see, e.g. "supercuts", Punch *et al.*, 1991). Including a dependency to image size and shower telescope geometry can improve the separation power of the cuts (Aharonian *et al.*, 1997; Krawczynski *et al.*, 2006).

Estimation of the height of the shower maximum in stereoscopic IACT systems allows for a severe reduction of low-energy background. Isolated muons crossing the telescope at an impact parameter of several tens of meters produce images similar to low-energy gamma rays (Maier and Knapp, 2007). Such events are reconstructed to occur at the height of $\sim$ 2–3 km, while the bulk of gamma rays with such size has a shower maximum that is much higher (see Fig. 3.9). Therefore, stereoscopic reconstruction of the showers can easily remove the muon-dominated events, greatly improving the performance of stereo IACTs at the lowest energies.

3.3.2. *Reconstruction of the shower geometry*

Cherenkov light induced by a gamma-ray shower can be seen as a glowing column in the upper atmosphere (Hillas, 1996), its direction aligned with the direction to the source. By viewing such a column of light simultaneously by more than one IACT from different angles, simple three-dimensional geometry can be applied to reconstruct the properties of a shower (Kohnle *et al.*, 1996). Every point in the camera corresponds to a given direction in the sky — hence, the main axis of an image represents the directions of the shower axis in the camera plane. The same shower seen by a second telescope from another angle results in a different orientation of the main axis of the image (see Fig. 3.7 solid and dotted line and ellipse). The crossing point of the two main axes corresponds to the only shared direction of the two, i.e. the direction of the shower axis, which is the estimation of the direction of the primary particle. The method loses its power when the shower impact point is aligned with the position of the two telescopes. In a case when more than two images are used, a weighted average of the intersection points can be utilized, with the weight dependent on the crossing angle of the main axes and/or image quality (Hofmann *et al.*, 1999).

In a similar way, the shower impact point on the ground can be estimated by intersecting the main axes of the image starting from the telescope position (Hofmann *et al.*, 1999, see also Fig. 3.10). The distance from the shower axis to the telescope is usually called the impact parameter. Assuming that the most luminous part of the image in its centroid corresponds to the shower maximum, its height, H_{max}, can then be estimated using the shower impact parameter and the angular distance between the image centroid and the reconstructed arrival direction (see, e.g. Kohnle *et al.*, 1996). Multiple alternative methods are used to obtain the height of the shower maximum in IACT analysis. For example, it can be obtained as the altitude of a plane that minimizes the crossing point of the lines led from the individual telescopes in the direction of the centroid of the image (Sitarek *et al.*, 2018). In the case of the standard analysis of the MAGIC telescope data, two such lines are led from the direction of each of the telescopes in the direction of the center of gravity of its image. An additional third line is led from the reconstructed impact point along the reconstructed direction of the shower. Then, the height of the maximum is reconstructed as the height of a horizontal plane, in which the triangle formed by the crossing points with those three lines has the smallest circumference.

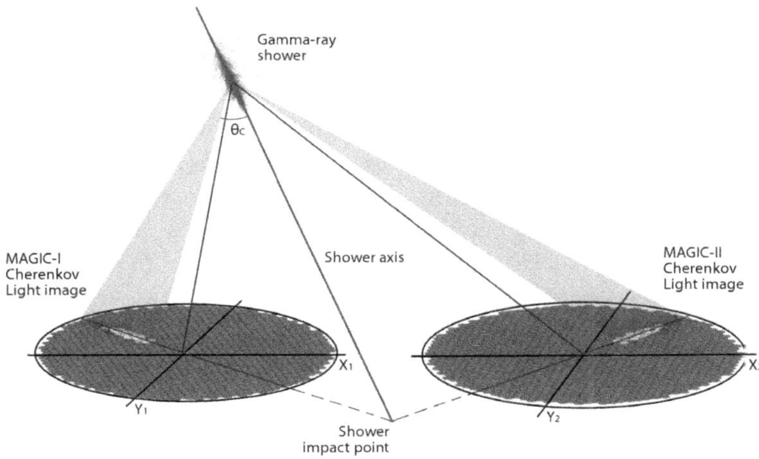

Fig. 3.10. Principle of the impact point reconstruction in stereo observations. Main axis of an image (ellipses in the figure) is projected to the ground (dashed lines). These lines cross in the impact point of the shower.

The above methods are obviously only directly applicable in the case of a stereo system. In the case of observations with a single telescope, the direction of the primary particle can be reconstructed based on the ellipticity of the image (Fomin *et al.*, 1994; Domingo-Santamaria *et al.*, 2005). An event with a small impact parameter will have a more roundish image, while farther events are more elongated. Even while this method provides an estimate of the arrival direction, the impact parameter and the height of the shower maximum cannot be unambiguously disentangled. For example, an event that is closer but lower in altitude will have the same distance between the source position and the centroid of the image as a farther but higher event. Part of this ambiguity can be solved using timing information. Due to the dependence of the Cherenkov angle on the altitude, there is a correlation between the impact parameter and the time gradient measured along the main axis of the image. The correlation was already noticed in the case of HEGRA telescopes (Heß *et al.*, 1999). The application of such a parameter into single IACT analysis was done for the MAGIC-I telescope, which greatly improved its performance (Aliu *et al.*, 2009a). In fact, the time gradient measured along the main axis of the image can also improve the performance of a stereo system, especially in a case of far-impact events (de la Calle Pérez and Biller, 2006; Stamatescu *et al.*, 2011; Aleksić *et al.*, 2012d).

3.3.3. *Multivariate methods*

As seen in the previous sections, the gamma/hadron separation can be based on multiple image and stereo parameters. Simultaneous cuts in multiple parameters are, however, difficult to optimize due to limited event statistics. In addition, the correlations of parameters could often provide additional gamma/hadron separation. Therefore, it is desirable to have a single parameter summarizing how likely a given event is, in fact, a gamma ray. Multivariate classification methods can be employed to provide such a parameter. While the multivariate classification methods were already tried for earlier Cherenkov instruments (see, e.g. Halzen *et al.*, 1991; Reynolds and Fegan, 1995; Dumora *et al.*, 2001), their popularity greatly increased with the third generation of IACTs (Bock *et al.*, 2004; Albert *et al.*, 2008a; Krause *et al.*, 2017). In practice, such a parameter is based only on the image shape and stereo parameters, with the background rejection based on the arrival direction of events treated independently.

3.3.3.1. *Multivariate decision trees*

A type of a multivariate classification algorithm, the multivariate decision tree, is a commonly used method in the third generation of IACTs. Random Forest (RF) (Breiman, 2001) is the implementation of multivariate decision trees used in the analysis of the MAGIC telescope data (Albert *et al.*, 2008a). Similar to the case of artificial neural network (ANN), the training of gamma/hadron separation with an RF requires samples of gamma-ray and background events and a list of image parameters calculated for each event. For training of a single tree, random sub-samples of gamma rays and background events are bootstrapped (i.e. randomly sampled with repetitions, such that about $\sim 2/3$ of events are selected from each sample; the use of such a method for parameter estimation is sometimes referred to as "bagging"). At each step of the tree generation, a number of image parameters are selected as possible cut parameters. Typically, the number of parameters selected at each step is equal to the square root of the total number of parameters used in the training. The optimal cut parameter and value are selected and two new nodes are created. Branching of the tree finishes once the end nodes contain only particles of a single type[d] (see Fig. 3.11). The selection of the best cutting parameter and its value at each

[d]Alternatively, to avoid over-training, the branching can stop earlier, once the node content is smaller than a given number of events.

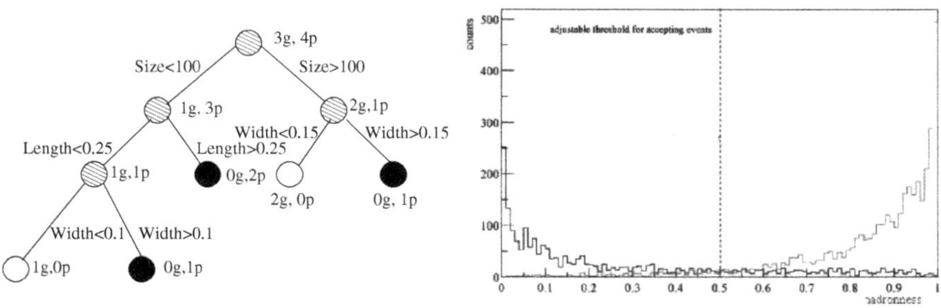

Fig. 3.11. *Left*: Schema of gamma/proton classification with a RF tree. Topmost node contains three gamma rays and four background events. A cut in size parameter divides it into further nodes (dashed circles). Further cuts in width and length parameter divide the nodes until they contain only gamma rays or protons (white and black circles). *Right*: Distribution of the final classification parameter of RF for gamma rays (black) and background events (gray). Image reproduced from Albert *et al.* (2008a)

step is done with the help of the Gini index (Breiman *et al.*, 1984). The Gini index is defined as a scaled variance of a binomial distribution (Albert *et al.*, 2008a), and for a node with N_g gamma rays and N_b background events is equal to

$$Q_{\text{Gini}} = 4N_gN_b/(N_g + N_b)^2. \tag{3.1}$$

It spans the range of values from 0 (sample of only one type of particle) to 1 (sample with equal contribution of both particle types). The cuts at each step are therefore selected to minimize the average Gini index of the two newly created nodes. In a typical case, hundreds of thousands of events are being used in the training and the tree contains tens of thousands of nodes. The procedure is then repeated about a hundred times over newly bootstrapped sub-samples to construct individual trees.

In order to apply RF classification to an independent event, it has to be propagated through the cuts stored in the RF tree to find out the terminal node to which it corresponds. The final classification is the average response of the terminal nodes corresponding to the event in all the trees of the RF (see Fig. 3.11).

The RF method can be used not only to provide classification between a number of event classes, but also to estimate continuous variables, such as the energy of an event. The first method for doing so is to simply divide the continuous variable into bins, treat each bin as a single class of events and then apply the aforementioned procedure. Alternatively, the binning can be avoided by using as a selection criterion the generalization of the

Gini index to a continuous variable, i.e. the variance of the distribution in the bin (Albert *et al.*, 2008a).

The RF can also be exploited to estimate the arrival direction of the events (Aleksić *et al.*, 2010c). Contrary to ANN described in Section 3.3.3.2, RF can only estimate a single continuous variable, rather than the full 2D position on the camera. Therefore, in this approach, the distance of the source position from the center of gravity of the image, the "DISP" parameter, is estimated with RF. The source position is assumed to lie on the extension of the main axis of the image. Ambiguity between the "head" and "tail" of the shower can be solved by using the asymmetry of the image or exploiting stereo information. For a stereoscopic system, estimation of the arrival direction with the DISP method can be done for each image separately and the average position from multiple images is calculated (see, e.g. Aleksić *et al.*, 2012d).

The RF method is currently used as the standard gamma/hadron separation and arrival direction reconstruction method in the MAGIC Collaboration (Aleksić *et al.*, 2016b). The method has also been used in the Monte Carlo studies of the future MACE telescope (Sharma *et al.*, 2017) and CTA (Hassan *et al.*, 2017). The analysis of the data from the VERITAS telescopes employs a similar method called Boosted Decision Trees (BDT) (Krause *et al.*, 2017). The first tree in the BDT method is produced according to the same principle as an RF tree. Before each next-generated tree, however, the classification of events is checked and the misclassified events (signal events classified as background or the other way around) have their weight increased. Such re-weighting forces the BDT method to focus on the training of events that are difficult to classify. BDT implementation of the multivariate decision trees has also been used for the analysis of the H.E.S.S. telescope data (Ohm *et al.*, 2009).

3.3.3.2. *Artificial neural networks*

An Artificial Neural Network (ANN) (Crick, 1989; Reynolds and Fegan, 1995; Bishop, 1995) is a mathematical construct built to emulate the properties of biological neurons in a brain and can be exploited for pattern recognition tasks. ANNs are composed of nodes imitating individual neurons. The signal propagated jth neuron is a weighted sum of signals from all the neurons: $n_j = \sum_i w_{i,j} n_i$ (see, e.g. Reynolds and Fegan, 1995). The values of the weights have to be trained for the search of a particular pattern. To give the ANN the possibility to learn complex, nonlinear patterns in input variables, the output of nodes is transformed with a Sigmoid function[e] The nodes can be either organized into a single layer, or into multiple layers, with each one taking as input the output of the previous one. In particular, for the purpose of gamma/hadron separation, the input layer can be fed with the low-level data of the signal registered in each pixel, or partially processed event data such as image parameters (see, e.g. Reynolds and Fegan, 1995), and the final layer consists of a single separation parameter. A number of "hidden" layers can be placed in between with often gradually decreasing numbers of nodes. To train the ANN, samples of gamma-ray and background events are used. After each pass of the ANN, the classification error of the training sample is computed and the values of weights are adjusted using a given learning rule.

ANN can be applied not only for gamma/hadron separation, but also for energy and arrival-direction estimation (see, e.g. Dumora *et al.*, 2001; Sembroski and Kertzman, 1991). In the case of energy estimation, the output node value can be associated with the energy of the event. As the ANN can have a nearly arbitrary number of nodes in the final layer, for the estimation of the arrival direction, it can be composed of two nodes directly representing the direction of the event in the camera.

Despite their high potential, ANNs have not been commonly used in the analysis of the third generation of IACT data. Analysis of Whipple telescope data with various MC methods showed that while an ANN provides similar performance to supercuts, the latter method is preferred due to robustness (weaker dependence on small, instrument-related changes in the distributions of input parameters) (Moriarty *et al.*, 1997). ANNs have also been used for the data analysis of a still-operating TACTIC telescope, however only for the energy estimation (Koul *et al.*, 2007). A comparison of various multivariate classification methods (Bock *et al.*, 2004) showed that, depending on the particular implementation of an ANN, its performance for gamma/hadron separation gives results over a wide range, between the very best and the rather mediocre.

3.3.3.3. *Deep learning*

In recent years, deep learning (Lecun *et al.*, 2015; Goodfellow *et al.*, 2016) algorithms have become very popular. Their objective is to exploit the full information about a given image, not dependent on any parametrization.

[e]Sigmoid is an S-shaped function such as, e.g. $(1 + \tanh(x))/2$.

This means that the charge information from all the pixels has to be used, resulting in a very large number of input parameters. For dealing with such a task, a special type of ANN (see Section 3.3.3.2), a Convolutional Neural Network (CNN), can be applied. In CNNs, the input parameters are processed through many hidden layers of nodes. Each node takes information from a small group of nodes of the previous layer. The weights are selected such as to form "filters" that are shared between different bunches of input nodes. This allows us to look for image features not dependent on their location or orientation.

The application of deep learning algorithms in IACT data analysis so far has been rather limited, but has increased in popularity over recent years. The VERITAS collaboration showed that deep learning algorithms can be used for analyses of very characteristic muon images (Feng and Lin, 2017). A study using MC simulations of the future CTA telescopes showed that deep learning applied on individual telescope images can be used for gamma/hadron separation (Nieto *et al.*, 2017). The performance of the method, however, was worse than the BDT method, probably due to a lack of stereoscopic information in this deep learning approach. On the other hand, the analysis of H.E.S.S. data with deep learning showed improvement with respect to Shilon *et al.* (2019). In this case, the method was applied to 2-4 images simultaneously. Despite possible improvements in performance, one should note that deep learning algorithms are very sensitive to any data/MC mismatches (Shilon *et al.*, 2019).

3.3.4. *Model and template analyses*

One of the best-performing data analysis methods for VHE gamma-ray astronomy is a "model" or "template" analysis. At its core, it is a maximum-likelihood method of parameter estimation applied to Cherenkov images. Its implementation for IACTs goes back to the HEGRA and CAT second-generation IACTs (Ulrich, 1996; Le Bohec *et al.*, 1998), and had been refined and exploited by the H.E.S.S. experiment some years after its inception (de Naurois and Rolland, 2009; Parsons and Hinton, 2014). For each event, the most likely incident particle parameters are estimated by comparing the pixel-wise camera data to the expected average Cherenkov photon distributions on the camera plane. These can be generated by means of semi-analytical shower models or using Monte Carlo (MC) air-shower simulation templates. Both model and MC templates are multi-dimensional over incident particle parameters such as direction, energy, impact distance and shower-maximum depth in the atmosphere. Usually, only gamma-ray models are used. The shower parameters that best fit the data in the camera are found by maximizing the likelihood over this multi-dimensional model space. The strength of the method resides in the fact that the likelihood is computed for all pixels in the camera, foregoing the image cleaning step, therefore using a greater amount of information than Hillas-based methods. The results of the fit can be used for both gamma/hadron separation and (most importantly) direction and energy estimation. The angular resolution achieved with this method is better than a factor of two with respect to the standard Hillas-type box-cut analysis. Energy resolution is lower than 10% around 1 TeV, and overall better than the standard one by about 50%. The overall sensitivity is improved by a factor of two. Furthermore, this reconstruction method is highly insensitive to the intensity of the night sky background light and to the presence of non-working pixels, and can better reconstruct shower images located at the edge of the camera.

3.3.5. *Background estimation*

Neither of the gamma/hadron separation methods can exclude the background completely.[f] In the case of weak sources, the gamma-ray excess can be as small as a few percent of the total number of events after all the cuts. Hence, it is crucial to be able to accurately estimate the background. Historically, the ON–OFF observation mode was used. The observations of a possible emitter of gamma rays were followed by observations of a dark patch of the sky from which no point-like source was expected. The background (both residual hadronic as well as diffuse gamma-ray) was expected to be the same in both observations. ON–OFF observations have two major problems. First, it is very important that ON and OFF observations are taken in the same conditions. While the proper selection of the dark patch[g] allows one to obtain the same Zenith and Azimuth angle distributions, the varying atmospheric conditions can produce mismatches between ON and OFF observations. Those, if not taken into account, could lead to fake detections of the source. The second disadvantage of ON–OFF observations is that they require a twice longer observation time. To achieve

[f]Except in the energy range of tens of TeV, where nearly background-free gamma-ray astronomy is possible.

[g]Typical solution is to take a point in the sky with the same declination but somewhat different right ascension than the source of interest.

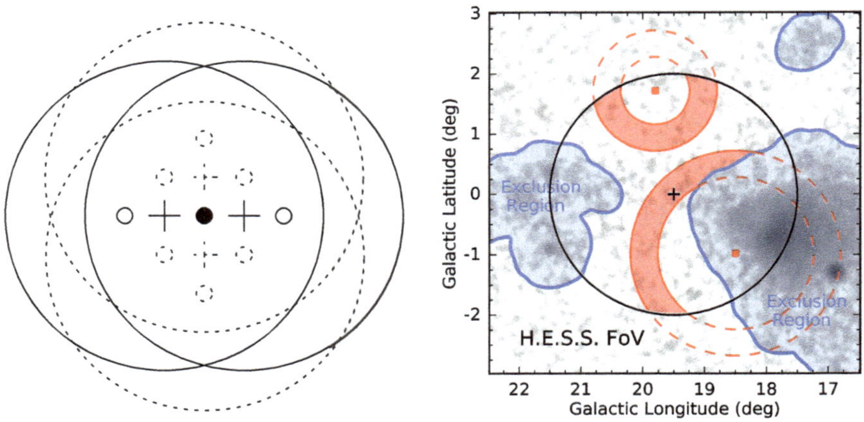

Fig. 3.12. *Left*: Principle of wobble observations. Large circles with a cross in the center show the camera position in the sky; filled circle, the investigated source; empty circles, the background estimation regions. Solid lines are used for the case of two wobble positions, dotted show the additional positions in case four wobble positions are used. Right: The right panel explains the same principle with a more realistic field of view that includes other sources (the exclusion region). Reproduced from (Abdalla *et al.* 2018j).

optimal statistical uncertainty on the background estimation, the time of OFF observations should be at least comparable to the ON observations.

An alternative is the wobble observation mode (or the false-source tracking) (Fomin *et al.*, 1994). In wobble observations, the telescopes point not to the source of interest, but slightly (typically $\sim 0.5°$) off (see the left panel of Fig. 3.12). In the camera plane, the source is then located at such an offset from the camera center. The point in the sky located at the same offset, but on the other side of the camera center, can then be used to estimate the background. To counteract inhomogeneities in the cameras of IACTs, every few tens of minutes, the telescope repoints such that the place of the source region and background estimation region swap. More then two wobble pointings can be used (see the solid and dotted lines in the left panel of Fig. 3.12); this allows one to estimate, for each wobble position, the background from multiple regions. The advantage of the wobble observations is that source and background region data are taken at the same time, both saving time and decreasing the influence of systematic uncertainties. The main caveat of wobble-mode observations is that the gamma rays from the source are also visible in the background estimation region at a larger θ^2 value (see Exercise 3.6). For sources that are either point-like or only slightly extended, this is not a problem — however, largely extended sources require either a larger wobble offset or a different observation scheme. When the wobble offset is too large, the smaller acceptance, which worsens the sensitivity, needs to be correctly taken into account. It is also possible that a different source is located in the background estimation region of the source of interest (see, e.g. Aleksić *et al.*, 2012c). The observation time of each wobble position also has to be properly tuned. Due to the Earth's rotation, the source position in the case of wobble observations will slowly rotate in the camera — hence, if the time is too long, after moving to the next wobble position, the new region of the background will not be in exactly the same camera position as the location of the previously observed source region. Additionally, with the long duration of individual wobble observations, it is much more difficult to reach the equal fraction of observations in each wobble position. On the other hand, too-short wobbling times result in a relatively large overhead of re-pointing to the new position and stabilization of the telescope (tracking and the trigger, in particular).

Sources with complicated morphology or regions with multiple sources are a challenge to IACT analysis due to complicated background estimation. The background can be estimated in rings along the source position (see the right panel of Fig. 3.12) — however, for such a procedure not to introduce large systematic uncertainties, either the background acceptance must be homogeneous in the whole FoV, or it has to be homogenized (Da Vela *et al.*, 2018). Additionally, the parts of the rings affected by neighboring sources should be excluded from the background estimation region (see the right panel of Fig. 3.12 and Abdalla *et al.*, 2018j). Instead of excluding the region with other sources, one can also add them into the background model and perform a global fit of all of the sources in the FoV (Vovk *et al.*, 2018). A similar procedure has been used for some time in wide FoV instruments (Mattox *et al.*, 1996). Finally, instead of using gamma-like background events in the regions devoid of sources, one can use non-gamma-like events in the whole FoV. Such events can be used to obtain a background template describing the relative acceptance in each part of the sky (Rowell, 2003).

3.4. Performance

A number of performance parameters can be calculated for an IACT system. The most important one is sensitivity, i.e. the minimal flux of the source that is detectable in a given observation time. For studies of morphology of the VHE sources, angular resolution is an important parameter. Finally, observations of spectral features (such as cutoffs) profit from IACTs with good energy resolution.

When examining a source, one of the main observables is how it behaves as a function of energy. All three performance parameters mentioned before depend on the energy and degrade very fast at low energies due to the lack of Cherenkov photons reaching the reflector surface. In this sense, it is worth having a reference number that tells at which energy the performance of a given IACT system degrades significantly. Energy threshold has been historically used for that and to compare different instruments. It is commonly evaluated as a peak of energy distribution weighted to a particular spectrum, usually a power law with spectral index −2.6. Another definition (used by H.E.S.S.) is the energy at which the energy bias is equal to 10%. It should be noted that it is possible for an IACT to perform measurements below the threshold; however, those measurements are typically burdened with large systematic uncertainties.

All of the performance parameters can be evaluated using MC simulations. Some of the parameters (in particular, sensitivity and angular resolution) can be additionally validated using real data. For this purpose, observations of the Crab Nebula, the standard candle of VHE gamma-ray astronomy, are often used. The validation of performance with real data is important, as IACTs are complex instruments, hence small simplifications and imperfections of the Monte Carlo simulations are common. It should be noted that the performance of IACTs depends on observational conditions, such as the transmission of the atmosphere, zenith angle at which the observations are performed or the distance of the source from the optical axis of the telescope (see, e.g. Aleksić *et al.*, 2016b). The performance parameters (especially the energy threshold) also degrade in cases of observations under strong moonlight (see, e.g. Ahnen *et al.*, 2017d).

3.4.1. *Angular resolution*

The angular resolution describes the response of an IACT system to a point-like source of gamma rays. It is also sometimes referred to as the gamma-ray PSF, which should not be confused with the optical PSF of the IACT mirror. The angular resolution of the IACT system can be relatively easily evaluated using the θ^2 distribution of the excess events for a strong point-like source (see Fig. 3.6). θ is the angular distance between the reconstructed arrival direction of an event and the nominal position of the source. The distribution of θ^2 is more common, as for isotropic arrival directions (as it happens for the cosmic-ray background events), it is flat. Two complementary methods are often used to evaluate the angular resolution. In the first method, a given shape (often 2D Gaussian) is assumed to describe the gamma-ray PSF of the instrument. Assuming radial symmetry, i.e. $f_{PSF,2D}(\theta)d\theta_x d\theta_y$, one can integrate it in rings of θ and calculate the θ^2 distribution (see Exercise 3.6). According to this method, if the $f_{PSF,2D}$ is a two-dimensional Gaussian function, its standard deviation can be taken as a measure of the angular resolution of an IACT. In the second method, the angular resolution is defined as the angular distance containing a given fraction (e.g. 68%) of the gamma rays. Even while both methods evaluate the capability of an IACT system for observations of extended sources, the numerical values of the angular resolutions computed according to them will be different. The latter method is more sensitive to tails of the distribution.

The angular resolution of the third generation of IACTs is of the order of 0.1°, much worse than high-resolution imaging X-ray, optical and radio instruments. It improves slowly with energy of the gamma rays, as the higher-energy showers are better reconstructed. Nevertheless, nearly all extragalactic[h] and a large fraction of Galactic sources are point-like to the third generation of IACTs. The angular resolution gives the accuracy of the location of individual gamma rays. Since a large number of gamma rays are observed from a source, their location can be estimated with higher precision, which can be an order of magnitude better than the angular resolution value. In fact, the precision in the localization of sources without a counterpart in other wavelengths is often limited by the systematic uncertainties in the pointing precision of IACT telescopes.

3.4.2. *Energy resolution*

In the analysis of IACT data, each event has an assigned estimated energy E_{est}. Using Monte Carlo events of simulated energy E_{true}, we can construct a distribution of $(E_{\text{est}} - E_{\text{true}})/E_{\text{true}}$. The standard deviation of such a distribution can be interpreted as the energy resolution of the experiment.

[h]Notable exception is the detection of extended emission from the jet of Cen A (Abdalla *et al.*, 2020c) by H.E.S.S.

Similarly, the mean of the above distribution is the bias of the energy estimation. When reconstructing the spectrum of a source observed with an IACT, it is important to take into account the migration of energies from E_{true} to E_{est} either by simple folding with a migration matrix, or by an unfolding procedure (see, e.g. Albert *et al.*, 2007c).

The energy resolution of the third generation of IACTs is of the order of 25%, just above their energy threshold. It improves with energy, reaching close to 15% at an energy $\sim 1\,\text{TeV}$. The highest-energy behavior depends strongly on the method used for the energy estimation. It should be noted that, due to the lack of a test beam for IACTs, the energy resolution can only be calculated using MC simulations. It is, however, possible for stereo systems to partially validate the energy estimation method by comparing the outputs of the energy estimation from individual telescopes (Aleksić *et al.*, 2016b).

3.4.3. *Sensitivity*

The sensitivity of an IACT system is the minimum flux of gamma-ray emission that can be detected with such an instrument in a given observation time. The source is typically considered as "detected" if its gamma-ray excess has significance calculated according to Eq. (17) of Li and Ma (1983) of $\geq 5\sigma$ in 50 h of observation time. Under certain conditions (nearly perfect estimation of the average background), the condition can be simplified to $N_\gamma/\sqrt{N_{bgd}} = 5$, where N_γ and N_{bgd} are the number of gamma-ray and background events after 50 h of observations. Since both N_γ and N_{bgd} scale linearly with observation time, and N_γ scales with the source flux, the sensitivity scales as the inverse of the square root of the observation time — i.e. to detect a twice-lower flux, one needs a four-times-longer observation time. Such simple scaling holds only in the middle-energy range (see the left panel of Fig. 3.13). At the highest energies, the background is nearly completely rejected by very-efficient gamma/hadron separation and excellent angular resolution. In such a background-free regime, however, N_γ is usually low, and source detection is only claimed if $N_\gamma > 10$. This results in the sensitivity scaling inversely with observation time. On the other hand, at the lowest energies, the N_γ and N_{bgd} statistics are large, but the gamma-ray signal is only a small fraction of the background. Such a signal can be mimicked (or hidden) by systematic effects affecting the background rejection. Therefore, the detection of a

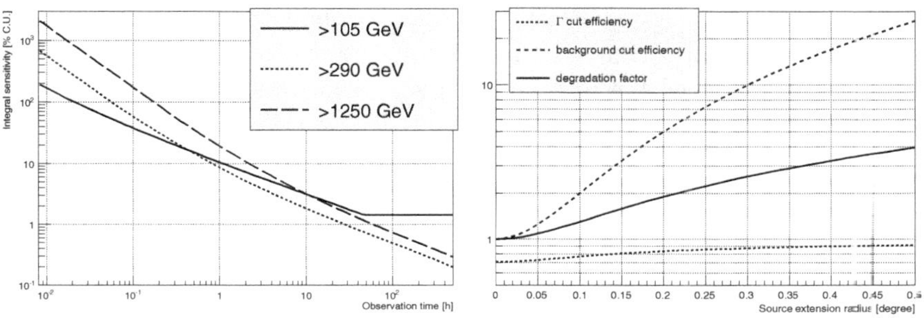

Fig. 3.13. *Left*: Integral sensitivity of the MAGIC telescopes above a few energy thresholds (see legend) as a function of the observation time. *Right*: Sensitivity-worsening factor of an extended source of a given radius with respect to a point-like source sensitivity for an IACT with a PSF of $\sim 0.1°$ (solid line). The corresponding increase of the background and acceptance of the gamma rays are shown with dashed and dotted lines, respectively. Both panels reproduced from Aleksić *et al.* (2016b).

source can be claimed usually only if $N_\gamma/N_{bgd} > 5\%$.[i] Thus, at the lowest energies, the sensitivity saturates and cannot be improved further by longer observations.

The sensitivity can be calculated as either integral or differential. In the case of integral sensitivity, N_γ and N_{bgd} are calculated above a given energy threshold. In the case of differential sensitivity, only events between E_1 and E_2 are used. The larger the bin size, the better the sensitivity value (note that integral sensitivity can be considered as the $E_2 = \infty$ case). Typically, five bins per decade in energy are used as a balance between possible spectral features and broadband spectra. The sensitivity can be expressed in absolute flux units; however, it is also often computed in units of the Crab Nebula flux.

The sensitivity also depends on the extension of the source. An extended source requires a broader angular cut for accepting the gamma-ray events from the source. Typically, the cut is of the order of $\theta_{\text{cut}} = \sqrt{\theta_0^2 + \theta_s^2}$, where θ_0 is the cut for a point-like source and θ_s is the angular radius of the source. Since the number of accepted background events scales with θ_{cut}^2 and sensitivity scales with the square root of the background, in the

[i]It is also common to perform various "sanity checks" of the detection, such as the flatness of the θ^2 distribution in the off-source region, or the variance of the background.

first approximation, the sensitivity for an extended source is $\sim \theta_{\text{cut}}/\theta_0$-times worse for the extended than for the point-like sources. Note that the above calculations are only valid for sources with extents much smaller than the size of the IACT camera. For larger sources, the dependence of the sensitivity on the offset from the camera center has to be taken into account. Moreover, for sources with sizes comparable to the size of the IACT camera, the background estimation is problematic and burdened with large systematic uncertainties, preventing observations of weak, very-extended sources.

3.4.4. *Physical limits of the performance*

With the improvements in both the hardware and analysis methods of IACTs, the performance of the instruments is naturally improving. There are, however, some performance limitations intrinsic to the IACT technique. Separation of gamma rays from the more abundant hadronic background in IACTs is based on differences between electromagnetic and hadronic showers. A large fraction of the primary proton energy can, however, be transferred into a single π^0 particle in the early generations of the shower. The π^0 will decay then into two gamma rays, forming an electromagnetic cascade nearly indistinguishable from a primary gamma ray (Maier and Knapp, 2007), in particular, if only one electromagnetic component falls into the FoV of the IACT (Sobczynska, 2007). Such "false gamma" events by definition create exactly the same EAS as primary gamma rays. The only way to partially reduce the effect of such a background is by exploiting the fact that, due to the prior hadronic interaction, "false gamma" events start deeper in the atmosphere and hence also have deeper shower maximums. The maximum separation of the gamma-ray and "false gamma" distributions by the height of the shower maximum can, however, only improve the sensitivity by a factor of ~1.35 due to shower-to-shower fluctuations. Most of this improvement is already exploited by the current analysis methods used for the MC simulations of the future CTA instrument (Sitarek *et al.*, 2018). A similar effect also occurs due to the cosmic-ray electrons. As they also develop an electromagnetic cascade in the atmosphere, very similar to the one initiated by a gamma ray, they are nearly indistinguishable from gamma-ray showers. The spectrum of cosmic-ray electrons is steeper (see, e.g. Aharonian *et al.*, 2009c) than the one of the cosmic-ray protons, hence one would naively expect that their effect will be most pronounced at the lowest energies. At the lowest energies, however, the rate of detected cosmic-ray electrons is still much smaller than the rate of the protons (see, e.g. Cortina and González, 2001), which — combined with poor rejection of hadronic background — results in negligible effects of the electrons. On the other hand, at energies of hundreds of GeV, most of the hadronic background is rejected, and electrons can constitute up to a few tens of percent of the remaining background (Actis *et al.*, 2011; Aleksić *et al.*, 2012d; Sitarek *et al.*, 2018).

The energy resolution of the IACTs is limited by the fluctuations of the density of the Cherenkov photons on the ground. The fluctuations have different origins; at the lowest energies, close to the threshold of IACTs, the Poissonian fluctuations in the number of obtained ph.e. in the camera are important. Those can be counteracted by building telescopes with larger mirror areas, and/or more sensitive light detectors, as well as combining information from multiple telescopes. There are additional shower-intrinsic fluctuations that further limit the energy resolution of IACTs (see, e.g. Chitnis and Bhat, 1998). Even within one event, the density of light at different impact parameters will fluctuate in a non-Poissonian way due to correlations (the same electrons produce multiple Cherenkov photons on their paths).

Sampling the local density of Cherenkov photons with multiple telescopes can lower the influence of such correlations. Finally, the amount of observed light depends not only on the energy of the primary gamma ray, but also on the height in the atmosphere at which the first pair production happened. Therefore, the estimation of the height of the shower maximum is very important in the estimation of the energy of the primary gamma ray. It should be noted that, in general, the Cherenkov photon density fluctuations are larger for hadronic showers than for gamma-ray showers.

The performance parameter for which we are still relatively far from intrinsic limitations is angular resolution. MC studies of an ideal detector with a large detection efficiency of Cherenkov photons can reach an angular resolution down to $\sim 0.01^\circ$ a factor of a few better than the third-generation IACTs (Hofmann, 2006). The simulations of large arrays of IACTs show a strong dependence of the achievable angular resolution on the number of telescopes seeing a given event, with an additional partial dependence on the pixelization of the camera (see Krennrich, 2009 and references therein).

3.5. Calibration of IACTs and Systematic Uncertainties

A characteristic feature of IACTs is the rather large systematic uncertainties of their measurements. This is partially due to the fact that, contrary to satellite experiments operating in GeV range, the response of an IACT to a shower cannot be calibrated with a man-made particle beam. For example, the claimed systematic uncertainty of the *Fermi*-LAT is 2% (Ackermann *et al.*, 2012c), while for the MAGIC and VERITAS telescopes it is $\lesssim 15\%$ (Aleksić *et al.*, 2016b) and 20% (Abeysekara *et al.*, 2015), respectively. The uncertainty in the light collection of an IACT by a given factor, in the first approximation, leads to a similar factor of uncertainty in the energy scale. Let us assume that an uncertainty of the energy scale is $x\%$, and the source spectrum is a power law of $dF/dE \propto E^{-\alpha}$ type. The change of the energy scale both shifts the spectrum in the X direction, and also changes the bin size of the differential flux. As a result, its effect on the systematic uncertainty of the reconstructed flux can be roughly estimated as a factor of $(1 - x\%)^{-(\alpha-1)}$ to $(1 + x\%)^{-(\alpha-1)}$. Therefore, such an uncertainty of the energy scale of $\pm 15\%$, for a source with a Crab-like spectrum with a spectral index $\alpha = 2.6$, would lead to a propagated flux uncertainty of $^{+30\%}_{-20\%}$. Similar values (uncertainty on the flux at the level of 20%) have also been estimated for the H.E.S.S. telescopes (Aharonian *et al.*, 2006d). It is therefore clear that for strong and/or soft sources, the systematic uncertainty can exceed the statistical uncertainties (see Exercise 3.6). They are particularly important to take care of (by energy-dependent cross-calibration) in the case of joint campaigns of observations of the same source with different IACTs (see, e.g. Acciari *et al.*, 2020b).

The total systematic uncertainty on the energy scale can be estimated by summing up all the contributions of all the possible sources of systematic uncertainties (see, e.g. Aleksić *et al.*, 2012d). Notably, one of the largest contributions comes from the uncertainties in the atmospheric transmission of the Cherenkov light. In the third generation of IACT telescopes, various methods for decreasing (and evaluating) systematic uncertainty on the energy scale are used. The systematic uncertainty on the energy scale of the MAGIC telescopes was validated by comparisons of the spectrum of the same source observed with different analysis thresholds (Aleksić *et al.*, 2016b). The systematic uncertainties of the measurement can also be validated by comparing the measurement of the same source (normally the Crab Nebula is used for this task) with different IACTs (see Fig. 3.14) or with a satellite instrument (Meyer *et al.*, 2010). In the energy range, around $\sim$1 TeV, the agreement between different IACTs is at the level of $\sim$30%. At the lowest and highest energies, the uncertainties are larger.

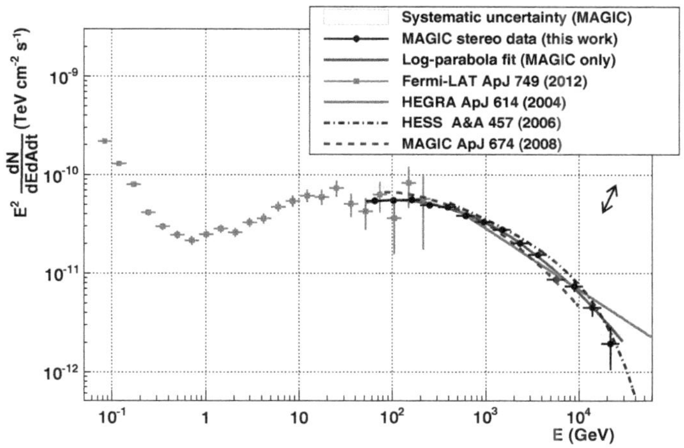

Fig. 3.14. Comparison of the Crab Nebula spectrum observed by different IACTs (see legend) and the *Fermi*-LAT. Image reproduced from Aleksić *et al.* (2015a).

3.5.1. *Calibration methods*

It is clearly desirable to lower the systematic uncertainties of IACT systems, and multiple methods have been devised for such a task. The first step is a cross-calibration of individual telescopes in an array of a stereo IACT system. In the first approximation, the density of the Cherenkov light is radially symmetric in the plane perpendicular to the shower axis (i.e. it depends mainly on the impact parameter)[j]. Therefore, the comparison of the reconstructed light yield of different telescopes seeing the same event at a similar impact parameter can allow us to cross-calibrate the telescopes (Hofmann, 2003). Similarly, instead of telescope light yield, the telescope-wise energy estimation can be used (Aleksić *et al.*, 2016b). Also, the rates of cosmic rays can be used for relative calibration of IACT observations taken under varying conditions (Lebohec and Holder, 2003).

The absolute calibration of the IACTs is much more difficult. The most common method, used by all the third-generation IACTs, is exploiting the images of individual muons with their direction crossing the reflector of the telescope (Vacanti *et al.*, 1994). Such events produce characteristic, ring-like images since, close to the ground, the whole cone of the Cherenkov light

[j]Note that the effect of the geomagnetic field produces an asymmetry, spreading the Cherenkov light pulse in the East–West direction (see, e.g. Cocconi, 1954).

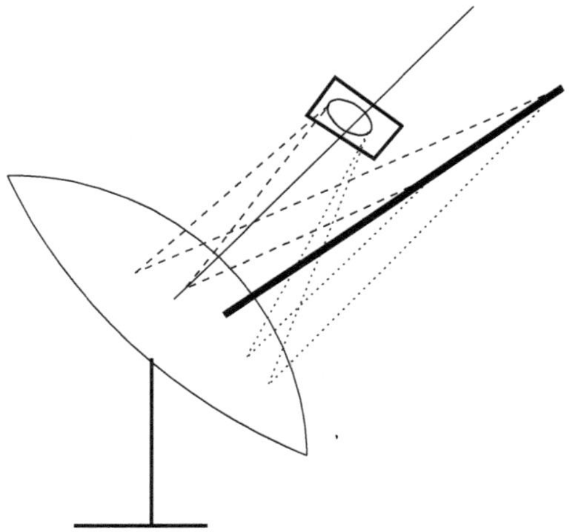

Fig. 3.15. Reflection of the Cherenkov light from a single muon. Thin line: optical axis of the telescope, thick line: path of the muon. Dotted and dashed lines show the photons emitted in two different directions. The cone of the emitted Cherenkov photons is represented as a circle in the camera (thick square).

will hit the reflector (see Fig. 3.15). The total light yield of a single muon can be predicted and thus compared with an analytical model. It should be noted, however, that the wavelength and emission height distribution of the Cherenkov light from muons is different from that of gamma-ray showers. Single muons produce most of their light within 2 km of the telescopes — hence, their Cherenkov light is less absorbed and thus bluer.[k] Therefore, the application of the muon calibration needs to employ full MC simulations to correct for such effects, and is also burdened by its own systematic uncertainties (see, e.g. Daniel *et al.*, 2015).

3.5.2. *Atmospheric monitoring and corrections*

One of the main ingredients increasing the systematic uncertainty of the IACTs is the atmosphere. First, the average atmospheric properties over the telescope site (in particular, the density of atmospheric nuclei that determines the evolution of an EAS and the height-dependent content of aerosols affecting the absorption of the Cherenkov light) are only known with certain accuracy. Moreover, the atmospheric properties show seasonal

[k]The atmospheric absorption is the strongest in the blue range due to ozone.

variation (see, e.g. Puerto-Giménez *et al.*, 2013). Finally, passing clouds can affect the performance of the telescopes on timescales of minutes. Therefore, as in the case of other types of telescopes, the systematic uncertainties of the IACT technique can be partially lowered by the use of atmospheric monitoring.

Various instruments are used for the monitoring of atmosphere for the purpose of IACT observations (see Hahn *et al.*, 2014; Hui, 2008; Will, 2017), differing in complexity of the analysis methods and applicability. A pyrometer is a device that measures the integral thermal radiation from the sky in its line of sight. The presence of clouds increases the measured temperature, allowing observers to infer "cloudiness" during the IACT observations.

Even while a pyrometer can provide information about the cloud coverage of a given sky region at a given time, it is not sensitive to the height at which the cloud resides. Clouds can occur below or above the shower maximum, and they will naturally absorb only the light emitted above their height. Therefore, a desirable piece of information is the transmission of the atmosphere in a given direction from different heights. Such information is provided by the elastic LIght Detection And Ranging (LIDAR). A LIDAR emits a narrow beam of constant-frequency laser light along the line of sight of the telescope. The light is scattered in the atmosphere and, in particular, in clouds, reflected back to the LIDAR, and registered there after a delay corresponding to a double path to a given height of the atmosphere. A cloud will show up as an increase in the expected light from the height of the cloud (the reflected light) and a decrease of the light emitted above the cloud (see Fig. 3.16). Using such information, the transmissitivity of the atmosphere from a given height can be computed (Fruck and Gaug, 2015). It should be noted that the calculations are not trivial, as both the back-scattering coefficient and transmission have to be computed from the same information, which can only be done with extra assumptions as to their relation.

Disentangling the back-scatter and absorption can be done with the so-called Raman LIDAR. The Raman LIDAR measures reflected light in a few different frequencies, allowing one to disentangle the elastic and non-elastic (i.e. shifted in frequency) scattering channels. It is expected to reduce the systematic uncertainties of the molecular and aerosol components of the atmosphere to below 5%, and hence a Raman LIDAR is going to be used in the next-generation IACTs: the CTA (Vasileiadis *et al.*, 2017).

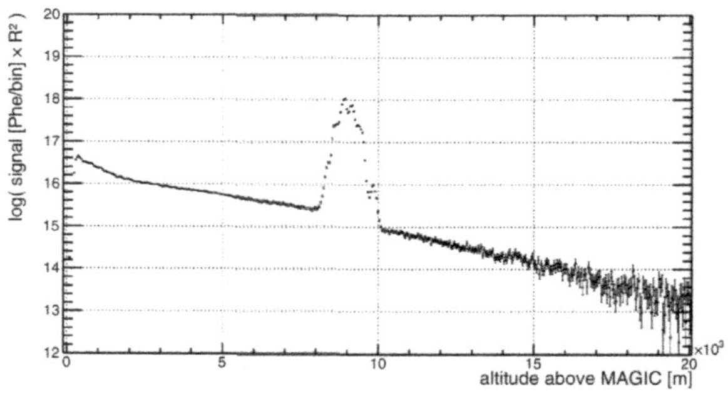

Fig. 3.16. Example of a range-corrected LIDAR response for a cirrus cloud at 9 km altitude. Image reproduced from Fruck and Gaug (2015).

It should be noted here that atmospheric information can be used twofold. First, it is used for rejection of the data recorded during adverse atmospheric conditions that would mar the performance of the telescopes. Second, the information about the additional absorption can be applied to the data to correct the data affected by deficient atmospheric conditions. This will either recover the unusable data, or reduce the systematic uncertainties.

Pyrometers are used by H.E.S.S. (Hahn *et al.*, 2014), MAGIC (Will, 2017) and VERITAS.[1] LIDAR information is used in the analysis of the data from MAGIC (to both select and also to correct data, Fruck and Gaug, 2015), VERITAS (Hui, 2008) and H.E.S.S. (Hahn *et al.*, 2014).

3.6. Summary

As can be seen from this chapter, the third generation of IACTs brought about multiple innovations, both in the applied technologies and analysis methods. For the first time, arrays of a few relatively large telescopes were built. The constant improvement of light detectors, mirrors, trigger-logic algorithms and fast-sampling electronics strongly aided the sensitivity of those instruments. The contemporaneous progress in the analysis methods — in particular, the use of machine-learning algorithms — allowed the H.E.S.S., MAGIC and VERITAS Collaborations to exploit the power of those instruments for observations of gamma rays. The fact that, during the $\sim$ 15 years of the operations of the third generation of IACT

[1]see http://www.physics.utah.edu/~kieda/Atmosphere/.

instruments, the number of known VHE gamma-ray sources increased by over an order of magnitude shows how successful the used approaches are. The hardware and analysis solutions used in the third generation of IACTs laid foundations for the next generation of the Cherenkov telescopes, the CTA, which will start pushing the IACT technique to the physical limits of its power. It should also be stressed that, due to the greatly improved sensitivity of the third-generation IACTs, more and more often, the systematic uncertainties play a larger role in the observations of gamma-ray sources than statistical uncertainties. The efforts to estimate and correct such systematic uncertainties caused, for instance, by atmospheric transmission are also going to be a part of the third-generation IACT heritage.

Ideas for Exercises

(1) Declination of the Crab Nebula is $+22°$, at which lowest zenith angle is it observable by H.E.S.S., MAGIC and VERITAS?

(2) Calculate the time spread of spherical and parabolic mirrors of diameter 12 m and $f/d = 1$.

(3) Consider wobble observations with offset of $0.5°$ and gamma-ray PSF of $\sim 0.1°$. In which range of θ^2 distribution of the background estimation region do you expect to see the effect of the gamma rays from the source region? Perform the calculations for two cases: two opposite pointing positions, or four pointing positions with their centers forming a square in camera coordinates.

(4) Gamma-ray PSF of an IACT is described in a first approximation as a two-dimensional Gaussian; show that the distribution of a θ^2 for gamma rays is exponential.

(5) Assume that signal distribution in pixels is a Gaussian with a mean $= 2$ ph.e. and $\sigma = 1$ ph.e., estimate a chance probability (or average number of cases) that in a given event two neighboring pixels are found with charge > 5 ph.e. if camera has 1000 pixels and each pixels has 6 neighbors.

(6) Calculate (maybe numerically) how much larger is 68% containment of PSF than sigma of a two dimensional Gaussian.

(7) $N_{\mathrm{ON}} = 150$, $N_{\mathrm{OFF}} = 100$ (for one OFF position) in a given time $t = 1$ h, calculate what is the $exc/\sqrt{bgd} = 5$ sensitivity and what Li and Ma significance we would get with 1 and 5 OFF positions from a source with a flux equal to such sensitivity.

(8) Source has a flux of 10% of Crab, and a 2.6 spectral index. The instrument has sensitivity of 1% of Crab and dominating systematic uncertainty in the energy scale of 15%. For which observation time will the systematic uncertainties be equal to the statistical one?
(9) Why does a muon make a ring in the camera plane? What determines how much of the ring is seen and where its center is in the camera?
(10) Consider a shower generated directly above a telescope, what does the light distribution in the focal plane look like and from what part of the shower is the first light received? How does this change for a shower with an impact parameter of 200 m?
(11) Assume that in 1 ns you observe 20 photons of Cherenkov light from a shower per m^2. What magnitude star has the same photon flux? Would such a star be visible? Why can't we see the Cherenkov flash?

https://doi.org/10.1142/9789813275720_0004

Chapter 4

The Status of Very-High-Energy Gamma-Ray Astronomy

Robert Daniel Parsons[*,§], Nahee Park[†,¶] and Elisa Prandini[‡,‖]

[*] *Humboldt University ät zu Berlin*
Berlin, Germany
[†] *Queen's University*
Kingston, Ontario, Canada
[‡] *Università degli Studi di Padova*
Padua, Italy
[§] *daniel.parsons@physik.hu-berlin.de*
[¶] *npark7@wisc.edu*
[‖] *elisa.prandini@unipd.it*

4.1. The Very-High-Energy Gamma-Ray Sky

Very-high-energy (VHE) ($E > 100\,\text{GeV}$) gamma rays are produced by the interaction of high-energy (HE) particles with the matter and radiation fields around their source regions. Because gamma rays are neutral particles, their directional information is preserved and they can be used to study the acceleration and diffusion of HE particles around the source regions. Since the first detection of VHE gamma-ray emission from the Crab Nebula in 1989 by the Whipple Observatory (Weekes *et al.*, 1989), VHE gamma-ray astronomy has rapidly progressed. At the time of publication, the number of VHE gamma-ray sources detected by IACTs has reached 188 in around nine broad source classes. Since the beginning of observations with the third generation of IACTs (high-energy stereoscopic system (H.E.S.S), major atmospheric gamma-ray imaging cherenkov (MAGIC) and very energetic radiation imaging telescope (VERITAS)), the total source count has increased by nearly a factor of ten. This trend will continue as the next-generation telescope, the Cherenkov telescope array observatory (CTAO),

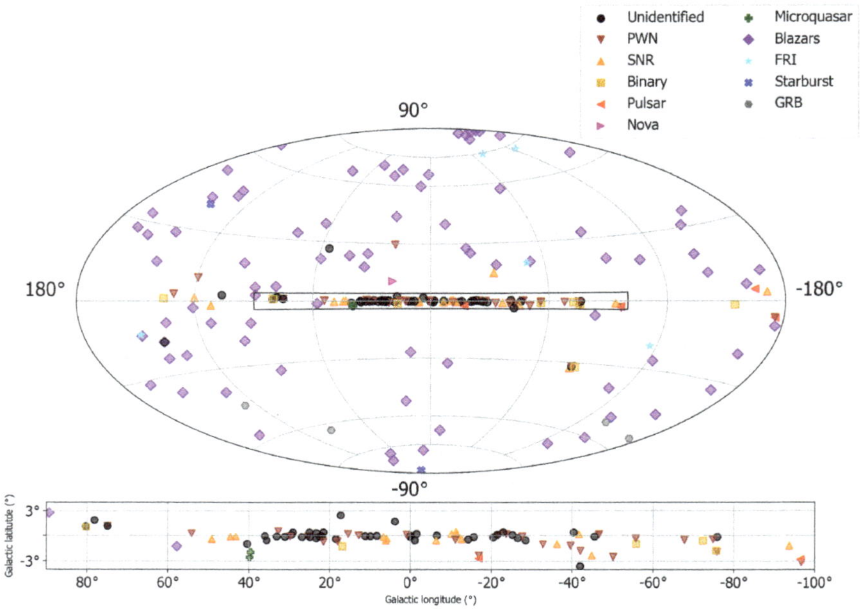

Fig. 4.1. Distribution of currently detected VHE gamma-ray sources shown in Galactic coordinates, split into broad source classes. Lower panel: close-up of the marked Galactic plane region. Sources with multiple possible counterparts are included under the label "unidentified".

comes online. In this chapter, we will focus on the properties of detected gamma-ray sources and describe the source populations.

Figure 4.1 shows the positions of all known VHE gamma-ray sources in Galactic coordinates;[a] within this source distribution, the Galactic plane can be clearly seen. In total, 104 sources are either confirmed, associated or likely connected to Galactic stellar objects, namely SuperNova Remnants (SNRs), Pulsar Wind Nebulae (PWNe), binaries and pulsars (see Table 4.1). Outside of our Galaxy, the Large Magellanic Cloud (LMC), located at a distance of 50 kpc, is the only other galaxy in which individual VHE gamma-ray sources can also be resolved, with four discrete sources detected. For sources located further away, we see only the brightest part of the Galaxy as a point-like source. We have detected a total of 78 sources

[a]We have generated a catalog of acrshortvhe gamma-ray sources for this study based on the information from the TeVCAT and GammaCAT catalogs. Overlapping sources and unclear source associations are sorted out based on our own criteria.

Table 4.1. List of VHE gamma-ray source classes.

Classes	Location	Source number	Variability?
Unidentified	Galactic + LMC + extragalactic	44/1/1	Not detected
Pulsar Wind Nebulae	Galactic + LMC	30/1	Not detected
Supernova remnant	Galactic + LMC	19/1	Not detected
Binaries	Galactic + LMC	7/1	Orbital phased
Pulsar	Galactic	4	Phase dependence
BL Lac	Extragalactic	73	Variable
FSRQ	Extragalactic	9	Variable
Radio galaxy	Extragalactic	5	Variable
Starburst galaxy	Extragalactic	2	Not detected
GRB	Extragalactic	4	Transient events
	Nova Galactic	1	Variable
	Microquasar Galactic	2	Variable

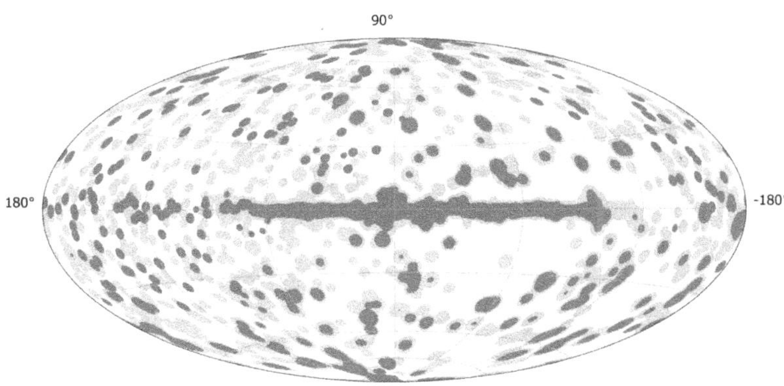

Fig. 4.2. Approximate sky coverage by the current generation of VHE gamma-ray telescopes. Areas with more than 30 minutes of exposure are shown in light gray, and those deeper than 10 hours, in dark gray.

that are considered extragalactic; more than 90% of these extragalactic sources are blazars. In general, these extragalactic sources are located in the local Universe, lying at a redshift of ~0.9.

The catalog of gamma-ray sources presented here, however, is incomplete. It is clear from the map of observation positions for the current generation of instruments shown in Fig. 4.2 that the sky coverage in VHE gamma rays is highly non-uniform and rather patchy. This non-uniformity is caused by the relatively small field of view (FoV) (<5°) of gamma-ray telescopes, combined with their low duty cycle (around 10%) and low sensitivity in comparison with instruments at other wavelengths. With current instruments, around 25 h of observation (of roughly 1000 available hours per year) are required to detect a source with 1% of the Crab Nebula flux. The total sky coverage at the 1% Crab flux level is therefore smaller than even shown in Fig. 4.2 and typically strongly grouped along the plane of the Galaxy and a few well-known extragalactic objects.

An additional important constraint on the source catalog completeness is the correlation of VHE gamma-ray sensitivity with angular source size. This reduction in sensitivity makes the largest (lowest surface brightness) sources more difficult to detect. Such a restriction will have the largest effect when searching for sources close to the Earth and may bias the measured distribution of sources.

The measured spectra for most VHE gamma-ray sources can be described by a simple power-law model. For around 15 sources, an exponential cutoff must be added to the power-law model in order to adequately describe the VHE spectrum. The distributions of spectral index and integral flux over 700 GeV for all sources are shown in Fig. 4.3. The distribution of spectral indices shows a clear bimodality between Galactic and extragalactic sources, with the extragalactic sources (dominated by blazars) having the wider distribution, with a peak at around 3.5. On the other hand, Galactic source classes show a tighter distribution of indices, generally peaking between 2.0 and 2.5 — a notable exception to this being the spectra of the two detected pulsars, both of which show a softer spectrum. All of the sources with a soft spectral index (>3) were detected in the energy range from 100 GeV to a few hundred GeV, and this produces the correlation between the index and the integral flux shown in Fig. 4.3. For sources with a spectral index harder than 3.0, there is less of a correlation. The distribution of fluxes from these sources varies by a factor of more than 300.

4.1.1. *Connection between IACT observations to GeV and multi-TeV gamma-ray observations*

As of 2018, the sensitivity of gamma-ray instruments spans a very wide energy range. The *Fermi*-LAT has continuously monitored the GeV energy sky since its launch in 2008. While the energy range of *Fermi*-LAT spans

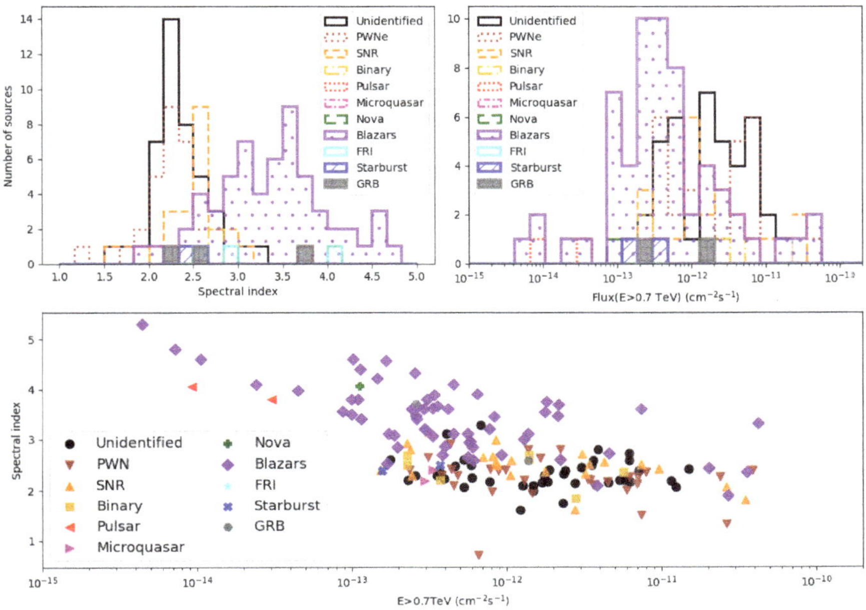

Fig. 4.3. Distributions of spectral properties of the major VHE gamma-ray source classes.

from a few tens of MeV up to the TeV energy range, its peak sensitivity lies around a few GeV due to its small effective area. The typical energy coverage of IACTs provides good measurements, from ~100 GeV up to tens of TeV. Combining these observations has provided spectrum measurements over six orders of magnitude in energy. This coverage is important for studying the details of gamma-ray sources. As shown in Fig. 4.4, there are clear changes in the gamma-ray spectrum from the GeV energy range to the TeV energy range, which may reflect the gamma-ray emission and absorption mechanisms, as well as the environment of the sources. A complementary measurement of VHE gamma-ray observations is provided by particle array detectors. The High-Altitude Water Cherenkov (HAWC) Observatory is the most sensitive operational particle array detector and has operated with its full detector array since 2015. While the instantaneous sensitivity of HAWC is worse than that of the IACTs, it provides daily high-duty-cycle observations of a large portion of the sky.

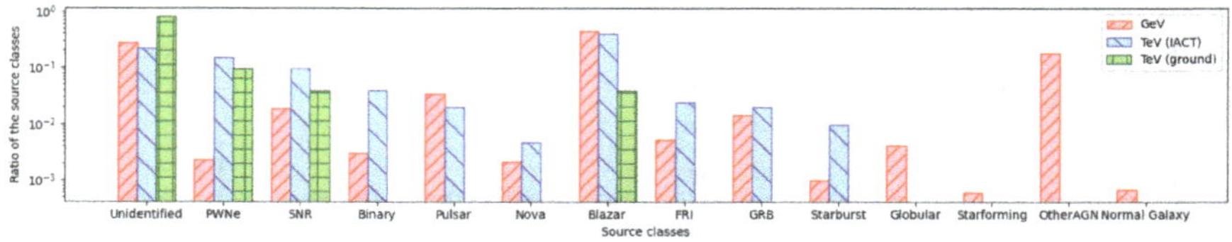

Fig. 4.4. Population comparisons between GeV gamma-ray sources detected by *Fermi*-LAT and TeV gamma-ray sources detected by IACTs. TO BE UPDATED? Add GRB class, update FSRQ and BL Lacs also. Galactic.

Above 10 TeV, HAWC surveys a large portion of the sky with a sensitivity similar to 50 hours of IACT observation in 5 years. Additionally, due to their large FoV, it is easier for particle array detectors to study very extended sources, especially those that are much larger than the FoV of IACTs.

Figure 4.4 shows the number of sources detected for each source class by *Fermi*-LAT and the current IACTs. Blazars are the most common source class in both GeV and TeV gamma rays, followed by unidentified sources, which are most likely associated with Galactic sources. It is also notable that there is a larger population of pulsars in the GeV sky, simultaneous with larger populations of PWNe, binaries and FRIs in the TeV sky. This is likely due to the differences in the spectral energy distribution (SEDs) between source classes. For example, pulsars are generally bright around the GeV energy range, and the flux rapidly decreases at higher energy. There are a few source classes that are detected only in the GeV energy range, such as globular clusters, novae and star-forming regions.

In the following sections, we will discuss the properties, characteristics and populations of the primary source classes from both galactic and extragalactic observations.

4.2. The Galactic Very-High-Energy Gamma-Ray Sky

4.2.1. *Galactic observations*

4.2.1.1. *Survey*

Strategies for observing the Galactic sky can be largely divided into two categories: sky surveys and pointed observations. Sky surveys provide unbiased observations of the sky, typically by tiling observations in a grid across a given region of the sky, and continuing observations until

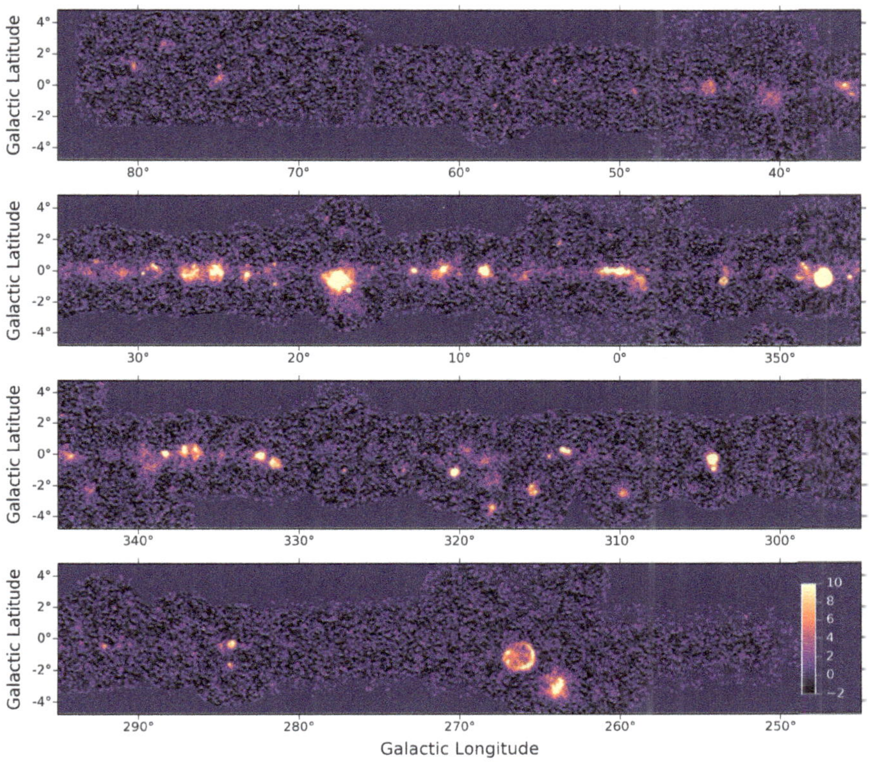

Fig. 4.5. Significance map of the Galactic plane consisting of the combination of the survey regions observed by H.E.S.S. and VERITAS.

a minimum level of sensitivity is reached. Such surveys are particularly important for source population studies as well as for searching for sources that can be revealed only by VHE gamma-ray observations. The constraint of limited observation time requires that surveys cover a carefully selected region of the Galaxy. As a result, VHE surveys by IACTs have been mainly focused on the regions close to the Galactic plane, or where high star-formation activity has been reported. Figure 4.5 shows the combined significance map of the Galactic plane from two surveys: the H.E.S.S. Galactic Plane Survey (GPS) and the VERITAS Cygnus Survey.

The H.E.S.S. GPS: For the first ten years of H.E.S.S. operations, a significant fraction of the observation time was spent performing the

H.E.S.S. GPS — the purpose of which was to provide the first deep and unbiased survey of the Galactic plane as seen from the southern hemisphere. In total, the H.E.S.S. GPS consists of over 2700 h of data split between source observations and a grid of survey observations. The survey regions cover the Galactic plane from 250 degrees to 75 degrees in Galactic longitude and **val** in latitude, reaching a sensitivity of at least 10% of the Crab Nebula flux over the entire survey region and significantly deeper in most locations.

In addition to producing and distributing maps of the Galactic plane, the H.E.S.S. Collaboration also used this dataset to produce a catalog of Galactic gamma-ray sources, independent of observations at other wavelengths. Overall, the H.E.S.S. GPS independently identified 78 sources in its survey region; around half of these sources (47) have no firmly identified counterpart at other wavelengths, typically due to source confusion. The remainder of the sources can be identified as known PWNe (12), SNRs (8) or binary systems (3).

VERITAS Cygnus survey: VERITAS carried out a survey in the Cygnus region, which contains many systems with recent star formation (Abeysekara *et al.*, 2018). Between April 2007 and December 2008, VERITAS spent 135 h observing a $\sim 15^\circ$ by $\sim 5^\circ$ portion of the Cygnus region centered on a Galactic longitude of 74.5° and a Galactic latitude of 1.5°. After including targeted and follow-up observations totaling 174 h and focusing on three important sources in this region (γCygni SNR, TeV 2032 and MGRO J2019+18), a total of 4–5 sources were detected. The sensitivity of the observations in the survey region is somewhat uneven, with an average sensitivity of $\sim 3\%$ of the Crab Nebula flux. In several regions, however, significantly deeper observations were made — given the sensitivity difference between the H.E.S.S. GPS, the Cygnus survey and a rough expectation based on extrapolating the *Fermi*-LAT sources, the number of detected VHE gamma-ray sources in this survey is in reasonable agreement with the number of sources detected in the H.E.S.S. GPS.

4.2.1.2. *Pointed observations of stable sources*

Pointed observations provide the most efficient way to search for gamma-ray emission from selected sources with high sensitivity. Due to the limited observation time, sources are prioritized based on their likelihood to be VHE particle accelerators. Generally, pointed observations are carried out by using the *wobble* observation method, in which the telescope is pointed with a fixed angular offset from the target in different directions. The offset is selected based on the size of the source, which is usually estimated based on other wavelength data, including radio, X-ray and GeV gamma-ray observations. Due to the size of the FoV and declining sensitivity with the offset angle, the typical size of a source detected via pointed observations is smaller than 0.5° in radius. ON/OFF observations are carried out for sources that are larger than the FoV of the telescope. In this observation mode, OFF-source data are collected from a region where no VHE sources are expected, in order to estimate the expected background in the ON-source region. In general, we attempt to match the observational conditions (such as elevation angle and night-sky background brightness) of the OFF region as closely as possible to the ON region. Although it requires longer observations, this method has yielded fruitful scientific results for very rich and extended fields, i.e. the Galactic center (GC) region.

4.2.2. *Source classification*

Figure 4.6 shows the population of Galactic sources. There are largely five different source classes: unidentified, SNRs, PWNe, binaries and pulsars.

In order to identify the counterparts and study the nature of VHE gamma-ray sources, it is necessary to compare the position and spectrum of each source to as much multi-wavelength data as possible. Typically,

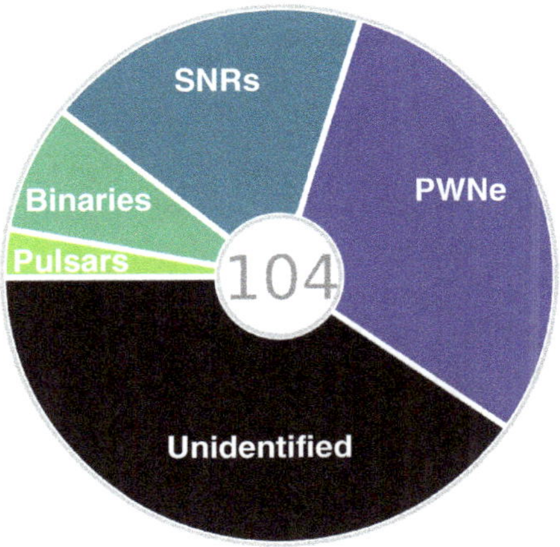

Fig. 4.6. Galactic source population.

this includes studies of non-thermal emissions in the radio, X-ray and GeV gamma-ray bands spatially coincident with the VHE source. In the case of variable sources, such as binary systems, correlated variability with the known period of the source provides conclusive proof of the source association. Because the radio, optical and X-ray bands have superior spatial resolutions compared to VHE gamma-ray observations, non-periodic sources often have multiple potential counterparts for the observed VHE gamma-ray emission, especially for the extended sources. In these cases, model-based studies and source population studies can be used to identify the counterparts, though they are often inconclusive. It is also possible that the source luminosity is higher in VHE gamma rays than it is at lower energies, making it hard to find the counterparts in the lower energy band. Due to these effects, there are still many sources that fall into the unidentified category.

4.2.3. *Source positions and extensions of the Galactic sources*

Since the exposure and the observational strategies vary from source to source for VHE gamma-ray observations, it is hard to make a solid conclusion based on the distributions of their measured properties. Still, it is worthwhile to talk about the general properties of the sources we have observed. Figure 4.7 shows the position of VHE gamma-ray sources overlaid on a simple two-dimensional top view of our Galaxy. The positions of the sources are generally well-correlated with the locations of the Galactic arms. It is also evident that we have detected more nearby sources than farther ones.

Figure 4.8 shows the positions of VHE gamma-ray sources in Galactic latitude and longitude, together with the *Fermi*-LAT source positions and 2HWC source positions. A total of 97 out of 104 Galactic VHE gamma-ray sources are located between the latitudes of $-5°$ and $5°$. The latitude distribution of the sources located within $5°$ from the Galactic plane has a mean value of $-0.17°$ and RMS of $1.29°$. The *Fermi*-LAT distribution is derived from an unbiased survey of the entire sky, while the longitudinal distribution of the 2HWC sources clearly shows the source detections being dominated by the part of the sky covered by HAWC. Compared to the *Fermi*-LAT's latitude distribution, the distributions of IACT-detected sources and 2HWC sources have more of a concentration around zero latitude.

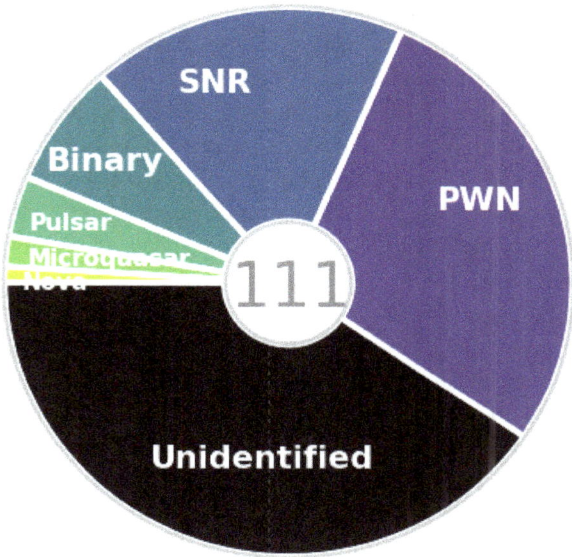

Fig. 4.7. Location of sources within our Galaxy.

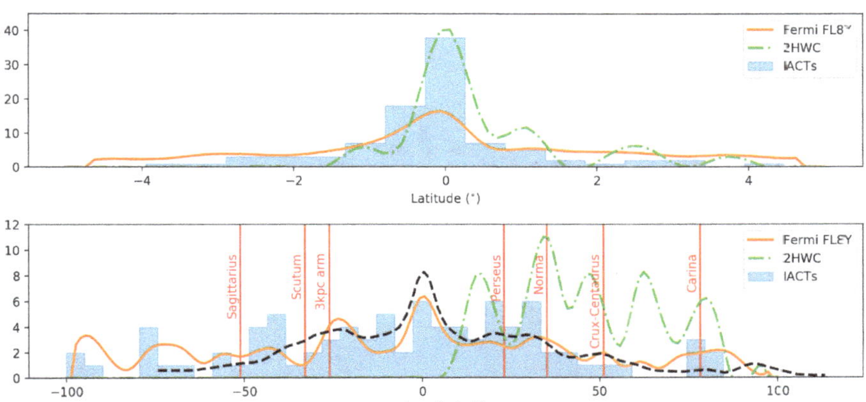

Fig. 4.8. Galactic latitude and longitude distribution of Galactic sources. The distributions of Fermi FL8Y sources, 2HWC sources and CO distributions are normalized to match the area of the source distributions measured by IACTs.

Among Galactic sources, there are 41 sources that are point-like — including sources that are point-like in all wavebands, such as pulsars and binary systems, and sources that are point-like because the angular resolution of the current IACTs is not sufficient to resolve the morphology. The rest of the 63 sources are clearly detected to be extended. Except for nine SNRs with shell-like morphology, the extension of the sources is assumed to have a two-dimensional Gaussian morphology. There are ~20 sources that are reported to have asymmetric spatial distributions, and the morphology of these sources is generally modeled with a two-dimensional asymmetric Gaussian.

Figure 4.9 shows the comparison between the detected source extension and the flux measured above 700 GeV. For simple plotting purposes, the average value of the major axis and minor axis of the asymmetric Gaussian are reported for sources that are asymmetrically extended. Excluding the point-like sources, the distribution of source extensions has an average of 0.21° and standard deviation of 0.16°. The median of the distribution is 0.17°. While the observation time for each source varies greatly, the correlation between the extension and the source flux is still evident. As the extension of a detected source becomes larger, the flux required for the

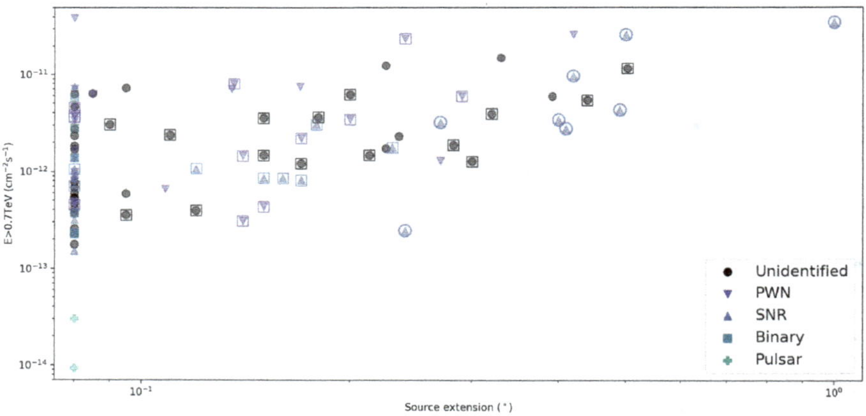

Fig. 4.9. Galactic source extension vs. flux. Sources with shell morphologies are marked with an additional empty circle marker around the existing marker, with the extension defined as the radius of the shell. Sources with multiple-Gaussian morphology or asymmetric Gaussian morphology are marked with an additional empty square around the existing marker, and the extension is defined as the average of the Gaussian σs. The extensions of the rest of the sources are reported with 2D symmetric Gaussians, and the 1σ value of this Gaussian is drawn in the plot.

detection also gets larger. This, as well as the median values of the extension for detected sources, is influenced by the FoV of current instruments and analysis methods. The largest extended source detected is Vela Junior with a radius of 1°; one of the brightest sources, based on its integral flux above 0.7 TeV.

4.2.3.1. *Supernova remnants*

The list of SNRs detected by IACTs and the properties of these sources are shown in Table 4.2. We have detected a total of 19 confirmed SNRs. Two related sources, HESS J1800-240A and HESS J1800-240B, produce VHE gamma rays due to the interaction of particles accelerated by the SNR W28 with nearby molecular clouds. In addition, there are two unidentified sources that are associated with SNRs, HESS J1745-303 and VER J2019+407.

Table 4.2. List of TeV SNRs. A total of 21 sources are confirmed to be related to SNRs and two sources are associated with SNRs. Associated sources are located at the bottom of the table.

SNR name	Age (kyr)	TeV morphology	TeV spectrum
Tycho, SNR	0.441	Point-like	PL
LMC N132D	1.3–3.5	Point-like	PL
IC 443	3–30	Shell	PL
Vela Junior	1.7–4.3	Shell	ECPL
RCW 86	1.82	Shell	ECPL
HESS J1457-593		Point-like	
SN 1006	1.01	Shell	PL
HESS J1534-571	?	Shell	PL
HESS J1614-518	?	Shell	PL
RX J1713.7-3946	1.6	Shell	ECPL
CTB 37B	5	Extended	PL
CTB 37A	1.65?	Extended (mol. cloud)	PL
SNR G349.7+0.2	1.8	Point-like (mol. cloud?)	PL
HESS J1731-347	27	Shell	PL
HESS J1800-240B	N/A	Extended (mol. cloud)	PL
W 28	33–36	Extended (partial shell)	PL
HESS J1800-240A	N/A	Extended (mol. cloud)	PL
W 49B	5–6	Point-like (mol. cloud?)	PL
HESS J1912+101	?	Shell	PL
W 51C	18–30	Extended (mol. cloud + PWN)	PL
Cassiopeia A	0.3	Point-like (shell)	ECPL
HESS J1745-303	>10	Extended (mol. cloud?)	PL
VER J2019+407	8–16	Extended (shell?)	PL

The VHE SNRs divide largely into two main groups: SNRs with emission from the SNR shell, and SNRs with emission from interactions with molecular clouds. Nine SNRs clearly demonstrate shell-type morphology and fall into the first group. Among these, four shell-type SNRs (HESS J1534-571, HESS J1614-518, HESS J1731-347 and HESS J1912+101) were detected by VHE gamma-ray observations in blind searches. There are a handful of point-like SNRs, such as Cassiopeia A and Tycho SNR, where the size of the shell detected at lower energies is smaller than the angular resolution of the current IACTs. The rest of the sources are connected to the interactions of particles with molecular clouds located near the SNRs. These sources are identified by the spatial coincidence of gamma-ray emission with the molecular clouds. These sources are also determined by their offset from the center of the remnant or because the VHE emission overlaps only with a part of the shell instead of the entire remnant.

The distribution of the extensions for SNR-associated VHE gamma-ray sources is shown in Fig. 4.10 in comparison to the size of SNRs measured in radio. Compared to radio-detected SNRs, the number of VHE gamma-ray detected SNRs decreases rapidly as the extension of the source increases.

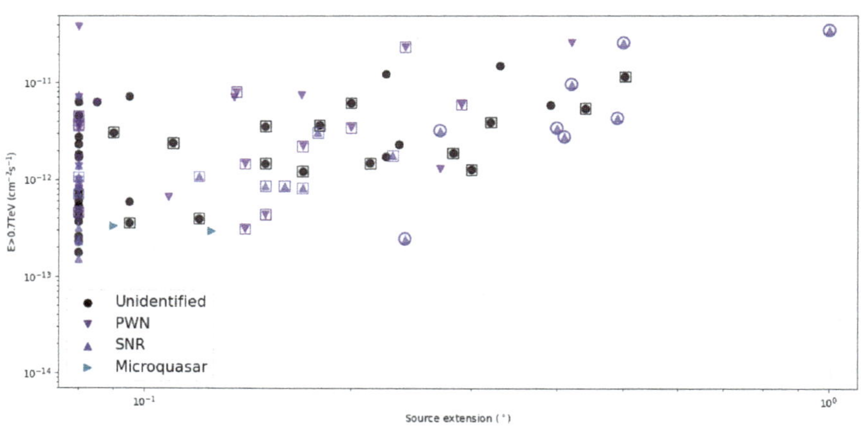

Fig. 4.10. SNR extension distributions for radio observations and IACT observations. Empty histogram: extension of the radio shell or composite-type SNRs from Green's SNR list. The extension for radio sources is calculated to be half of the reported size. Filled histogram: extension of the TeV SNRs.

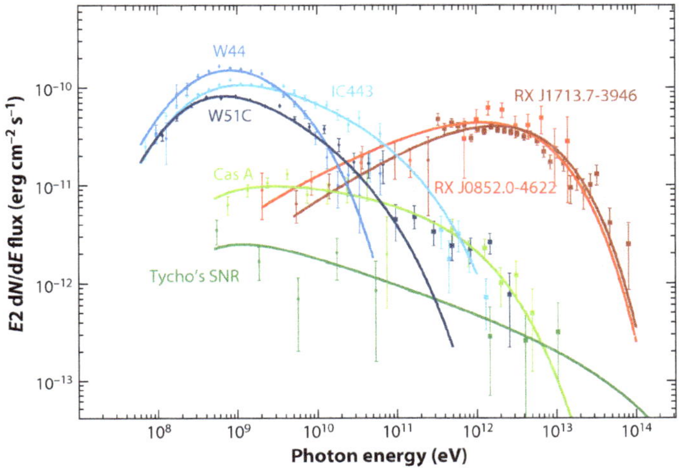

Fig. 4.11. SEDs of the most prominent SNRs with associated gamma-ray emission. Young SNRs are illustrated in red. The older shell-type SNRs are displayed in blue color while the middle-aged SNRs can be found in cyan. Figure taken from Funk (2015).

This may be due to the size-dependent bias of current observations as described in the previous section (also see Fig. 4.9).

Figure 4.11 shows the SEDs of all of the detected and associated SNRs. There are four sources whose spectral distribution follows a Exponential Cutoff Power Law (ECPL), while the rest of the sources are well-modeled by simple power laws. A number of SNRs — i.e. Tycho, W51C and W49B — have softer spectral indices than their GeV counterparts, indicating that a larger number of gamma-ray SEDs would follow an ECPL in a wider energy range.

TeV gamma-ray emission from SNRs is expected to be the result of both hadronic and leptonic interactions. Depending on the environment and evolution of the SNR, either the hadronic or leptonic component may dominate the observed TeV gamma-ray emission. It is possible to use TeV measurements to study the environments of SNRs, which are typically not well measured. Recent SNR population studies based on H.E.S.S. GPS data reported the typical ambient density around shell-type SNRs as well as the electron-to-proton energy faction above 10 TeV (Abdalla *et al.*, 2018f).

4.2.3.2. *Pulsar wind nebulae*

The largest portion of Galactic VHE gamma-ray sources are associated with PWNe. There are 30 identified PWNe in our Galaxy and one identified in the LMC. Among the unidentified sources, there are three sources with potential associations with PWNe. The full list of these sources is shown in Table 4.3. This table contains only the PWNe detected by IACTs.

There are few other PWNe detected in the VHE range that are not included in Table 4.3 because they are detected only by particle array

Table 4.3. List of 31 PWNe with three unidentified sources that are potentially connected to PWNe and are listed at the bottom of the table.

TeV name	Canonical name	Associated pulsar	log $\dot{E}$	τ_c (kyr)	d (kpc)	$L_{1-10TeV}$ (10^{33} erg s^{-1})
VER J0006+729	CTA 1	J0007+7303	35.65	13.9	1.40	0.71±0.10
3C 58	3C 58	J0205+6449	37.43	5.37	2.00	0.23±0.06
HESS J0537-691	N 157B	J0537-6910	38.69	4.93	53.7	760±80
Crab Nebula	Crab Nebula	B0531+21	38.65	1.26	2.00	32.1±0.7
Geminga	Geminga	J0633+1746	34.32	342	0.25	
HESS J0835-455	Vela X	B0833-45	36.84	11.3	0.28	0.83±0.11
HESS J1018-589		J1016-5857	36.41	21.0	8.0	8.1±1.4
HESS J1026-582		J1028-5819	35.92	90	2.33	5.3 ±1.6
HESS J1119-614	G292.2-0.5	J1119-6127	36.36	1.61	8.40	23±4
HESS J1303-631		J1301-6305	36.23	11.0	6.65	96±5
HESS J1356-645		J1357-6429	36.49	7.31	2.5	14.7±1.4
HESS J1418-609	Kookaburra(Rabbit)	J1418-6058	36.69	10.3	5.0	31±3
HESS J1420-607	Kookaburra(K2)	J1420-6048	37.00	13.0	5.61	44±3
HESS J1458-608	SNR G327.1-1.1	J1459-60?	35.95	64.7	4	
HESS J1514-591	MSH 15-52	B1509-58	37.23	1.56	4.40	52.1±1.8
HESS J1616-508		J1617-5055	37.20	8.13	6.82	162±9
HESS J1632-478		....	36.48	20	3.0	
HESS J1640-465		J1640-4631	36.64	3.35	12.8	210±12
HESS J1708-443		B1706-44	36.53	17.5	2.60	6.6±0.9
HESS J1718-385		J1718-3825	36.11	89.5	3.60	7.2±0.9
HESS J1747-281	G0.9+0.1	J1747-2809	37.63	5.31	13.3	46±7
HESS J1837-069	G25.5+0.0	J1838-0655	36.74	22.7	6.6	204±8
HESS J1813-178		J1813-1749	37.75	5.60	4.70	19.0±1.5
HESS J1818-154	SNR G15.4+0.1.	J1818-1541?	33.36	9	7.8	
HESS J1825-137		B1823-13	36.45	21.4	3.93	116±4
HESS J1833-105	G21.5-0.9	J1833-1034	37.53	4.85	4.10	2.6±0.5
HESS J1837-069		J1838-0655	36.74	22.7	6.60	204±8
HESS J1846-029	Kes 75	J1846-0258	36.91	0.728	5.80	6.0±0.7
HESS J1849-000	IGR J18490?0000	J1849-0001	36.99	42.9	7.00	12±2
VER J1930+188	G54.1+0.3	J1930+1852	37.08	2.89	7.00	5.5±1.8
VER J1952+293	DA 495	....		20.0	1.0	
VER J2016+371	CTB 87	....				
HESS J1831-098						
VER J2019+368	Dragonfly Nebula	J2021+3651	36.53	17		
TeV J2032+4130		J2032+4127				

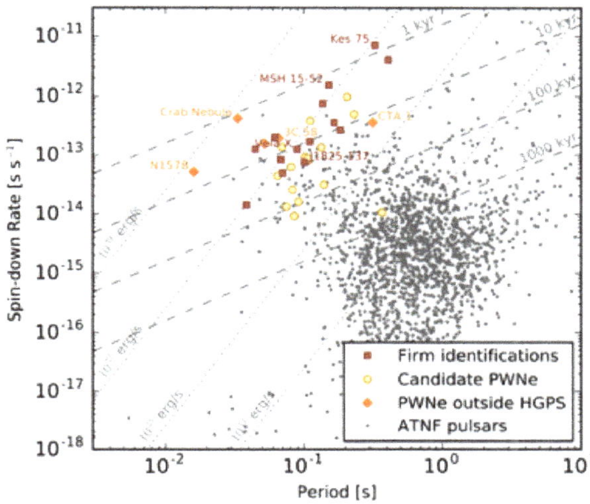

Fig. 4.12. Spin-down rate compared to the period of the ATNF pulsars (Abdalla *et al.*, 2018k). VHE-detected PWNe are plotted with larger markers.

detectors such as Monogem (Abeysekara *et al.*, 2017a). These sources are usually located in the very nearby Universe and are observed to have angular extensions larger than the typical FoV of the current IACTs.

A recently published PWN population study by H.E.S.S. (Abdalla *et al.*, 2018k), as well as previous studies, pointed out that TeV PWNe observed by the current generation of IACTs are associated with young pulsars with high spin-down luminosities. Figure 4.12 shows the relationship between the spin-down rate and the pulsar period for the ATNF pulsars. Pulsars associated with TeV PWNe are drawn with larger markers. As can be seen from this plot, all of the firmly identified TeV PWNe are associated with pulsars with spin-down rates higher than 10^{-15} s^{-1}, spin-down luminosities larger than 10^{35}erg s^{-1} and ages less than 100 kyr.

The sizes and morphologies of TeV PWNe vary depending on the evolutionary stage of the PWNe and the environment. The pulsar motion, SNR reverse shock and particle diffusion can all change the distribution of VHE gamma rays in PWNe. Abdalla *et al.* (2018k) presented the most complete TeV PWNe population study, with a simplified, time-dependent model for the evolution of the VHE electron population and TeV emission from PWNe. Figure 4.13 shows some representative characteristics of observed TeV PWNe shown in that study. The left panel shows the relation

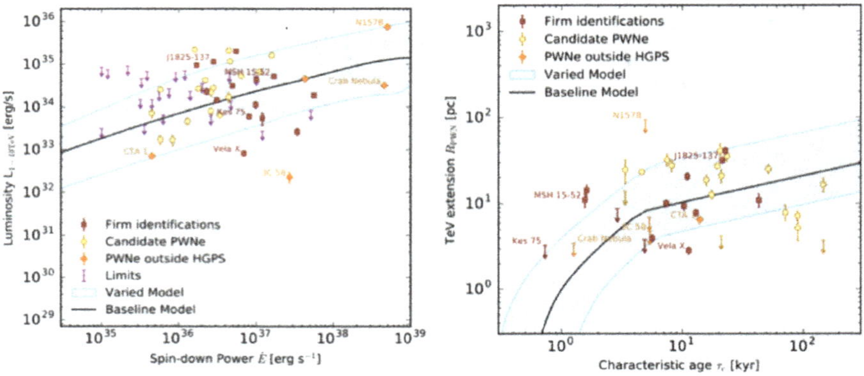

Fig. 4.13. PWNe population plot (Abdalla *et al.*, 2018k).

between TeV luminosity and the spin-down power of the pulsar. While the TeV luminosity scatters relatively widely around the baseline model, there is generally a mild positive correlation between the luminosity and the spin-down power. The relation between TeV source extension to the pulsar's characteristic age is shown in the right panel of Fig. 4.13. Generally, the models predict PWNe to go through fast, free expansion in the initial stages. This expansion rate is expected to slow down as PWNe start interacting with the surrounding medium and the SNR reverse shock. Measurements of TeV extension generally agree with this trend of increasing PWNe size with time.

TeV PWNe are generally considered to be stable sources. *Fermi*-LAT, however, reported several flaring episodes from the Crab Nebula in the GeV energy range, with a peak flux exceeding the average flux by up to a factor 6 (Mayer *et al.*, 2013). No corresponding flux changes have been observed in the VHE range from this flare (Aliu *et al.*, 2014a) or from any other time period.

4.2.3.3. *Binaries*

There are eight binary systems known to emit VHE gamma rays in our Galaxy: PSR B1259-63, LS 5039, LS I +61 303, HESS J0632+057, 1FGL J1018.6-5856, Eta Carinae, PSR J2032+4127/MT91 213 and RS Ophiuchi. Outside of our Galaxy, only one binary system has been detected: P3 in the LMC. All except two systems, i.e. Eta Carinae and RS Ophiuchi, are composed of a massive star and a compact object — either a pulsar or black hole — orbiting around the common center of mass. The identity of the compact object in these systems is unknown except in the case of the PSR B1259-63 (Johnston *et al.*, 1992) and PSR J2032+4127/ MT91 213 (Lyne *et al.*, 2015) systems, where the compact object for both cases is known to be a pulsar. Theories explaining the particle acceleration in binary systems can be largely divided into two categories: acceleration in an accretion-driven relativistic jet or in a shock interaction between the stellar wind and a pulsar wind. It is generally accepted that the second scenario is the one explaining the VHE emission of these binary systems, usually referred to as gamma-ray binaries since their SED peaks in the gamma-ray energy band. The following are only exceptions:

- Eta Carinae, which is a colliding wind binary system composed of a massive luminous blue variable star with a O- or B-type companion star.
- RS Ophiuchi, which is a recurrent nova detected by both H.E.S.S. (Abdalla *et al.*, 2022) and MAGIC (Acciari *et al.*, 2022) up to 1 month after its 2021 outburst. This detection signs the establishment of novae as a new class of gamma-ray VHE emitters.

All binary systems are point-like sources. Due to the change in environment as the compact object orbits the massive star, we see a variation in VHE gamma-ray emission depending on the orbital phase. The periods of known VHE gamma-ray binaries vary from ∼4 days (in the case of LS 5039) to ∼40–50 years (in the case of PSR J2032+4127/MT91 213). The latter, given its long orbital period, could be the only exception to the point-like behaviour: a PWN-like extended emission can temporarily form in some orbital phases.

Table 4.4 shows the orbital period for each of the VHE gamma-ray binaries.

Table 4.4. List of VHE gamma-ray binary systems.

Name	Companion star	Compact object	Orbital period
PSR B1259-63	LS 2883 (Be star)	Pulsar	∼3.4 years
LS 5039	? (O6.5V star)	?	3.91 days
LS I +61 303	? (B0 Ve star)	?	26.5 days
HESS J0632+057	MWC 148 (Be star)	?	315 days
1FGL J1018.6-5856	?(O6V star)	?	16.6 days
PSR J2032+4127	MT91 213 (Be star)	Pulsar	∼30 years
Eta-Carinae	? (O or B star)	luminous blue variable	5.5 years
LMC P3	? (O5 star)	?	10.3 days
RS Ophiuchi	red giant	white dwarf	454 days

Table 4.5. Characteristics of different binary systems.

Name	Comments
PSR B1259-63	GeV flare at ±15 days around periastron
LS 5039	Anti-correlation between GeV & TeV, PL+cutoff at INFC, PL at SUPC
LS I +61 303	Weak correlation between X-ray & TeV, Anti-correlation between GeV & TeV, TeV flares
HESS J0632+057	Correlation between X-ray & TeV, TeV/X-ray flare on 2018
1FGL J1018.6-5856	X-ray, GeV, TeV max at similar phase
PSR J2032+4127	
Eta-Carinae	Maximum GeV flux & TeV detection at periastron
LMC P3	GeV & TeV emission out of phase
RS, Ophiuchi	recurrent nova TeV emission during optical outbursts

The orbital phase dependence of the VHE gamma-ray flux differs from system to system because of their different environments and geometrical properties, with respect to the line of sight. Table 4.5 summarizes the characteristics of the known VHE binary systems. PSR B1259-63, LS 5039, 1FGL J1018.6-5856 and the LMC P3 systems all show highly predictable orbit-dependent flux changes (Abramowski *et al.*, 2015a; Abdalla *et al.*, 2018c). Meanwhile, orbit-to-orbit variability has been reported for LS I +61 303, which may be associated with a super-orbital variation with a longer period (Ahnen *et al.*, 2016d). HESS J0632+057 was believed to have repeating behavior until the X-ray and TeV gamma-ray fluxes showed a flaring episode around 2017 (Mirzoyan and Mukherjee, 2017). In systems with longer orbital periods (PSR J2032+4127, Eta Carinae, and PSR B1259+63), TeV emission has been detected only around periastron.

Figure 4.14 shows observations of LS 5039 as an example of the wealth of data we can get from binary systems. The characteristics of this system have been well studied, thanks to its short orbital period of 3.91 days and the high VHE gamma-ray luminosity.

4.2.3.4. *Pulsar*

While there are a handful of pulsars observed by IACTs (Aliu *et al.*, 2015; Archer *et al.*, 2015), only three pulsars have been detected: the Crab pulsar (Aliu *et al.*, 2008, 2011a), Vela pulsar (Gajdus *et al.*, 2015) and Geminga pulsar. Among these, the only pulsar that has been detected above 100 GeV is the Crab pulsar — a young pulsar with a high spin-down

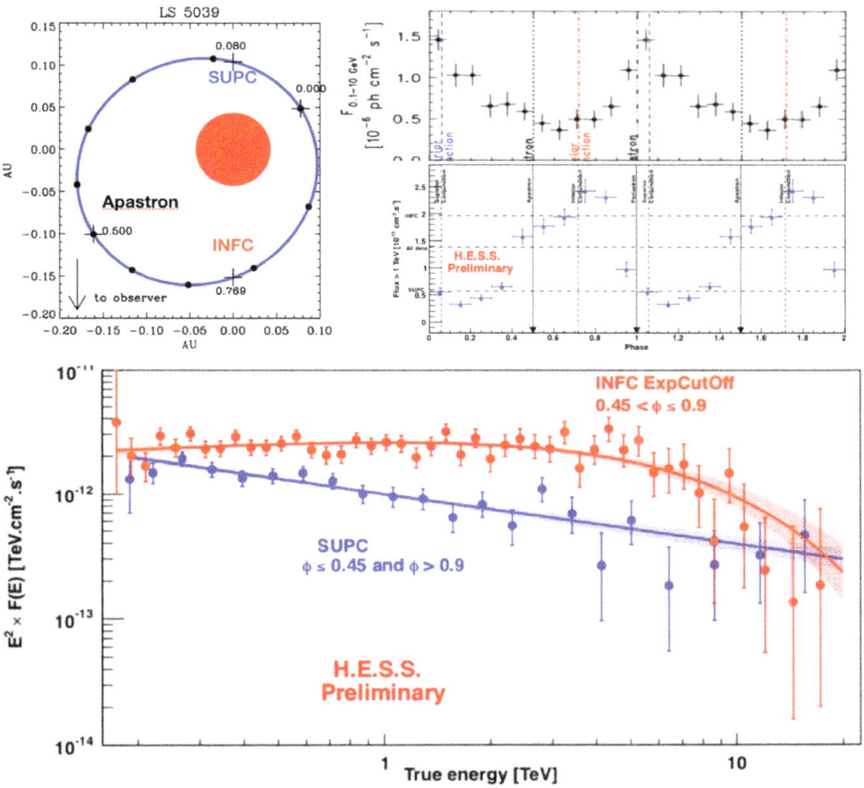

Fig. 4.14. Binary system diagram, flux dependency, and the SEDs of LS 5039.

luminosity of 4.6×10^{38} erg s^{-1}, and its rotational period is 33 ms. As shown in Fig. 4.15, TeV emission from the Crab pulsar after phase folding shows two clear peaks. The pulse phase between the two main pulses is defined as the "bridge", and TeV emission from this region has also been reported (Ansoldi *et al.*, 2016).

4.2.3.5. *Galactic center region*

The Galactic center (GC) region represents one of the most complex and interesting regions of bright non-thermal emission in the sky. Table 4.6 shows the list of sources that are detected within 200 pc around the GC.

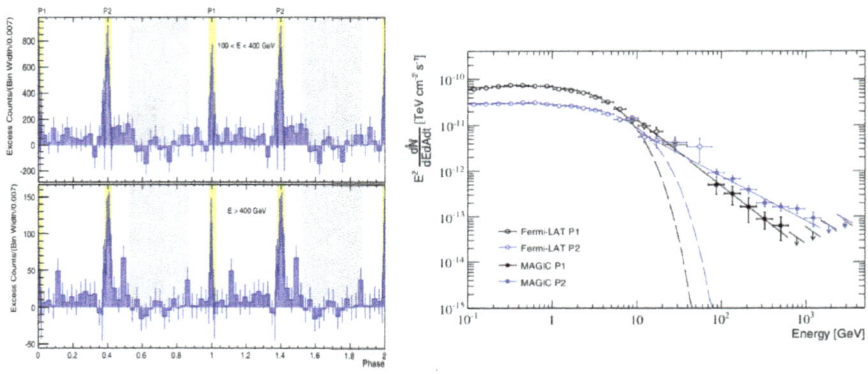

Fig. 4.15. Example plot showing the detection of the Crab pulsar in the VHE band (Ansoldi *et al.*, 2016).

Table 4.6. List of sources around the Galactic center.

Name	G.Lat (°)	G.Lon (°)	Class	Comments
Galactic center	−0.05	359.9	unid	Sgr A⋆, HESS J1745-290, VER J1745-290
Galactic center ridge	−0.05	359.9	unid	Correlated with Central Molecular Zone
Arc source	−0.15	0.05	unid	VER J1746-289, HESS J1746-285, MAGIC J1746.4-2853
SNR G0.9+0.1	0.09	0.86	PWN	HESS J1747-281,VER J1747-281
HESS J1745-303	−0.58	358.9	unid	Associated with G359.1-0.5, molecular cloud interaction

Sgr A⋆: Sgr A⋆ is a bright and variable source of non-thermal radio and X-ray emission lying at the heart of our Galaxy. It is associated with the accretion of material onto the supermassive black hole at the center of the Milky Way. A point-like source of VHE gamma rays (HESS J1745-290) with a flux around **FLUX** Crab has been observed by all current-generation IACTs at a position coincident with Sgr A⋆. The spectrum, position and variability of this source have been studied by all three of the current IACTs in comparison to observations at other wavelengths. Even though significant

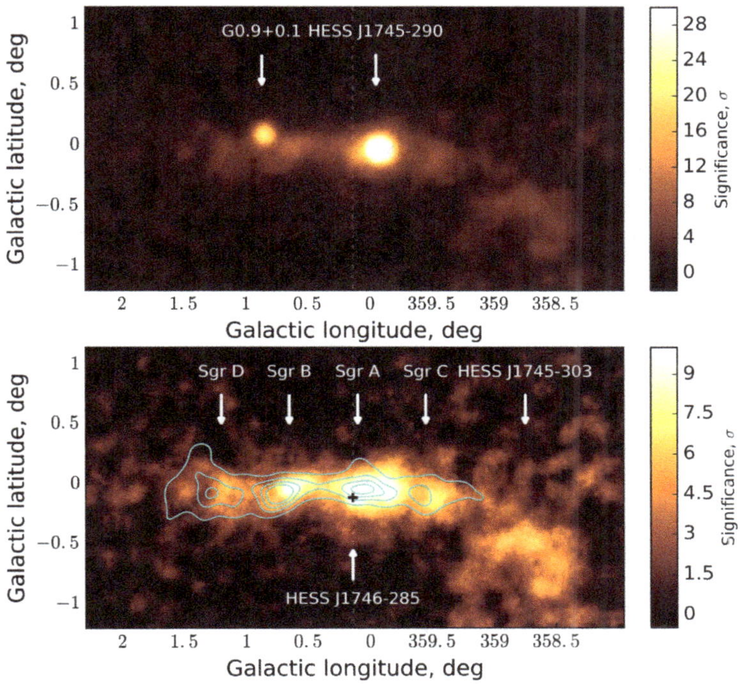

Fig. 4.16. Significance map of H.E.S.S. observations of the GC region, with the emission of the central point source and G 0.9+0.1 subtracted.

efforts have been made to reduce the uncertainties on these measurements, due to the complex nature of this source region, the reconstructed position of HESS J1745-290 is compatible with numerous potential counterparts.

Although variability in Sgr A⋆ has been detected at both radio and X-ray wavelengths, as of yet, no associated variability has been observed in the gamma-ray flux. The detection of such variability would not only conclusively prove the association of HESS J1745-290 with Sgr A⋆, but would also significantly aid in understanding the mechanisms and timescales of particle acceleration in this region.

Galactic center ridge: Surrounding Sgr A⋆ lies a region of diffuse gamma-ray emission extending several degrees along the Galactic plane. This emission is typically associated with the interaction of hadronic cosmic rays (CRs) with the dense gas in this region. When one combines the observed gas density in this region with the Galactic average CR density, however, the predicted gamma-ray emission is insufficient to account for the GC ridge. This implies that the local CR density in this region is higher than the Galactic average, likely due to a nearby CR accelerator (probably HESS J1745-290). The H.E.S.S. collaboration has made a number of studies of this region, discovering that no cutoff is seen in the gamma-ray spectrum up to 50 TeV. This implies the presence of protons at PeV energies that originate from the region surrounding the central accelerator.

4.2.3.6. *Unidentified sources*

There are 41 unidentified sources that are considered to be of Galactic origin. Among these, three are associated with PWNe and two are associated with SNRs. Unidentified sources around the GC region have been described in the previous section. The counterparts and the emission mechanisms of the rest of the unidentified sources are unclear.

The unidentified source category covers more than half of the entire population of VHE Galactic sources. Table 4.7 shows the full list of unidentified sources that were not mentioned in the previous section. The most common reason for the lack of a firm source identification is "source confusion", which occurs when several potential counterpart candidates are seen in the VHE source region. This confusion is exacerbated by the

Table 4.7. List of unidentified sources.

Source name	Extension(°)	gamma	Source name	Extension(°)	gamma
MAGIC J0223+403	point-like	3.1 ± 0.31	Westerlund 2	0.18	2.58 ± 0.19
HESS J1427-608	0.04 × 0.08	2.16 ± 0.14	HESS J1503-582	0.28	2.4 ± 0.4
HESS J1507-622	0.15	2.49 ± 0.18	HESS J1626-490	0.07 × 0.1	2.18 ± 0.12
HESS J1634-472	0.11	2.38	HESS J1641-463	point-like	2.07 ± 0.11
Westerlund 1	0.50	2.19 ± 0.08	HESS J1702-420	0.30 × 0.15	2.07 ± 0.08
HESS J1708-410	0.06 × 0.08	2.58 ± 0.10	HESS J1729-345	0.12	2.24 ± 0.15
HESS J1741-302	point-like	2.3 ± 0.2	HESS J1804-216	0.2	2.72
HESS J1808-204	0.09	2.3 ± 0.2	HESS J1809-193	0.53 × 0.25	2.2 ±0.1
HESS J1813-126	0.21	1.99 ± 0.14	HESS J1826-130	0.17	—
HESS J1828-099	point-like	2.25 ± 0.12	HESS J1831-098	0.15	2.10 ± 0.1
HESS J1832-085	point-like	2.38 ± 0.14	HESS J1832-093	point-like	2.6 ± 0.4
HESS J1834-087	0.09	2.45	HESS J1841-055	0.41 × 0.25	2.41 ± 0.08
HESS J1843-033	point-like	2.15 ± 0.05	HESS J1844-030	poine-like	2.48 ± 0.12
HESS J1848-018	0.32	2.8 ± 0.2	HESS J1852-000	point-like	2.17 ± 0.1
HESS J1857+026	0.11 × 0.08	2.39 ± 0.08	HESS J1858+020	0.08 × 0.02	2.17 ± 0.12
MGRO J1908+06	0.44	2.2 ± 0.1	MGRO J2019+37		1.75 ± 0.08
VER J2227+608	0.27 × 0.18	2.29±0.33	HESS J1746-308	point-like	3.27 ± 0.22
30 Dor C	point-like	2.6 ± 0.2	**Terzan 5**	0.16	2.5±0.3

often dense grouping of potential sources around regions of massive star formation, combined with the limited angular resolution of the current IACTs. The remainder of the unidentified sources are the "dark" sources, in which the VHE gamma-ray emissivity exceeds that of any potential low-energy counterparts, which is likely due to the source environment.

Finally, there are two sources that are associated with a unique source class. 30 Dor C is associated with a superbubble, and Terzan 5 is associated with a globular cluster. As these are the only sources associated with each particular source class, we categorize these as unidentified sources here and list them in bold font in Table 4.7.

4.2.3.7. *Undetected Galactic source classes*

Although the current generation of IACTs has made huge strides in the discovery of VHE gamma-ray-emitting source classes, there still remain a large number of proposed source classes that have yet to be discovered. Table 4.8 lists the classes that have been searched by IACTs, but have not yet yielded any conclusive detection.

4.2.3.8. *Galactic diffuse emission*

In addition to the numerous distinct point-like and extended sources detected in the Galactic plane, there also exists a large-scale diffuse emission component covering almost all Galactic longitudes and extending at least one degree above and below the plane. This diffuse emission can be seen in all of the current generation of instruments, but is clearest in the central galactic regions observed by H.E.S.S. This diffuse emission originates from two major sources: unresolved and undetected sources, and the interaction of hadronic CRs with ambient material in the Galactic disk. The relative importance of these two sources is not fully understood; however, these diffuse measurements can provide limits on the CR density in the Galactic disk.

Table 4.8. List of undetected source classes.

Source class	Searched sources
Microquasar	GRS 1915+105, Cyg X-3, Cyg X-1, SS 433
Young stellar objects	
Magnetars	4U 0142+61, 1E 2259+586
Low mass X-ray binary	V 404 Cyg, 4U 0115+634

4.3. The Extragalactic Very-High-Energy Gamma-Ray Sky

The extragalactic VHE gamma-ray sky counts to date 78 objects detected in the last 30 years. The first 20–25 years were mainly dedicated to the discovery of new extragalactic sources needed to populate the VHE gamma-ray sky. In the past 5 years, we instead passed from a discovery science plan to an astrophysical one, requiring deeper observations on specific source states or multi-year monitoring campaigns.

The current generation of IACTs is characterized by a limited FoV and usually does not operate in scanning mode. These instruments need an *a priori* knowledge of the region of the sky where a potential emitter is located. A fundamental characteristic of the extragalactic VHE gamma-ray sky is its variability, with a significant number of sources detected only during an enhanced state of emission. The limited FoV, in combination with the variable nature of the extragalactic emitters, bears the consequence that IACTs have to work in close connection with other facilities monitoring the sky (or at least a sample of interesting objects) that may trigger VHE gamma-ray observations. Observations at VHE gamma rays have also often been triggered by enhanced emission states observed in frequencies other than gamma rays.

Table 4.9 summarizes the different trigger criteria (third column) currently adopted by the running experiments. Band-dependent strategies (second column) have been set up during these decades. They are strongly

Table 4.9. Summary of the trigger criteria currently adopted by IACTs.

Band	Strategy	Trigger criterion	Alert delay
Optical	Monitoring programs with dedicated facilities	Brightening of the host-galaxy subtracted flux	1 day
Optical polarization	Monitoring programs (via proposal submission) and ATels	Fast rotations of EVPA and/or increase of the polarization degree	days
X-ray	SWIFT: GCN/ATel alerts and monitoring programs. MAXI:ATel or private communication	Count rate increase	days
HE gamma rays	Internal analysis chain and private communications	Both flux level and spectral shape changes are considered (extrapolation)	1 day
VHE gamma rays	IACTs + HAWC agreement	Enhanced flux	~hour

related to the existing instrumentation and its monitoring capabilities. The last column of the table reports the average delay of the triggered alert in each band.

For the optical band, IACTs have often profited from a dedicated optical telescope mainly devoted to AGN monitoring, such as ATOM in case of H.E.S.S. (Abramowski *et al.*, 2014b) and KVA (Nilsson *et al.*, 2018) in case of MAGIC. Optical triggers have been fundamental to discovering a number of sources now populating the VHE gamma-ray sky, in particular before the advent of *Fermi*-LAT (observing the full sky at higher frequencies, in gamma rays). Optical telescopes, however, do not have a full-sky coverage and may offer a biased sample of targets. The typical delay of the issued alert is one day.

Optical polarization measurements, which evaluate the percentage of polarized light observed and the reconstructed EVPA, are sometimes performed in combination with optical photometric observations. The percentage of polarized light depends on the mechanism of emission and is at the level of a few percent in blazars. Recently, polarization measurements have been adopted for triggering VHE gamma-ray observations. In this case, monitoring programs via proposal submission have been employed. The typical delay of the alert is few days.

In the X-ray band, the IACT community has profited from monitoring programs and private communications with astronomers performing automatic analysis of the available data. X-ray telescopes also do not have full sky coverage, hence they may offer a biased sample of targets. Typical delays in the alert are of the order of 1–2 days.

In the HE gamma-ray band, since *Fermi*-LAT went into orbit in August 2008, a large number of successful triggers based on HE gamma rays were sent to the VHE gamma-ray colleagues. *Fermi*, in fact, operates a continuous monitoring of the gamma-ray sky and, in the case of an enhanced signal from a particular source promptly detected with a daily analysis of the data, the community is alerted through several channels, such as GCN circulars and Astronomical Telegrams (ATels). In addition, for faster triggers, direct communication channels with the *Fermi*-LAT team members have been opened, reducing the delay of the issued alert to one day.

Finally, at VHE gamma rays, there is an agreement between the running IACTs and HAWC, and an alert is sent whenever the flux from a particular source from a common list reaches a certain threshold. Since, during data-taking the data are analyzed on site with a fast analysis, in the case of exceptionally high states, the alert is circulated via email almost simultaneously.

The main limitation in the maximum distance at which we can actually see a VHE gamma-ray emitter is set by the interaction of gamma-ray photons with the optical and IR photons permeating the Universe: also called extragalactic background light (EBL), discussed in detail in Chapter 7 of this book. The gamma-ray flux above tens of GeV, in fact, is suppressed as a function of the energy of the photon and its traveled distance, in analogy with the highest energy CRs that are suppressed at the highest energies due to the interaction with the cosmic microwave background. The VHE gamma-ray Universe is therefore characterized by a horizon at redshift $z \sim 1$, which confines the detectable sources within this distance.

We now briefly describe the *main observables* usually considered when studying the extragalactic sources at these extreme energies. Given the limited angular resolution of VHE gamma-ray detectors when compared to other facilities such as radio or optical telescopes, IACTs are not used for imaging extragalactic sources. The main studies of the emission are focused on the characterization of the VHE gamma-ray energy spectrum and its evolution with time.

Figure 4.17, right panel, shows the spectrum of Mkn 501 from Acciari *et al.* (2011b). This source is one of the brightest and best-studied extragalactic TeV-emitters. The differential energy spectrum, dF/dE, represents the number of photons at a given energy emitted by a source and detected in a given area and time interval. If we multiply the differential energy spectrum by E^2, we get the SED, $E^2dN/dE = \nu F_\nu$. The broadband SED of Mkn 501 built with archival data from the SSDC data center

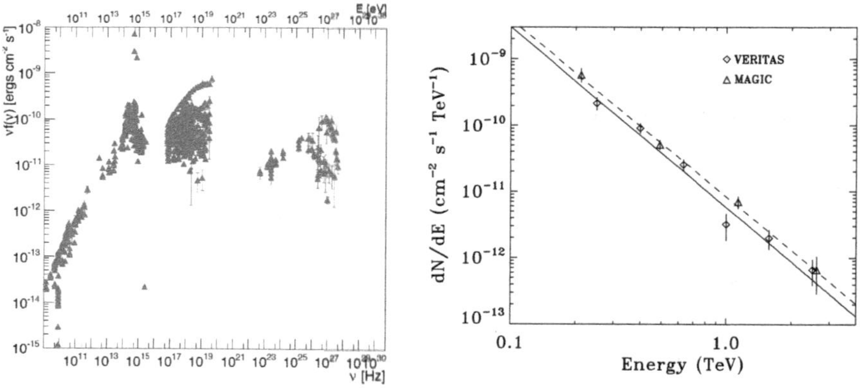

Fig. 4.17. *Left*: Broadband spectral energy distribution of Mkn 501, adapted from the SSDC website. *Right*: Differential energy spectrum measured in the VHE gamma-ray band by MAGIC and VERITAS in 2009, from Acciari *et al.* (2011b).

is displayed in Fig. 4.17, left plot. The x axis represents the increasing frequency, in Hz. The SED covers the entire electromagnetic spectrum, from radio to VHE gamma rays ($\nu > 10^{26}$ Hz). In this representation, we can appreciate the energy density of the emitted radiation and compare the amount of energy emitted at different energy bands. Different points at a given frequency refer to different measurement of νF_{ν}. The source is clearly variable, in particular in X-rays.

As opposed to optical and X-ray spectra, gamma-ray spectra are featureless, containing no emission or absorption lines. They are usually well fitted by power-law or log-parabola functions. In general, we study these spectra because they probe the physical conditions where particles are accelerated; offer clues on the nature and energy of these particles; and allow us to infer some basic information on the fields surrounding the emitting region.

The second kind of information that is extracted from VHE gamma-ray data collected from extragalactic objects is the timing evolution of the integral spectrum, i.e. the flux observed above a certain energy threshold. Some objects, such as the blazars, are characterized by a highly variable emission across the different bands of the electromagnetic spectrum. The study of the time evolution can give valuable insights into the acceleration region. In particular, timing information is crucial to constrain the size of the emitting region. Since, for causality, the variability timescale cannot be smaller than the crossing-light time, the size of the emitting region can be inferred once this timescale is measured. The range for this variability timescale spans from a few minutes up to several days. One of the most representative light curves collected at VHE gamma rays is shown in Fig. 4.18, from Aharonian *et al.* (2007).

Timing and spectral information can be combined to provide a time-resolved model of the emission.

4.3.1. *Extragalactic populations*

The 78 sources detected up to now that are above 100 GeV are illustrated in Fig. 4.19: there are three non-AGN sources, the starburst galaxies M82 and NGC 253, and the GRB 190314. Out of the remaining sources, five are radio-galaxies (FRI, also known as misaligned blazars), and 68 are blazars, with seven FSRQs and 61 BL Lac objects.

For comparison, the (non-GRB) extragalactic sources detected with *Fermi*-LAT between 110 MeV and 300 GeV and reported in the 3LAC catalog (Ackermann *et al.*, 2015b) are 1591, a factor of twenty larger than those observed above 100 GeV by IACTs. Of these sources, 40% are

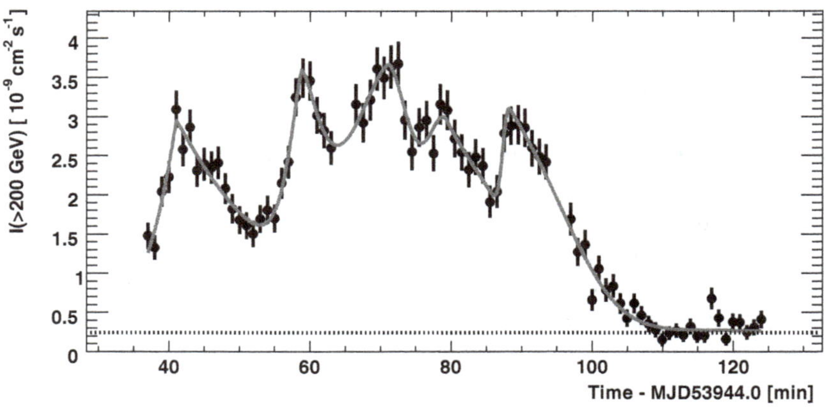

Fig. 4.18. Time evolution of the integral flux above 200 GeV from the BL Lac object PKS 2155-304 recorded by H.E.S.S. during an exceptional flare in 2006, from Aharonian *et al.* (2007).

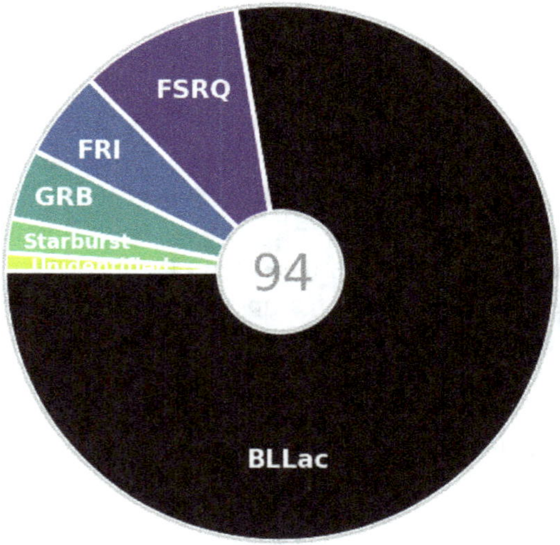

Fig. 4.19. Pie chart illustrating the different populations of extragalactic sources detected at VHE gamma rays.

BL Lac objects, 29% are FSRQs, another 29% are blazars of uncertain type, while only 2% are non-blazar AGN (including 14 radio galaxies). The main reasons for this large difference in numbers are *Fermi*-LAT full-sky coverage; the fact that VHE gamma-ray photons require extremely energetic conditions to be produced, which are more rare, and the EBL absorption effect.

The 3-year HAWC catalog (Albert *et al.*, 2020b) includes only two bright extragalactic sources (Mkn 501 and Mkn 421) due to its limited sensitivity. For these two sources, however, HAWC has the unique possibility of an unbiased monitoring of the emission thanks to the duty cycle close to 100%, thus offering important information on the time evolution over different scales, almost inaccessible by IACTs.

The aperiodic variations of the large majority of TeV-emitting extragalactic sources, combined with the redshift-dependent spectral absorption at the highest energies due to the EBL, make any attempt of population study with the current catalog of extragalactic VHE sources very difficult. The situation is even worse when considering the non-uniform exposure of the sky, with most sources observed only during active emission states.

This is a challenge for the next generation of IACT and instruments — that with their improved sensitivity and, in the case of IACTs, wider FoV will certainly open the possibility of population studies with TeV extragalactic sources.

In the following subsection, we will briefly describe the main observational properties of the objects of the classes of TeV-detected sources and outline the open questions.

4.3.1.1. *Blazars*

Blazars constitute a large majority of the extragalactic TeV emitters. They are AGNs, i.e. galaxies with an intense emission from the core region, characterized by a jet of particles closely aligned with the line of sight of the observer. Particles in the jet are accelerated, either by shock waves or processes involving magnetic reconnection, up to ultra-relativistic speed. The radiation emitted at all bands is therefore Doppler-boosted. This enhancement explains why blazars are the most numerous VHE gamma-ray emitting extragalactic objects detected.

There are two major sub-classes of blazars: BL Lac objects and Flat-Spectrum Radio Quasars (FSRQs).

FSRQs are blazars characterized by strong emission lines in the optical spectrum (equivalent width > 5 Å).

The SED is complex, due to the superposition of several radiation fields, the jet emission, a narrow and a broad-line emitting region, the hot accretion disk (emitting in X-ray) and a warm IR-emitting torus of dust.

The jet emission is non-thermal and presents a double-bump shape, with a low-energy bump peaking at radio to IR frequencies and due to synchrotron emission by electrons accelerated in the jet. The HE bump peaks in the MeV to GeV domain, and is attributed to external Compton emission, with synchrotron photons upscattered when interacting with one of the external radiation fields (the broad-line region of the torus). The luminosity of FSRQs is very large, and the jet emission is largely variable. Seven FSRQs have been detected to now at VHE gamma rays, listed in Table 4.10. The sources span a wide range of redshift, from ~0.19 to 0.99. All of these sources were discovered during a state of enhanced activity and their emission is highly variable in all the energy bands. The time evolution of the VHE gamma-ray emission of PKS 1222+216 is displayed in the left panel of Fig. 4.20. This measurement was used to constrain the size of the emitting region.

An example of spectral energy distribution from the distant FSRQ PKS 1441+25 is shown in the right panel of Fig. 4.20. The Compton peak luminosity spans several order of magnitudes. From the figure, it is evident that the VHE gamma-ray band in these sources covers the last part of the SED and becomes detectable, in general, only during flares given the sensitivity limit of the current generation of IACTs. An external Compton model (continuous line) nicely fits the simultaneous data of the flare. One of the open questions, which has been partially addressed with recent observations in VHE gamma rays, is *where* the particle acceleration takes place in the jet — whether it is close to the central black hole or in the outer region.

Table 4.10. TeV-emitting FSRQs (Adapted from TeVCat[a]).

Source name	Class	Distance z
PKS 0346-27	FSRQ	0.991
S3 0218+35	FSRQ	0.954
PKS 1441+25	FSRQ	0.939
TON 0599	FSRQ	0.725
B2 1420+32	FSRQ	0.682
3C 279	FSRQ	0.536
PKS 1222+216	FSRQ	0.432
PKS 1510-089	FSRQ	0.361
PKS 0736+017	FSRQ	0.189

[a]Website: http://tevcat.uchicago.edu

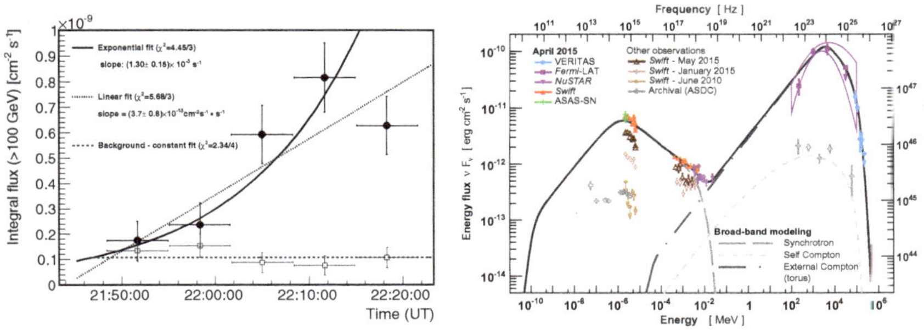

Fig. 4.20. *Left*: Six-minute binning light curve above 100 GeV of the FSRQ PKS 1222+216, courtesy of the MAGIC Collaboration. *Right*: Spectral energy distribution of PKS 1441+25, one of the few FSRQs detected at VHE gamma rays, courtesy of the VERITAS Collaboration.

BL Lac objects (BL Lacs) are blazars whose optical spectrum displays very faint (equivalent width < 5 Å) or no emission lines. In BL Lacs, the SED is fully dominated by the jet emission and is characterized by two broad bumps covering the entire electromagnetic spectrum. These objects are, on average, less luminous than FSRQs, but the peaks of the SED are shifted toward higher frequencies. For this reason, there are several BL Lacs that are detected not only during an enhanced state of emission but also in a low/quiescent state. Experimentally, this allows us to study the general properties of these objects independently on the emission state. Depending on the position of the first synchrotron peak of the SED, BL Lacs are further divided into low-, intermediate- and high-frequency-peaked BL Lac objects (LBL, IBL and HBL, respectively). In general, the second peak also moves toward higher frequencies when passing from LBLs to HBLs. For this reason, HBLs, having the second peak located in the VHE gamma-ray regime, are the ideal target for IACT observations. It is therefore not surprising that the large majority of the extragalactic objects belongs to this sub-class. In a few cases, the source classification is uncertain. The source HESS J1943+213 represents an interesting case. This source is a point-like TeV emitter close to the Galactic plane detected by the H.E.S.S. Collaboration (Abramowski *et al.*, 2011). Due to its location, the nature of the source was debated. Multi-wavelength observations exploring the classification of the source are converging toward its identification as an HBL (Archer *et al.*, 2018).

Due to the absence of optical emission lines in their optical spectrum, many BL Lacs lack a precise redshift determination. In the TeV-selected sample listed in Table 4.11, 14 sources, ~20%, are affected by this problem. Knowing the distance of BL Lacs is important for two reasons: to evaluate the overall intrinsic bolometric luminosity, and to estimate the intrinsic, EBL-corrected spectrum in the VHE range. The last generation of optical instruments has been recently employed for this task. One of the

Table 4.11. TeV-emitting BL Lac objects (Adapted from TeVCat[b]) ordered according to their distance. (last update March 2019).

Source name	Class	Distance z	Source name	Class	Distance z
PKS 0903-57	IBL	0.695	B2 1811+31	IBL	0.117
1ES 0033+595	HBL	0.467	PKS 2155-304	HBL	0.116
S4 0954+65	BL Lac	0.370	1ES 1312-423	HBL	0.105
PKS 0447-439	HBL	0.343	W Comae	IBL	0.102
1ES 0502+675	HBL	0.340	SHBL J001355.9-185406	HBL	0.095
3C 66A	IBL	0.340	1ES 1741+196	HBL	0.084
TXS 0506+056	Blazar	0.336	RGB J0152+017	HBL	0.080
OT 081	LBL	0.322	PKS 2005-489	HBL	0.071
OJ 287	BL Lac	0.306	BL Lacertae	IBL	0.069
1ES 0414+009	HBL	0.287	PKS J0648.7+1516	HBL	0.069
PKS 0301-243	HBL	0.266	PGC 2402248	HBL	0.065
1RXS J023832.6-311658	HBL	0.232	1ES 1727+502	HBL	0.055
MS 1221.8+2452	HBL	0.218	1ES 2037+521	HBL	0.053
1ES 1011+496	HBL	0.212	AP Librae	LBL	0.049
RBS 0723	HBL	0.198	1ES 1959+650	HBL	0.050
MRC 0910-208	HBL	0.198	TXS 0210+515	HBL	0.049
RBS O413	HBL	0.190	Markarian 180	HBL	0.045
1ES 0347-121	HBL	0.188	1ES 2344+514	HBL	0.044
1ES 1101-232	HBL	0.186	Markarian 501	HBL	0.034
1ES 1218+304	HBL	0.182	Markarian 421	HBL	0.031
RX J0648.7+1516	HBL	0.179	IC 310	HBL	0.019
MAGIC J2001+435	HBL	0.174	PKS B0625-354	AGN	0.055
1ES 2322-409	HBL	0.174	PG 1553+113	HBL	—
1RXS J081201.8+023735	HBL	0.172	1ES 1440+122	HBL	—
H 2356-309	HBL	0.165	RGB J2056+496	Blazar	—
1RXS J101015.9-311909	HBL	0.143	RGB J2243+203	HBL	—
1ES 0229+200	HBL	0.140	H 1722+119	HBL	—
PKS 1440-389	HBL	0.138	RGB J0136+391	HBL	—
1ES 0806+524	HBL	0.138	KUV 00311-1938	HBL	—
S3 1227+25	IBL	0.135	HESS J1943+213	HBL	—
RX J1136.5+6737	HBL	0.134	PKS 1424+240	HBL	—
1ES 1215+303	HBL	0.131	VER J0521+211	IBL	—
H 1426+428	HBL	0.129	S5 0716+714	IBL	—
TXS 1515-273	HBL	0.128	S2 0109+22	IBL	—
RGB J0710+591	HBL	0.125	GB6 J1058+282	BL Lac	—
1RXS J195815.6-301119	HBL	0.119	1ES 0647+250	HBL	—
B3 2247+381	HBL	0.119			

[b]Website: http://tevcat.uchicago.edu

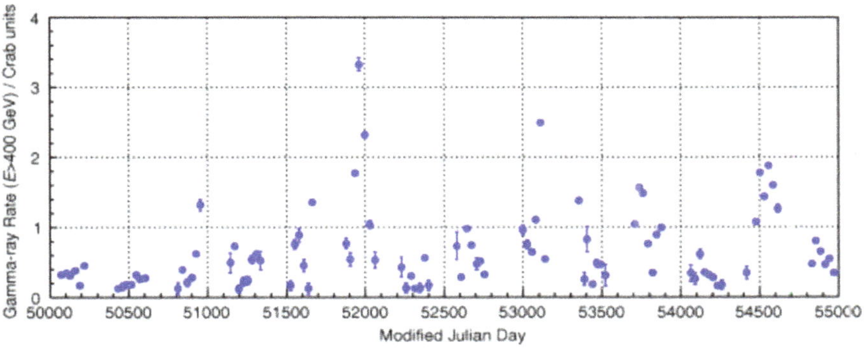

Fig. 4.21. Multi-year TeV light curve of the BL Lac object Mkn 421, from Acciari *et al.* (2014).

most important determinations was that of the redshift of TXS 0506+650 (Paiano *et al.*, 2018), the neutrino-blazar (see Chapter 9).

In BL Lacs, the variability timescale spans from minutes to several months. For example, Fig. 4.21 illustrates the multi-year light curve of the BL Lac object Mkn 421.

Main open questions on the study of BL Lac objects regard the role of the magnetic field in the particle acceleration and the role of hadrons in the overall emission (hence the neutrino flux).

4.3.1.2. *Radio galaxies*

Radio galaxies can be considered the misaligned counterpart of blazars; they are, in fact, jetted AGNs with the jets almost perpendicular to the observer's line of sight.

The five known TeV-emitting radio galaxies have all been discovered in this energy band by current-generation IACTs. The main characteristics are a hard spectrum extending to several TeV and a rapid variability down to day-scale. They are relatively nearby, as shown in Table 4.12. The reason

Table 4.12. TeV-emitting radio galaxies.

Source name	Class	Distance z
PKS 0625-35	FRI	0.05
3C 264	FRI	0.02
NGC 1275	FRI	0.02
M 87	FRI	0.004
Centaurus A	FRI	0.002

is that in these objects the jets are misaligned, so the flux is not strongly enhanced by Doppler effect like in blazars. While in gamma rays, radio galaxies cannot be spatially resolved, observations in the other bands allow for a clear identification of the emitting zone. When studying the source emission, this must be taken into account since in low-energy bands — such as radio, optical and X-ray — we have spatially resolved spectra and light curves. At higher energies, instead, we have a superposition of different, unresolved, emitting zones. On the other hand, the coordinated and simultaneous monitoring of the emission in several bands might help in localizing the VHE emission site. This was the result of a joint M87 VERITAS, MAGIC and H.E.S.S. campaign with VLBA (radio band) carried out in 2008. During this campaign, a strong flare at VHE gamma rays was detected, followed by an ejection of a radio-emitting component from very close to the black hole (Acciari *et al.*, 2009c). The light curve is reported in Fig. 4.22. The combined observations were used to constrain the origin of the VHE gamma-ray emission as close to the central engine.

The broadband VHE displays a two-bump shape, similar to what was observed in BL Lac objects. The measure of an unexpected spectral hardening in Centaurus A (Abdalla *et al.*, 2018i) when passing from the HE to the VHE gamma-ray regime might point to an additional component in the spectrum, i.e. a hadronic component. According to several theories, radio galaxies might also be responsible for the observed flux of ultra-high-energy CRs and a hadronic component would therefore be expected.

The origin of the VHE gamma-ray emission, however, is still largely debated: a multitude of scenarios — relying on either hadronic and/or leptonic radiation processes and operating on different scales — have been developed and applied to the data. Future observations with more precise and sensitive instruments will help us to better characterize these nearby, extreme accelerators.

4.3.1.3. *Starburst galaxies*

The class of starburst galaxies is less represented in the VHE gamma-ray sky: only two sources are known to date be VHE gamma-ray emitters and are listed in Table 4.13. Both sources were discovered by the current generation of IACTs after deep exposures of more than 100 hours, collected in multi-year campaigns (Acciari *et al.*, 2009a; Acero *et al.*, 2009a). Their VHE gamma-ray spectra do not show any flaring activity and connect

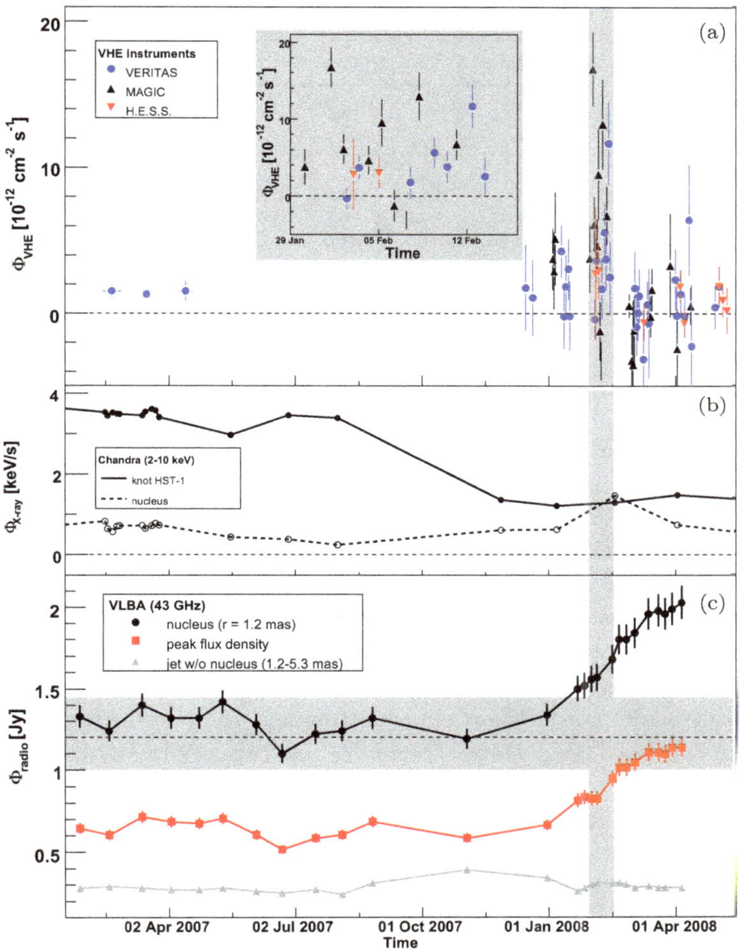

Fig. 4.22. Light curve of M87 in VHE gamma rays (*top panel*), X-rays (*central panel*) and radio (*bottom panel*). From Acciari *et al.* (2009c).

smoothly with the *Fermi*-LAT spectrum, suggesting the same origin for the emission mechanism (Abramowski *et al.*, 2012).

In these objects, the origin of the HE emission is believed to be related with the stars of the galaxies. In particular, energetic CRs produced by massive stars during supernova explosions and interacting with the interstellar gas and radiation are mainly responsible for this emission. Combined TeV

Table 4.13. TeV-emitting starburst galaxies.

Source name	Class	Distance z
M 82	Starburst	3900 kpc
NGC 253	Starburst	2500 kpc

and GeV spectral measurements allowed an initial evaluation of the CR energy density in those galaxies; further details are reported in Chapter 5.

Future observations with the CTAO should increase the number of starburst galaxies detected at TeV energies.

4.3.1.4. *Gamma-ray bursts*

Gamma-ray bursts (GRBs) are transient phenomena characterized by a fast flash of electromagnetic radiation mainly peaked in the gamma-ray regime. Of extragalactic origin, GRBs are considered the most energetic phenomena of the Universe. They are classified according to their duration, and may last just a second ("short" GRBs) or tens of seconds ("long" GRBs).

Since 2017, we know that at least in a fraction of GRBs the radiation is accompanied, or anticipated, by the emission of gravitational waves (Goldstein *et al.*, 2017). This detection confirmed for the first time that at least a fraction of GRBs occur due to binary compact mergers.

The *Fermi*-GBM, since beginning operations in mid 2008, has autonomously detected over 2000 GRBs in the few keV-tens of MeV energy range. A fraction of these events has also been detected by *Fermi*-LAT at higher energies. H.E.S.S., MAGIC and VERITAS have observed hundreds of GRBs since the early 2000s, as illustrated in Fig. 4.23. Over time, the automatic, quick response of IACTs improved significantly until the first claim, in 2019, of a VHE gamma-ray signal from the GRB 190114C by the MAGIC collaboration. The observation started at about 50 s after *Swift*-BAT alert. Catching a GRB at VHE gamma rays was one of the main objectives of MAGIC that devoted large efforts for the construction and maintenance of a fast-repositioning telescope system. Also, this detection paved the road for the future CTAO, and in particular for the Large-Size Telescopes (LSTs), whose low energy threshold will open a much deeper Universe to the search for VHE-emitting GRBs.

There are four GRBs currently detected at VHE, all belonging to the "long-GRB" class and related to the explosion of massive stars. In addition, two hints of detection have been published (see Table 4.14). Interestingly,

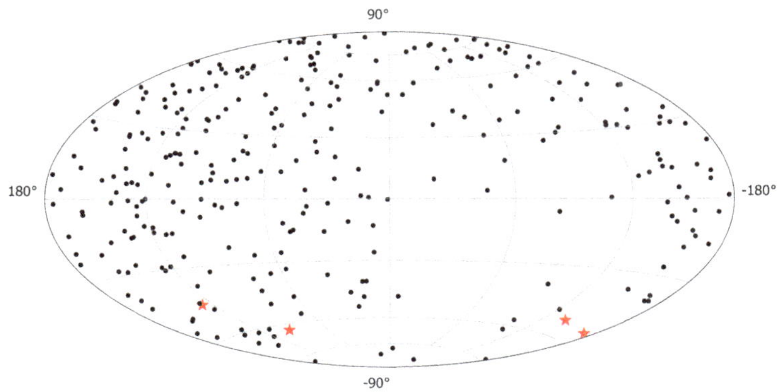

Fig. 4.23. Galactic projection of the positions of the GRBs followed up by the current generation of IACTs. The position of GRB 190114C is marked by the large triangle.

Table 4.14. TeV-emitting GRBs (Adapted from TeVCat[c]).

Source name	Type	Distance z
GRB 201216C	lGRB	1.1
GRB 180720B	lGRB	0.65
GRB 201015A	lGRB candidate	0.43
GRB 190114C	lGRB	0.42
GRB 160821B	sGRB candidate	0.16
GRB 190829A	lGRB	0.08

[c]Website: http://tevcat.uchicago.edu

one of the hints refers to a "short-GRB", an event related to a star merger that might be associated with gravitational wave emission.

Current open questions on GRB studies at VHE energies are related to pinpointing the emission mechanism causing such energetic radiation (synchrotron or inverse Compton emission).

4.3.1.5. *Search for fast radio bursts*

A type of transients that has not yet been detected by current-generation IACTs, despite a number of efforts, is the class of Fast Radio Bursts (FRBs). These phenomena are bright flashes observed typically at radio (GHz) frequencies with short, millisecond durations. Their nature remains unknown and thus motivates searches for counterparts at other wavelengths, up to the VHE gamma-ray band.

More than 800 FRBs are known to date and their origin is very likely extragalactic. The FRB Catalog[b] presents an up-to-date list of FRBs. Among the known sources, only two have been detected more than once after their discovery, the FRB 121102 (the name of the event is given according to its first detection, as in the case of GRBs or other transient events), and in January 2019 the FRB 180814.J0422+73 (Amiri *et al.*, 2019). In the case of FRB 121102, the repeating nature of the burst allowed the identification of the host of the source, a very distant galaxy about three billion light years from Earth in an extreme environment (Chatterjee *et al.*, 2017). VERITAS and MAGIC observed this source during both its persistent and burst states, and no detection was found in both cases (Acciari *et al.*, 2018b; Bird *et al.*, 2017).

4.3.2. *Multi-wavelength context*

Given the variable nature of extragalactic VHE gamma-ray emitters, in particular that of AGNs, the multi-wavelength approach was, since the beginning, utilized by IACT collaborations.

The existence of dedicated optical facilities aimed at a dense monitoring of interesting AGNs, which helped substantially to identify interesting objects, especially before the continuous monitoring done with *Fermi*-LAT. Refined over the years, the trigger strategies outlined in Section 4.3 were adopted and led to the discovery of a remarkable number of new objects. Moreover, these strategies allowed for simultaneous data-taking between telescopes operating in different energy bands, which is the only viable way to build a consistent SED to be modeled and interpreted. Simultaneous data have also been used for cross-correlation analysis to search for correlations and delays during active states.

In the last decade, multi-band simultaneous observations have become the standard approach for the study of the most extreme, persistent extragalactic accelerators of the Universe. We concentrate on the possible connection between VHE gamma rays and the other bands, as follows:

Radio band: TeV-radio correlation studies are very powerful tools used to study and localize the emitting region in radio galaxies, thanks to the excellent resolution of the radio arrays of telescopes and the favorable viewing angle. In blazars, on the contrary, TeV and radio

[b]https://www.wis-tns.org/.

bands are not always strictly correlated, because the radio emission can be generated in an outer region with respect to the VHE gamma-ray emission. In some cases, however, an ejection of a new radio component is observed simultaneously with a flaring state of a blazar (Ahnen *et al.*, 2017c).

Optical band: Optical-TeV correlation studies are crucial for the identification of the electron population(s) responsible for the VHE gamma-ray emission in all AGN classes. If a flaring state is reported simultaneously in two bands, it probes the common origin of the emission. In FSRQs, increasing attention is given to the role of optical polarimetry, since a number of electron polarization vector plane rotations have been detected in relation with flares.

X-ray band: This band has a role similar to that played by the optical band, in particular in IBL and HBL, which emit a significant part of their electromagnetic power at X-ray frequencies. Remarkably, a non-correlation between an X-ray and a VHE gamma-ray flare could be explained by hadronic processes at work.

gamma-ray band: The low-energy counterpart of the emission studied at the highest energies offers the unique possibility to test the origin of the emission and constrain the shape and continuity of the spectrum.

4.3.3. *Properties of extragalactic sources*

4.3.3.1. *The redshift distribution of TeV-detected AGNs*

The redshift distribution of the different extragalactic emitters belonging to the AGN family is represented in Fig. 4.24. In this plot, we have excluded the starburst galaxies since they belong to a completely different family and are only two sources.

The redshifts span from a minimum of 0.002 to a maximum value of 0.95. The peak of the distribution is between $z = 0.1$ and 0.15, where a large number of sources are located. The average value of the three populations of sources reflects intrinsic differences in the source classes, and is $\langle z \rangle = 0.59$ for FSRQs, 0.17 for BL Lac objects and 0.02 for radio galaxies. It must be noted, however, that a large fraction of BL Lacs, ~15%, still lacks a redshift determination and could be located at large distances.

Due to the EBL absorption effect, only a few, very bright emitters have been detected at $z > 0.6$. All of these sources are FSRQs: objects that exhibit strong flares and an exceptional intrinsic luminosity.

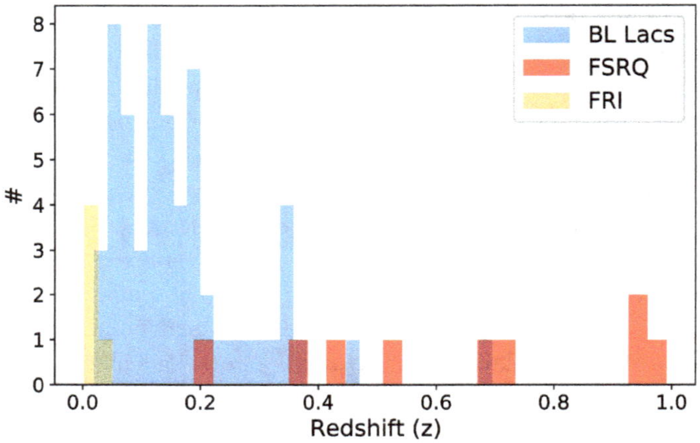

Fig. 4.24. Histogram with the redshift of the jetted AGNs detected at VHE gamma rays per population type.

4.3.3.2. *AGN variability at TeV energies*

While starburst galaxies are believed to be constant emitters of VHE gamma rays, jetted AGNs and in particular blazars are, by their own nature, variable sources.

AGN variability in the VHE gamma-ray energy band has been observed at different timescales, from years down to minutes. The amplitude of TeV flares has also been detected in a wide range of values, with a maximum flux level from 1.5 to 2 times its value in the quiescent state up to tens, even hundred times its value.

Due to non-uniform and often only sparse sampling of the sources, a systematic study of the time variability of all the sources has not yet been possible for IACTs. Moreover, some sources were detected above 100 GeV only during particular active states of emission, and their emission level during a non-active (or quiescent) state is still unknown. This is, for example, the case for all known VHE gamma-ray emitting FSRQs — except one, PKS 1510+089 (Acciari *et al.*, 2018).

As mentioned above, short-term variability of AGNs is a key observation useful to directly probe the size of the emitting region. Causality implies that the shorter the variability timescale, the smaller is the emitting region. This is particularly relevant in the VHE gamma-ray domain, as the imaging capabilities of IACTs are by far too limited to resolve the emitting region in extragalactic sources. Short-term variability in AGNs is also a very powerful tool for testing Lorentz invariance violation, as will be further discussed in Chapter 8.

The VHE gamma-ray integral flux variability at any scale has been extensively used to constrain some physical properties of the emitting region when compared with the variability at other wavelengths. As a first approximation, in single-zone SSC models, we expect a strong and clear correlation between different bands, in particular, among bands related to the same electron population. For example, in HBLs, we expect it in X-ray (direct synchrotron emission by HE electrons) and gamma rays (upscattered X-ray emission by HE electrons). In a multiple zone SSC model, the emitting zone is more than one and the situation is more complicated. In each band we might have the contribution of one or both emitting zones and the connection of the variability between zones becomes more complex.

Yet another case is represented by objects whose HE emission is due to the external Compton process of target photons on the accelerated electrons responsible for the low-energy synchrotron emission. This is the typical case observed in FSRQs where there are strong external fields surrounding the jet, as mentioned above. In this case, like in the one-zone SSC model, there is a connection between high- and low-energy bands, since the emission is related to the same electrons.

All the cases considered up to now are based on leptonic models. In the case of hadronic models, the relation between high- and low-energy bands can be completely lost. The effect is the possibility of having "orphan flares" such as the one observed from 1ES 1959+650 on June 4, 2002 (Aharonian *et al.*, 2003). Also, in this kind of model, it is more difficult to accommodate fast flares, due to the larger mass of the proton with respect to that of the electron. A statistical variable very commonly used for the study of multi-band variability is the $F_{\rm var}$.

4.4. Status and Prospects of Gamma-Ray Astronomy

4.4.1. *The Cherenkov Telescope Array Observatory*

The next major construction project in the world of VHE gamma-ray astronomy is the Cherenkov Telescope Array Observatory (CTAO). The design for the two CTAO arrays are expected to allow a point-like source sensitivity that is a factor 5 to 10 larger than the current generation of facilities, with significantly improved angular resolution and increased FoV ($\sim 8°$). This combination of features will allow the CTAO to significantly increase the number of sources detected, potentially reaching almost thousand. In order to facilitate source discovery, the CTAO will conduct three major source surveys (Acharya *et al.*, 2019), the GPS, an unbiased scan of 25% of the extragalactic sky and the deepest ever scan of the 10° around the GC.

4.4.2. *The CTAO Galactic Plane Survey*

One of the main key science projects of CTAO is the GPS extending over the entire Galactic plane (Remy *et al.*, 2021). CTAO aims to provide survey coverage in this region, improving the point-like sensitivity on the currently available H.E.S.S. GPS data by about one order of magnitude (2 mCrab). Additionally, the improved angular resolution of the CTAO will help in the identification of the large number of currently unidentified sources by its ability to better determine their location. Improved resolution will also help in the disentangling of multiple overlapping sources in the galactic plane. Significant challenges will still remain, in the identification of sources, however, with the increased sensitivity of the CTAO potentially detecting many more sources in the most complex regions.

4.4.3. *The CTAO extragalactic survey*

In addition to the survey of the galactic plane region, CTAO plans to conduct — for the first time — an unbiased search for VHE gamma-ray sources of one-forth of the extragalactic sky. This will be the most sensitive scan of a large extragalactic region at VHEs: the expected sensitivity is of about 5 mCrab. This survey will, for the first time, be able to fully assess the populations of extragalactic sources, without biases introduced from observing objects seen with bright non-thermal emission at other wavelengths.

https://doi.org/10.1142/9789813275720_0005

Chapter 5

Cosmic-Ray Physics

Sabrina Casanova*,§, Daniele Gaggero†,¶ and Fabio Zandanel‡,‖

*Instytut Fizyki Jadrowej
Krakow, Poland
†Universidad de València and CSIC
Paterna, Spain
‡Universiteit van Amsterdam
Amsterdam, Netherlands
§sabrina.casanova@ifj.edu.pl
¶dgaggero@pi.infn.it
‖f.zandanel@uva.nl

5.1. The Cosmic-Ray Spectra at Earth

The all-particle spectrum of cosmic rays (CRs), spanning over many orders of magnitude in energy and flux (see left panel in Fig. 5.1), is a power law in energy characterized by a spectral index of roughly -2.7 up to several PeV (PeV $= 10^{15}$ eV), where the slope steepens to -3.0 (Apel *et al.*, 2009). This prominent feature in the all-particle CR spectrum is called the "knee". Another significant change in the slope is present at even higher energies, around $10^{18.6}$ eV, and is usually called the "ankle".

CRs with energy below the knee are supposed to be accelerated at Galactic sources, whereas the decrease in CR flux beyond the knee is usually interpreted as the convolution of the High-Energy (HE) cutoff of different species, which should be located at the same rigidity, but shifted in kinetic energy due to the different charge (Hörandel, 2003). For instance, if the maximum energy of hydrogen nuclei is E, for helium it is $2E$, for carbon it is $6E$ and so on.

In order to confirm this scenario and other competing explanations, it is very important to pinpoint the knee in single-particle spectra as well — a very challenging task from the experimental point of view. The KASCADE

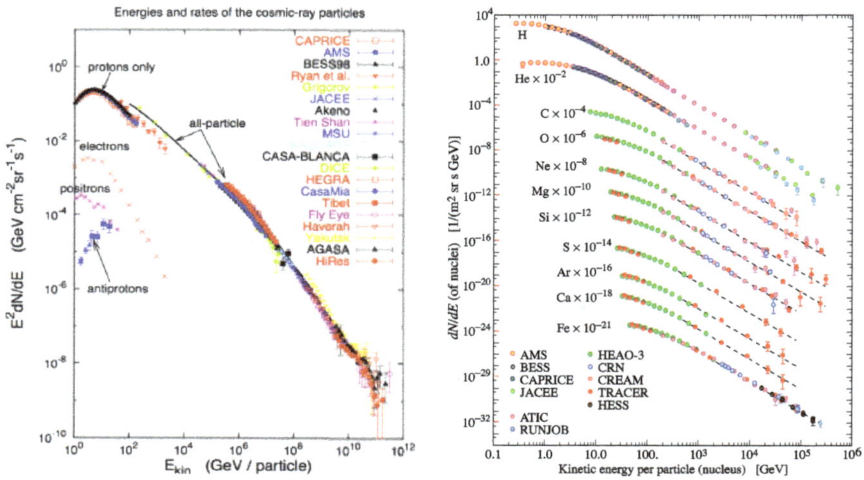

Fig. 5.1. (Left): All-particle spectra taken from Hillas (2006). The local energy density in CRs amounts to $w_{CR} = 1\,\mathrm{eV\,cm^{-3}}$. (Right): Zoom of the spectra of different nuclei up to PeV energies. The fluxes of nuclei of the primary cosmic radiation in particles per energy-per-nucleus are plotted vs. energy-per-nucleus (Tanabashi *et al.*, 2018).

collaboration reported evidence for a light element knee at several PeV (Antoni *et al.*, 2005), and KASCADE-Grande reported a further steepening at $\simeq$ 80 PeV, which corresponds to a knee-like structure in the heavy component of CRs (Apel *et al.*, 2013); on the other hand, the ARGO-YBJ experiment detected the light element (hydrogen + helium) knee at lower energy $\simeq$ 400–500 TeV (Bartoli *et al.*, 2015).

Thanks to the increasing precision of CR measurements, additional minor features are emerging in the CR spectra of the different nuclei shown in the right panel of Fig. 5.1. For instance, at about 200 GeV/n in rigidity, the spectra of protons and heavier nuclei get slightly harder and the different CR species show discrepant spectral hardening indices, with this trend confirmed up to 100 TeV (Adriani *et al.*, 2011; Aguilar *et al.*, 2015a, 2015b). These features might be the result of acceleration processes, such as the contribution from discrete sources close to the Solar System or re-acceleration in weak SN shocks, or caused by transport effects, which will be discussed in the detail in Section 5.2.

The region between the knee and the ankle in the CR spectrum is believed to mark the onset of CRs of extragalactic origin. This assumption implies that the Galaxy must contain CR factories accelerating particles up to the knee at energies close to PeV (Berezinskii *et al.*, 1990; Gaisser, 1990). These highly efficient particle factories are named "PeVatrons", as they are supposed to accelerate particles up to several PeVs. Since it is impossible to isolate all Galactic CR sources acting now or in the past in the Galaxy, a common approach to the question of the origin of CRs has been to focus on a specific population of candidate cosmic-ray sources, such as SuperNova Remnants (SNRs), and to understand whether this population is able to provide CRs at the measured injection rate and can explain the features of the CR spectrum upto the knee. Section 5.4 and beyond will describe the state of the art with regards to the origin of Galactic CRs.

The CR leptonic component (mainly electrons and positrons), although contributing to only 1% of the total energy in CRs at GeVs, is of particular importance in the field of CR physics. Here we summarize the main aspects; we provide a more detailed discussion of this in Section 5.3.

According to the conventional picture, electrons are expected to be primary CRs while positrons are expected to be produced as a secondary product of CR proton spallation onto interstellar gas. The spectra of these particles is shaped by both diffusion and energy losses, thus revealing critical information about CR sources and transport.

As a matter of fact, electrons and positrons cool very efficiently in the Galactic magnetic and radiation fields, with their cooling time being $t_{\rm cool} \approx 3 \times 10^8 (E/1\,{\rm GeV})^{-1}$ yr in the $3\,\mu$-Gauss Galactic magnetic field. Thus, the highest energy electrons and positrons detected at Earth can be, at most, hundreds of thousands of years old and accelerated within one kilo parsec from Earth. The question of the origin of the most energetic electrons detected at Earth has been recently revived thanks to the detection of electrons above TeV by the main IACT facilities: H.E.S.S. (Aharonian *et al.*, 2008c), MAGIC (Borla Tridon *et al.*, 2011) and VERITAS (Staszak, 2015; Kerszberg *et al.*, 2017), the *Fermi*-LAT experiment (Abdollahi *et al.*, 2017) and dedicated experiments in space, namely, CALorimetric Electron Telescope (CALET) (Adriani *et al.*, 2018) and DArk Matter Particle Explorer (DAMPE) (Ambrosi *et al.*, 2017). In fact, few very close astrophysical sources, such as local pulsars, can accelerate electrons up to these energies (Aharonian *et al.*, 1995; Profumo, 2012). Alternatively, the transport of particles in the nearby Galaxy has to proceed differently from the average diffusion in the Galaxy. In particular, the detection of 20 TeV electrons by H.E.S.S. suggests the existence of an electron TeVatron (a factory accelerating electrons up to TeV energies) located extremely close to Earth, such as Geminga (Aharonian *et al.*, 1995;

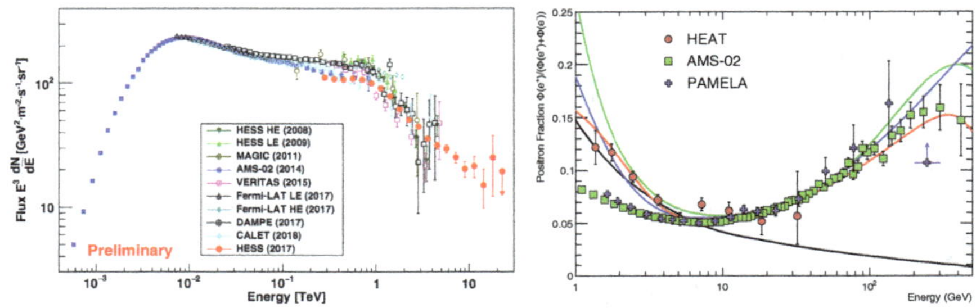

Fig. 5.2. *Left*: Spectrum of cosmic-ray electrons measured up to 20 TeV by different experiments, both space-based detectors and IACTs (Kerszberg *et al.*, 2017). The measurements reported are from H.E.S.S. (Aharonian *et al.*, 2008c; Kerszberg *et al.*, 2017), MAGIC (Borla Tridon *et al.*, 2011) and VERITAS (Staszak, 2015; Kerszberg *et al.*, 2017), the *Fermi*-LAT experiment (Abdollahi *et al.*, 2017) and dedicated experiments in space, namely, *CALET* (Adriani *et al.*, 2018), *DAMPE* (Ambrosi *et al.*, 2017) and AMS-02 (Aguilar *et al.*, 2014). *Right*: Positron fraction from Tanabashi *et al.* (2018).

Atoyan *et al.*, 1995). It is, however, challenging to extract the properties of this electron factory from the featureless HE electron spectrum. A break in the spectrum would help constrain the distance and age of the source and the nature of particle transport in the nearby Galactic environment. We will discuss in more detail the implications of the current data in Section 5.3.

In parallel, the unexpected discovery of a significant excess above $\simeq$10 GeV in the positron spectrum with respect to the expected secondary flux (Adriani *et al.*, 2011; Aguilar *et al.*, 2014; Abdollahi *et al.*, 2017) added an extra piece to the puzzle of the origin of the HE leptons. This feature triggered an intense debate in the astrophysical community; despite the interpretations in terms of dark matter annihilation or decay,[a] the excess seems to point toward a population of astrophysical objects that accelerate and release primary electron+positron pairs (Aharonian *et al.*, 1995; Profumo, 2012; Di Mauro *et al.*, 2014; Cholis *et al.*, 2018).

5.2. Distribution and Propagation of CRs

Let us turn our attention to the long journey that CRs undertake in the InterStellar Medium (ISM) of our Galaxy before reaching our detectors at

[a]The DM interpretation is quite challenging on the model-building side and in tension with several constraints coming from other channels (see, e.g. Gaggero and Valli, 2018 and references therein for a recent review)

Earth and in space. In order to characterize the properties of CR transport in the Galaxy, a useful starting point is the following set of observations regarding the CR flux measured at Earth:

- The high degree of isotropy of the incoming flux ($\sim 10^{-3}$ from $\sim$TeV to $\sim$PeV), recently measured by several different experiments (Abdo *et al.*, 2009c; Antoni *et al.*, 2004; Amenomori *et al.*, 2017; Aglietta *et al.*, 2009; Aartsen *et al.*, 2013; Abbasi *et al.*, 2012);
- The higher abundances of several light elements — e.g. Lithium, Beryllium and Boron — with respect to the Solar System elements. This piece of information implies that the "primary" species (e.g. hydrogen, helium and heavier nuclei) have interacted with a non-negligible column density of interstellar material, which can be quantified as $\simeq 3$ g/cm^2. Given the average density of the interstellar environment and the typical spallation cross sections, such a "grammage" implies a residence time as large as $\mathcal{O}(10$ Myr) and eventually implies that the primary species have often crossed the Galactic disk; and
- The observation of a large-scale diffuse gamma-ray emission across the entire Galactic disk, accurately measured from $\sim$1 MeV to $\sim$100 GeV energies. The presence of this non-thermal radiation was predicted more than 50 years ago (Pollack and Fazio, 1963), first measured in the late 1960s by the OSO-3 satellite (Kraushaar *et al.*, 1972). Nowadays, the *Fermi*-LAT instrument has provided extremely accurate maps in the 30 MeV–300 GeV range. In the TeV domain, particle array detectors such as Milagro and HAWC as well as IACTs played a crucial role. This section will particularly emphasize the importance of the latter category in the determination of CR transport properties.

These pieces of information contribute to suggest a "conventional scenario" for the propagation of cosmic particles in the Galaxy. We refer to the renowned textbooks Ginzburg and Syrovatskii (1964); Berezinskii *et al.* (1990) for a comprehensive presentation of this picture and to the more recent review by Gabici *et al.* (2019) for a critical discussion in light of the most recent data. The main feature is the presence of a smooth "diffuse sea" of CRs in the Galaxy resulting from an efficient confinement mechanism, causing a complicated random walk of CRs within the Galaxy. A basic theoretical framework that describes this transport mechanism is the quasi-linear theory (QLT) of pitch-angle scattering on magnetic inhomogeneities, elaborated on in the 1960s (Jokipii, 1966; Jokipii and Parker, 1968).

In the following sections, we will present in more detail the current CR transport scenarios, with particular focus on the role of HE gamma-ray observations, mainly provided by IACTs, in the understanding of the underlying physical processes.

5.2.1. *The CR transport equation*

The phenomenon of Magneto Hydro Dynamic (MHD) turbulence is ubiquitous in astrophysics; it has a preeminent role, in particular, in the dynamics of the ISM and is considered the ultimate origin of the Galactic CR diffusive regime up to ~PeV energy scales.

The turbulent power is associated with magnetic inhomogeneities and is present over a remarkably wide range of length scales, spanning from the Astronomical Unit (AU) all the way up to kpc scales (see, e.g. Armstrong *et al.*, 1995), on top of a coherent large-scale magnetic field structure, $\vec{B}$, whose study is today a thriving field of research (Jansson and Farrar, 2012; Terral and Ferrière, 2017). The turbulent cascade couples different modes and ultimately transfers energy from large- to small-length scales. The QLT of pitch-angle scattering is based on a simplified description of the cascade, which is considered isotropic and made up of "Alfvén waves"; such perturbations, named after Hannes Alfvén, are dissipationless, incompressible magneto-hydrodynamic oscillations of the magnetized plasma, traveling in the direction of the magnetic field with speed

$$v_A = \frac{B}{\sqrt{4\pi\rho}}, \tag{5.1}$$

in Gauss units, where B is the magnetic field intensity and ρ, the plasma density.

Alfvén waves are characterized by a motion of the ions transverse to the direction of propagation; the restoring force is provided by "magnetic tension".

The turbulent cascade generates perturbations of smaller and smaller wavelengths. It is believed to be initiated at scales as large as $\sim 10^2$ pc, for instance by large supernova explosions or other mechanisms (Elmegreen and Scalo, 2004). Alternatively, and especially at smaller scales, the formation of Alfvénic perturbations can be triggered by CRs themselves via a non-linear process called "streaming instability", which has been outlined and studied in detail for the first time in Skilling (1975); Cesarsky (1980). The key idea behind this process is that streaming CRs generate forward Alfvén waves in the direction of the streaming with the same wavelength as the particle Larmor radius; the growth rate of this instability is proportional to the CR gradient.

Alfvén waves scatter CRs and are responsible for their random walk across the Galaxy. In the following paragraphs, we will briefly recap the main aspects of the QLT. For a full derivation of the theory, we refer to Berezinskii *et al.* (1990); for a more recent review on QLT and other non-trivial aspects of CR transport and acceleration, a useful and comprehensive reference is Blasi (2013).

The key process described by the theory is the resonant interaction between a charged CR (carrying momentum $\vec{p} = m\vec{v}$) and an isotropic cascade of MHDwaves $\delta\vec{B}$ featuring small amplitude at the scale of interest (compared to the regular field $\vec{B}$). Under these hypotheses, it is possible to write a stochastic equation in terms of a quantity usually called "pitch angle",[b] which is characterized by a "resonance condition": CRs only interact with the perturbations with a wavelength $\sim$s$2\pi/k$ that matches the gyroradius r_L. This is the most relevant feature of QLT; as an example, let us mention that particles carrying $\mathcal{O}(GeV)$ momenta are scattered by waves on the astronomical unit scale, while on the very-high-energy side, particles with $\mathcal{O}$(PeV) momentum can interact with waves on the parsec scale.

The interaction can be written in the spatial domain in the Galaxy rest frame (see, e.g. Blasi, 2013; Gaggero and Valli, 2018 for a simplified derivation): In this frame, the interaction can be described by a "spatial diffusion" term describing a random walk along the direction of the regular field, characterized by a coefficient D, and a "diffusion term in momentum space", characterized by a coefficient usually called D_{pp}.

These operators describe the most important processes characterizing the phenomenon of Galactic CR transport, and the two associated coefficients obey the following relation (Skilling, 1975):

$$DD_{pp} \propto v_A^2\, p^2, \tag{5 2}$$

for $\vec{B} = B\,\hat{z}$. In general, when doing accurate phenomenology, a general transport equation that builds on some key results of QLT is considered. Such an equation features, as a key process, diffusion in position and

[b]Defined by $\mu = \cos(\vec{B} \times \hat{p})$.

momentum space, as shown above, and contains a wider set of operators, as well as describing other physical processes.

The general equation is usually written as follows:

$$\begin{aligned}\frac{\partial N}{\partial t} &+ \vec{\nabla}\left(\vec{u}N\right) - \frac{1}{3}\frac{\partial}{\partial p}\left[p\left(\vec{\nabla}\cdot\vec{u}\right)N\right] - \vec{\nabla}\left(D_{\vec{x}\vec{x}}\vec{\nabla}N\right) \\ &- \frac{\partial}{\partial p}\left[p^2 D_{pp}\frac{\partial}{\partial p}\left(\frac{N}{p^2}\right)\right] + \frac{\partial}{\partial p}\left(\frac{dp}{dt}N\right) \\ &= Q_0 + Q_{\text{sec}} - \frac{N}{\tau_N},\end{aligned} \tag{5.3}$$

where $N(\vec{x}, p, t)$ and $Q(\vec{x}, p, t)$ are the CR density and CR source term, respectively.

It is important to point out, when looking at the left-hand side, that the diffusion operator is typically modeled — in the phenomenological context — as "homogeneous and isotropic", i.e. fully described by a scalar, space-independent coefficient (see, for instance, the recent paper Cerri *et al.*, 2017, for a discussion of the more complex phenomenology associated with anisotropic transport). In the simplified context of isotropic diffusion, the spatial diffusion coefficient is written as

$$D = \frac{c r_L}{\mathcal{F}(k)}, \tag{5.4}$$

where $\mathcal{F}(k)$ is defined as the (normalized) power associated with the turbulent modes with wave number $k \propto 1/p$ resonating with the particles carrying momentum p. Given the power associated with the turbulent cascade scales as a power law in k, the rigidity dependence of the coefficient is usually parametrized as

$$D = D_0\left(\frac{p}{p_0}\right)^{\delta}, \tag{5.5}$$

where D_0 and δ become free parameters and correspond to large-scale averaged transport properties, fixed by comparison against data (especially of secondary-to-primary ratios).

The average diffusion coefficient in the Galaxy in the GeV–TeV range, as estimated by statistical analyses based on secondary CR data (Trotta *et al.*, 2011; Evoli *et al.*, 2015; Yuan *et al.*, 2017), roughly comprises between 10^{28} and 10^{29} cm^2/s (for convenience, we recall that $3 \cdot 10^{29}$ cm^2/s $\simeq$ 1 kpc^2/Myr).

The spatial dependence of these parameters, which is expected to originate from the spatial variations of the properties of turbulence, is neglected in almost all analyses, with some relevant exceptions (Evoli *et al.*, 2012; Gaggero *et al.*, 2015a, 2017; Cerri *et al.*, 2017; Feng *et al.*, 2016; Tomassetti, 2015; Johannesson *et al.*, 2016).

Other relevant terms in the transport equation are as follows:

- An advective operator, characterized by the bulk velocity of the plasma in the lab frame (Galactic winds affecting CR motion may be described by such a term, together with adiabatic energy losses involving velocity gradients even of $\mathcal{O}(10^2)$ km s^{-1} kpc^{-1} perpendicularly to the Galactic disk);
- The primary source term Q_0, representing the primary accelerators of CRs at work in a class of Galactic sources (or in several different classes — see above);
- The secondary source term $Q_{\rm sec}$, describing the production of a given CR nucleus by spallation of a heavier one on the interstellar gas;
- An energy loss term, characterized by the continuous rate dp/dt, particularly relevant in some cases, e.g. at large energies for light CR leptons (Strong *et al.*, 2007; Evoli *et al.*, 2017);
- A catastrophic loss term, capturing inelastic collisions characterized by an interaction time τ_N.

In order to solve the full CR transport equation for all species, several numerical codes are available. These packages are equipped with different numerical techniques and contain different astrophysical ingredients. We mention here (in chronological order): `GALPROP` (Strong and Moskalenko, 1998; Moskalenko and Strong, 1998), `DRAGON` (Evoli *et al.*, 2008, 2017, 2018b; Gaggero *et al.*, 2013; and `PICARD` (Kissmann, 2014; Werner *et al.*, 2015)). A semi-analytical approach instead is followed by `USINE` (Maurin *et al.*, 2001).

Our comprehension of MHD turbulence has greatly improved during the latest decades. In the currently accepted scenarios (Sridhar and Goldreich, 1994; Goldreich and Sridhar, 1995), MHD turbulence is characterized by an "anisotropic" cascade of Alfvén waves and an isotropic cascade of fast magnetosonic modes; this picture is theoretically demonstrated and numerically confirmed by simulations. Due to the anisotropy of the Alfvénic cascade, the scattering efficiency on this kind of wave becomes very low, especially at small scales (Chandran, 2000) far from the injection point. This fact poses a challenge to the framework presented above. Among others, a possible way out of the problem (Yan and Lazarian, 2002, 2008; Fornieri *et al.*, 2021) is that magnetosonic waves actually dominate the gyro-resonance interaction. In summary, the actual microphysics underlying CR transport is still far from being completely understood and the very nature of CR confinement in the Galaxy is still a matter of debate.

Another key process that is expected to play a major role at GeV scales is the nonlinear feedback between Alfvénic turbulence and the cosmic-ray flux, briefly mentioned above. If CRs stream faster than the Alfvén speed, they can amplify waves (naturally of the correct shape for scattering) through the "resonant streaming instability" (Wentzel, 1974; Skilling, 1975; Cesarsky, 1980; Farmer and Goldreich, 2004). As the waves amplify, the CR diffusion coefficient gets lower (more confinement) and the streaming velocity is reduced. For this process of self-confinement to operate, the excitation rate of the Alfvén waves by CRs themselves must exceed the sum of all wave-damping rates.

Nonlinear CR transport can have observable signatures in the local CR data: for instance, the spectral feature discovered by *PAMELA* in the proton flux, and confirmed by *AMS* with even larger accuracy in several CR species at $\simeq$200 GV, can be interpreted as the transition between CR diffusive propagation in self-generated and pre-exisiting turbulence (Blasi, 2013). Recently, the physics of non-linear self-confinement has been widely debated in the literature (see Aloisio *et al.*, 2015 for a recent review) in the context of several other anomalies detected in CR data.

Despite all these complications and caveats, however, the QLT of pitch-angle scattering can still be considered a quite useful guideline. In particular, Eq. (5.3) is usually taken as a reference and provides an effective phenomenological tool to investigate the complexity of CR and gamma-ray data, keeping in mind our poor knowledge of the microphysics behind the phenomenon of CR confinement.

5.2.2. *What we have learned from gamma-ray data so far*

The analysis of the gamma-ray diffuse emission — mainly due to π^0 decay, inverse Compton scattering and Bremsstrahlung — is a crucial tool to infer extremely useful information about the distribution and spectral properties of the smooth "sea" of CRs across our Galaxy.

In the GeV–TeV domain, the *Fermi*-LAT collaboration has provided the most accurate maps so far. We point out that, in this energy range, and at least at first order, there is a remarkable agreement between data and model predictions (Ackermann *et al.*, 2012e).

Current models are mostly based on a scenario mainly featuring the following:

- A population of SNRs, distributed according to current catalogs of pulsars, OB stars and other tracers of star formation, as main CR accelerators in the GeV–PeV domain (although, as mentioned above, other classes of sources can contribute);
- An isotropic and homogeneous steady-state diffusive propagation of CRs in the Galaxy, whose average properties are tuned on local CR data (the long escape time due to this process originates a "smooth diffuse sea" of CRs in the GeV–TeV domain); and
- Gamma-ray emitting targets, namely the gas and interstellar radiation field, whose distribution is modeled by exploiting a wide set of data in different wavelengths.

Several anomalies have been recently outlined, however, and their characterization can help to shed light on the big questions discussed above regarding the ultimate origin of CRs and the physics of their transport. The most relevant are the following:

- A progressive hardening of the CR proton flux toward the inner Galaxy (Gaggero *et al.*, 2015a; Yang *et al.*, 2016; Acero *et al.*, 2016; Pothast *et al.*, 2018), which can have very different interpretations. For example, the scenario presented in Cerri *et al.* (2017) is based on the presence of a poloidal component of the regular Galactic magnetic field and on the harder scaling of the parallel (with respect to the regular field) diffusion coefficient with rigidity inferred from simulations (DeMarco *et al.*, 2007; Snodin *et al.*, 2016). The hardening would then originate from anisotropic CR transport — as predicted by QLT (see above) — that shows a progressive transition from perpendicular to parallel (vertical) escape from the Galactic plane. On the other hand, in Recchia *et al.* (2016), the key ingredient is CR self-confinement due to streaming instability. As mentioned in the previous section, GeV CRs can trigger the growth of Alfvén waves that are in turn responsible for their confinement. Since this effect is stronger in the presence of larger CR gradients, the diffusion coefficient in the inner regions of the Galaxy should be much lower, advection can take over all the way up to large energies and the propagated spectrum should be closer to the injected spectrum and therefore harder;
- A hard emission extending out to $\simeq 40^\circ$ in longitude and $\simeq 50^\circ$ above and below the Galactic center called Fermi bubbles (Su *et al.*, 2010).

This emission seems associated with the signal — featuring a similar morphology — detected in WMAP data (Finkbeiner, 2004);

- A discrepancy between the predicted CR gradient, computed by means of numerical codes based on simple transport scenarios, and the gradient inferred from gamma-ray emissivity along the Galactic plane (Strong *et al.*, 2004; Evoli *et al.*, 2012). This anomaly can point toward different transport properties in different regions of the Galaxy; and
- A roughly spherical excess in the inner few degrees from the Galactic Center and peaking at an energy of a few GeV (Vitale *et al.*, 2009; Goodenough and Hooper, 2009; Hooper and Goodenough, 2011; Calore *et al.*, 2015; Ackermann *et al.*, 2017). This excess, originally claimed to be a possible case of dark matter indirect annihilation, seems to have an astrophysical explanation, e.g. emission from a population of unresolved pulsars (Bartels *et al.*, 2016; Lee *et al.*, 2016; Manconi *et al.*, 2021); we remark that its characterization has triggered a debate on the current modeling of the inverse Compton emission from the Galaxy on the distribution of CR sources in the inner Galaxy (Gaggero *et al.*, 2015b; Carlson *et al.*, 2016) and on CR transport in molecular clouds (de Boer *et al.*, 2017).

The multi-TeV domain plays a crucial role in this context and in most of these issues. Given the lower fluxes, we do not have in this case a detailed mapping of the diffuse emission along the whole Galactic plane. A remarkable — and progressively more significant — diffuse emission from a massive and dynamic complex of giant molecular clouds called the Central Molecular Zone (CMZ), however, has been observed during the latest decade by the H.E.S.S. collaboration as well as the VERITAS experiment (Archer *et al.*, 2014, 2016). The features and implication of this emission will be discussed in more detail in the following subsection.

5.2.3. *A closer look at the Galactic center*

5.2.3.1. *The GC gamma-ray diffuse emission and the Pevatron claim*

The Galactic Center of the Milky Way is an extremely active region of our Galaxy. It contains about ten million stars within one parsec, in particular red giant stars. It also harbors a population of massive supergiants and Wolf-Rayet stars, which are relicts of a star-formation event that happened one million years ago. A supermassive black hole of about $4 \times 10^6 M_\odot$ at the

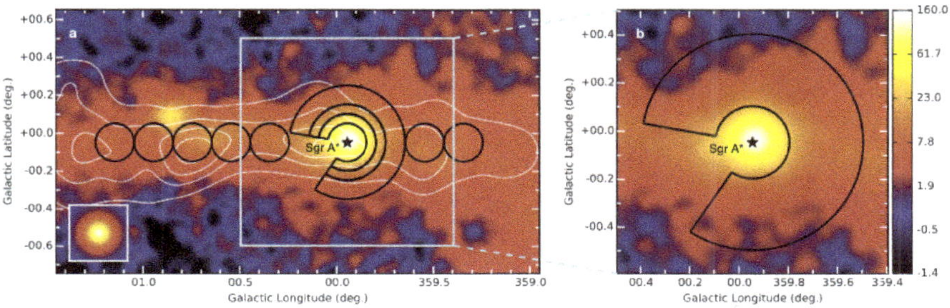

Fig. 5.3. *Left*: H.E.S.S. TeV map of the Galactic Center region. The regions where the cosmic-ray energy density was evaluated are shown in black. The contours of the density distribution of molecular gas traced by its CS line emission are shown in white. *Right*: Contour of the region used to extract the spectrum of the diffuse emission is shown in black.

Galactic Center powers the Sagittarius A* (Sgr A*) compact radio source. The Galactic Center is also the closest center of a spiral Galaxy to us and is a unique laboratory for testing processes that might happen in other galaxies containing active nuclei. Within a region of radius $\lesssim$400 pc from the Galactic Center, the CMZ contains around 5% of the total Galactic molecular gas (Terral and Ferrière, 2017; Tsuboi *et al.*, 1999). While dense gas and dust prevents observations of this region in the optical and ultra-violet wavelengths, the Galactic Center is extremely bright in radio, infrared, X-rays and gamma rays.

The Galactic Center at TeV energies includes a steady point-like source, HESS J1745-290, at the gravitational Center of the Galaxy, coincident with the positions of the supermassive black hole Sgr A* and the pulsar wind nebula G359.95-0.04 (at a projected distance of 0.4 pc to Sgr A*). The TeV sky map also includes an extended region of VHE gamma-ray emission, spatially correlated with the CMZ region presented in Fig. 5.3 (Tsuchiya *et al.*, 2004; Aharonian *et al.*, 2004c; Albert *et al.*, 2006; Aharonian, and others, 2009; Acero *et al.*, 2010b; Abramowski *et al.*, 2016; Archer *et al.*, 2016; Abdalla *et al.*, 2018a).

While the spectrum of HESS J1745-290 in Fig. 5.4 is a power law with a cutoff at several TeV, the diffuse emission has a power-law spectrum with a spectral index −2.3 and very remarkably does not show any evidence of cutoff; this extended emission with power-law energy distribution is likely to be emitted by PeV protons.

The distribution of the protons responsible for this radiation can be obtained by combining the gamma-ray and gas maps. From this correlation

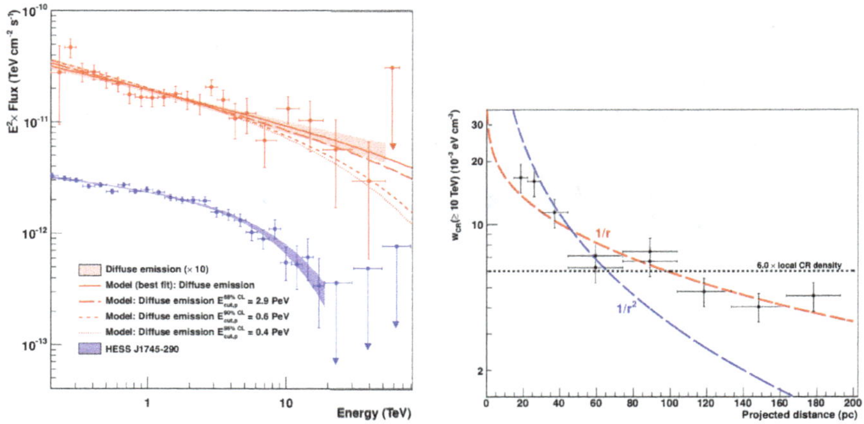

Fig. 5.4. *Left*: Energy spectra of HESS J1745-290 and of the diffuse gamma-ray emission, multiplied by a factor of 10, are shown. The red lines show the spectra of emission predicted from the decay of neutral pions produced by proton–proton interactions assuming different cutoffs in the parental energy distribution. *Right*: Combining the information from the morphology and luminosity of the TeV diffuse emission and from the gas distribution, it is possible to determine the energy density distribution of the emitting particles and determine the location and acceleration history of the particle factory. From the H.E.S.S. analysis we find that the energy density distribution of multi-TeV protons in the Galactic Center region is ten times higher than in the local Galaxy. Furthermore, at different projected distances from the Galactic Center follows a 1/r dependence. A consequence of this proton distribution is that the proton factory source (located somewhere within 10 pc of the Galactic Center) accelerates protons continuously and the accelerated particles propagate diffusely in the CMZ.

study we find that the distribution of protons is spherically symmetric and drops as $1/r$ with the distance r with respect to the Galactic Center (see the right panel of Fig. 5.4). This proton distribution implies that the proton factory — which should be located somewhere within 10 pc of the Galactic Center — accelerates protons continuously and the accelerated particles propagate diffusely in the CMZ. As a matter of fact, the distribution of particles undergoing isotropic diffusion out of a steady-state point-like accelerator is given by

$$w_{\rm CR}(E,r,t) = \frac{\dot{Q}_p(E)}{4\pi D(E) r}\,{\rm erfc}\left(\frac{r}{r_{\rm diff}}\right), \tag{5.6}$$

where D(E) is the particle diffusion coefficient (Aharonian and Atoyan, 1996) and $\mathrm{r_{diff}} \equiv \sqrt{4D(E)t}$ is the diffusion radius for timescales larger than $\mathrm{t_{diff}} = \frac{R^2}{6D(E)} = 2 \times 10^3 \frac{D(E)}{D_0}$ (where $R = 200$ parsec and $D_0 = 10^{30}$

$cm^2\ s^{-1}$ is the characteristic Galactic diffusion coefficient for multi-TeV protons). Since the Taylor expansion of the complementary error function is $\mathrm{erfc}(x) = 1 + \frac{2}{\sqrt{\pi}}x + \mathcal{O}(x^3)$, the radial distribution near a steady-state source simply scales as $1/r$. On the other hand, the distribution of particles diffusing out of a bursting source would be

$$w_{\rm CR}(E, r, t) = \frac{\dot{Q}_p(E)}{(4\pi D(E)t)^{3/2}} \exp\left(-\frac{r^2}{4D(E)t}\right), \tag{5.7}$$

which is approximately constant in the vicinity of the source given the Taylor expansion $\exp(x^2) = 1 + O(x^2)$. Therefore, due to the $1/r$ dependence, the H.E.S.S. measurements disfavor a burst-like acceleration event, which would imply a constant profile for the w_{CR}. An advective phenomenon, which would imply a $1/\mathrm{r}^2$ profile, is also not compatible with the data.

As far as energy budget is concerned, the level of the measured gamma-ray luminosity implies that the power in accelerated protons is $\dot{Q}_p(E)$ $(E > 10\,\mathrm{TeV}) = 4 \times 10^{37}$ $(\mathrm{D}/\mathrm{D}_0)$ erg/s. Assuming that $\mathrm{t_{diff}} > R/c$, $\dot{Q}_p(E)$ $(E > 10\,\mathrm{TeV}) < 1.2 \times 10^{38}$ erg/s.

Since the gamma-ray emission shows no evidence of a spectral cutoff, the spectrum of the parent protons — with a spectral index close to 2.4 — should extend to energies close to 1 PeV. Assuming a cutoff in the parent proton spectrum, the corresponding secondary gamma-ray spectrum deviates from the H.E.S.S. data at 68%, 90% and 95% confidence levels for cutoffs at 2.9 PeV, 0.6 PeV and 0.4 PeV, respectively, which makes this discovery the first robust claim of detection of a Galactic cosmic PeVatron.

Integrating the CR distribution over the whole diffuse emission region yields $\mathrm{W_{CR}} = 10^{49}$ erg. The astrophysical source associated with the proton PeVatron could be a supernova remnant, a pulsar wind nebula, a compact cluster of massive stars or the supermassive black hole, Sgr A*. The particles could be in this case accelerated either in the accretion flow close to the black hole or at the termination of an outflow. If the supermassive black hole, Sgr A*, is the PeV factory responsible for the H.E.S.S. diffuse emission from the Galactic center, then Sgr A* cannot be the only PeV factory in the Galaxy unless it was more powerful over the last 10^{6-7} years. Its current bolometric luminosity of 10^{39} erg is two or three orders of magnitude lower than the one needed to sustain the CR population, and would constitute a significant (as large as 1%) fraction of the current accretion power of the supermassive black hole. As testified by X-ray outbursts and giant outflows from the Galactic center, Sgr A* was likely more active over the last 10^{6-7} years and could have accelerated then a significant fraction of the CR population close to the knee. Alternatively, the extended TeV emission from the Galactic center could have been the combined result of acceleration within three powerful star clusters: the Arches, the Quintuplet and the Nuclear clusters. In this case, the continuous injection over millions of years, required by the $1/r$ distribution of the diffuse emission, could result from the characteristic ages of massive clusters, roughly 10^6 years (Aharonian *et al.*, 2019). The hypothesis that clusters of massive stars and star-forming regions accelerate particles up to PeV energies will be further discussed in Section 5.5.1 Finally, the extended emission proceeding up to tens of TeV without a cutoff could be produced by a population of relativistic electrons. The expected high radiative losses in the intense magnetic and radiation fields in the Galactic center region, however, cool down electrons very quickly and would prevent them from diffusing over the entire CMZ region.

5.2.3.2. *The role of the diffuse CR sea*

Let us now focus on the possible implications related to CR propagation, since this bright diffuse emission poses a challenge to the long-standing quest for CR accelerators and may allow us to shed more light on CR transport in the inner Galaxy.

In particular, we point out an alternative scenario (with respect to the interpretations that invoke the presence of a HE accelerator in the center) proposed in Gaggero *et al.* (2017), which connects the H.E.S.S. detection to all the issues related to CR transport discussed in the previous section and to the anomalies in the GeV gamma-ray data mentioned above.

The idea stems from three main hypotheses: (1) A smooth CR sea is present across the Galaxy all the way up to PeV energies as the result of the history of CR acceleration in different regions and diffusive-advective escape from the Galaxy. (2) The CR sea is harder in the inner Galaxy as the *Fermi*-LAT data seems to suggest. (3) The spectral feature detected by *PAMELA* and *AMS* in the primary (and, most likely, secondary) local CR spectra is present in the Galactic center region as well.

Under these hypotheses, the diffuse emission identified by H.E.S.S. at the Galactic ridge can be interpreted as π^0 emission from the smooth and hard CR sea that fills up the inner Galaxy interacting with the CMZ massive cloud complex.

This scenario can be corroborated by the fact that the gamma-ray spectrum in the Galactic ridge, as measured by both *Fermi*-LAT and H.E.S.S., appears quite featureless and smooth from GeV all the way up to multi-TeV energies, as well as that the local CR spectrum appears almost featureless up to the knee. We emphasize that this picture is agnostic on the very nature of the PeVatrons and essentially stems from the assumption that a previous history of CR acceleration up to PeV energies — taking place in some class of sources distributed across the Galaxy, chiefly in star-forming regions — has contributed to creating a steady-state CR sea with a harder spectrum in the inner Galaxy. As already mentioned, the scenario naturally connects to the hardening inferred by *Fermi*-LAT data at lower energies, and — if confirmed — could point toward an explanation of that feature in terms of harder scaling of the diffusion coefficient in the inner Galaxy due to parallel escape along the vertical magnetic field component; on the other hand, any explanation of that hardening that invokes a low diffusion coefficient due to streaming instability could not explain the hard feature all the way up to multi-TeV gamma-ray energies.

Many caveats exist; in particular, the role of unresolved sources needs to be quantified with sufficient accuracy. The future challenge is to disentangle the contribution of the diffuse sea from the sea emerging from local accelerators that are expected to be present in the CMZ: the central supermassive black hole, then the supernova events associated with the central star clusters and finally the massive star winds mentioned above. In all cases, a careful analysis of the energy budget of the required diffusion coefficients and the detailed analysis of the competition between the timescale associated with the acceleration and the (very short) escape time from the CMZ environment is in order.

Needless to say, more data are needed and the Cherenkov Telescope Array (CTA) has the opportunity to study in great detail the morphology of the signal with particular focus on the outer regions of the CMZ (e.g. the Bania clump), which may help to extricate different scenarios and confirm the validity of the inferred $1/r$ profile up to large Galactocentric radii.

5.3. The Spectrum of CR Electrons

The most relevant energy loss processes in the TeV domain are inverse Compton scattering off the diffuse Galactic radiation field and synchrotron emission due to the interaction with the Galactic magnetic field; both processes are characterized by a rate that follows this scaling with the particle energy

$$\frac{\mathrm{d}E}{\mathrm{d}t} \equiv b(E) = -\beta E^2. \tag{5.8}$$

The interplay between these energy loss mechanisms and the diffusive propagation has been studied analytically in Bulanov and Dogel (1974) in cylindrical symmetry. A key role is played by the "energy loss length" λ, i.e. the distance through which a CR electron can travel before losing a significant fraction ($\sim$1/2) of its energy. This quantity is given by

$$\lambda^2(E) = \int \frac{D(E^2)}{b(E)} \mathrm{d}E^2 = \frac{D(E)}{(1-\delta)\beta E}, \tag{5.9}$$

where δ is the scaling of the diffusion coefficient with respect to the particle energy. This length scale becomes as low as $\sim$1 kpc at $\simeq$100 GeV and $\sim$200 pc at 10 TeV. The impact of the energy loss term is therefore very relevant, especially at high energy.

If a continuous distribution of sources is considered, a CR electron spectrum $\propto E^{-\alpha}$ injected into the ISM is steepened by the combined effect of diffusion and energy losses in different ways depending on the value of λ.

- If the loss length is larger than the half-height of the diffusive halo $\lambda > H$ (this condition corresponds to energies below 0.1 GeV), losses are not relevant and the spectrum is shaped by diffusion only. The propagated spectrum becomes $\propto E^{-\alpha-\delta}$.
- If the loss length is smaller than the half-height diffusive halo, but larger than the half-height of the disk where CR sources are located, the spectrum is steepened by the combined effect of losses and diffusion and becomes $\propto E^{-\alpha-\frac{\delta}{2}-\frac{1}{2}}$.
- If the loss length is smaller than the half-height of the source disk (this condition corresponds to energies above $\sim$100 TeV), the spectrum is further steepened to $E^{-\alpha-1}$.

Above 1–10 TeV, however, given the small values of the loss length scale, the continuous approximation breaks down and the discrete nature of CR sources has to be taken into account; in that energy range, the CR electron spectrum can significantly depend on the properties of the individual nearby sources.

The electron spectrum measured by CR observatories shows some anomalies as well. The low-energy portion of the spectrum is often overlooked and seems to point toward a significant hardening below $\simeq$10 GeV of unknown origin confirmed by independent analyses of radio data (Di Bernardo *et al.*, 2013; Orlando and Strong, 2013). The HE portion, especially above the TeV, is also critical. As explained above, this energy range probes the "local" interstellar environment (within $\sim$1 kpc) and should be particularly sensitive to the presence of individual accelerators in the solar neighborhood. In the pulsar scenario the multi-TeV flux should be dominated by the emission from a few nearby young or middle-aged pulsars (Aharonian *et al.*, 1995; Profumo, 2012) such as Geminga or Vela. The softening detected by H.E.S.S. and *DAMPE* is the key feature in this domain and its origin is still unknown. The compatibility between this feature and the scenarios based on a significant contribution from pulsars to the HE part of both the electron and positron spectra is still under debate (Genolini, 2018; Yuan and Feng, 2018). On the other hand Lipari (2018) proposes an alternative picture that invokes a common, purely secondary origin for both the HE positrons and the HE antiprotons without any contribution from pulsars. This scenario requires a much shorter CR "residence time" in the Galaxy and would therefore significantly change the conventional paradigm of CR transport if corroborated by further evidence. Indeed, since the positron and antiproton spectra are surprisingly close to the spectra expected to be injected in the local interstellar medium by proton spallation on interstellar gas, they should not be affected significantly by either rigidity-dependent diffusion or energy losses, a fact that implies stringent constraint on their residence time in the Galactic halo. In this context, the spectral break detected by H.E.S.S. would be interpreted as a "cooling break" naturally originating at the energy corresponding to $t_{\text{cool}} \sim t_{\text{age}}$, where the age of the CR electrons reaching us would be equal to the residence time, hence $\sim$1 Myr in this scenario (compared to a conventional estimate larger than $\sim$10 Myr).

The pulsar interpretation of the local CR electron and positron spectra up to the highest energies still seems favored by the HAWC detection (Abeysekara *et al.*, 2017a) of the inverse Compton emission associated with a population of highly confined electrons in the vicinity of nearby pulsars (Profumo *et al.*, 2018). This measurement triggered a relevant debate about the nature of CR transport in the vicinity of pulsars (Hooper *et al.*, 2017; Di Mauro *et al.*, 2020; Recchia *et al.*, 2021; Profumo *et al.*, 2018; Evoli *et al.*, 2018a) and about a possible very relevant contribution of these halos to the overall multi-TeV emission (Hooper and Linden, 2018; Sudoh *et al.*, 2019).

In conclusion, we stress once again the importance of the spectral features in the context of the physics of CR transport and the necessity to test alternative scenarios of CR propagation against all possible data given the many unclear aspects of the microphysics behind CR propagation and the numerous puzzles provided by the most recent data.

5.4. The Supernova Paradigm for the Origin of CRs

A role of SNs in the acceleration of CRs in the Galaxy was first proposed in 1934 (Baade and Zwicky, 1934; Ter Haar, 1950; Morrison, 1957). The current hypothesis that CRs might originate from SNRs located in our Galaxy was formulated by Ginzburg and Syrovatskii (1964). The SNR paradigm is based on the energetic consideration that the supernova explosions in the Galaxy can provide the power needed to sustain the CR population. If the CR confinement region is roughly a disk of 20 kpc in radius and about 400 pc in height, then the confinement volume, V, is about 10^{67} cm^3. Assuming that the locally measured energy density is representative of the CR energy density everywhere in the Galactic disk, the CR production rate is $\dot{W} = V w_{\rm CR}/t_{\rm conf} \approx 10^{41}$ erg/s, where the CR confinement time, $t_{\rm conf}$, is $\approx 10^7$ yr. This is roughly the power that CR sources need to provide to sustain the Galactic CR population. Typically, in the whole Galaxy, an SN event happens every 30 years and the total energy released in the explosion is about 10^{51} erg. If SNRs are the Galactic CR factories, then about 10% of their typical explosion energy has to be released in non-thermal accelerated particles.

The other fact that supports SNRs as major CR contributors is that shell-type supernova remnants provide adequate conditions to have efficient CR acceleration. The current paradigm is that CRs up to PeV energies are accelerated in the shells of young SNRs through a mechanism known as diffuse shock acceleration (DSA) in the presence of strong magnetic fields. In the shells of SNRs, suprathermal particles would be injected into the acceleration process at shocks propagating into the ISM at speed >1000 km/s where collisionless processes transfer to them the kinetic energy of the magneto-hydrodynamic shocks. DSA is not only a viable theory of CR acceleration in SNR shocks — it also predicts that 10% of the kinetic energy available in supernova explosions is transferred to CRs in accordance with the energetic requirement. There is a vast literature on DSA in SNRs (see, for instance, Krymskii, 1977; Blandford and Ostriker, 1978; Axford, 1981; Bell 1978a, 1978b; Blandford and Eichler, 1987; Ellison *et al.*, 1997; Lucek and Bell, 2000; Malkov and Drury, 2001).

Within a DSA scenario, SNRs accelerate the highest-energy CRs (up to at least a few PeVs) at the transition between the free expansion and the Sedov phase, which typically happens a few hundred years after the supernova explosion. During the Sedov phase, the SN shock slows down and the magnetic field intensity decreases so that the shock can no longer confine the most energetic particles that escape the SNR. This means that an SNR can accelerate particles at the highest energies close to PeV for a relatively short time. Considering the rate of SN explosions in the entire Galaxy (about three per century), the chances to observe an SNR when it still accelerates up to PeV energies (when it is a PeVatron factory) are thus very low, and any proof of emission from PeV CRs even from young SNRs is challenging to find.

5.4.1. *Testing the supernova paradigm*

To recap, the SNR paradigm implies that SNRs accelerate CRs up to the knee and that, on average, each SNR provides roughly 10^{50} ergs in accelerated particles. The observational evidence for the SNR paradigm has been long thought to come from radiation signatures of proton TeVatrons and PeVatrons in the very-high-energy gamma-ray band, from GeV to TeV (Drury *et al.*, 1994; Drury, 2012). For CRs close to PeV energies, the dominant emission processes are the decay of neutral pions, π^0, produced when CR hadrons collide with ambient gas in the ISM (commonly called the "hadronic production mechanism"), and inverse Compton scattering of CR electrons off radiation fields (called the "leptonic production mechanism"). As a rule of thumb, either process will produce gamma rays of around 10% of the energy of the parent CR, i.e. several tens to hundreds of TeV for CRs close to PeVs. Typically, the spectrum of the hadronic radiation at TeV mimics the parental CR spectrum where the proton energy is shifted to lower energy by a factor of 20–30.

Searching for the GeV-to-TeV radiation from proton TeVatrons and PeVatrons and testing the SNR paradigm for the origin of CRs have been among the top priorities for all IACT facilities over the last 20 years. Studies of the GeV-to-TeV spectra of the remnants, morphology and correlation of the GeV-to-TeV images of the SNRs with maps of the ambient gas and X-ray radiation are the three most powerful instruments to extract the population of accelerated particles in SNRs.

Several other shell-type SNRs have been observed at TeV energies. Refer to Section 4.2.3.1 for a complete list, among which young SNRs such as Cas A (Aharonian *et al.*, 2001; Albert *et al.*, 2007b; Acero *et al.*, 2010a; Ahnen *et al.*, 2017a), Tycho (Acciari *et al.*, 2011a; Archambault *et al.*, 2017a), RX J1713.7-3946 (Aharonian *et al.*, 2004a; Abdalla *et al.*, 2018e) and middle-aged SNRs such as IC433 (Albert *et al.*, 2007a; Acciari *et al.*, 2009b), W28 (Aharonian *et al.*, 2008; Hinton and Hofmann, 2009) and W51C (Ferrand and Safi-Harb, 2012) are listed. In these latter cases, the

TeV radiation is thought to be produced by runaway CRs interacting with the dense gas of nearby molecular clouds.

In Section 5.4.1.1, the most extensively studied SNR at TeV energies, RX J1713.7-3946, will be used to illustrate how spectral, morphological and correlation studies at TeV help constrain the population of particles accelerated in the SNR.

5.4.1.1. *The young shell-type SNR, RX J1713.7-3946*

RX J1713.7-3946 is a young shell-type SNR (about 1000 years old) that is an extremely powerful non-thermal X-ray and TeV gamma-ray emitter (see Fig. 5.5). Contrary to most young shell-type SNRs, RX J1713.7-3946 shows weak radio emission and no thermal X-radiation. The first evidence of VHE emission from the shell of RX J1713.7-3946 has been reported by CANGAROO (Muraishi *et al.*, 2000).

X-ray-TeV morphological correlation: The shell morphology of RX J1713-3946 after 10 years of H.E.S.S. data, together with the *XMM*-Newton contours, are shown in the top left panel of Fig. 5.5 (Aharonian *et al.*, 2004a; Uchiyama *et al.*, 2007; Tanaka *et al.*, 2008; Abdalla *et al.*, 2018e). The spatially resolved remnant has a shell morphology at TeV energies similar to that seen in X-rays. Non-thermal hard X-rays are commonly interpreted as synchrotron radiation of hundreds of TeV electrons. These same electrons also produce TeV radiation through inverse Compton scattering very efficiently. This proves that VHE electrons up to hundreds of TeV are accelerated in the shells of RX J1713-3946. Gamma rays, however, extend radially beyond the X-ray emitting shell (see Fig. 5.5). The differences in the radial extent of the gamma-ray and *XMM*-Newton X-ray (Cassam-Chenaï *et al.*, 2004) shells do not depend on energy and the same behavior is evident for $E < 1\,\mathrm{TeV}$, $1 < E < 3\,\mathrm{TeV}$ and $E > 3\,\mathrm{TeV}$. The main shock position and extent are visible in the X-ray data, and the gamma-ray emission extending further is either due to accelerated particles escaping the acceleration shock region or particles accelerated in the shock precursor region.

GeV–TeV energy spectrum: The HE part of the spectrum of RX J1713.7-3946, from X-ray to TeV, can be used to infer the population of particles producing this emission. The gamma-ray spectra of this young SNR is hard, with spectral indices close to -2 (as predicted by DSA), and

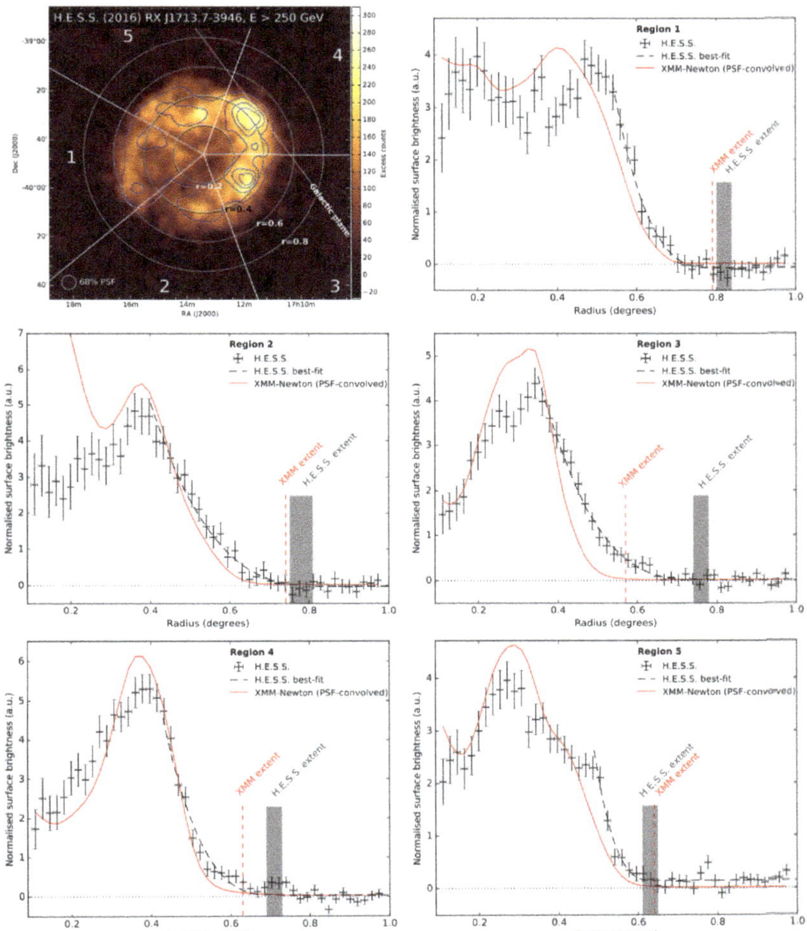

Fig. 5.5. The VHE gamma-ray excess map and radial profiles of RX J1713.7-3946 above 250 GeV obtained with 170 hr observations by the H.E.S.S. telescope (Abdalla *et al.*, 2018e). The H.E.S.S. gamma-ray count map ($E > 250$ GeV) with *XMM*-Newton X-ray contours (1–10 keV, smoothed with the H.E.S.S. PSF) overlaid, is shown on the top left. The five regions used to compare the gamma-ray and X-ray data are indicated along with concentric circles (dashed gray lines) with radii of 0.2–0.8 and centered at the SN location. The radial profiles of region 1 to region 5 are shown in the top right panel and in the middle and bottom panels. The profiles are extracted from the H.E.S.S. maps (black crosses) and from the *XMM*-Newton map convoluted with the H.E.S.S. PSF (red line). The relative normalization between the H.E.S.S. and *XMM*-Newton profiles is chosen such that for regions 1, 2, 4 the integral in [0.3,0.7] is the same, for regions 3, 5 in [0.2,0.7]. The gray shaded area shows the combined statistical and systematic uncertainty band of the radial gamma-ray extension. The vertical dashed red line is the radial X-ray extension.

shows a clear cutoff in the few TeV range. This tells us that RX J1713.7-3946 is an efficient particle accelerator only up to about 100 TeV ($=10^{14}$ eV). In other words, this young SNR is an efficient TeVatron (Abdalla *et al.*, 2018b, 2018e). Because of the cutoff at a few TeVs in the observed gamma-ray spectrum, however, RX J1713.7-3946 does not seem to be currently accelerating the bulk of CRs up to PeV energies. In other words, we have no observational evidence that RX J1713.7-3946 is currently a Galactic PeVatron.

Protons are likely accelerated in the supernova shell through the same mechanism accelerating electrons, but it is not clear what fraction of energy input of the supernova explosion goes into the acceleration of electrons vs. protons. Both protons or electrons can be, in fact, responsible for the TeV emission from RX J1713.7-3946 ("leptonic-hadronic degeneracy"). In Fig. 5.6, the measurements of the shell emission carried out by *Fermi*-LAT (Abdo *et al.*, 2011c), H.E.S.S. (Abdalla *et al.*, 2018e) and Susaku (Tanaka *et al.*, 2008) are compared to leptonic (left) and hadronic (right) model predictions by Zabalza (2015). The total energetics in accelerated electrons and protons in the relevant leptonic and hadronic models of gamma rays can be estimated by assuming a distance to the source of about 1 kpc. The required budget in electrons is determined only by the reported gamma-ray fluxes, $W_{\rm e} \simeq 1.2 \times 10^{47}$erg, while the total energy budget of protons in hadronic models depends on the uncertain ambient gas density, $W_{\rm p} \simeq 5 \times 10^{49}(n/1{\rm cm}^{-3})^{-1}$erg. The theoretical curves shown in Fig. 5.6 correspond to the model predictions calculated assuming that the GeV and TeV gamma-ray regions fully overlap and assuming a homogeneous emitter within the shell. The leptonic model in the left panel corresponds to the electron inverse Compton scattering off CMB photons and an infrared radiation field with energy density 0.415 eV cm^3 and temperature $T = 26.5$ K.

The leptonic model requires a break in the electron spectrum at 2.5 TeV, where the index of the electron energy distribution changes from $\Gamma = 1.7$ to $\Gamma = 3$ beyond the break. If is due to synchrotron cooling, this spectral break in the electron spectrum implies a magnetic field of about 140 μG. Such a high value of the B-field in the shell is in contradiction with the X-ray measurements that predict an average magnetic field of about 15 μG. An additional photon field of energy density of 140 eV cm^{-3}, about 100 times larger than in all previous estimates, would be alternatively needed to explain the spectral shape of the gamma-ray emission from RX J1713.7-3946 within a leptonic scenario and a single power-law electron spectrum.

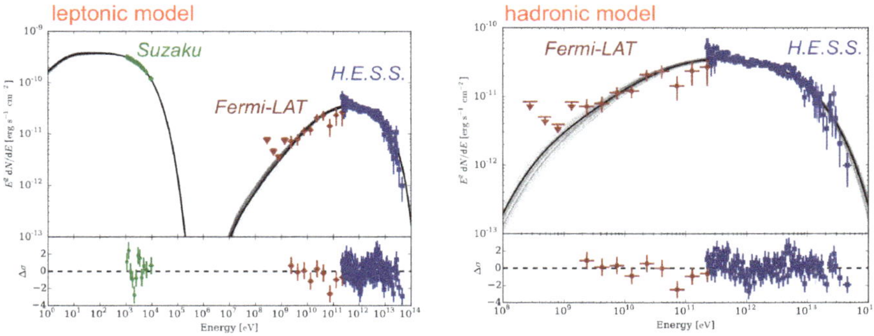

Fig. 5.6. Spectral energy distribution of RX J1713.7-3946 based on the Suzaku (Tanaka *et al.*, 2008), *Fermi*-LAT (Abdo *et al.*, 2011c) and H.E.S.S. data (Abdalla *et al.*, 2018e). The H.E.S.S. spectrum of the full SNR using an extraction radius of 0.6° around the central location of the SNR shell is shown. The line (red) shows the best-fit power law with index $\Gamma = 2$ and exponential cutoff at 12.9 TeV, which is preferred at a confidence level of 13.5 σ over the pure power-law model. RX J1713.7-3946 emits photons up to 50 TeV, the highest-energy H.E.S.S. data points being in the cutoff region. Although the TeV spectra of this SNRs is "hard", with $\Gamma = 2$, as predicted by DSA, it shows a clear cutoff in the few TeV range, which tells us that this SNR is currently an efficient accelerator only up to about 100 TeV (Abdalla *et al.*, 2018e). The theoretical hadronic and leptonic gamma-ray spectra are taken from Zabalza (2015).

The hadronic π^0-decay gamma-ray spectrum is produced by a population of protons with a spectrum index of $\Gamma = 1.5$ up to a spectral break at about 1 TeV, where the index becomes $\Gamma = 1.9$, and an exponential cutoff at 79 TeV. The spectral break in the proton population spectrum can be explained if the hadronic emission is produced mostly in dense clumpy regions where low-energy protons are efficiently excluded during the approximate 1000 years since the SNR explosion. Penetration in clumpy regions can also explain the hard gamma rays detected by *Fermi*-LAT (Fukui *et al.*, 2012; Gabici and Aharonian, 2014; Inoue *et al.*, 2012). Furthermore, it can explain the lack of thermal X-ray emission from the shell, which implies a gas density as low as 0.1 cm^{-3}, which could speak against a hadronic origin of the emission (Ellison *et al.*, 2010; Fukui *et al.*, 2012; Gabici and Aharonian, 2014). Contrary to the leptonic mechanism, gamma-ray production in gas condensations implies narrow angular distribution of the radiation, which is beyond the reach of the current generation of IACTs but could probably be tested with the upcoming CTAO with an angular resolution of about 1–2 arcmin (Zirakashvili and Aharonian, 2010; Ambrogi *et al.*, 2018).

Finally, X-ray synchrotron emission with a variability timescale of a few years has been observed in the shell of this SNR (Uchiyama *et al.*, 2007). The variability of X-ray images on year timescales shows that electron acceleration in RX J1713-3946 takes place in a strongly magnetized environment with amplification of the magnetic field by a factor of more than 100 (Uchiyama *et al.*, 2007). These amplified magnetic fields, which cool down extremely fast electrons accelerated up to hundreds of TeV, can be produced by CR-induced instabilities. The X-ray spectrometric measurements of RX J1713.7-3946 show also that electron acceleration proceeds in the most effective way, namely, in the Bohm-diffusion regime. Together with the amplification of the magnetic field, this result is additional evidence that the acceleration of protons and nuclei in the shell of this young SNR must proceed simultaneously with electron acceleration (Uchiyama *et al.*, 2007).

The bottom line of the investigation of the spectral features of RX J1713.7-3946 is that both hadronic and leptonic models do satisfactorily explain the spectral points in Fig. 5.6, but both models still present weak aspects. Furthermore, no observational evidence that this SNR accelerates particles beyond 100 TeV has yet been found.

5.4.1.2. *Very-young SNRs*

Very-young SNRs are primary candidates to be PeVatron sources because of high shock speed, not yet slowed down by the interaction with the interstellar medium. They are optimal candidates to study SNR development, eventually looking for the start of the PeVatron-emission phase.

The youngest known SNR in the Galaxy is G1.9+0.3, which is only 110 years old. While its shock speed of 10000 km/s, estimated through X-ray measurements, is ideal to produce efficient diffusive acceleration (Borkowski *et al.*, 2017), H.E.S.S. could not detect any evidence of TeV emission and thus of efficient particle acceleration in this very-young SNR. Although the efficiency of particle acceleration might be low in the very first phases of SNR evolution, this result confirms the difficulty in testing the SNR paradigm. The youngest SNR in the local group of galaxies occurred 30 years ago in 1987: the core-collapsed SN 1987A. Located in the LMC, it is visible from the H.E.S.S. site. The shock wave of SN 1987 has reached and is interacting with the equatorial ring. Since the gas density in the equatorial ring ranges from 10^3 cm^{-3} to 3×10^4 cm^{-3}, the H.E.S.S. upper limit of 5×10^{-14} ph cm^{-2} s^{-1} at a 99% C.L. translates into an energy input in CRs, $W_{pp} < 9 \times 10^{48}$ erg. This implies that only 1% of the explosion energy of SN 1987 is carried by accelerated CR nuclei (Abramowski *et al.*, 2015b).

5.4.2. *Testing the supernova paradigm: Final remarks*

To recap, the GeV-to-TeV radiation from young shell-type SNRs, which has been long thought to deliver a solution with regard to the origin of CRs, can be either of hadronic or leptonic origin (leptonic–hadronic degeneracy). Typically, a 1 TeV gamma-ray photon is emitted either by an electron or a proton of about 10 TeV. The cooling time for a 10 TeV electron is $\approx 5 \times 10^4$yr (for scattering off the CMB photons), while for a proton the cooling time is $5 \times 10^7 (n/1\,\text{cm}^{-3})^{-1}$yr (Aharonian, 2004). Thus, the ratio of the luminosity in inverse Compton gamma rays to π^0-decay gamma rays is of the order of $10^3 \times (\frac{W_e}{W_p}) \times (\frac{n}{1\text{cm}^{-3}})^{-1}$. The leptonic contribution to the emission is thus dominant over the hadronic contribution unless $\frac{W_e}{W_p} << 10^{-3}$, which could be true in SNRs because of the short synchrotron cooling time of electrons (Malkov and Drury, 2001). Alternatively, the gas density inside the shell has to be $n \gg 1\,\text{cm}^{-3}$, which would however have the drawback of slowing down the shock wave and thus would prevent efficient acceleration. Finally, in the presence of the magnetic field B exceeding $10\,\mu$G, the electrons accelerated within the SNR are cooled predominantly via synchrotron radiation and thus only a small fraction, $w_{MBR}/w_{\text{B}} \approx 0.1(\text{B}/10\mu\text{G})^{-2}$, is released in inverse Compton gamma rays. In fact, an amplified magnetic field exceeding $10\,\mu$G, which can be induced by cosmic-ray streaming instability in the region upstream the shock, has been observed in the shell of young SNRs detected at TeV energies.

The power in accelerated protons within SNRs derived from gamma-ray observations depends on the highly uncertain local density n as $1/n$ and on the magnetic and radiation fields within the remnant. Thus, it is not possible to prove that the dominant contribution to the observed TeV emission from these supernova remnants is produced by accelerated protons and that the population of SNRs in the Galaxy injects the necessary power (about 10^{50} erg per supernova event) to sustain the CR population.

5.4.3. *Molecular cloud–SNR interaction*

Molecular clouds are regions of the Galaxy, typically a few tens of parsecs in radius, where the density of cold molecular gas is often orders of magnitude higher than elsewhere in the diffuse ISM. Stars are believed to be born in these clouds. Radio observations of the rotational 1− > 0 line emission of carbon monoxide are mainly used to trace the distribution of molecular gas.

The gamma-ray emission from molecular clouds depends only upon the total mass of the cloud, M; its distance from the Earth, d; and the CR flux within the cloud. The CR flux Φ_{CR} in the cloud is thus determined as $\Phi_{\mathrm{CR}} \propto \frac{\Phi_\gamma d^2}{M_{cl}}$, where Φ_γ is the gamma-ray flux from the cloud. There is a caveat in this approach. The diffusion of CRs into molecular clouds depends upon the highly uncertain diffusion coefficient. CRs can penetrate clouds if the diffusion coefficient inside the molecular clouds is the same as the average diffusion coefficient in the Galaxy, as derived from spallation measurements. CRs are effectively excluded if the diffusion coefficient is suppressed (Gabici *et al.*, 2007). Generally, CRs at TeV energies are believed to diffuse into even the densest parts of molecular clouds.

Giant molecular clouds enhance the gamma-ray emission produced through the hadronic CR channel. As a result, the relative contribution of the IC leptonic component of the VHE emission compared to the π^0-decay gamma rays from specific dense regions is significantly reduced because of the enhanced gas density. Finally, the presence of a massive cloud along the line of sight distance from a given direction in the Galaxy, on the one hand, enhances the gamma-ray emission from that direction and, on the other hand, tells us where the emission is mostly produced along the line of sight distance, i.e. at the location of the massive clouds. In this way, molecular clouds can provide a sort of "tomography" — a three-dimensional view of the CR distribution (Issa and Wolfendale, 1981; Aharonian and Atoyan, 1996; Aharonian, 2001; Casanova *et al.*, 2010b; Neronov *et al.*, 2017; Peron *et al.*, 2021; Baghmanyan *et al.*, 2020; Aharonian *et al.*, 2020).

Although an SNR can act as a PeVatron for a relatively short time, the observation of HE gamma rays from dense regions in the ISM close to the candidate SNR, caused by the interaction of the escaped PeV CRs with the dense gas (Aharonian and Atoyan, 1996; Aharonian, 2001), can be interpreted as proof of the existence of PeVatrons in the past. This scenario was proposed and explored for several SNRs. If CRs escape their injection sources, they can collide with the ambient atomic and molecular gas. Usually, CR sources are associated with regions of high-gas-density molecular clouds (Montmerle and Cesarsky, 1979; Issa and Wolfendale, 1981). This is fortunate for us, as escaping CRs can then interact strongly with the dense gas, producing bright gamma-ray sources. Enhanced gamma-ray emission from molecular clouds located close to young SNRs can thus provide the first evidence of the parent population of runaway CRs. The emission produced by these runaway CRs could be crucial to trace back acceleration up to PeV energies. Gamma rays, in association with dense molecular clouds close to SNRs, have been detected both at GeV and TeV energies. Depending on the location of massive clouds, acceleration history and the timescales of the particles' escape into the interstellar medium related to the diffusion coefficient, a broad variety of energy distributions of gamma rays is produced from very hard spectra (much harder than the spectrum of the SNR itself) to very steep spectra (Aharonian and Atoyan, 1996; Gabici *et al.*, 2009; Casanova *et al.*, 2010a).

At TeV energies, MAGIC and VERITAS detected extended emission from the vicinity of the SNR IC 443, with its brightest regions coincident with the dense cloud material and maser emission. The H.E.S.S. observations have revealed VHE gamma-ray sources in the field of the SNR W28 that positionally coincide well with molecular clouds studied with the Nanten telescope (Aharonian *et al.*, 2008). Such emission correlates quite well with the position of dense and massive molecular clouds and thus it is often interpreted as the result of hadronic cosmic-ray interactions in the dense gas. If W28 is located at 2 kpc, a hadronic scenario for the TeV emission implies a CR density between 10 and 30 times the local CR density. Constraints on the CR diffusion coefficient in the region can be obtained under the assumption that the CRs responsible for the gamma-ray emission have been accelerated in the past at the SNR shock and subsequently escaped in the surrounding medium. In this scenario, gamma-ray observations can be explained only if the diffusion coefficient in the region surrounding the supernova remnant is significantly suppressed with respect to the average value in the Galactic region (Gabici *et al.*, 2010). A re-analysis of *Fermi*-LAT data outside the middle age SNR, W44, revealed clouds of runaway CRs escaping collectively and anisotropically from the shell along the magnetic field of the remnant (Peron *et al.*, 2020) as predicted by some theorists (Malkov *et al.*, 2013; Giacinti *et al.*, 2013) (See Fig. 5.7).

Finally, gamma-ray emission from Giant Molecular Clouds (GMCs) is proved to be a powerful tracer of the level of the background CRs at GeV energies (Ackermann *et al.* 2012a, 2012f; Aharonian, 2001; Aharonian *et al.*, 2020, Baghmanyan *et al.*, 2020; Casanova *et al.*, 2010b). Upper limits on the emission from local GMCs at TeV energies set with HAWC (Albert *et al.*, 2021) encourage similar measurements with both imaging (H.E.S.S., VERITAS, MAGIC and CTAO) and large field-of-view detectors at TeV energies (LHAASO and SWGO).

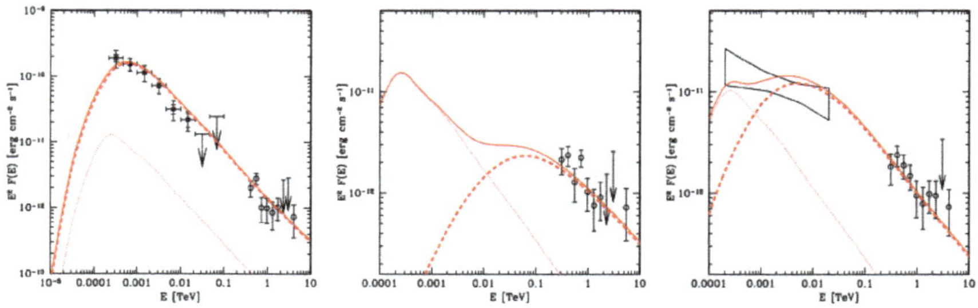

Fig. 5.7. Broadband fit *Left*: IC443. *Right*: Broadband fit to the gamma-ray emission detected by *Fermi*-LAT and H.E.S.S. from the sources HESS J1801-233, HESS J1800-240 A and HESS J1800-240 B (left to right). Dashed lines represent the contribution to the gamma-ray emission from CRs that escaped W28, dotted lines show the contribution from the CR Galactic background and solid lines show the total emission. Distances to the SNR center are 12, 65 and 32 pc (left to right). *Fermi*-LAT and H.E.S.S. data points are plotted in black. No GeV emission has been detected from HESS J1800-240 A (Gabici *et al.*, 2010).

5.5. Beyond the SNR Paradigm

The bottom line of all the studies of SNRs conducted at TeV energies is that SNRs have been proven to be efficient accelerators of CR electrons up to 100 TeV. That protons up to 100 TeV are accelerated in young SNRs seems a reasonable assumption, despite remaining unproved. The hypothesis that acceleration of CRs within the shells of SNRs proceeds up to PeV energies, though, has been rejected in all known young SNR-emitting TeV photons, as all SNRs detected by IACTs show clear cutoffs in the gamma-ray spectra at several TeV (see Section 5.4.1.2).

Since CRs with energies up to a few PeV (=10^{15} eV) — the knee in the CR spectrum — are believed to be of Galactic origin, one or more sources able to accelerate particles up to PeV energies, PeVatrons, must inhabit our Galaxy. Some other classes of astrophysical sources (different from SNRs), such as superbubbles, star-forming regions (see Section 5.5.1) or remnants of GRBs in our Galaxy and, in particular, the Center of our Galaxy have long been proposed as effective accelerators of particles and major contributors to the population of Galactic CRs up to PeV energies (Casse and Paul, 1980; Voelk and Forman, 1982; Cesarsky and Montmerle, 1983; Torres *et al.*, 2004; Reimer *et al.*, 2006; Bykov and Toptygin, 2001; Parizot *et al.*, 2004; Higdon and Lingenfelter, 2005; Atoyan *et al.*, 2006).

The first evidence of a breakthrough in understanding the origin of the highest-energy CRs in the Galaxy has been the controversial discovery of a powerful PeVatron in the Center of the Milky Way (see Section 5.2.3), which

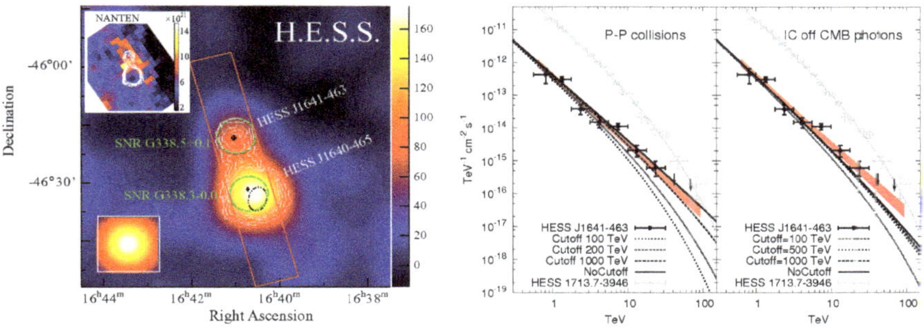

Fig. 5.8. *Left*: H.E.S.S. emission from the region of the hard spectrum source, HESS J1641-463, hosting the nearby bright SNR, SNR G338.3-0.6, coincident with HESS J1640-465. In the bottom insert, the H.E.S.S. PSF is presented, while the upper insert shows the gas distribution in the region according to the NANTEN CO survey (Onishi *et al.*, 2004). CRs leaking from G338.3-0.6 and interacting with dense molecular clouds is the most likely scenario to explain the TeV emission detected from HESS J1641-463, which has one of the hardest spectra ever detected in TeV astronomy, extending up to 30 TeV without evidence of a cutoff (see the right panel). *Right*: Differential VHE gamma-ray spectrum of HESS J1641-463 for energies between 0.64 TeV and 100 TeV, together with the expected emission from p-p collisions (left) and inverse Compton off CMB photons (right). The pink area represents the 1-σ confidence region for the fit to a power-law model, the red data points the H.E.S.S. measured photon flux (1-σ uncertainties), the arrows the 95% C.L. upper limits on the flux level, and the black curves the expected emission from the models, assuming different particle energy cutoff values. For comparison, the gray data points and curve represent the archival spectrum and the corresponding best-fit model, respectively, of the young SNR RX J1713.7-3946 (Abramowski *et al.*, 2014a).

has been associated with either the radio source at the Center of our Galaxy, Sgr A*, or three powerful star clusters in the Central Molecular Zone (CMZ): the Arches, the Quintuplet and the Nuclear clusters (Abramowski *et al.*, 2016; Abdalla *et al.*, 2018a; Aharonian *et al.*, 2019).

Unidentified gamma-ray sources, such as HESS J1641-463 and HESS J1702-420A, extending up to several tens of TeV without evidence of a cutoff in the gamma-ray spectrum, have been discovered by H.E.S.S. in different regions of the Galactic Plane (Abramowski *et al.*, 2014a; Abdalla *et al.*, 2018j, 2021a). The absence of any break in the spectrum up to several tens of TeV (see Fig. 5.8) and the frequent correlation with molecular clouds suggests that the dominant mechanism for the emission from these unidentified sources could be hadronic, and in this case, the parental proton spectrum should extend to energies close to the knee. In Fig. 5.8, we show how the parental particle population can be deduced from the TeV emission.

The search for PeVatron sources has been given a new impulse by the survey of the Galactic Plane at higher photon energies carried out with the HAWC Observatory. HAWC has observed several gamma-ray sources emitting above 56 TeV, with three of them showing emission continuing to 100 TeV and beyond (Abeysekara *et al.*, 2020; Albert *et al.*, 2020, 2021; Abeysekara *et al.*, 2021). While one of these sources is likely associated with the SNR G106.3+2.7, evidence of powerful particle acceleration close to PeV energies has emerged from the Cygnus star-forming region (Abeysekara *et al.*, 2021) (more about this discovery in Section 5.5.1.1) and in the vicinity of a 2MASS young star cluster (Albert *et al.*, 2021).

The LHAASO Collaboration reported recently the detection of more than 530 photons at energies above 100 TeV from 12 ultra-high-energy gamma-ray sources and up to 1.4 PeV from the Cygnus region of the Galaxy (Cao *et al.*, 2021). The photon energies reported by the LHAASO Collaboration are so high that the attenuation of these HE gamma rays, due to pair production of gamma rays interacting with background photons from both the cosmic microwave background (CMB) and interstellar radiation fields (ISRFs), has been taken into account for the first time for radiation sources located within the Galaxy.

The Galactic PeVatrons discovered by HAWC and LHAASO likely represent the prototype of some source populations contributing significantly to the Galactic cosmic-ray flux up to the knee. In addition to the Galactic Center PeVatron, many PeVatrons should, in fact, be currently active to explain the measured CR flux up to PeV energies.

Another crucial piece of information to clarify the origin of the highest CRs are the HAWC and Tibet Air Shower Array measurements of diffuse emission up to hundreds of TeV from the Galactic Plane. This radiation, compatible with expectations from a hadronic emission scenario in which gamma rays originate from the decay of pions produced through the interaction of protons with the Galactic ISM, is strong evidence that CRs are accelerated beyond PeV energies in our Galaxy and spread over the Galactic disk (Nayerhoda *et al.*, 2019; Amenomori *et al.*, 2021).

5.5.1. *Star-forming systems*

Collective stellar winds and SNR shocks in clusters and associations of massive stars have been long suggested as possible alternative or additional protons to TeVatrons and PeVatrons. As a matter of fact, CR acceleration could take place close to the stars (Casse and Paul, 1980; Voelk and Forman,

1982; Cesarsky and Montmerle, 1983). Core-collapse SN progenitor stars and colliding wind binaries evolve in giant molecular clouds and mostly remain close to their birthplaces in groups of loosely bound associations or dense stellar clusters. The winds of multiple massive stars in such systems can collide and form collective cluster winds that drive a giant bubble, also called a superbubble, filled with a hot ($T = 10^6$ k) and tenuous ($n < 0.01$ cm^{-3}) plasma. At the termination shock of the stellar cluster wind, turbulences in the form of MHD fluctuations and weakly reflected shocks can build up (Torres *et al.*, 2004; Reimer *et al.*, 2006; Bykov and Toptygin, 2001; Parizot *et al.*, 2004; Higdon and Lingenfelter, 2005). Turbulence in superbubble interiors can accelerate particles to very-high energies, not only through the first-order but also via the second-order Fermi mechanisms (Bykov and Toptygin, 2001). Supernova explosions of massive stars in thin and hot superbubbles can also produce efficient particle acceleration at the boundary of the superbubbles or at MHD turbulence and further amplify existing MHD turbulence (Ferrand and Marcowith, 2010). It has also been recognized that multiple shocks can result in efficient acceleration beyond PeV energies (Klepach *et al.*, 2000). The interaction of the accelerated particles with the ambient medium, including often dense molecular clouds or electromagnetic fields, leads to the efficient production of VHE gamma rays.

The TeV observations of massive star-forming regions and stellar clusters, such as 30 Doradus from the LMC (Abramowski *et al.*, 2015b), Westerlund 1 (Abramowski *et al.*, 2012b) and the Cygnus region (Aharonian *et al.*, 2002; Abdo *et al.*, 2007; Konopelko *et al.*, 2007; Albert *et al.*, 2008b; Abdo *et al.*, 2009d; Bartoli *et al.*, 2012; Aliu *et al.*, 2014b; Abeysekara *et al.*, 2018, 2021; Cao *et al.*, 2021), support the hypothesis that star-forming regions are sites of HE particle acceleration and give new impulse to the gamma-ray research in this field. These observations point to constrain the fraction of mechanical stellar-wind energy transferred into relativistic particles and, hence, gamma rays study particle acceleration and propagation in Galactic stellar clusters and superbubbles. Furthermore, HE phenomena are getting increasing attention in relation to the life cycle of interstellar matter and star-formation processes. The rate and efficiency of the star-formation process depends, in fact, on the balance between the self-gravity of dense molecular cores and countervailing forces that act to support the clouds. The most important of these are likely to be thermal pressure, turbulence and magnetic fields. In order for magnetic support to be effective, a population of ionized particles must be present in the core. Since molecular clouds are opaque to ultraviolet radiation from stars, the

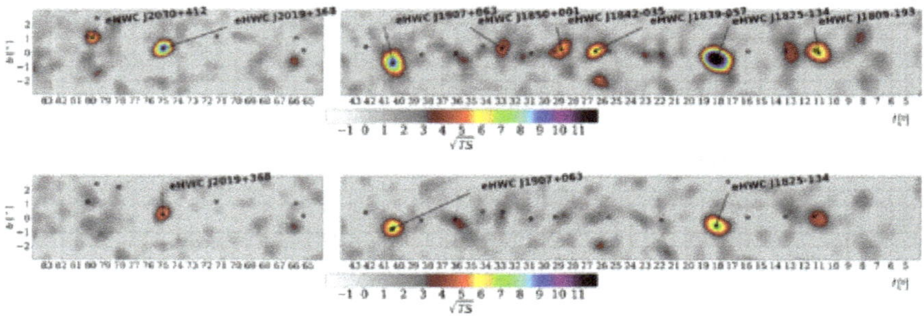

Fig. 5.9. TS map of the Galactic plane for photon energies above 56 TeV (top) and 100 TeV (bottom). A disk of radius 0.5° is assumed as the morphology. Black triangles denote the HE sources. For comparison, black open circles show sources from the second HAWC catalog (Abeysekara *et al.*, 2017c, 2020, 2021; Albert *et al.*, 2020, 2021).

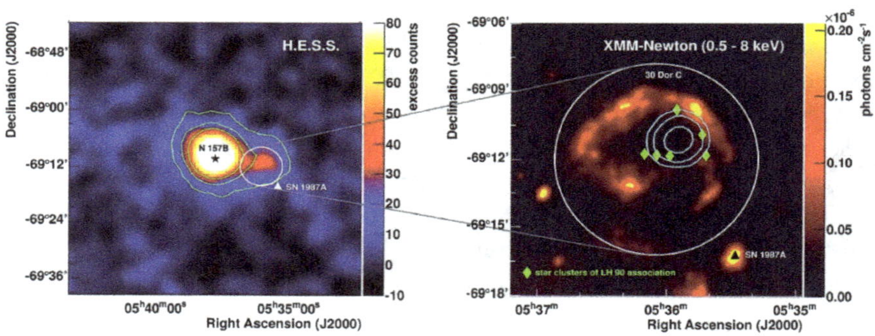

Fig. 5.10. *Left*: VHE gamma-ray emission in the region around N 157B within the LMC. The green lines represent contours of 5, 10 and 15 σ statistical significance of the gamma-ray signal. *Right*: *XMM*-Newton X-ray flux image of the region of 30 Dor C. The superimposed cyan lines represent contours of 68, 95 and 99% confidence levels of the position of the gamma-ray source. Diamonds denote the positions of the star clusters of the LH 90 association. Figures from Abramowski *et al.* (2015b).

main ionizing agent is thought to be low-energy CRs, and the magnetic support of the cloud is critically dependent on this factor.

The Large Magellanic Cloud (LMC), a satellite Galaxy of the Milky Way, hosts a factor of five higher star-formation rate-per-unit mass than the Milky Way. Within the LMC, H.E.S.S. discovered VHE emission from the largest star-forming region in the local group of galaxies, 30 Doradus (30 Dor), a superbubble in the starburst region of the Tarantula nebula (see Fig. 5.10). The TeV emission, which provides an unambiguous detection of

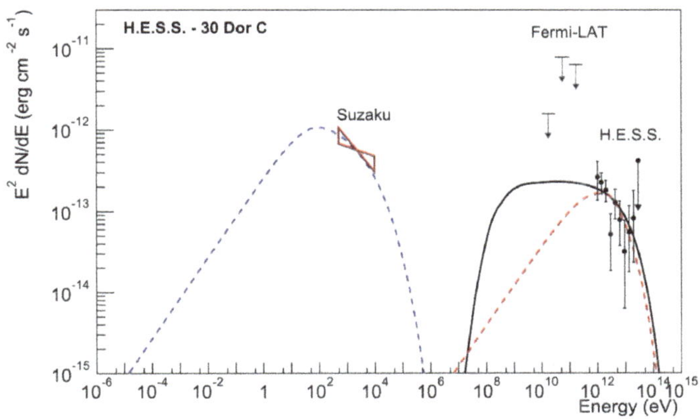

Fig. 5.11. Spectral modeling of MWL data from 30 Dor C. The 30 Dor C X-ray data are taken from Yamaguchi *et al.* (2010). Both leptonic (dashed lines) and hadronic (solid lines) models are shown. Figure from Abramowski *et al.* (2015b).

a superbubble in VHE gamma rays, can be explained by the production of neutral pions due to collisions of hadronic CRs with the background plasma. In Fig. 5.11, the spectral hadronic and leptonic modelings of multi-wavelength data from 30 Dor C are shown.

The hadronic scenario requires an energy in CRs of $W_{pp} = 0.7 \times 10^{52}(\frac{n_{H1}}{\text{cm}^{-3}})^{-1}$ erg, which can be explained considering that 30 Dor C probably experienced several supernova explosions, likely providing about 5×10^{50} erg in CR energy. In order to explain the energy accumulated in protons, W_{pp}, the average gas density, should be $n > 20\,\text{cm}^{-3}$, which is higher than the density estimate of $n = 0.1$ to $0.4\,\text{cm}^{-3}$ based on the X-ray thermal emission. Locations of high densities may be present, however, if the X-ray thermal emission comes from smaller radii than the dense outer shell or if cool, dense, clumped gas survived inside the otherwise rarefied interior of the bubble. Furthermore, the hadronic scenario puts constraints on the local CR diffusion coefficient because the diffusion length scale should be smaller than the radius of the shell, with D(10 TeV) $\leq 3.3 \times 10^{16}(\text{t}/10^6\text{year})^{-1}\text{cm}^2\text{s}^{-1}$, which, given an age for the superbubble of a few million years, gives a much smaller diffusion coefficient than the typical Galactic diffusion coefficient. This small diffusion coefficient requires magnetic-field amplification combined with turbulent magnetic fields, as hypothesized by Bykov (2001). Moreover, the gamma-ray and X-ray observations suggest active particle acceleration by a very large, fast-expanding shell. This may provide the right conditions for accelerating some protons to energies exceeding 3×10^{15} eV, namely, very turbulent magnetic fields and a sufficiently large extension to contain VHE particles for up to millions of years.

In the Milky Way, the most closely related object to 30 Dor C is the stellar cluster, Westerlund 1, the most massive stellar cluster in our Galaxy with a lower limit on its mass of 10^5 solar masses. The H.E.S.S. observations have revealed an emission region 20 times the size of the cluster. The detailed spectral and morphological studies demonstrated that a significant part of the emission may very well originate from Westerlund 1 and that protons are most likely responsible for the emission. The apparent size of the emission region also supports the idea that particles are accelerated in the cluster, escape, and later interact with the surrounding material and a nearby HII region. This scenario is further supported by the morphology of the HI gas in this region, which shows a bubble-like structure where dense gas partly overlaps with regions of strong TeV emission.

Finally, Aharonian *et al.* (2019) argued that three ultra-compact clusters located in the Galactic center power the H.E.S.S. diffuse emission from the CMZ.

The efficiency of conversion of kinetic energy of powerful stellar winds can be as high as 10%. This implies that the population of young massive stars can provide production of CRs at a rate of up to 10^{41} erg/s, which is sufficient to support the flux of Galactic CRs without invoking other source populations. The Galactic center PeVatrons, together with the other massive star-forming regions, such as Westerlund 1 and Cyg OB2, would represent the major factories of Galactic CRs (Aharonian *et al.*, 2019).

5.5.1.1. *Cosmic-ray acceleration in the Cygnus star-forming region*

The Cygnus region hosts some of the most remarkable star-forming regions in the Milky Way, including Cygnus X at only 1.5 kpc from the Sun with a total mass in molecular gas of a few million solar masses, which is at least ten times the total mass in all other close-by star-forming regions — such as Carina or Orion — and a total mechanical stellar wind energy input of 10^{39} erg s^{-1}. This corresponds to several percent of the kinetic energy input by SNe in the entire Galaxy. Cygnus X hosts many young star clusters and several groups of O- and B-type stars called OB associations. One of these associations, Cygnus OB2, contains 65 O stars and nearly 500 B stars. These super stars have likely created cavities filled with hot, thin

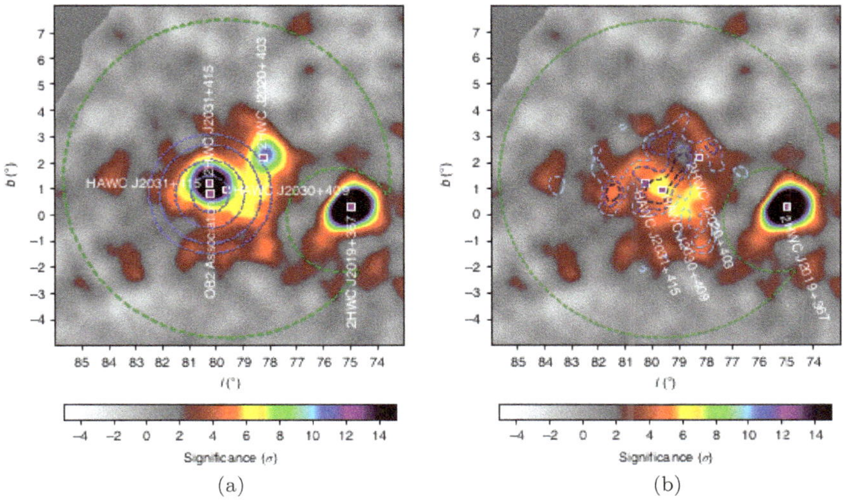

Fig. 5.12. *Left*: Significance map of the Cocoon region before subtraction of the known sources in Galactic coordinates, latitude, b, and longitude, l. The blue contours are four annuli centered at the OB2 association, where the CR density was estimated. The green contour is the ROI used for the study, which masks the bright source 2HWC J2019+367. *Right*: Significance map of the Cocoon region after subtracting HAWC J2031+415 (PWN) and 2HWC J2020+403 (γ-Cygni). The light blue, medium blue and dark blue dashed lines are contours for 0.16, 0.24 and 0.32 photons per 0.1 deg x 0.1 deg spatial bin, respectively, from *Fermi*-LAT Cocoon.

gas surrounded by ridges of cool, dense gas where stars are now forming, which strongly emit at GeV energies, called the Fermi Cocoon superbubble (Ackermann *et al.*, 2011).

The TeV counterpart of the Fermi Cocoon has been studied with the ARGO detector up to about 10 TeV, with the HAWC observatory up to at least 200 TeV and LHAASO up to 1.4 PeV (Bartoli *et al.*, 2014; Abeysekara *et al.*, 2021; Cao *et al.*, 2021). CRs up to PeV energies, accelerated in the enclosed star cluster Cyg OB2, likely produce the detected emission. The spectral energy distribution shows a significant softening at a few TeV; this is revealed by the comparison between the ARGO-YBJ, HAWC and *Fermi*-LAT data. This break in the gamma-ray spectrum hints at a cutoff in the injected CR spectrum or can possibly be explained as due to suppressed diffusion in the high-turbulence environment of the Cygnus superbubble, which would confine low-energy particles as higher-energy particles escape. An analysis of the CR density as a function of the distance from the star cluster has revealed at *Fermi*-LAT energies a distinct signature of

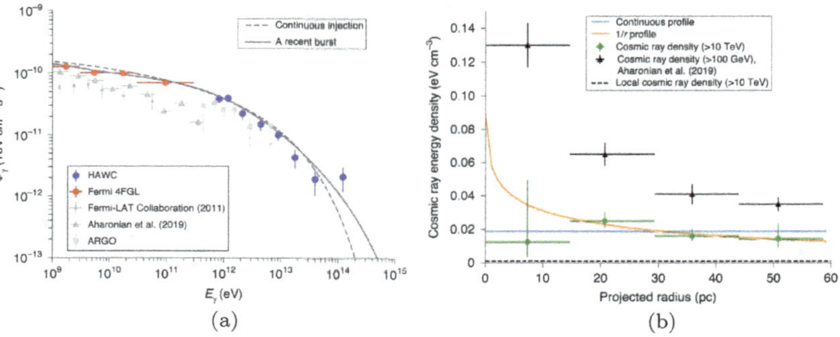

Fig. 5.13. *Left*: Spectral energy distribution of the gamma-ray emission at the Cocoon region. HAWC errors on the flux points are the 1 sigma statistical errors. At low TeV energy, HAWC data agree with the measurements by the ARGO observatory (Bartoli *et al.*, 2014). The red and gray circles are *Fermi*-LAT flux points published in Abdollahi *et al.* (2020) and Ackermann *et al.* (2011), respectively. The gray triangles are from the *Fermi*-LAT analysis in Aharonian *et al.* (2019). The gray solid and dashed lines are the spectra derived from the hadronic modeling of the region. *Right*: CR energy density profile calculated for four rings (0–15 pc, 15–29 pc, 29–44 pc and 44–55 pc) centered at the OB2 association. The green circles are the cosmic-ray densities derived above 10 TeV using HAWC gamma-ray data. The y errors are the statistical errors and the x error bars are the width of the x bins. The orange and blue lines are the $1/r$ profile for the case where the particles are continuously injected, and the constant profile (signature of the burst injection), respectively. The black dashed line is the local cosmic ray density above 10 TeV based on the Alpha Magnetic Spectrometer measurements. The black triangles are the cosmic-ray densities above 100 GeV from Aharonian *et al.* (2019).

.

continuous injection of CRs over the cluster lifetime and following diffusion through the ISM (Aharonian *et al.*, 2019). The $1/r$ decrement of the CR density with the distance from the star cluster is, in fact, a distinct signature of continuous injection of CRs over the cluster lifetime and their diffusion through the ISM. At TeV energies, the CR energy density as a function of the distance from the center of the emission, roughly coincident with the location of the Cyg OB2 cluster, is shown in Fig. 5.13. The CR density as resulting from TeV measurements is significantly higher than the CR energy density at Earth. It could, however, be produced by both a continuous injection or by a burst-like event happening in the cluster tens of millions years ago (Abeysekara *et al.*, 2021).

5.6. Future Perspectives

The crucial test of the SNR-as-PeVatron paradigm, or of any competing paradigm, has to come necessarily from high-significance spectra of multi-TeV photons (10–100 TeV photons) produced by CRs close to the knee energy colliding with the ambient gas. Multi-TeV photons tell us whether the parent CR spectrum extends all the way up to the PeV knee energy or if there is evidence of a cutoff in the parent proton spectrum, which produces a cutoff in the gamma-ray spectra. The spectrum of highest-energy gamma rays thus contains important information about the shape of the proton spectrum around the cutoff, crucial for the identification of acceleration mechanisms in SNRs and for understanding the role of different processes responsible for the formation of the knee in the CR spectrum.

Until now, the approach to the question of the CR origin has been to select a possible class of astrophysical sources and decide if this class can be a CR factory. Different astrophysical sources, however, might contribute at different energies and different levels to the CR population. Therefore, ideally, one would like to make a census of all possible CR contributors. An unbiased survey of the Galactic Plane in the crucial multi-TeV up to hundreds of TeV energy range could help in understanding the origin of CRs. Such surveys are currently being undertaken with particle array detectors such as HAWC (Malone *et al.*, 2017; Abeysekara *et al.*, 2021; Peron *et al.*, 2021; Albert *et al.*, 2021), LHAASO (Cao *et al.*, 2021) and in the future by the SWGO. A map of the Galactic plane above 56 TeV as seen by HAWC is shown in Fig. 5.9. Studies of PeVatrons conducted with detailed spectroscopy, higher sensitivity above 10 TeV, and, in particular, higher angular resolution, are furthermore key scientific goals of the upcoming Cherenkov Telescope Array. In the search for PeVatrons in the Galaxy with respect to current imaging Cherenkov telescopes, CTAO will benefit from the wider energy range extending up to 100 TeV in gamma rays, a realm where the hadronic mechanism is believed to be strongly dominant over the leptonic mechanism, and on the better angular and energy resolution, which is crucial in order to establish correlation with the gas maps.

Hadronic collision of protons off the ambient gas also produces X-rays, the result of synchrotron radiation of secondary electrons — the products of $\pi^{\pm}$-decays. The X- and gamma-ray fluxes depend on the present content of total energy of accelerated protons accumulated in the CR source and on the density of the ambient matter. Approximately the same fraction of energy of the parent protons is transferred to secondary electrons and gamma rays. Since the energy of sub-TeV electrons is not radiated away effectively, however, the direct (π^0-decay) gamma-ray luminosity exceeds the synchrotron luminosity. The L_X/L_γ ratio depends on the proton spectrum as well as on the particle injection history; typically it does not exceed 0.2–0.3. In the search for the proton PeVatrons, these X-rays from secondary electrons could be treated as a "prompt" radiation emitted simultaneously with gamma rays — valuable complementary information to constrain the acceleration efficiency of the PeVatron candidates.

5.7. The Extragalactic CRs

While the exact energy where the transition between Galactic and extragalactic CRs happens is highly debated, the Ultra-High-Energy Cosmic Ray (UHECR) above the ankle, i.e. with energies above 10^{18} eV, is certainly of extragalactic origin (e.g. Aloisio *et al.* 2012, 2018). The simplest explanation for this is that a proton with $E > 10^{18}$ eV under the typical magnetic field of our Galaxy, i.e. about 3 μG, has a Larmor radius $r_{\rm L} > 1$ kpc and therefore cannot be confined in the Galaxy and must be of extragalactic origin. In this context, two topics are important in relation to the contribution of IACTs to this field: the astrophysical sources generating extragalactic CRs and the transport of those CRs across the intergalactic space. We refer the reader to Aloisio *et al.* (2018) for a recent review.

5.7.1. *Propagation, interactions and secondary gamma rays*

The propagation of extragalactic CRs through intergalactic space is governed by their interaction with the astrophysical photon backgrounds and the possible presence of the poorly known Inter-Galactic Magnetic Fields (IGMFs). UHECRs interact with the photon backgrounds via pair production, photo-pion production and photo disintegration. The photon backgrounds are the well-known Cosmic Microwave Background (CMB) and the Extragalactic Background Light (EBL). The propagation of protons due to the energies involved in the problem is mainly governed by the CMB, while the EBL is important only for the photo-disintegration processes of nuclei and the production of secondary particles. IACTs study both the EBL and IGMFs through the observations of AGNs. These topics are covered in depth in Chapter 6 of this book and therefore will not be further discussed here. The impact of the IACT studies on the EBL and IGMFs in the context of CR physics is so far limited, but will be of growing importance with CTAO (Acharya *et al.*, 2019).

Before moving on to the astrophysical sources generating extragalactic CRs, we want to briefly mention the production of secondary gamma rays due to the interaction of UHECRs with the astrophysical photon backgrounds. These gamma rays are diffuse in nature and therefore contribute to the extragalactic gamma-ray background (see e.g. Fornasa and Sánchez-Conde 2015 for a review). This background is made by all of the extragalactic gamma-ray sources that are not resolved by current instruments — some of which will be by themselves sources of UHECRs (see the next subsection)–and by other diffuse components, such as the aforementioned secondary gamma rays from UHECR propagation. It follows, then, that observations of the extragalactic gamma-ray background can help in studying the origin and propagation of UHECRs. The current generation of IACTs have limited constraining power as observations of large portions of the extragalactic sky are needed for such studies. These studies have been so far done exclusively using observations by *Fermi*-LAT (e.g. Ackermann *et al.* 2015a, 2016a). CTAO, however, will survey a significant portion of the sky and will be able to complement the *Fermi*-LAT results with higher-energy information (Doro *et al.*, 2013; Acharya *et al.*, 2019).

5.7.2. *Astrophysical sources*

The study of gamma rays from extragalactic astrophysical sources is important for CRs for two main reasons: the origin of the UHECRs and the physics of CRs at the source. The separation can be easily understood by thinking that studying the CR acceleration processes in extragalactic astrophysical sources is complementary to what is done in our own Galaxy, but on different scales, and that CRs with enough energy will escape those extragalactic accelerators and become the UHECRs measured on Earth. These two topics are obviously strictly connected. and it is often very difficult to clearly separate them given that the CR population in extragalactic sources and their escape are not well known. An additional important part of the study of the sources of UHECRs is the cross-correlation of known extragalactic gamma-ray sources with the arrival directions of UHECRs, which is, however, very difficult given the poor angular resolution of UHECR experiments (e.g. Abraham *et al.* 2007; Moskalenko *et al.* 2009). Nevertheless, recently the Pierre Auger collaboration was able to detect a dipolar anisotropy in the arrival direction of UHECRs with energies above 8 EeV (Aab *et al.*, 2017) that sparked a number of works trying to relate this with local extragalactic gamma-ray sources (e.g. Joshi *et al.* 2018; Eichmann *et al.* 2018).

The basic requirement for an astrophysical source to be a source of UHECRs is obviously the ability to accelerate protons to energies above the ankle. Assuming a given accelerator has size R, this requirement translates to the fact that for particles to reach energies E their Larmor radius has to be less than the accelerator size $r_{\rm L} < R$. This concept is the basis of the Hillas plot already introduced in Chapter 3. Several sources can meet this requirement for protons (note that the acceleration of nuclei is easier than that of protons thanks to their larger electric charge). In particular, we are referring to clusters of galaxies, AGNs, GRBs and neutron stars. We want to add to the list of sources of interest here star-forming galaxies, already discussed in Chapter 3 and in Section 5.5.1 of this chapter. While these galaxies cannot be the sources of UHECRs, their study is complementary to that of our own Galaxy for the sake of understanding CR physics and acceleration. Additionally, it is worth mentioning the more exotic hypothesis for the origin of UHECRs, such as dark matter annihilation or decay, discussed in detail in Chapter 8. We will discuss thoroughly the clusters of galaxies below in the separate Section 5.2.1, as they are not discussed elsewhere in this book. Again, here, we will briefly discuss the other mentioned sources — AGNs, GRBs and neutron stars. Both AGNs and GRBs are introduced and discussed in detail in other chapters of this book.

For clusters of galaxies, we will talk about non-relativistic shocks, for AGNs and GRBs, we are instead referring to relativistic shocks. For particles to gain significant energy, they have to undergo multiple shock crossings; in other words, the probability that particles return to the shock from the downstream region has to be substantial. This is a non-trivial condition for relativistic shocks because both particles and shocks are moving with relativistic speeds, implying a large anisotropy in the distribution of particles at the shock. In this case, the maximum energy attainable by CRs critically depends on the details of the unknown physical processes in the region. Alternatively, neutron stars present some advantages over AGNs and GRBs as UHECR sources. Such fast-spinning magnetized stars can have very large electric fields that can therefore accelerate CRs to very-high energies. Neutron stars are naturally rich with heavy elements that can accommodate the latest Pierre Auger results, and are energetically viable sources of UHECRs, with only a small fraction of them needed to contribute to the observed UHECR flux. We refer the reader to Aloisio *et al.* (2018) for a more complete discussion of these three classes of sources. While GRBs have not yet been detected by IACTs, dozens of AGNs have been detected and monitored by H.E.S.S., MAGIC and VERITAS (see Chapter 3). In this sense, particularly relevant for

UHECRs are powerful local sources such as Centaurus A (Aharonian *et al.*, 2009b) and M87 (Aharonian *et al.*, 2006c; Acciari *et al.*, 2008; Aleksić *et al.*, 2012). Neutron stars have been observed so far by IACTs only in our own Galaxy: in the form of pulsars and pulsar wind nebulae, and as part of binary systems (see Chapter 3). The impact of such observations on the study of extragalactic CRs is to be searched in a multi-wavelength and multi-messenger context, as was well demonstrated recently by the gravitational-wave event GW170817 associated with a binary neutron star merger (Abbott *et al.*, 2017d) (see also Chapter 8), and will become of paramount importance in the upcoming CTAO era.

5.7.2.1. *Clusters of galaxies*

According to the commonly accepted hierarchical model of structure formation, clusters of galaxies are the latest and largest gravitationally bound objects to form in the Universe (Voit, 2005). They form and grow through the merging of smaller objects into larger ones, and through the accretion of matter onto an already formed cluster. During this formation process, particles can be (re)accelerated by turbulence and non-relativistic shocks rising from plasma motions due to the violent formation processes (see Brunetti and Jones, 2014 for a review). Both the accretion shocks in cluster outskirts and the merging shocks propagating through the intracluster medium can accelerate particles to very-high energies and potentially generate UHECRs (e.g. Kang *et al.* 1996, 1997; Jones 2004; Murase *et al.* 2008; Kang and Ryu 2013; Vannoni *et al.* 2011).

The quest for the possible gamma-ray emission coming from the hadronic interaction of CR protons with the protons of the intracluster medium have been going on for about two decades now (e.g. Gabici and Blasi, 2003; Pfrommer and Enßlin, 2004; Pfrommer *et al.*, 2008; Pinzke and Pfrommer, 2010; Zandanel *et al.*, 2014). Despite the fact that no clusters of galaxies have been firmly detected in gamma rays so far (e.g. Reimer *et al.*, 2003; Ackermann *et al.*, 2014b, 2016b), IACT observations have significantly contributed to constraining the CR content and acceleration properties in these environments. These include the early Whipple and VERITAS observations of the Perseus and Coma clusters (Perkins *et al.*, 2006; Perkins, 2008; Galante and for the VERITAS Collaboration, 2009; Acciari *et al.*, 2009d; Arlen *et al.*, 2012), the H.E.S.S. observations of the Coma, Abell 85 and 496, and Hydra A clusters (Aharonian *et al.*, 2009a, 2009d; Domainko *et al.*, 2009; Abramowski *et al.*, 2012a), the CANGAROO-III observations of the Abell 3667 and 4038 clusters (Kiuchi *et al.*, 2009) and the MAGIC observational campaign on the Perseus cluster (Aleksić *et al.*, 2010d, 2012a; Ahnen *et al.*, 2016a). Here, we will focus on summarizing the results of the observations of the Coma and Perseus clusters. These are among the better-studied clusters of galaxies across the whole electromagnetic spectrum. Their study is complementary, with Coma being a merging cluster and Perseus, a relaxed cluster. They are also among the most interesting targets for non-thermal studies, as both of them host prototypical diffuse radio emission filling a large volume of the cluster (see Feretti *et al.*, 2012 for a review). The presence of such a diffuse radio emission originally motivated the hadronic model — where secondary electrons from hadronic interactions would generate this emission — and, therefore, gamma-ray observations. Nowadays, thanks to gamma-ray observations, we know that the diffuse radio emission in the Coma cluster cannot be of hadronic origin and turbulent re-acceleration scenarios are preferred (Brunetti *et al.*, 2012; Zandanel and Ando, 2014; Brunetti *et al.*, 2017). The quest for protons, however, continues.

The MAGIC observational campaign on Perseus provided the tightest constraints on CR physics in clusters among all IACT observations, thanks both to the long duration of observations — about 250 h — and the characteristics of Perseus itself: being a relaxed cool-core cluster, it has a very centrally peaked gas density (i.e. the target protons) favoring a centrally peaked hadronic-induced signal, rather than a more diffuse one that presents more difficulties for current IACT observations. The MAGIC observations of Perseus are also comparable in constraining power to *Fermi*-LAT results. This is due to the presence in Perseus of the two bright AGNs NGC 1275 (Aleksić *et al.*, 2012b) and IC 310 (Aleksić *et al.*, 2010b), which imply that *Fermi*-LAT cannot efficiently study the diffuse emission in Perseus due to its poor angular resolution at low energies and small statistics at high energies where the angular resolution would be better. We show in Fig. 5.14 the Perseus field-of-view as seen by MAGIC and the constraints on the CR content as obtained by the MAGIC upper limits on the possible diffuse hadronic-induced gamma-ray emission. Considering that both the spectral and spatial distribution of CRs is unknown — the latter being strictly connected with the unknown propagation and transport properties of CRs in the intracluster medium — the main outcome of the MAGIC campaign, according to their latest results (Ahnen *et al.*, 2016a), can be summarized as follows. Assuming a CR-proton spectral index of $\alpha = 2.2$ (as in the right panel of Fig. 5.14), the CR-to-thermal pressure for the whole cluster is $\lesssim$1%–2%. In the case of a radially flat

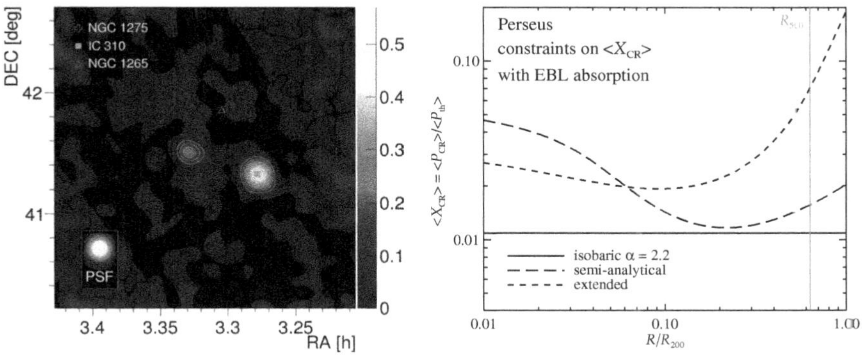

Fig. 5.14. *Left*: MAGIC Perseus sky map above 250 GeV. The color scale represents the signal-to-background ratio, and the symbols indicate the positions of the three brightest AGNs of the cluster. *Right*: Perseus volume-averaged CR-to-thermal pressure as constrained by the MAGIC upper limits for models that assume different underlying distributions of CRs. Figures taken from Ahnen *et al.* (2016a).

CR spatial distribution, however, the CR-to-thermal pressure constraint worsens significantly to $\lesssim$20%. Note also that the softer the CR-proton spectral index, the looser the constraints on the CR-to-thermal pressure. This result implies that contrary to the CR population in our Galaxy, CRs are not dynamically important in clusters of galaxies.

The H.E.S.S. (Aharonian *et al.*, 2009a) and VERITAS (Arlen *et al.*, 2012) observations of the Coma cluster are not as constraining for CR physics as the aforementioned Perseus clusters, both for the limited amount of observation time, of about 8 and 18 h, respectively, and because the possible hadronic-induced gamma-ray emission in Coma is expected to be much more diffuse than in Perseus (e.g. Pinzke and Pfrommer, 2010). In particular, the VERITAS observations constrain the CR-to-thermal pressure in Coma to be $\lesssim$16% for a centrally peaked CR spatial profile and $\alpha = 2.2$. The VERITAS observations, however, are at the center of a controversial debate still ongoing at the time of writing this book. In fact, authors not belonging to the VERITAS collaboration used the published Coma sky map obtained by VERITAS to claim the detection of a ring-like gamma-ray emission around the Coma cluster interpreted as correlated with the accretion shock in the cluster outskirts (Keshet *et al.*, 2017). While we consider it fair to mention this work, we also must note that both the IACT and the cluster non-thermal community consider this result, at the time of writing this book, as inconclusive and thus disregard its interpretation. Nevertheless, there have been a number of predictions for gamma-ray signals induced by CRs accelerated at accretion shocks in the outskirts of clusters of galaxies (e.g. Keshet *et al.*, 2003; Kushnir and Waxman, 2009; Inoue *et al.*, 2005; Vannoni *et al.*, 2011). This is also strictly connected with CRs given that accretion shocks can accelerate particles at higher energies with respect to the slower merging shocks, and that these CRs are more likely to leave the cluster environment with respect to those accelerated at merging shocks, which are more strongly confined within clusters with escape energies actually unknown.

The future of such observations is represented by the key science project of CTAO on clusters of galaxies (Acharya *et al.*, 2019). In this context, a long observational program of Perseus is envisioned with the goals of providing the final test on the hadronic interpretation of its diffuse radio emission, and, more importantly, of limiting the CR-to-thermal pressure below 5% independently from the underlying CR-proton spectral and spatial characteristics.

Acknowledgments

D.G. acknowledges the support from Generalitat Valenciana through the plan GenT program (CIDEGENT/2021/017).

S.C. acknowledges the National Science Center, Poland, grant no. UMO-2017/27/B/ST9/02272.

F.Z. acknowledges the supported by the Netherlands Organization for Scientific Research (NWO) through a Veni grant.

https://doi.org/10.1142/9789813275720_0006

Chapter 6

Astrophysics

Emma de Oña Wilhelmi[*,‡], Daniel Mazin[†,§] and Andrew Taylor[*,¶]

[*] *Deutsches Elektronen-Synchrotron (DESY)*
Zeuthen, Germany
[†] *University of Tokyo*
Tokyo, Japan
[‡] *emmadeona@gmail.com*
[§] *mazin@mpp.mpg.de*
[¶] *andrew.taylor@desy.de*

6.1. Introduction

The recent scientific results achieved with Cherenkov telescopes have transformed the field of high-enerygy (HE) gamma-ray astronomy from being considered a part of the particle (or rather astroparticle) physics domain to becoming acknowledged as a full-fledged domain in astrophysics. The astrophysical scenarios from which energetic gamma rays emerge imply, in general, violent phenomena involving shocks that develop at different speeds in a broad variety of different-density plasma. Such plasmas are embedded with magnetic fields that vary from μG to 10^{12} G, in the most extreme cases. Observations in the TeV regime, then, allow for testing a range of astrophysical parameters that are otherwise inaccessible in other wavelengths.

The basic processes can be well-described according to the Lorentz factor or speed of the particle flow ($\beta\Gamma$) and the magnetization parameter (see Fig. 6.1). The magnetization of a given medium represents the ratio of the Poynting flux to the particle energy flux, namely, $\sigma = \mathrm{B}^2/4\pi\rho c^2$ for a medium with a magnetic field B, a mass density $\rho = n \times m$, composed of particles of mass m and proper density n. In case the flow is moving at relativistic velocity, the magnetization parameter becomes $\sigma = B^2/4\pi\rho\Gamma^2\mathrm{c}^2$ for a medium with a four-velocity (Γ, $\beta\Gamma$) and Lorentz factor

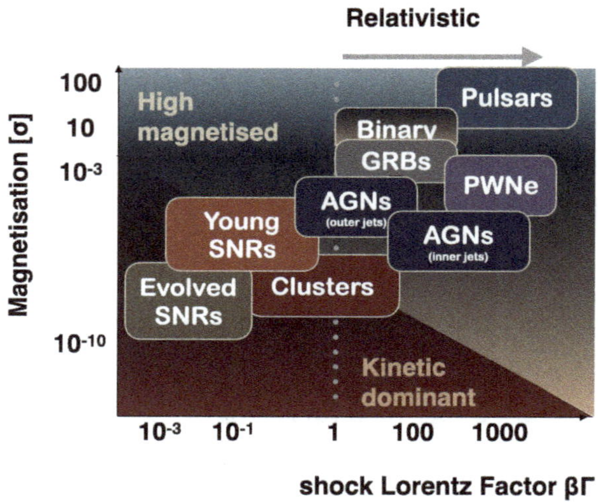

Fig. 6.1. Sketch of the relation between the magnetization parameter, defined as the ratio between the magnetic energy and the kinetic energy, vs. shock velocity, represented as $\beta\Gamma$, for different very-high-energy (VHE) gamma-ray sources.

$\Gamma = (1 - \beta^2)^{-1/2}$. The interplay between these two aspects defines the type of acceleration, propagation and radiation that can occur in an astrophysical source. To supersede thermal heating — required to explain the non-thermal radiation observed in the TeV regime — either high magnetization in relativistic flows or efficient Fermi shock acceleration is needed. The field of TeV astrophysics allows us to investigate different scenarios for a wide range of parameters. Highly magnetized compact galactic objects and their magnetospheres, as well as large-scale extragalactic objects, are optimal laboratories to study relativistic flows at different densities, whereas explosive shocks created by massive stars, for example, provide conditions to understand behaviors of collisionless plasma.

The novelty of our approach is to focus on the astrophysical problem rather than the source-by-source phenomenology. In the following chapter, we attempt to provide a global view of the astrophysical advances in the field of TeV astronomy. We describe the different relevant mechanisms dictating the acceleration, propagation and energy losses from particles in the different regimes outlined above (Sections 6.2–6.4). We then put the results (or open questions) from TeV observations into context considering the lessons learned from the last 30 years of gamma-ray observations. This chapter aims to provide information related to the acceleration, propagation and emission processes obtained in different astrophysical scenarios, in which similar phenomena (at, however, very different scales in relevant parameters such as magnetic field or size) can be used to derive conclusions of the same process. Surprisingly, many conclusions are common to different object classes, e.g. the failure of the equipartition assumption in objects as different as Pulsar Wind Nebulae (PWNe) and blazars. We finally conclude with a list of open problems that we believe will be addressed in the near future.

6.2. Acceleration of Particles to the TeV Regime

Observations in the TeV regime probe particle acceleration to energies above a few hundreds of GeV — that is, mechanisms that can supersede thermal heating are required. Several mechanisms are behind the TeV observations dominating the acceleration of particles, depending on the relativistic (or not) regime and magnetization of the medium. In the following section, we review the most relevant processes and their impact on the spectral shape and maximum possible energy achieved.

Source class	Shock speed ($\beta\Gamma$)	B-field (G)	Chapter section
Pulsar magnetosphere	$10^2 - 10^7$	$10^{12} - 10^{14}$	6.5.1
Pulsar wind nebula	$10^{3.5} - 10^{4.5}$	$10^{-5} - 10^{-3.5}$	6.5.2.1, 6.5.2.2, 6.5.2.3
AGN-inner jet (∼ pc)	$10^1 - 10^2$	$10^{-2} - 10^{-0.5}$	6.5.2.1, 6.5.2.2, 6.5.2.4
AGN-outer jet (∼ kpc)	$10^{0.5} - 10^{1.5}$	$10^{-5} - 10^{-3.5}$	6.5.2.1, 6.5.2.2, 6.5.2.4
SNR (young)	$10^{-2.5} - 10^{-1.5}$	$10^{-6} - 10^{-3}$	6.5.4
SNR (middle-aged)	$10^{-3.5} - 10^{-2.5}$	$10^{-6} - 10^{-5}$	6.5.4
Gamma-ray binaries	$10^0 - 10^1$	$10^{-3} - 10^0$	6.5.3
Gamma-ray burst	$10^{0.5} - 10^{2.5}$	$10^{-3} - 10^0$	6.5.3.2
Starburst galaxies	$10^{-3.5} - 10^{-2.5}$	$10^{-3} - 10^{-6}$	6.5.4

6.2.1. *Fermi shock acceleration*

Gamma-ray production within a source is a secondary phenomenon, which demands firstly that particles are accelerated by their source, and the accelerated particles thus undergo energy losses and radiate their energy away in gamma-ray radiation. Of particular concern for probing the efficiency of the acceleration process is the determination of two key parameters: the acceleration time, $t_{\rm acc}$, and the maximum energy the source can accelerate the particles to, $E_{\rm max}$. We focus our discussion here on the determination of these two values.

Insight regarding the physics of the acceleration process can be provided by considering the collision of a fast-moving fluid component with a slower fluid component, for the case of particles with thermal velocities much lower than the bulk velocity difference between the two colliding flows. In such a case, a shock front (whose velocity we denote as $\beta_{\rm sh}$) is set up at the interface between these differing velocity flows. Assuming that the Coulomb collisional scattering times of the particles are longer than the dynamical time of the system, the thermal distribution of the plasma is not maintained by such collisions. Instead, for such collisionless systems, magnetic fields are responsible for scattering and isotropizing particles on either side of the shock, allowing the possibility for a subset of the particles to gain energy from the velocity difference of the two flows and giving rise to the energization of a fraction of the plasma particles into a non-thermal pool.

Non-thermal particles, being considerably more energetic than the thermal particle pool they are elevated from, possess large velocities and are subsequently able to propagate considerable distances through the background plasma before scattering. This advantage allows the particles to act as a large-scale viscous agent at any boundary between differing velocity flows. Conversely, non-thermal particles in return may be energized by these large-scale scatterings, with the energy they drain from the fluid passing into the non-thermal particle population.

Continuing the assumption of negligible collisional interactions, the non-thermal distribution of particles produced within the source couples weakly to the background plasma via scattering interactions with its (frozen-in) magnetic field. Through these scatterings, the non-thermal particles diffuse (with diffusion coefficient D) through the background plasma, scattering on the turbulence present (δB) within the background magnetic field (B), within a scattering length ($l_{\rm mfp}$) that quasi-linear-theory predicts to be (Schlickeiser, 1989)

$$\frac{3D}{c} = l_{\rm mfp} = \left(\frac{B^2}{\delta B_\lambda^2}\right) r_{\rm lar} = \eta r_{\rm lar}, \tag{6.1}$$

where λ is the wavelength of the relevant turbulence mode, and only in the mode satisfying the condition $\lambda = r_{\rm lar}$ (i.e. resonant with $r_{\rm lar}$ being the Larmor radius) is scattering considered.

For extreme astrophysical accelerators, the term $\eta = \left(B^2/\delta B_\lambda^2\right)$ in the above expression is close to 1. This is referred to as the "Bohm scattering" regime, with particles scattering every Larmor period on average. Such a situation requires effective production of turbulence at the gyration-length scale of the particle in order to ensure effective disturbance of the particles by the particular component of turbulence in the resonant magnetic field.

More generally, the energy density in a particular mode is $\delta B_\lambda^2 \approx \delta B^2(\lambda/\lambda_{\rm max})^{q-1}$, where $\delta B^2 = \int \delta B_\lambda^2 d\ln\lambda$ describes the total energy density summed over all the turbulent modes (λ), and q describes the spectral index of the turbulence power spectrum. Therefore, $\eta = (B^2/\delta B^2)(r_{\rm lar}/\lambda_{\rm max})^{1-q}$, indicating that scattering lengths become significantly larger than $r_{\rm lar}$, for $r_{\rm lar} \ll \lambda_{\rm max}$.

Assuming that these non-thermal particles are isotropized after every interaction, starting with the particles with an initial delta distribution at the origin, their ensemble distribution at later times can be described by a spreading Gaussian (i.e. random-walk result) with $dN/dr \propto e^{-r^2/4Dt}$. Thus, a fiducial diffusive distance for particles with a scattering length $\sim D/c$ on a timescale t is anticipated of $r \approx (4Dt)^{1/2}$.

In contrast to this, for particles in an advective flow started in a similar initial configuration, their distribution at later times will be $dN/dr \propto \delta(t - r/v_{\rm adv})$, where $v_{\rm adv}$ is the advection velocity, i.e. the particles will be located at $r = v_{\rm adv}t$. A competition between diffusive and advective transport, therefore, will invariably be dominated by the latter process at sufficiently late times.

In this way, the bulk of the non-thermal particle distribution in the shock region (see Fig. 6.2(a)) will eventually drift downstream out of the region on a timescale dictated by the time required for advective transport to take over from diffusive transport, namely, $t_{\rm tran} = D/(\beta_{\rm adv}c)^2$. Within this timescale, however, the particle will have on average crossed the shock front $1/\beta_{\rm sh}$ times. Thus, the timescale $\langle \Delta t_{\rm cyc}\rangle \approx D/\beta_{\rm sh}c$ is associated with the average time required to perform one cycle of crossing the shock front and returning back during the epoch before advective transport takes over.

As appreciated from Fig. 6.2(a), in the rest frame of the fluid on either side of the shock front, the plasma on the other side of the shock front is always approaching (though with differing magnitudes of velocity).

For a particle in the downstream region with a pitch angle μ_1, after being boosted into the moving upstream frame, its pitch angle in this frame is

$$\mu_2' = \frac{(\mu_1 - \beta)}{(1 - \beta\mu_1)}. \tag{6.2}$$

After boosting the particle from this boosted frame back into the original frame, its new pitch angle will be μ_2 and its energy, E_2, which are related

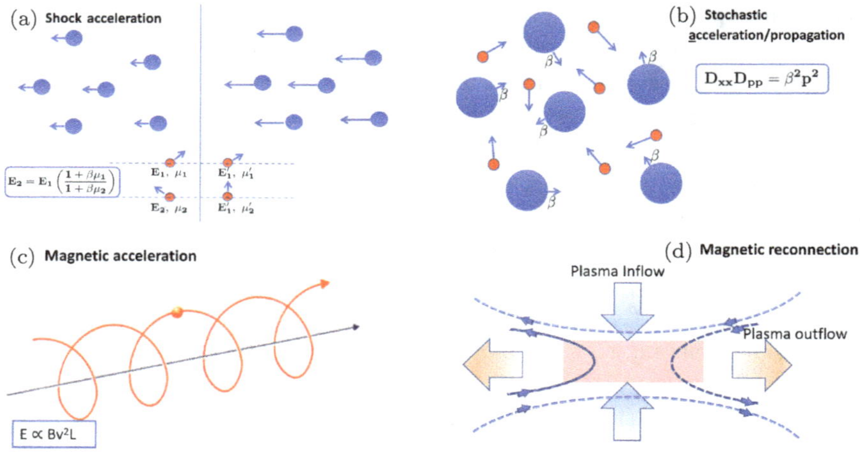

Fig. 6.2. Diagrams sketching out four different astrophysical acceleration processes: (a) shock-acceleration process, (b) stochastic acceleration, (c) bead-on-a-wire magnetic acceleration and (d) magnetic reconnection.

by

$$E_2 = \Gamma^2 E_1(1 - \beta\mu_1)(1 + \beta\mu_2') \tag{6.3}$$

$$= E_1 \left(\frac{1 - \beta\mu_1}{1 - \beta\mu_2}\right) \tag{6.4}$$

Thus, only if the particle's direction changes pitch angle (i.e. the particle scatters) is it able to gain energy from the flow during a crossing cycle. This scattering might occur in either relativistic shocks or non-relativistic shocks:

- **Non-relativistic shocks:** In the case of non-relativistic shocks, the energy-gain-per-crossing is $\langle \Delta E_{\mathrm{cyc}} \rangle \approx E\beta_{\mathrm{sh}}$. Thus, the acceleration timescale for a particle repeatedly crossing the shock front is

$$t_{\mathrm{acc}} \approx E\frac{\langle \Delta t_{\mathrm{cyc}} \rangle}{\langle \Delta E_{\mathrm{cyc}} \rangle} \sim \frac{\eta r_{\mathrm{lar}}}{\beta_{\mathrm{sh}}^2 c}. \tag{6.5}$$

  If, following each cycle, the change in the energy of the particles is $d \ln E = (1 + \beta)$ and the change in the number of particles is $d \ln N = (1 - \beta) \approx 1/(d \ln E)$ after n collisions, the energy of the particles and the number of particles in the system will be $E(n){=}E_0(1 + \beta_{\mathrm{sh}})^n$, $N(n){=}N_0(1 - \beta_{\mathrm{sh}})^n \approx N_0E_0/E(n)$. The subsequent spectrum in the acceleration region is therefore $dN/dE \propto (N/E) \propto E^{-2}$. The maximum energy achievable by such an accelerator is dictated by the balance between the acceleration time (Eq. (6.5)) and the diffusive escape time ($t_{\rm diff} = t_{\rm esc}$) from the source $t_{\rm esc} = R^2/4D \approx R^2/\eta c r_{\rm lar}$, leading to $r_{\rm lar} = \beta_{\rm sh} R/\eta$. This expression can be re-expressed as

$$E_{\rm max} = \frac{\beta_{\rm sh}}{\eta} eBR, \tag{6.6}$$

which is the well-known Hillas criterion (Hillas, 1985b).

- **Relativistic shocks:** Following on from the discussion above and the consideration of Eq. 6.4, assuming full turnaround of the particle after crossing the shock (i.e. $\mu_1 = 1$ and $\mu_2 = -1$ in Eq. 6.4) leads to a large energy gain for the case of relativistic shocks, $\langle \Delta E_{\rm cyc} \rangle \approx E\Gamma^2_{sh}$. For relativistic shocks, the diffusive angular spread of particles that have crossed the shock front may not be isotropic by the time advection takes over from diffusive transport on the timescale $t = D/(\beta_{\rm sh} c)^2$. In this case, the full Γ^2 energy boosting per shock crossing cycle may not be achieved, considerably increasing the acceleration timescale beyond the naive extrapolation from the non-relativistic shock case, i.e. acceleration times longer than $t_{\rm acc} \sim D/c^2$. Following the same line of reasoning as for the non-relativistic case,

$$E_{\rm max} = \frac{1}{\eta} eBR, \tag{6.7}$$

where $E_{\rm max}$, B and R are all quantities measured in the observer's frame. We note here that the factor η for a relativistic shock is of particular importance due to the challenge of the production of large-scale turbulence by relativistic sources.

Generally, the steady-state spectrum produced within the source by the acceleration process is a power law of the form $EdN/dE = E^{-(t_{\rm acc}/t_{\rm esc})}$. As discussed above, however, for the case of shock acceleration, with particle escape being dominated by advection, $t_{\rm acc}$ and $t_{\rm esc}$ are equivalent, leading to universal spectra of the form $dN/dE \propto E^{-2}$. This is only true, however, for the particular cases in which particle escape and acceleration timescale are connected, while other acceleration processes might occur in other environments for which $t_{\rm acc} \neq t_{\rm esc}$, leading to a departure from the expectation of an E^{-2} spectrum being produced.

6.2.2. *Stochastic acceleration*

Magnetic turbulence within a source leads to cosmic scattering and its subsequent diffusion in space. If the scattering centers giving rise to this spatial diffusion are themselves moving, e.g. fast-moving turbulent magnetic waves with velocity $\beta_{\rm Alf}$, energy boosts and energy losses will be experienced during particle scattering. Since the non-thermal particles may collide head-on or rear-on with such scattering centers, the average energy gain per scattering is $\langle \Delta E_{\rm scat}\rangle \propto E\beta_{\rm Alf}^2$. With a scattering timescale of $t_{\rm scat} \approx D/c^2$, the subsequent acceleration time achieved by this process is

$$t_{\rm acc} \approx E\frac{\langle \Delta t_{\rm acc}\rangle}{\langle \Delta E_{\rm scat}\rangle} \sim \frac{\eta r_{\rm lar}}{\beta_{\rm Alf}^2 c}. \tag{6.8}$$

It is important to note that, interestingly, the expression for the particle acceleration rate for stochastic acceleration compared to shock acceleration (i.e. Eqs. (6.5) and (6.8)) differs only in the Replacement of $\beta_{\rm sh}$ with $\beta_{\rm Alf}$ (Jones, 1994). Subsequently, the maximum energy expression achievable is Eq. (6.6), with $\beta_{\rm Alf}$ substituted instead of $\beta_{\rm sh}$. The resulting particle spectrum produced by stochastic acceleration, however, is generally quite different from that produced by shock acceleration.

6.2.3. *Magnetic-driven acceleration*

Compact objects possessing strong magnetospheres exert a force on the surrounding plasma, causing it to co-rotate (see diagram (c) in Fig. 6.2 for a representation). The field lines of these systems can subsequently be considered locally rigid, forcing the surrounding plasma to be accelerated along the local field lines out in the radial direction (Rieger and Mannheim, 2000). This acceleration occurs until the point where the Lorentz force on the particle eventually balances the Coriolis force of the rotating field. Note that the field lines of the magnetosphere are themselves unable to maintain a radial direction, since they would be required to move faster than the speed of light at distances beyond the light cylinder radius.

For a particle with an initial Lorentz factor Γ_0, injected at a point in a rotating field with velocity $\beta_{\Omega 0}$, the characteristic acceleration time is $t_{\rm acc} \approx T_{\rm orbit}/[\Gamma_{\Omega 0}(\Gamma/\Gamma_0)^{1/2}]$, where $\Gamma_{\Omega 0} = (1-\beta_{\Omega 0}^2)^{-1/2}$. Since $\Gamma_{\Omega 0} \approx 1$ can be expected, the acceleration time of mildly relativistic particles will be the rotation time of the field lines. This acceleration time will subsequently decrease for higher-energy particles, making the acceleration process more effective as particles are accelerated up to higher and higher energies.

6.2.4. *Magnetic reconnection*

The merging of plasma regions, each with their own magnetic field lines, can lead to differing magnetic field structures being brought into close proximity (see Fig. 6.2 (d)). Due to Ohmic resistivity in the plasma, magnetic field-line wandering provides the opportunity for differently oriented field-line structures to be brought together and merged, with the subsequent release of kinetic energy into the plasma. A naive estimate of the maximum energy achievable in such merging flows follows on dimensional grounds from the velocity of the merger flows $v_{\rm Alf}$ and the length scale L that the reconnection process occurs over, with $E_{\max} = \beta_{\rm Alf} eBL$. A comparison of this result with Eq. (6.6) indicates that unless the length scale L can approach the macroscopic size of an astrophysical system, the maximum energy achievable by this process will be dwarfed by that achievable by diffusive shock or stochastic acceleration processes.

6.2.5. *Final comments on acceleration processes*

The different processes described are studied in astrophysical sources in which different regions are ruled by the interaction of plasma, magnetic-field and non-thermal particles present. Indeed, the gamma-ray emission observed in the TeV regime from the ensemble of TeV sources originates in a diverse range of different conditions and length scales. For instance, in the surroundings of *compact objects*, namely, pulsars and black holes, their extreme electromagnetic activity can affect the environment from a few kilometers (surface on neutron stars) to several kpc (outer jets of AGNs) as a result of their extreme magnetic field and the relativistic particle flows they created around them. The magnetic field strength in these objects reaches $\sim 10^{9-13}$G and $\sim 10^{2-3}$G for pulsars and black holes, respectively, whereas the particle flows can reach a high Lorentz factor of $\sim 10^{5-6}$ and $\sim 10^{1-2}$, respectively. These rich environments in which different processes operate — including highly magnetized plasma, regions of accreting mass, radial relativistic flows and magnetic sheets — translate into a plethora of different source regions with unique physical conditions in which particles are accelerated (see Figs. 6.3 and 6.4).

6.3. Energy Losses, Escape and Propagation

Following the energization of particles in the acceleration zone, their escape and propagation away from this zone is considered next. Here it must be kept in mind that outside the source, the subsequent regions

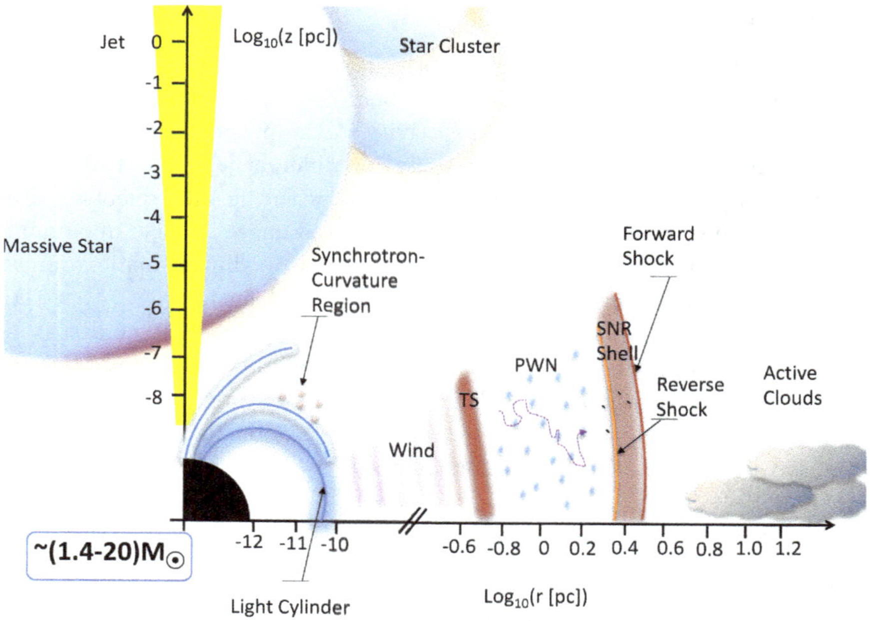

Fig. 6.3. Sketch of different Galactic sources or regions emitting VHE gamma rays.

that the particles propagate through will be predominantly low-density environments. Once particles have escaped into such regions from their extreme source environment, they are expected to propagate large distances without interacting directly with other particles in the plasma, interacting instead mostly via energy and momentum exchange with the radiation and magnetic field present. Quantitatively speaking, after being accelerated to very high energies, particles propagate either diffusively or advectively in the medium, occasionally losing energy by different radiation processes or adiabatically in the expanding flow. Despite these energy-loss processes, many — and at some energies, most — of the accelerated particles escape and do not interact, particularly at the highest energies. We see them as the cosmic ray (CR) flux, and the comparison between the measured CR flux and the gamma-ray flux gives clues about the processes of emission and propagation the particle suffered on its way to us. A longer discussion about CRs can be found in Chapter 5.

The emission observed from the fraction of particles that suffer radiative losses keeps an imprint of the manner in which they propagate. These

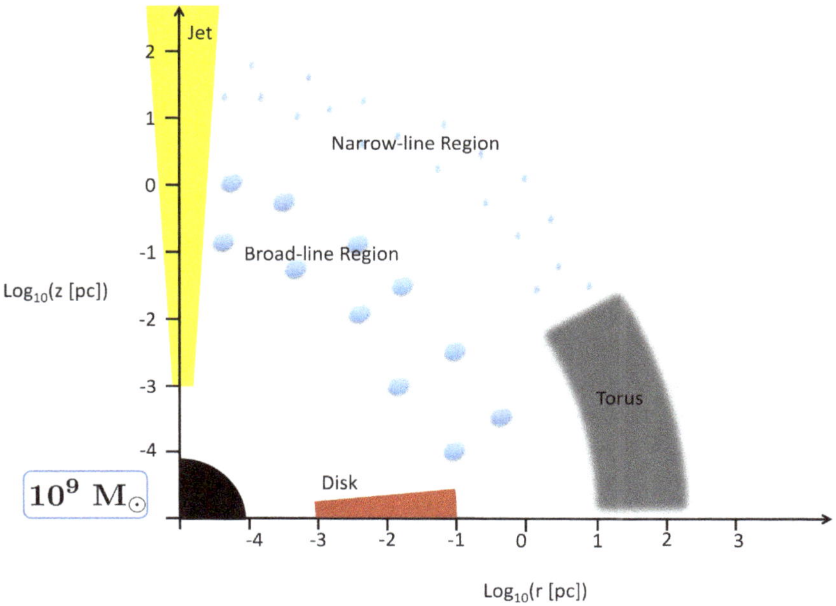

Fig. 6.4. Sketch of different extragalactic sources or regions emitting VHE gamma rays.

particles can be transported in either a diffusive or advective manner. Accounting for the transport mechanism and the particle escape from the system, the resultant particle spectrum is described by the transport equation

$$\frac{df}{dt} = -\vec{\nabla}\cdot[\underbrace{\vec{v}f}_{\text{advective current}} - \overbrace{D\vec{\nabla}f}^{\text{diffusive current}}] - \underbrace{\frac{\partial}{\partial E}\left[-\frac{E}{t_{\text{loss}}(E)}f\right]}_{\text{radiative losses}} - \overbrace{\frac{1}{E^2}\frac{\partial}{\partial E}\left[(\vec{\nabla}\cdot\vec{v})\frac{E^3}{3}f\right]}^{\text{adiabatic losses}}$$
$$- \underbrace{\frac{f}{t_{\text{esc}}(E)}}_{\text{escape}} + \overbrace{\frac{Q(E)}{E^2}}^{\text{source}}, \qquad (6\,9)$$

where $f = dN/d^3xd^3E$ (note that we assume here that all particles are relativistic such that E and p are equivalent).

We can thus describe the resultant steady-state spectrum of particles as that produced from a flow of particles injected (accelerated) by the source with spectrum $Q(E)$, propagating via diffusion on a timescale $t_{\text{diff}} = R^2/4D$

or via advection on a timescale $t_{\rm adv} = R/v_{\rm adv}$, experiencing energy losses that operate on a timescale ($t_{\rm loss} = E/(dE/dt)$), and eventually escaping the propagation zones on a timescale $t_{\rm esc}$. Insight can be provided into the above expression by considering the competition of specific terms.

First, we consider how energy-dependent transport (i.e. the "diffusive current" terms in Eq. (6.9)) can introduce deviation of the spectrum in the propagation zone away from the injected (accelerated) source spectrum. *Diffusive propagation* is described by the diffusion coefficient D, which characterizes the escape time from a region of size R by $t_{\rm esc} \approx R^2/4D$. Ignoring spatial information of the particle distribution and energy losses (i.e. the first three terms on the right-hand side (RHS) of Eq. (6.9)), the steady-state spectrum ($dN/dt = 0$) obtained is $dN/dE = Q(E)t_{\rm esc}(E)$. This result indicates how diffusion can lead to a buildup of particles, particularly at low energies where escape times are expected to be longest, and a resultant steepening of the particle energy spectrum within the diffusion region occurs.

In a similar vein, it is enlightening to next consider how diffusive transport (i.e. the second term in Eq. (6.9)) alters the *spatial distribution of the particles* away from the injection site. Assuming that advective transport and particle escape can be ignored, a steady-state spatial distribution can still be achieved (but only for diffusive transport in more than two dimensions). In such a case, the steady-state spatial distribution of particles observed for an injection at r_0 is $dN/dr \propto Q(r_0)/[D(E)r]$. Note that for explosive/highly variable sources, e.g. SuperNova Remnants (SNRs), the Active Galactic Nucleus (AGN), Gamma-Ray Bursts (GRBs), etc., the steady-state spatial distribution may only be obtained for particles for which the diffusive timescale $R^2/4D$ is less than the variability timescale. This result has strong implications on the observed gamma-ray profile in extended gamma-ray sources (Aharonian *et al.*, 2019; Abramowski *et al.*, 2016; Yang *et al.*, 2018).

Next, we consider how *energy-dependent losses* (i.e. the third term on the RHS in Eq. (6.9)) can alter the propagated spectrum. The form of $t_{\rm loss}(E)$ for HE particles can be expressed by the sum of different cooling processes. For protons, this term would convey the sum of Bremsstrahlung and pion-production energy losses ($\propto$ constant function of the energy) and synchrotron radiation ($\propto E^{-1}$) (see Fig. 6.5(a)). For electrons, this term might instead convey Bremsstrahlung, adiabatic losses ($\propto$ constant function of energy), inverse Compton (external or self) and synchrotron

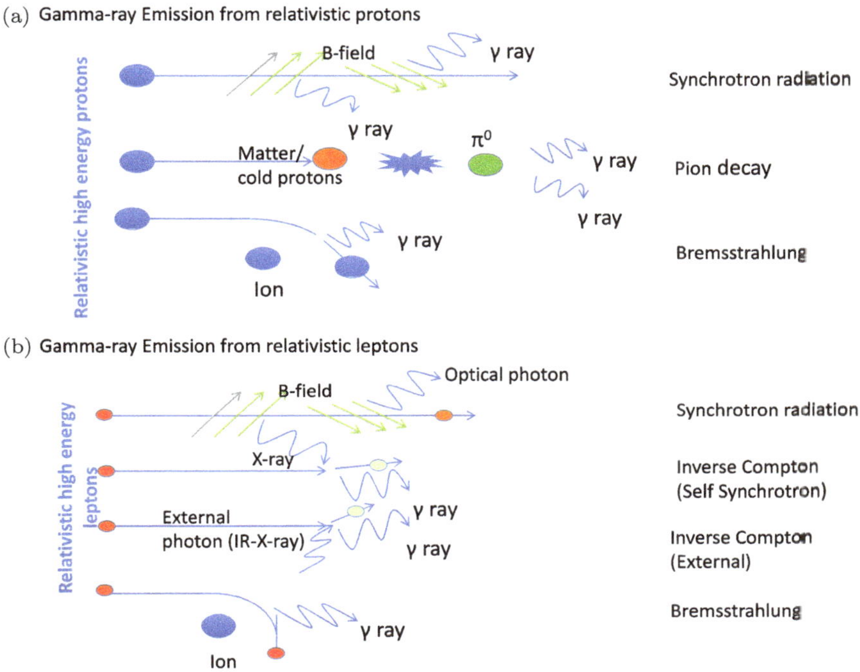

Fig. 6.5. Sketch of different emission processes relevant for VHE gamma rays.

radiation ($\propto E^{-1}$) (see Fig. 6.5(b)) (see Chapter 1 for more information). Considering an injection spectrum given by $Q(E) \propto E^{-s}$, (Eq. (6.9)) leads us to expect that the resultant spectrum depends only on the dominating cooling mechanisms, since in the steady-state regime

$$\frac{dN}{dE} = \frac{t_{\text{loss}}(E)}{E} \int_{E}^{E_{\max}} Q(E)dE. \tag{6.10}$$

For $s > 2$, this integral is dominated by its lower limit, resulting in $dN/dE \propto Q(E)t_{\text{loss}}(E)$. If Bremsstrahlung, proton–proton interaction or adiabatic loses dominate, the spectrum will mimic the one from the uncooled particle distribution, since τ_{loss} is approximately energy-independent for these cases. When an energy-loss process with $\tau_{\text{loss}} \propto E^{-1}$ dominates (e.g. synchrotron losses or inverse Compton losses in the Thomson regime), however, one would expect the steady-state energy spectrum (i.e. $dN/dt = 0$) to steepen the injected spectrum by one power, leading to $dN/dE \propto Q(E)t_{\text{loss}}(E) \propto E^{-(s+1)}$. For the case of $s < 2$, the integral in Eq. (6.10) is dominated by its upper limit and is therefore constant. The spectra in this case revert to $dN/dE \propto t_{\rm loss}(E)/E$.

Taking this discussion further, for the case of multiple processes being responsible for the achievement of steady state in the system, the correspondence of the change from one process dominating another will also be conveyed in the resultant particle spectrum. As a concrete example, we consider the steady-state spectrum brought about through the inclusion of the "source", "energy losses" and "escape" terms in Eq. (6.9). For the case where energy losses have a stronger energy dependence than escape, the HE region may be dictated by the former of these and the low-energy region by the latter. The resultant steady-state spectrum in this case is described by $dN/dE \propto Q(E)t_{\rm eff}(E)$, where $t_{\rm eff}(E) = (t_{\rm loss}(E)^{-1} + t_{\rm esc}(E)^{-1})^{-1}$, which effectively results in a broken power-law shape of the spectral energy distribution. In the same vein, for the case of energy and spatial transport terms in the transport equation (e.g. energy losses and advective transport), the steady-state spectrum at various distances away from the source will be different. For the simplified problem of a particle spectrum injected one time as a spectrum $dN/dE \propto E^{-s}$, the resultant spectrum at later times is

$$\frac{dN(t)}{dE} = \frac{dN(0)}{dE}\left(1 - \frac{t}{t_{\rm cool}(E)}\right)^{s} \tag{6.11}$$

Since in an advective flow, the age of a particle in a system maps simply to their distance away from the source via $t = R/v$, the spectrum as a function of distance away from the source follows by substituting this relation into Eq. (6.11). Thus, a telltale signature for particles trapped in advective flows can be provided from the evolution of the spectral index of their emission with their distance from the acceleration site.

Until here in the discussion we have ignored the *role played by the acceleration process itself in the resultant particle spectrum* (assuming it to be much faster than the other processes being discussed). At an energy scale at which acceleration balances with cooling or escape, a cutoff feature is introduced into the particle spectrum. In order to discuss this, it is necessary to scrutinize the acceleration process more closely by including an acceleration term in our transport equation

$$Q(E) = \vec{\nabla}_E \cdot D_{EE}\vec{\nabla}_E N(E) + \delta(E_0),$$

where we focus on the stochastic acceleration of particles and D_{EE} functions as the energy diffusion coefficient. Ignoring advective transport and energy losses for simplicity, and considering a system with spherical symmetry, the transport equation amounts to a second-order differential equation of the form

$$\frac{1}{E^2}\frac{\partial}{\partial E}\left[E^2 D_{EE}\frac{\partial f}{\partial E}\right] - \frac{f}{t_{\rm esc}(E)} + \delta(E_0) = 0. \tag{6.12}$$

Adopting a power-law function for escape, $t_{\rm esc} \propto E^{q-2}$, and for the energy diffusive acceleration term $D_{EE} \propto E^q$,

$$\frac{\partial^2 f}{\partial E^2} + \left[\frac{(2+q)}{E}\right]\frac{\partial f}{\partial E} - \frac{f}{k_0(E/E_0)^{(2q-2)}} = 0, \tag{6.13}$$

where $k_0 = D_{EE}(E_0)t_{\rm acc}(E_0)$.

The solution of such a differential equation amounts to a power law with a cutoff at high energies. The power-law slope in these results is dictated by the balancing of the first and second term, and the cutoff shape by the balancing of the first and third term. For the more general case — taking into account both energy losses and escape — the shape of the cutoff is found to follow the form $e^{-x^{\beta_e}}$, where $\beta_e = 2 - q - r$, where r is the exponent for the energy dependence of the particle energy losses (i.e. $r = -1$ for synchrotron losses) (Stawarz and Petrosian, 2008).

Putting the ensemble of results discussed together more broadly, one can anticipate the steady-state solution of the transport equation simply through a consideration of the various transport equation timescales of the problem. As an example, the left plot in Fig. 6.6 shows a potential set of such timescales (acceleration, escape and cooling). For this example

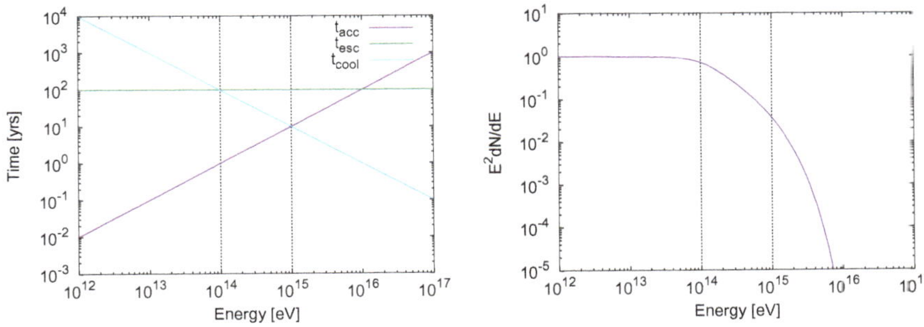

Fig. 6.6. *Left*: An example plot showing the acceleration, escape and cooling timescales as a function of particle energy. *Right*: The resultant steady-state electron spectrum obtained, assuming a source spectrum of the form $Q(E) \propto E^{-2}$. The two dashed vertical lines in the two plots indicate the energy scales at which transitions between the different transport timescales occur.

case, the right plot shows the resultant steady-state spectrum, revealing three regions within the spectra: the low-energy region in which the source spectrum is directly observed; the intermediate-energy region above the cooling-break energy; and the cutoff region formed through the competition between the acceleration and cooling processes. The separation of the three regions discussed is marked by vertical dashed lines.

Additionally, the position of the cutoff energy scale can itself provide information about the acceleration region. As an example case, one can derive the maximum energy that can be achieved (and thus tested) in the VHE regime around compact rotating objects, in which the acceleration of particles occurs by the effect of large rotating magnetic fields inducing enormous electric fields in gaps (Bednarek, 2012; Lyutikov, 2013). The maximum energy is given by the balance between the associated electric field E_F (which is a fraction of the magnetic field $E_F = \eta B$) and the cooling energy-loss rate. The cooling losses in the case of magnetically dominant environments depend on the curvature radius R_c as $dE/cdt = (E/m)^4 \alpha hc/R_\mathrm{c}^2$, where α is the fine structure constant and h is Planck's constant. Equating this with the energy-gain rate $dE/cdt = \eta B$ results in a maximum energy limit

$$E_\mathrm{max} = \left(\frac{\eta}{\alpha}\right)^{1/4} m \left(\frac{BR_\mathrm{c}^2}{hc}\right)^{1/4} \tag{6.14}$$

It should be noted that this result is general. Applying these results to the specific case of electrons whose curvature radius is their Larmor radius in a magnetic field leads to an expression of the maximum electron energy achievable in an accelerator with a fixed magnetic field strength

$$E_\mathrm{e,max} = \left(\frac{\eta}{\alpha}\right)^{1/2} m \left(\frac{B_\mathrm{crit}}{B}\right)^{1/2}, \tag{6.15}$$

where $B_\mathrm{crit} = 4.4 \times 10^{13}$ G is the Schwinger field strength (Schwinger, 1951).

6.4. Emission Processes

The non-thermal energy put into the accelerated particle population by the source will remain within the particle population for a timescale dictated by the energy-loss rate. In the extreme case in which the energy-loss time is sufficiently short compared to the particle transport time through the region, the particles will lose all of their energy within this environment. In such a case, the system is said to be acting as a "calorimeter" on this size scale. Such energy-loss processes give rise to the production and propagation of secondary non-thermal emission.

The resultant shape and intensity of the emission produced via energy-loss interaction depends on the type of accelerated particles and the environment in which they evolved. Gamma-ray emission can result from several different processes (see Chapter 1), e.g. inverse Compton upscattering of CMB and other soft photon fields by energetic electrons, or hadronic (pp) interactions of relativistic CR protons with the ambient gas decaying in gamma rays via π^0-decay. The physics of these heavy (hadronic) CRs is discussed at length in Chapter 5.

The gamma-ray spectrum produced via CR interactions with gas at high energies to first order repeats the spectral shape of parent *protons*, a power law with approximately the same power-law index and a cutoff at $E_\gamma \sim 0.1\ E_{0,\mathrm{p}}$, for a CR spectrum described with a standard power-law function plus exponential cutoff $E_{0,\mathrm{p}}$. A good knowledge of the material in the surrounding region of a hadronic accelerator provides direct information about the CR content and spectral distribution. The spectral energy distribution of the hadronic population can then be used as a diagnostic tool to understand the transport of particles (see previous section) or characteristics of the accelerator (continuous vs. impulsive accelerators, time evolution of the spectrum, etc.; see e.g. Aharonian, 2004; Gabici, 2013). Indeed, due to the energy dependence of diffusive escape time from the emission region, the CR spectra outside the accelerators may suffer significant modification, $dN/dE_\gamma \propto Q(E)t_{\mathrm{adv}}(E) \propto E_\gamma^{(\alpha+\delta)}$, where α is the power-law index characterizing the initial (acceleration) spectrum and Q is the proton injection rate (see discussion in Section 6.3).

In the case of gamma-ray emission originated from a *leptonic* population with energy E_e interacting with photons of energy E_γ, the process occurs in the Thomson regime (in which the cross section is constant) when electron recoil from the interaction is negligible. This happens when $E_e E_\gamma^{\mathrm{bg}} < m_e^2$.

Within this Thomson regime, the average interaction time of an electron with a target photon field is independent of energy. The average energy exchange between the electron and the target photon field, however, is energy-dependent, with

$$\frac{E_\gamma}{E_e} = \frac{4}{3}\left(\frac{E_e E_\gamma^{\mathrm{bg}}}{m_e^2}\right) = \frac{b}{3}, \tag{6.16}$$

where $b = 4E_e E_\gamma^{\mathrm{bg}}/m_e^2$ is the center-of-mass energy achieved in the electron–photon collision. The more general form of this expression, taking into account the effects of electron recoil in the interaction (the Klein–Nishina effect), which occurs when $b > 1$, is

$$\frac{E_\gamma}{E_e} = f(b)\frac{b}{1+b}, \tag{6.17}$$

where $f(b)$ is a factor of order unity describing the anisotropy of the cross section. In this case, the photon takes almost all of the electron's energy in the scattering and the energy exchange fraction becomes independent of the electron energy.

The resultant spectrum produced from the interaction of the non-thermal electron population with an isotropic photon field is determined by a convolution of the electron spectrum with that of the target photon field as follows:

$$\begin{aligned} E_\gamma \frac{dN}{dE_\gamma} &\propto \int E_e \frac{dN}{dE_e} d\ln E_e \int E_\gamma^{\mathrm{bg}} \frac{dN}{dE_\gamma^{\mathrm{bg}}} E_\gamma \frac{d\sigma(E_e, E_\gamma^{\mathrm{bg}})}{dE_\gamma} d\ln E_\gamma^{\mathrm{bg}} \\ &\approx \sigma_{\mathrm{T}} \int E_\gamma^{\mathrm{bg}} \frac{dN}{dE_\gamma^{\mathrm{bg}}} \left(\frac{E_\gamma}{\Gamma_e^2}\right) E_e \frac{dN}{dE_e} d\ln E_e \end{aligned} \tag{6.18}$$

where $E_\gamma^{\mathrm{bg}} dN/dE_\gamma^{\mathrm{bg}}$ is the target photon radiation field being scattered by the non-thermal electron population and $\Gamma_e = E_e/m_e$. Note that the term (E_γ/Γ_e^2) in the above assumes that the inverse Compton interactions only occur in the Thomson regime.

In the case of a peaked radiation target field distribution and a broad power-law electron spectrum distribution of the form $E_e dN/dE_e \propto E_e^{1-s}$, by approximating the target photon field as a delta-distribution of energy $E_{\gamma 0}^{\mathrm{bg}}$, $E_\gamma dN/dE_\gamma \approx n_\gamma \delta(E_{\gamma 0}^{\mathrm{bg}})$, Eq. 6.18 further simplifies to

$$E_\gamma \frac{dN}{dE_\gamma} \propto E_\gamma^{(1-s)/2} \tag{6.19}$$

This δ-approximation result, however, fails when the spectrum becomes sensitive to changes in the spectral slope such as in the cutoff part of the electron spectrum. Instead, in this case, the full expression in Eq. (6.18) must be used in order to allow the actual radiation field distribution to be taken into account. Adopting an electron spectrum with a cutoff of the form $E_e dN/dE_e \propto E_e^{-s} \exp\left[-(E_e/E_e^{\mathrm{cut}})^{-\beta_e}\right]$, and approximating a black body by $E_\gamma dN/dE_\gamma \propto E_\gamma^2 \exp\left(-E_\gamma/E_\gamma^{\mathrm{cut}}\right)$, the resultant inverse Compton spectrum takes the form

$$E_\gamma \frac{dN}{dE_\gamma} \propto \int_0^\infty \left(\frac{E_\gamma}{\Gamma_e^2}\right)^2 \exp\left(-\frac{E_\gamma}{\Gamma_e^2}\right) E_e^{1-s} \exp\left[-\left(\frac{E_e}{E_e^{\mathrm{cut}}}\right)^{\beta_e}\right] d\ln E_e. \tag{6.20}$$

At low energies (below the cutoff), the inverse Compton spectrum as before approximates to a power law of slope $(1-s)/2$. At higher energies, however, the spectrum has a cutoff. Overall, the resultant spectrum can be approximately described by

$$E_\gamma \frac{dN}{dE_\gamma} \propto E_\gamma^{(1-s)/2} \exp\left[-\left(\frac{E_\gamma}{E_\gamma^{\mathrm{cut}}}\right)^{\beta_\gamma}\right], \tag{6.21}$$

where β_γ describes the stretching parameters of the resultant cutoff in the photon spectrum, and for the case of inverse Compton interactions in the Thomson regime on an external radiation field, relates to that in the parent electrons via the relation $\beta_\gamma = \beta_e/(\beta_e + 2)$ (Fritz, 1989).

A set of example inverse Compton spectra are shown in Fig. 6.7. This figure shows the resultant inverse Compton spectrum produced by the electron spectrum shown in the right panel of the Fig. 6.6 scattering of a thermal photon distribution, at temperature kT, following the convolution expression given in Eq. (6.20). As appreciated from the right panel of Fig. 6.7, the difference in spectral shapes for the different values of kT is due to the onset of the Klein–Nishina suppression of the inverse Compton cross-section when $E_e E_\gamma^{\mathrm{bg}} > m_e^2$.

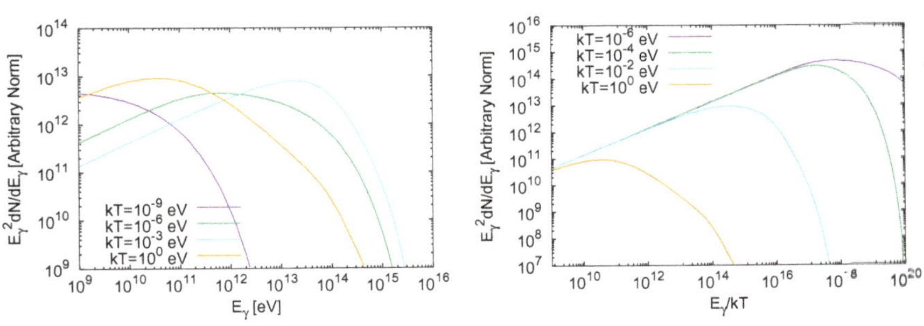

Fig. 6.7. Example inverse Compton emission spectra produced by the scattering of a steady-state electron spectrum (see right panel of Fig. 6.6) with a thermal photon distribution at temperature kT.

SSC scenarios: Alternative to the above result, the target radiation field for inverse Compton interactions may instead be the lower-energy non-thermal synchrotron radiation. In this case, the synchrotron spectrum inherits a spectrum and cutoff of the form described in Eq. (6.21). Consequently, to determine the inverse Compton spectrum produced, this new target photon cutoff must be convoluted with that of the electrons (i.e. similar to that in Eq. (6.20)). In this case, provided the subsequent inverse Compton interactions with the synchrotron photons occur in the Thomson regime, the subsequent cutoff in the inverse Compton photons relates to that in the parent electrons via the relation $\beta_\gamma = \beta_e/(\beta_e + 4)$ (Lefa *et al.*, 2012).

For all cases above, should electron energies be reached at which the onset of inverse Compton interactions in the Klein–Nishina regime occur, the subsequent cutoff in the photons simply reverts back to directly reflecting that of the parent electrons, i.e. $\beta_\gamma = \beta_e$ (Lefa *et al.*, 2012).

The *temporal* structure of the emission also encodes information about the nature of the emission process at work. As an example case, we consider the synchrotron and inverse Compton emission that follow from an increase in the number of accelerated particles in the system. It is here assumed that the emission from such particles is in quasi-steady state with the accelerated particle population. In such a case, there is rather distinct evolution of the synchrotron and inverse Compton peaks in the emitted spectra, depending on whether the inverse Compton emission is an SSC (inverse Compton scattering of the synchrotron radiation field) or an external Compton emission (inverse Compton scattering of the ambient radiation field present). In the former of these scenarios (SSC), the target radiation field for inverse Compton scattering increases proportionally to the increase in the population of the accelerated particles. Subsequently, a quadratic increase in the inverse Compton emission is expected. In the latter of these scenarios (external Compton), however, only a linear response of the brightening of the inverse Compton emission is expected.

Variability in the emission also encodes information about the maximum size of the emission zone. Assuming that such emission is produced by a relativistic emitter, the timescale of the variability relates to the maximum co-moving size of the emitter R′ via

$$R' < c\delta t_{\rm var}, \tag{6.22}$$

where $\delta = 1/\Gamma(1 - \beta\cos\theta)$. Note that for $\theta \approx 1/\Gamma$ (i.e. the emitter coming toward the observer, close to the line of sight of the jetted emission), $\delta \approx \Gamma$.

Thus, overall, the observed variability of the source for such an observer located at such a position is compressed relative to that of the emitter's intrinsic variability in its own reference frame. The fast variability can be attributed to either leptonic or hadronic emission (Barkov *et al.*, 2012; Bühler and Blandford, 2014). Precise measurement of the light curve can result in fine determination of (mean) magnetic field strengths and Doppler factors of the gamma-ray emitting blobs.

Finally, following the gamma-ray production by a source, the VHE photons produced may subsequently undergo pair-production interactions with the radiation field, either within the source or out in the extragalactic environment. Since this process has a threshold energy, it can only proceed when $E_\gamma E_\gamma^{\rm bg} > m_e^2$, where E_γ is the energy of the energetic photon, $E_\gamma^{\rm bg}$ is the energy of the target photon and m_e is the electron mass. Furthermore, with the cross-section for this process peaking at energies just above the threshold, a useful rule of thumb follows that a photon of energy E_γ predominantly interacts with a background target photon of wavelength $\lambda_{\mu m}^{\rm bg}$, such that $E_\gamma^{\rm TeV}/\lambda_{\mu m}^{\rm bg} \approx 1$.

6.5. Lessons Learned: From Detection of Gamma-ray Sources Above a Few Tens of GeV to Few Tens of TeV

The new generation of Cherenkov telescopes has unveiled a large number of gamma-ray sources related to both Galactic and extragalactic accelerators (see Chapter 2 for a compilation of these). These accelerators are related to late stages of massive stars, in the case of Galactic sources, and to the active nuclei of other galaxies giving rise to relativistic jets and/or large-star formation, in the case of extragalactic sources. Within these systems, several regions are susceptible to accelerating particles up to TeV energies, and subsequently brightening in gamma rays through the different processes outlined in the previous sections.

Galactic lessons: In the Milky Way, active regions of stellar formation supply a large reservoir of photon fields and promote the formation of objects with large magnetic fields such as *neutron stars* or *Galactic black holes.* They alter the InterStellar Medium (ISM) in the surroundings, powering relativistic shocks and inflating PWNe. These compact objects can also be found in *binary systems*, where the shock between the compact object outflow and companion wind creates favorable conditions for the acceleration of particles and production of gamma rays. These stellar regions also host shocks in which Fermi shock acceleration can take place, created by the ejecta of *supernova* explosions (*supernova remnants, SNRs*) or dense winds of massive stars. A schematic view of the identified Galactic gamma-ray sources is shown in Fig. 6.3.

In addition to observations of these objects within our own Galaxy, their observations in other galaxies have revealed further insights. Specifically, such objects have also been observed locally in the Large Magellanic Cloud (LMC) satellite galaxy (Abramowski *et al.*, 2015b). Several objects have been resolved in the LMC, including several SNRs, PWNs and a possible star cluster. This discovery allows the comparison between different structural evolution scenarios by means of comparing the gamma-ray emission from these systems.

Extragalactic lessons: HE emission from different types of AGNs has been identified, demonstrating a wide range of phenomenology with regards to the variability and spectral characteristics of the radiation. In fact, most of the extragalactic objects detected in the VHE gamma-ray domain are blazars (BL Lacs and, to a lesser extent, FSRQs), for which the jet points toward us, beaming the observed emission in the forward direction as well as boosting its observed energy. Besides blazars, a population of their unbeamed counterparts, *radio galaxies*, has also been observed to be bright in gamma rays. The number of the detected radio galaxies in gamma rays has increased up to four (maybe five) (Madejski and Sikora, 2016), all of them of FR I type. Contrary to their beamed counterparts, for which the emission seems to be clearly related to the jet activity, the large angle between the jet direction and the line of sight for radio galaxies poses doubts about the region where the gamma rays originate. The supermassive black holes in their center and their associated powerful jets are both prime candidates for the emission. The third extragalactic source class discovered in the gamma-ray domain is *starburst galaxies* (Acciari *et al.*, 2009a; Abramowski *et al.*, 2012). No powerful jets are present in these galaxies, suggesting that the gamma rays result instead from the ensemble of emission of sources embedded within them.

General remarks: In the following subsection, we will discuss the lessons learned during the past years from observing Galactic and extragalactic sources in the VHE regime and put them in context of the acceleration, propagation and emission summarized earlier. The structure is as follows: first, we review observations and results that provide clues on relativistic flows in low-density, highly magnetized plasmas (in Section 6.5.1). We discuss within this section the lessons learned from the study of spectral features, on the one hand, and temporal variability, on the other. Second, we assess the results concerning relativistic shocks in clumpy, magnetized plasmas (in Section 6.5.2). This scenario accounts for the optimal conditions to produce VHE radiation, therefore a larger dissertation on the spectral imprints is included, in addition to the variability imprints' subsection. Thus, in the spectral subsection, we included a general description of the state-of-the-art research lines, such as time-dependent modeling, magnetic evolution and the injection spectrum. We discuss separately an additionally relevant topic in the field: the study of distant AGNs as probes of the early Universe. In Section 6.5.3, we describe scenarios in which the relativistic shocks are believed to develop in high-density, weakly magnetized plasma at different space scales: Galactic and extragalactic. The last subsection on the lessons learned reviews non-relativistic shocks, which in general result in steady sources, for which we describe the astrophysical implications of several Galactic and extragalactic observations.

6.5.1. *Observations of relativistic shocks in low-density, highly magnetized plasma*

Ultra-relativistic cold flows are ejected by magnetized Galactic or extragalactic compact objects, which can accelerate particles to very high energies. Extragalactic active nuclei exhibit more complex dispositions, with various candidate regions in the relativistic jetted structures located either close to the black hole or extending up to Mpc scales far beyond that of the host galaxy. In particular, the region surrounding the black hole enclosing the black hole magnetosphere could provide conditions in which ultra-relativistic shocks propagate in a magnetized plasma (Rieger and Mannheim, 2000), similar to the phenomenology applied to neutron-star magnetospheres. In the following subsection, we will describe the different scenarios in which relativistic shocks interacting with low-density/highly magnetized plasma can be proven.

6.5.1.1. *Spectral imprints*

Magnetized objects are characterized by the large amount of the synchrotron component of their emitted photon spectrum. In our Galaxy, such an emission signature is observed from *pulsars* and *microquasars*. No firm detection has been found from the latest population using Imaging Atmospheric Cherenkov Telescopes (IACTs) despite extensive follow-up (Acero *et al.*, 2009b; Aleksić *et al.*, 2011a, 2010a). A large emission related to the jet of SS 433, however, has been recently reported by the HAWC Collaboration (Abeysekara *et al.*, 2018). SS 433 is a binary system containing a supergiant star whose Roche lobe is overflowing with matter accreting onto a compact object (either a black hole or neutron star). Two jets of ionized matter with a bulk velocity of $\sim 0.26c$ extend from the binary. The very-extended emission coincides with the extension of the jet at around 40 pc from the compact object. The low-density medium in which the jet evolves seems to lack enough molecular material to explain the emission by hadronic mechanisms given the total energetics of the microquasar. Leptonic emission seems to better explain the energetics and spectral distribution, assuming a magnetic field of $\sim 16\mu$G. In such a case, HE electrons may transport in a magnetized medium via diffusion or advection. The distance traveled via diffusion within the cooling time of electrons with energies of the order of PeV is of the order of 35 pc, while if one assumes an advecting propagation with the jet flow for a velocity of 0.26c, a much smaller distance of a few pc is obtained for the same energies. The two distances seem to be smaller than the one in which the TeV excess is located, suggesting that the acceleration of electrons might happen somewhere within the jets and not near the compact objects. Deeper and more detailed observations on the morphology and spectrum in different regions will provide more clues about how particles are accelerated and cooled in the surroundings of the microquasar.

In fast-rotating neutron stars, the curvature–synchrotron pulsed emission is usually explained by the magnetic acceleration of charged particles in low-density plasma regions or "gaps" created within the pulsar magnetosphere. The strong magnetic field pulled these charged particles (mainly electrons and positrons) from the star surface and they travel along the strong magnetic field lines (Cheng *et al.*, 1986a,b; Romani and Yadigaroglu, 1995; Takata *et al.*, 2006; Arons and Scharlemann, 1979). The dipolar magnetic field is disrupted in regions beyond the last closed field line, which defines the light cylinder, at which the velocity of the co-rotating plasma around the black hole would hit the limit of the speed of light. In those regions, the accelerated particles can escape, creating a low-density plasma in which an electrical potential difference is induced. In such gaps, the electrons and positrons are expected to create particle cascades, increasing the density of particles which are accelerated to relativistic velocities and subsequently injected into the surrounding medium. While moving in the high magnetic field medium, these particles lose energy via synchrotron–curvature processes (see, e.g., Torres, 2018 and references therein), quickly losing energy and hence displaying a sharp exponential cutoff in the photon spectrum at energies of a few GeV. This small volume, however, drives the particle flow that surrounds it, forming a relativistic wind that extends from a few thousand kilometers up to ~ 0.1 pc (10^{12} km!), eventually ending in a relativistic shock in which the bulk of the particles carried in the cold wind flow are thermalized. This cold ultra-relativistic wind extends from the light cylinder to the termination shock. As a result of the electromagnetic structure of an oblique magnetosphere, current sheets of particles might appear along the last closed magnetic lines and/or on the equatorial plane out at distances beyond the magnetosphere. These regions are, in theory, suitable places to accelerate particles via magnetic reconnection (Cerutti and Philippov, 2017; Mochol and Petri, 2015), resulting in a rapid transfer of magnetic energy to kinetic energy and giving rise to a fast, cold particle outflow. Up to a few years ago, the spectra of GeV pulsars were well-described with a sharp exponential cutoff at an energy scale of a few GeV, discouraging the study with Cherenkov telescopes at higher energies. The development of more sensitive instruments with lower-energy thresholds (Aleksić *et al.*, 2012d; Abdalla *et al.*, 2018d; Park *et al.*, 2015), however, coupled with deep observations of the most promising pulsar, finally resulted in an adequate sensitivity to access the sub-100 GeV region. The first detection of pulsars using Cherenkov instruments — in particular, the detection of the Crab, Geminga and Vela pulsars — made evident that the spectral shape above 1 GeV cannot be described by a super-exponential cutoff, as predicted by models that placed the emission region close to the surface of the pulsar, where large radiative losses are expected. Rather, a sub-exponential cutoff was clearly preferred, pointing to larger gaps or larger distances from the magnetized star at the particle acceleration site. Moreover, longer observations of bright GeV pulsars, namely, the Crab pulsar and the Vela pulsar (Aleksić *et al.*, 2015b, 2011c; Abdalla *et al.*, 2018d), have led to a general questioning of the previously established doctrine. These observations have revealed that, indeed, the pulsed emission extends up to energies of even a few tens of TeV (see Fig. 6.8 from Aleksić *et al.* 2015b). This unexpected result has drawn considerable attention in the astrophysics community and has triggered a large number of theoretical research projects aiming to unveil the mechanisms powering GeV/TeV-pulsed emission from pulsars. The electron population behind this extremely energetic radiation should have Lorentz factors greater than a few times 10^{6-7}. According to several authors (Bogovalov, 2014; Bednarek, 2012; Tchekhovskoy *et al.*,

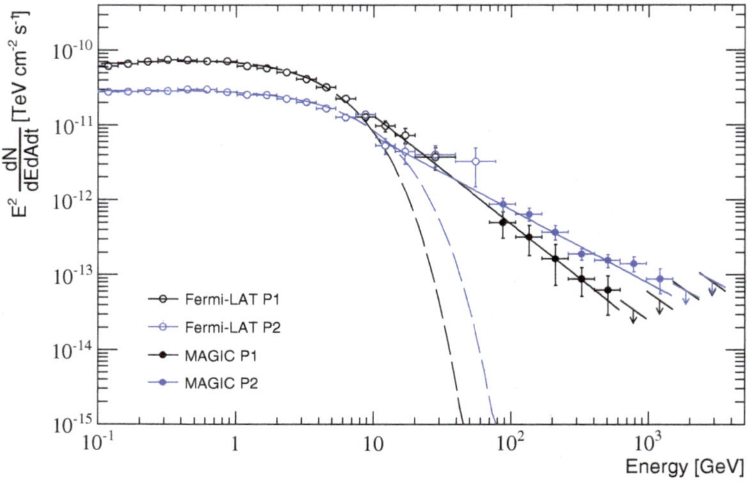

Fig. 6.8. Phase-folded spectral energy distribution of the Crab Peak 1 (black circles) and Peak 2 (blue circles) at high and very-high energies (open and filled circles, respectively). More information can be found in Ansoldi *et al.* (2016).

2013), several mechanisms (e.g. magneto centrifugal mechanisms and large near-the-light-cylinder gaps) can accelerate electrons to a Lorentz factor of more than 10^7 near the light cylinder in pulsars. The electron flow, in the form of Poynting flux, can be converted to gamma rays near the light cylinder via several mechanisms, namely, inverse Compton scattering of primary and/or secondary (even tertiary) electrons, and synchrotron–curvature radiation. A fraction of it can also be transferred to the cold ultra-relativistic wind in which particles are accelerated and radiate gamma rays via inverse Compton scattering on soft-pulsed X-ray photon fields or within the expanding sheets if a striped wind is formed. To achieve this HE observed in the synchrotron–curvature scenario — which is usually appealed to explain the gamma-ray emission up to a few GeV (Torres, 2018) — a curvature radius of $R_c \geq 200 R_{LC}$ is required (in the extreme case in which the magnetic-field-aligned electric field approaches the strength of the magnetic field, i.e. $\eta \sim 1$). In the case of the Crab pulsar, a very soft spectrum extending beyond TeV energies has been found. The spectrum seems to connect smoothly with the spectrum measured by gamma-ray satellites. If the TeV emission is attributed to curvature radiation on the magnetosphere, however, it would request a curvature radius in the emission region of several orders of magnitude larger than the typical radius, which is believed to be 0.3 to 2 times the light cylinder radius. A second possibility implies a second component due to inverse Compton emerging in the TeV regime, though still facing many challenges; which target photons are off-scattered, where this scattering takes place and how this emission manages to also avoid absorption are questions which still remain unresolved.

6.5.1.2. *Variability imprints*

The rotating beams ejected from compact, spinning magnetized objects result in variable sub-100 GeV and TeV emissions, which are usually associated with the interaction of those beams with the low density in the magnetosphere or ergosphere, or in the close surroundings, as explained above.

The different timescales measured and type of radiation point to different acceleration regions. Indeed, the fast variable and periodic emission measured, synchronized with the rotation period of the pulsar, demands an emission region very close to the rotating neutron star in the region dominated by the strong magnetic field of the order of 10^{8-14} G or in the wind. In fact, to avoid a time lag due to the expansion of the layers within the wind region or in different regions of the magnetosphere, the GeV and TeV emissions should arise from the same (or very nearby) locations. This region, however, cannot be too close to the pulsar magnetosphere, otherwise the electrons and positrons would be quickly absorbed before they can escape and radiate given the large particle density within it. Thus, the measurement of pulsed emission in the TeV regime gives important clues about the structure of the region in the vicinity of the magnetosphere from which this emission originates.

In a similar fashion to neutron stars, the fast variability found in some AGNs, in particular in the object IC 310 (Aleksić *et al.*, 2014a) down to minute timescales, has led some authors to describe the gamma-ray emission in a context similar to the mechanisms related to the neutron-star magnetosphere described above. Causality constrains the size of the emission region to be smaller than 20% of the gravitational radius of its central black hole if there is relativistic bulk motion. Ultra-fast variability can also be attributed to jet plasma moving with a speed close to the speed of light $\beta \sim 1$ within a small angle to the line of sight, leading to a bulk Lorentz factor $\Gamma > 1$ (shock-in-jet models). The relativistic boosting can actually affect the timescale of flux variations when the moving shock plasma producing the emission moves in the same direction as the emitted radiation, leading to a shortening of the observed variability

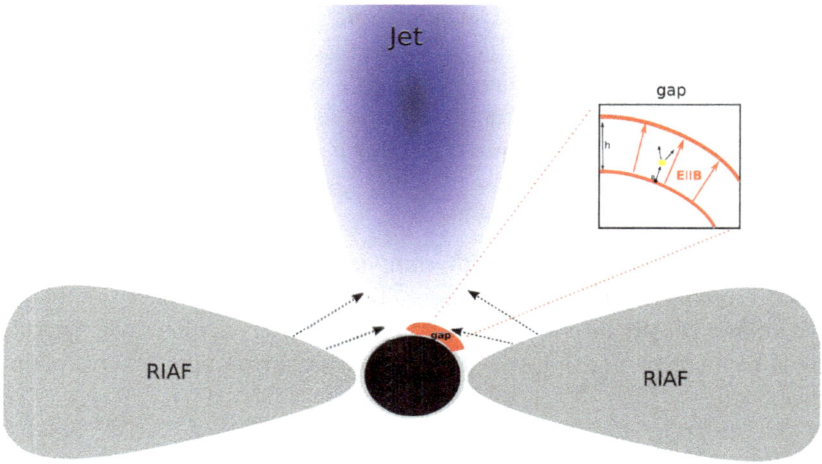

Fig. 6.9. Sketch of the unscreened electric field regions (vacuum gaps) that may form close to the horizon in under-dense black hole environments. From Rieger (2017).

timescale. IC 310, however, is a peculiar radio galaxy for which the orientation angle of the jet to the line of sight was found to be in the range $\theta \sim 10^{\circ}$ to 20°, facing difficulties to be explained using the shock-in-jet model. A different interpretation (see also the next section) involves jets shocking with dense material (Barkov *et al.*, 2012). In the magnetosphere interpretation (see Fig. 6.9), particles accelerating to ultra-high energies can initiate electromagnetic cascades, leading to a multiplication in the number of charge carriers in the region until their current shortcuts the gap, in a similar way to the gaps in the pulsar magnetosphere. These excess particles are then swept away with the jet flow until the gap reappears. The gaps in this case, however, are associated with the base of the radio jet. The potential drop across the gap is significantly larger than 10 TeV and the observed spectrum results from unsaturated electromagnetic cascading, implying a rather stable spectrum with a photon index close to 1.9. This attractive possibility may help in understanding jet formation in black holes.

6.5.2. *Observations of relativistic shocks in clumpy, magnetized plasma*

Relativistic flows injected in the medium by the compact objects discussed above propagate rapidly, reaching very-large-distance scales, hundreds of parsecs in the case of Galactic sources and Mpc in the case of extragalactic sources. Throughout this propagation, they evolve through media of different density depending on the complexity of the source. The different media and acceleration and emission region are reflected in the *variability* and *spectral characteristics* of the emissions. The study of the interaction of the relativistic flows with different media in different environments (both Galactic and extragalactic) through their time imprints and spectral shape then provides crucial information as to their astrophysical properties.

6.5.2.1. *Variability imprints in Galactic and extragalactic sources*

Fast variability is often related to relativistic jets that provide a comfortable energetic argument to the very often large and quick rise of the gamma-ray emission. In general, fast variability is usually associated with AGNs, whereas transitional behavior in Galactic sources is often related to a periodic or orbital nature. Variability in Galactic sources with relativistic flows that are generally spatially resolved, however, might provide essential information to explain the physics of jets in AGNs. For instance, Galactic flows in pulsars and jets in microquasars can reach from a few km to tens of parsecs, inflating magnetized bubbles in the low-density medium left behind by the hosting SNR explosion or in the ISM, in some cases forming a large nebula of magnetic fields and energetic particles around compact objects. These nebulae are also strongly affected by the evolution of the remnant left behind by the explosion and the medium in which this happens. In comparison to other sources of relativistic plasma, the surroundings of pulsars, and in particular PWNe, can be resolved in great detail Scales close to the gyro-radius of the highest-energy electrons in this system can be resolved by current radio, optical and X-ray telescopes, whereas the angular resolution reached by HE telescopes is good enough to observe their very-extended HE emission. These systems provide optimal test benches to study the behavior of relativistic outflows, which are thought to also be present in other HE sources, such as gamma-ray bursts or AGNs (Berger, 2014; Massaro *et al.*, 2015). Indeed, some of those PWNe also display rapid gamma-ray flares (Bühler and Blandford, 2014), which can be related to magnetic reconnection or shock accelerations. This remains an important topic for the next generation of Cherenkov telescopes (Acharya *et al.*, 2019).

In turn, the time domain in AGNs has uncovered perhaps the most revealing information to date about the nature of AGN emission. Presently, the shortest non-thermal variability timescale of an AGN observed at

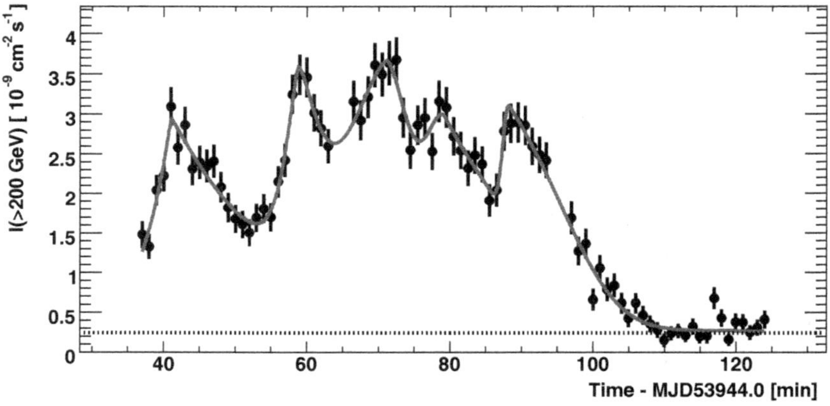

Fig. 6.10. The integral flux above 200 GeV observed from PKS 2155-304 on MJD 53944 vs. time. The data are binned in one-minute intervals. More details can be found in Aharonian *et al.* (2007).

VHE was that of the minute timescale variability observed of the BL Lac PKS 2155-304 in 2006 (Aharonian *et al.*, 2007) (see Fig. 6.10). Such variability places the constraint $R/\delta < 0.3$ A.U. As a means of comparison, the Schwarzschild radius for this system is estimated to be $R_{\rm Schwarz.} =$ 30 A.U. Thus, similar to the case discussed earlier, for reasonable values of the jet Doppler factor, the remarkable result is found that an emission site smaller than the Schwarzschild radius of the black hole appears to be required. Furthermore, beyond these general arguments, the variability timescale should ultimately connect to the competing timescales giving rise to the flare, e.g. the acceleration time ($t_{\rm acc}$) and the escape time ($t_{\rm esc}$). The detection of short timescale structure emission therefore conveys immense information about the nature of the acceleration site.

Another example is the fast variability exhibited by PKS 1222+21 (see Aleksić *et al.*, 2011b and Fig. 6.11). In contrast to BL Lac PKS 2155-304, PKS 1222+21 is an FSRQ with intense infrared emission from the broad-line region (BLR) surrounding the core on multi-parsec scales and leading to gamma-ray absorption of basically all TeV gamma rays produced inside the BLR, i.e. close to the black hole, suggesting the observed emission must come from a very small region outside of the BLR. Beyond the minimum variability time of AGNs, presently only a timescale set by the instrument sensitivity, the variability time structure of AGNs is also analyzed by power-spectrum analysis via the application of Fourier (or Laplace) transforms of the amplitude light curve of long-term data sets. Such an analysis can

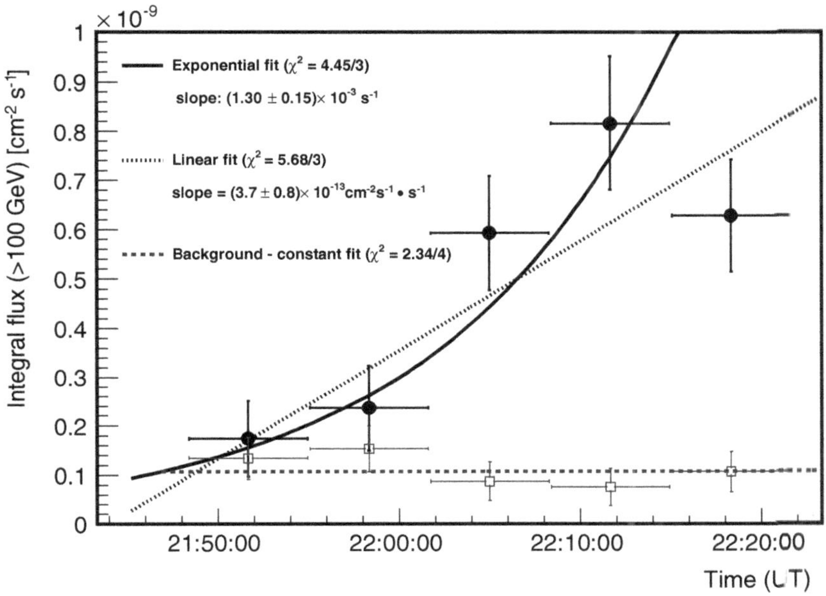

Fig. 6.11. PKS 1222+21 light curve above 100 GeV in six-minute bins (black filled circles). The observation was carried out on MJD 55364. The black solid line is a fit with an exponential function and the black dotted line is a fit with a linear function The gray open squares denote the fluxes from the background events and the gray dashed line is a fit with a constant function to these points. More details can be found in Aleksić *et al.* (2011b).

only be performed for well-studied AGNs over long timescales. At present, such studies at very high energies have only been carried out for the blazar PKS 2155-304 (Abdalla *et al.*, 2017a). The findings from this study indicate that the temporal structure of the variability of this blazar has a power-law form ($dP/df \propto f^{-\alpha}$), with a spectral slope consistent with that of pink noise structure ($\alpha = 1$) being found. Contrary to this general finding, however, during the giant outburst in 2006, this temporal structure changed and became red noise-like (i.e. $\alpha = 2$), with the light curves random walking during this period. Whether a broken power-law spectrum is implicit in the noise structure from this AGN or not, however, will require further observations. Regardless of whether such a structure exists, it is already apparent that no fundamental flare timescale exists within the variability of the gamma-ray emission from AGNs. It is therefore to be appreciated that the language of "flares" adopted by the community is used only to characterize a bump in the temporal emission rather than have a well-defined physical meaning.

The observation of fast variability imposes different constraints when observed in different source types. In blazars where the jet is strongly beamed toward the observer, the variability has been attributed to structures within the jets (Giannios *et al.*, 2010; Tavecchio and Ghisellini, 2008) or jet-cloud interactions, where the clouds may originate from stellar winds (Barkov *et al.*, 2012; Vieyro *et al.*, 2017). For flat-spectrum radio quasars (FSRQs) whose central region is known to harbor intense radiation fields (BLR, see above), further challenges of a short variability timescale are expected. As an example case, the flaring GeV emission of the FSRQ 3C 279 in 2015 exhibited minute timescale variability (Aleksić *et al.*, 2011b; Ackermann *et al.*, 2016b). Should the emission site, therefore, sit within the BLR, the VHE emission produced must traverse this intense radiation field, allowing for gamma–gamma absorption to occur. Estimates of the intensity of the radiation field within the BLR suggest that significant VHE gamma-ray attenuation should occur if their emission site sits at radii within this region (Tavecchio and Ghisellini, 2012). The spectrum emitted during flaring outbursts by FSRQs, therefore, provides additional information about the location of the central emitter, complementary to the probe provided in the temporal structure of the emission. The presence of strong radiation fields leading to large absorption in the gamma-ray spectrum might also affect radio quasars. As discussed earlier, the case of radio quasars is particularly interesting since no significant beaming is expected from radio galaxies in the direction of the observer. For cases like M87, a flux doubling time of 1–2 days in TeV gamma rays was reported by MAGIC (Albert *et al.*, 2008e) and VERITAS (Acciari *et al.*, 2008) as well as in joint H.E.S.S., MAGIC and VERITAS observations (Acciari *et al.*, 2009c), see Fig. 6.12. Again, as discussed earlier, the intrinsic volume in the jet of M87, in which the TeV gamma-ray emission takes place, must be similar to the regions of the fastest flares observed in blazars. This similarity to blazars — seen also in the non-thermal emission at lower energies — explains why some experts call radio galaxies seen at TeV gamma rays "misaligned blazars". Adding another piece to the puzzle, the collective spectral observation of the population of AGNs by the Fermi Large Area Telescope (Fermi-LAT) (in the sub-GeV regime) and by IACTs during flares has revealed a lack of evidence of internal absorption, which should be featured in the global spectral energy distribution. If the emission region is indeed external to the central accreting object, it could be due to the

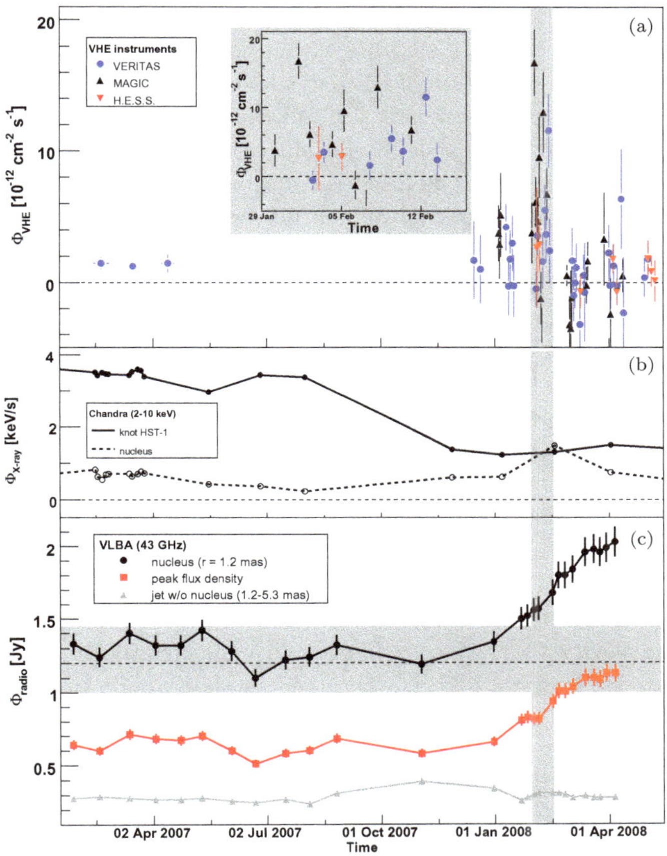

Fig. 6.12. Combined M87 light curves from 2007 to 2008. (a) VHE gamma-ray fluxes (E > 0.35 TeV, nightly average) showing the H.E.S.S., MAGIC and VERITAS data. The inlay shows a zoomed version of the flaring activity in February 2008; the time span is indicated by the gray vertical box in all panels. (b) Chandra X-ray measurements (2 to 10 keV) of the nucleus and the knot HST-1 (28). (c) Flux densities from the 43-GHz VLBA observations. The shaded horizontal area indicates the range of fluxes from the nucleus before the 2008 flare. Whereas the flux of the outer regions of the jet does not change substantially, most of the flux increase results from the region around the nucleus. See Acciari *et al.* (2009c) for details.

interaction between a hadronic jet and dense gas clouds traveling through the jet. In this scenario, the variability is explained based on emission produced by proton–proton (pp) interactions of accelerated protons located at the jet–cloud interface. This does not mean that hadronic scenarios are preferred; in leptonic scenarios, short flares can be explained by magnetic reconnection, which leads to particle acceleration and gamma-ray emission in smaller volumes than a disk filling the entire jet-opening angle.

Similarly, the acceleration of HE particles can occur in shocks produced by the collision of the relativistic jet and the wind of a star immersed in the jet's plasma. At these shocks, efficient particle acceleration can occur, with the relativistic electrons accelerated subsequently, efficiently cooling via inverse Compton scattering on the OB star's thermal radiation field. In this way, such objects can operate as very-efficient VHE gamma-ray sources.

Finally, one last consideration in explaining the rapid variability of TeV-emitting AGNs and avoiding absorption of the TeV photons by gamma–gamma pair production has been developed by dividing the relativistic jet into several sub-regions that move relatively to the main jet (Giannios *et al.*, 2010; Michel, 1969). The jets-in-a-jet approach ensures a small size of the emitting region since only a small fraction of the region is beamed into a narrow cone, boosting the Lorentz factor by scaling the two jets' relativistic moves.

Those different solutions should also be reflected in the spectral energy distribution of the observed flares (sometimes hampered by the lack of sensitivity to produce a high-quality spectrum). In the following subsection, the spectral features observed in the gamma-ray sources and the implications of propagation and emission are discussed.

6.5.2.2. *Spectral imprints in Galactic and extragalactic TeV sources*

General considerations: In general, many of the bright relativistic TeV sources observed, which are related to the clumpy medium, can be reasonably well explained by inverse Compton of energetic electrons upscattering soft photon background fields. A steady one-zone, power-law electron-injected spectrum model is often used for simplicity; this approach, however, usually fails when comparing the precise results obtained by the current IACTs with (simultaneously in in the case of AGNs) multi-wavelength data. Several trends are common to these efforts:

- Often, a power-law energy spectrum does not reproduce the spectral energy distribution, and instead, a broken power-law distribution has to be invoked, appealing to different arguments for different source types.
- Equipartition between magnetic and kinetic fields and homogeneous magnetic fields within the emission region are generally used when modeling

the observed gamma-ray emission. Current results show, however, for both extragalactic and Galactic sources, that this solution differs from reality, pointing to either a particle-dominant environment far from equipartition and/or an inhomogeneous magnetic field in the region.

- To the points above, we should add the need of including a time dimension to account for the continuous injection of particles in time (for steady sources that shine during tens of kiloyears) and to account for changes in the shape of the gamma-ray emission (in peak position, highest gamma-ray energy reached and flux level) in short (both minutes and years) timescales.

Time-dependent models in Galactic and extragalactic sources: Time-dependent leptonic models were found to be fundamental when studying the TeV observed on *Galactic* PWNe. In these magnetized plerions, particle acceleration and emission can be traced with the flow propagation time. To illustrate the effect of considering a continuous injection over a long time, in Fig. 6.13 we show an example of leptonic emission at different evolutionary stages of a PWN. As discussed in Section 6.3, the effect of the cooling losses is clearly displayed: the losses due to synchrotron radiation are

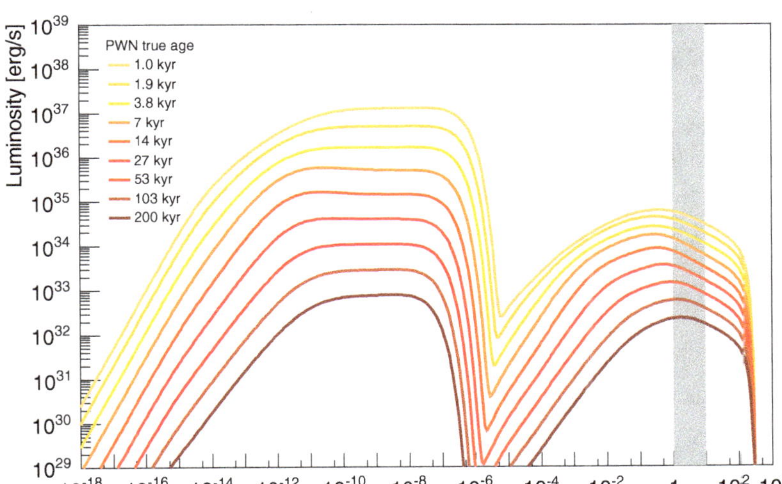

Fig. 6.13. Example of time-evolution modeling of a PWNe SED, ranging from 1 kyr to 200 kyr. From Abdalla *et al.* (2018k).

strongly reduced, as the magnetic field decreased in the expanding nebula, whereas the effect in the inverse Compton peak is not so drastic in the first years, as the radiation is ruled by the inverse Compton losses. PWNe are, in fact, one of the most efficient gamma-ray emitters in our Galaxy in terms of their efficiency in converting rotation energy into gamma-ray energy and constitute the largest population to date of identified Galactic sources (Abdalla *et al.*, 2018k). Spatially independent models in which the time evolution of the electron spectrum is taken into account (Martín *et al.*, 2012; Tanaka and Takahara, 2013; Fang and Zhang, 2010; Gelfand *et al.*, 2009) can satisfactorily fit the global spectrum of PWNe by assuming different injection electron spectra, target photon fields (namely, far-infrared, near-infrared and CMB) and magnetic-field strengths. Such models can unveil general trends, but fail to satisfactorily explain the particular details of the observed emission, in particular in X-rays and TeV energies. This is the case for the best-studied nebula, the Crab nebula, which exhibits the strongest flux for which dedicated models — applied to the latest results obtained with the MAGIC telescopes (Aleksić *et al.*, 2015a) — fail to reproduce the broadband shape.

Likewise, time-dependent models are also found to be important in explaining the GeV and TeV emissions of some evolved SNRs (Brose *et al.*, 2020; Schure *et al.*, 2010; Helder *et al.*, 2012). Recent detailed modeling efforts now also consider the time evolution of the CR spectrum in SNRs to account for the maximum energy reached, which is derived from the HE observations.

In the case of *extragalactic* sources, the large spectral variations found during the flaring activity suggest that the actual emission does not take place in a homogeneous magnetic field, and maybe not in a single emission region but in multiple zones, with the emission detected being a superposition of these. In fact, simultaneous observations with H.E.S.S., Chandra and the Bronberg Observatory were performed on a particularly bright outburst of the source PKS 2155-304, which allowed the direct comparison between the VHE flux and the X-ray flux at different flaring moments. It was found that the gamma-ray flux traces the variations in the X-ray flux far more than quadratically, namely, approximately as $F_{\mathrm{VHE}} \propto F_{\mathrm{X}}^3$ (see Fig. 6.14, from (Aharonian *et al.*, 2009f). Indication of a super-quadratic relation was also reported for Mkn 421 (Fossati *et al.*, 2008). These results disfavor even further a simple one-zone SSC scenario. The injected electron spectrum cannot change the shape and energy range freely (as one would model with snapshots of a one-zone SSC model),

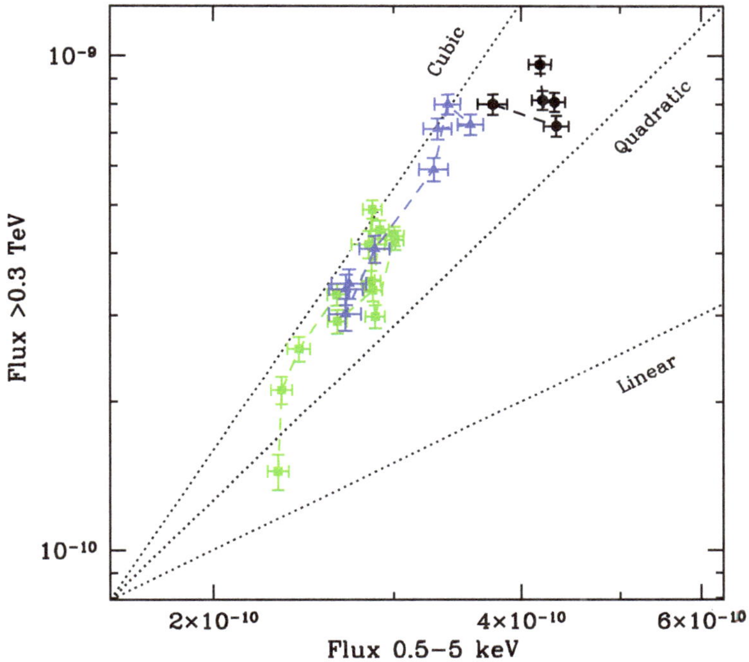

Fig. 6.14. Plot of the gamma-ray vs. X-ray flux correlation in a log–log diagram and units of erg cm^2 s^{-1}. For more details, see Aharonian *et al.* (2009f).

but these changes must be made self-consistent in a time-dependent manner. Although this was already clearly shown more than a decade ago (Krawczynski *et al.*, 2002), one-zone SSC models continue to remain popular. On the one hand, most of the data sets published arise from relatively short exposures and multi-wavelength data quality of limited statistical significance, which individually cannot rule out this model class. On the other hand, the parameter space opened with more complicated two/multi-zone models is, at the moment, too big for firm conclusions on the reasons for variability and spectral changes. More data with higher sensitivity and a better angular resolution are required to make significant progress in the domain of time-dependent modeling.

Another possibility is to self-consistently follow the electron distribution accelerated via Fermi-I (i.e. shock) or Fermi-II (i.e. stochastic) processes in the gamma-ray emission region. An example of the Fermi-II acceleration is developed by Asano *et al.* (2014), which follows the evolution of the electron

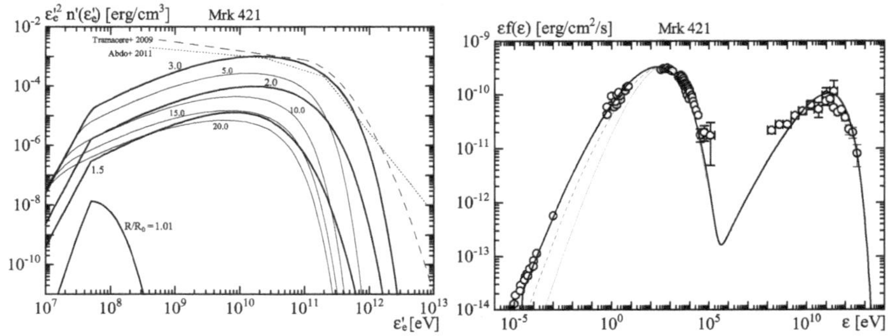

Fig. 6.15. Time-dependent leptonic model for gamma-ray emission from AGN Mrk 421: the electron injection (left) and resulting steady photon spectrum (right). See text for details. Figures from Asano *et al.* (2014).

energy distribution and photon production. The electrons are injected at a distance R_0 from the jet base into an emission region of the jet, and the evolution of electron population, as well as the emission photon spectrum, is followed as a function of time by dividing the jet into thin shells moving outwards with a Lorentz factor Γ. In such a description, the electron distribution is obtained with the effects of the injection, acceleration and energy losses being taken into account to determine the evolution of the energized particle population's energy distribution. An example of electron spectrum development inside the emission zone is shown in Fig. 6.15, left plot. The electron spectra during the acceleration process are denoted by thick lines, while those after the end of the acceleration are denoted by thin lines. The spectral shapes of other one-zone leptonic models are plotted (in arbitrary units) for reference. The thin dotted line is the DBP model in Abdo *et al.* (2011a) and the thin dashed line is the log-parabolic model in Tramacere *et al.* (2009). The resulting steady-state photon spectrum for the SSC model (i.e. without assuming external target photon fields for upscattering) is shown in Fig. 6.15, right plot. The data points are partially extracted from the data of the 4.5-month campaign (Abdo *et al.*, 2011a). Naturally, by varying environmental conditions along the jet axis as the electrons propagate, such a model can be used to describe AGN flares.

Consequence of spatial extent of emission region in Galactic and extragalactic sources

The simple modeling mentioned above may provide a poor description of reality considering the large spatial extension scales of the acceleration and emission regions. Even for the best angular resolution achieved using arrays of multiple IACT telescopes reaching less than 3′ (Hofmann, 2020; Acero *et al.*, 2010b), point-like objects located in the Galaxy could be as large as 3–8 parsecs, whereas if located further away, the dimension of the region within a few arc-minutes reaches thousands of parsecs. Nonetheless, the long cooling times (see above) involved in gamma-ray radiation processes result in very-large emission regions, which can be resolved if this region is close enough. The extended, resolved gamma-ray regions provide a unique scenario to investigate not only the propagation and re-acceleration of HE particles, but also the target or medium in which they propagate. Indeed, assuming that this target, which is scattered off in inverse Compton processes, is dominated by the synchrotron photon component may not be realistic for the majority of the detected gamma-ray sources. In general, the comparison between the VHE radiation attributed to inverse Compton and the synchrotron component related in principle to the same electron distribution result in a particle-dominated regime (kinematic jets in the case of blazars and kinematic winds in the case of relativistic winds in PWNe). In the case of *extragalactic* sources, the required magnetic field should be quite low, otherwise the particles would lose (by synchrotron cooling) their energy too quickly without having enough time to accelerate to sufficiently high energies. Consequently, 4–5 orders of magnitude more energy is contained in the kinetic energy of the particles compared to the energy in the magnetic field. This would imply that either the conversion from magnetic energy to the kinetic one is happening very fast (within the inner parsec scale), or that the TeV emission regions are located far away (kpc scale) from the central engine.

A similar comparison — maybe even more constraining, since a morphological comparison can be done for sufficiently extended sources comparing equal acceleration and emission regions — can be done for *Galactic* flows. In those winds, the observed radiation points to a turbulent medium in the nebula magnetic field, which could explain some of the features observed in TeV PWNe, including (in some objects) the extremely low magnetic field required to adjust the X-ray and TeV flux level (see, e.g. Abramowski *et al.*, 2012c). The morphological resolution allows in this case a two-fold approach that can provide further answers in the correct direction. In such an approach, a proper time-dependent radiative treatment is combined with MHD simulations, although the complexity of the morphology in evolved PWNe and computational expensive modeling is still a challenge. Basic questions — such as how and where the energy is transferred from the magnetic field to the particle kinetic energy — are still on hold. This is what is commonly referred to as the sigma problem: the pulsar wind, carrying most of the rotational energy lost by the pulsar, is Poynting-dominated (highly magnetized) at its base, while MHD modeling of PWNe requires it to become particle-dominated at the shock. The sigma problem has been discussed many times in the literature in light of the most recent results: 2D MHD simulations (Komissarov and Lyubarsky, 2003; Del Zanna *et al.*, 2006) somewhat alleviated the strong requirements of the (Kennel and Coroniti, 1984) model, and more recently, 3D MHD simulations (Porth *et al.*, 2017, 2013) have suggested that actually an equal amount of energy in particles and fields at the shock might be required to reproduce the Crab Nebula morphology. This latter conclusion, however, comes from the idea that, thanks to kink instabilities, the magnetic field becomes progressively more tangled and dissipates its energy into the bulk of the nebula, i.e. in the radiation region, rather than in the cold pulsar wind. Recently, a new ingredient for the Crab Nebula puzzle has been added, thanks to precise observations with angular resolution below $1'$ achieved with the H.E.S.S. telescopes. The extension of the nebula above $E > 0.7$ TeV was measured to be $\sim 50''$, comparable to the one measured in optical wavelengths and significantly larger than the one measured in the X-ray band. The comparison in the future with the size at other energy bands could strongly constrain results based on precise 2D and 3D MHD simulations (Volpi *et al.*, 2008).

Perhaps a better understanding of the cooling and transport of particles in PWNe can be achieved by studying middle-aged systems located at sufficiently close distances to resolve them. Given the slower cooling time of the inverse Compton emission, they appear to act as calorimeters, in which the life and cooling of the electron populations can be traced by looking at the energy-dependent morphology of their gamma-ray emissions. One of the best examples to study the effect of particle cooling within the nebula is HESS J1825–137 (see Fig. 6.16 from De Jager and Harding, 1992; Aharonian *et al.*, 2006b), which displays an extremely large size of 1°, which, for a 4-kpc distance, corresponds to 70 pc linear size of the gamma-ray production region. A detailed spectral and morphological analysis of HESS J1825-137 has revealed a steepening of the energy spectrum with increasing distance from the pulsar PSR J1823-13. This is also observed as a reduction of the nebula size for increasing energies. In the case of continuous injection and radiative lifetimes that are short compared to the age of the source, the spectral index α of the electrons steepens by

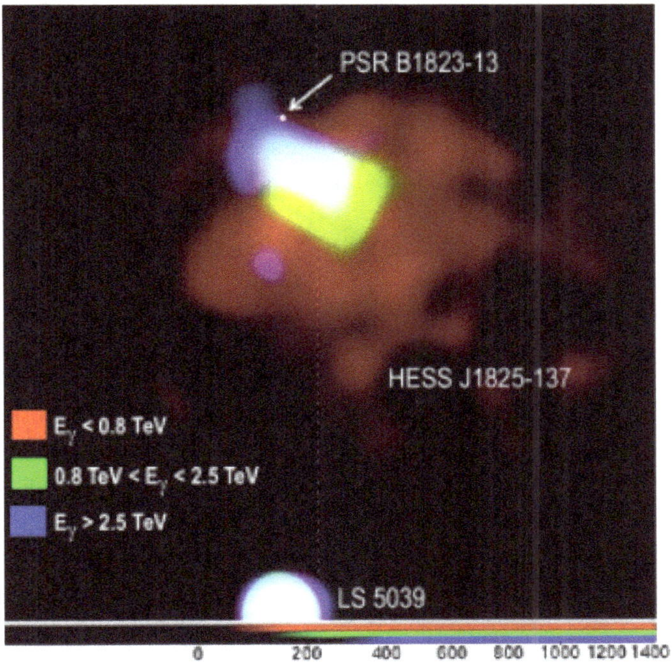

Fig. 6.16. Color image showing the energy-dependent morphology resolved in three basic colors as indicated in the legend. For details see Funk *et al.* (2007).

one unit, corresponding to a change of the photon index by half a unit, approximately matching the observed variation between the inner and outer regions of the nebula and pointing to synchrotron-dominated energy losses. Also, the large (and changing) size of the nebula allows for the study of different ways of particle propagation, advection, or diffusion. Efficient advection in PWNe is most likely limited by the inner region of the nebula, which may, however, extend significantly beyond the termination shock. The surface brightness across the source is roughly proportional to $1/\theta$, where θ is the angular distance from the accelerator. This dependence can be obtained by assuming different regimes of convection velocity or diffusion. For instance, this dependence is obtained when neglecting cooling effects and a constant radial convection velocity is considered, resulting in a $1/r^2$ density distribution (a surface brightness $\propto \theta^{-n}$ is obtained for spherical symmetry from a volume density $\propto r^{-n-1}$). That would imply, however, a rapid decrease of magnetic fields with distance from the pulsar, with very low fields at the edge of the PWN. A convection speed $\propto 1/r$ would allow instead a constant B-field resulting in a propagation time t$\propto r^2$. When applying the cooling losses, the size of the source follows $R \propto E_e^{1/2}$. Likewise, in the outer part, the nebula is bounded by the contact discontinuity (where the SNR limits its extension) where the plasma velocity should be non-relativistic and therefore favoring diffusive transport instead (Khangulyan *et al.*, 2018). If diffusion dominates, the size of the nebula can be estimated as $\mathrm{R} \propto (\mathrm{D(E)t})^{1/2}$, where $\mathrm{D(E)} \propto \mathrm{E}^{\delta}$ is the energy-dependent diffusion coefficient (see Section 6.2). A dependency of $\mathrm{R}\propto \mathrm{E_e^{1/2}}$ is again obtained when considering energy-independent diffusion ($\delta = 0$) and a propagation time given by the $\tau_{\rm rad}$ of radiating particles. Longer observations obtained with the H.E.S.S. telescopes — for which detailed spectral-morphological results were obtained — favor a description of the changing nebula extent with energy by an advection-dominated transport flow (Khangulyan *et al.*, 2018), with a possible contribution from diffusion in the outer part of the nebula. Still, the large size poses great problems when deriving the energy budget of the system. Different explanations, e.g. a faster rotation rate at earlier times, would be proven with the next generation of Cherenkov telescopes (Acharya *et al.*, 2019).

6.5.2.3. *PWNe: Galactic calorimeters tracing the electron lifetime*

One extragalactic object that also demonstrates a surprising extension of its spectrum up into the HE gamma-ray band is the local AGN called Centaurus A. This object is the most local radio galaxy to the Milky Way, at a distance of only 3.8 Mpc. The jets emitted by this AGN are not aligned with its line-of-sight direction to Earth, which instead provides the opportunity to view an AGN from an approximately side-on perspective. When viewed side-on, AGNs are referred to as radio galaxies, owing to the initial discovery of these objects through observations with radio instruments.

The spectral energy distribution from this Centaurus A exhibits a double-bump structure (see right plot in Fig. 6.17). The lower-energy bump in this spectrum is understood to originate from the synchrotron emission produced when the non-thermal electrons accelerated by Centaurus A interact with the magnetic field in the source region. The higher energy bump, instead, is considered to originate from inverse Compton emission produced when these same non-thermal electrons interact with the radiation fields present in the source environment.

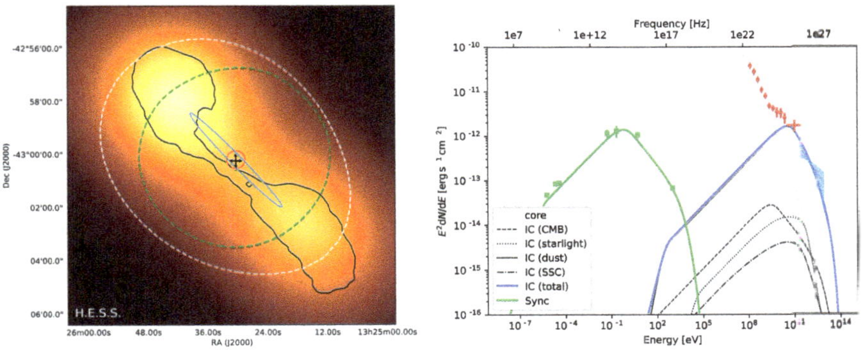

Fig. 6.17. *Left*: Extension of Centaurus A detected in the VHE band (the white oval indicates the detected extension, the green circle indicates the PSF and the blue ellipse indicates the extension inferred, taking into account the PSF of the instrument). *Right*: Spectral energy distribution of Centaurus A, showing both the synchrotron and inverse Compton emission components. Both plots are taken from Abdalla *et al.* (2020c).

Observations by the *Fermi*-LAT gamma-ray telescope of this object over the last decade at GeV energies have revealed the onset of an additional spectral component in the inverse Compton (HE bump) emission at approximately 10 GeV. Recent new insight into the nature of this new component has been provided by the discovery of the spatial extension of this object at very high gamma-ray energies by the H.E.S.S. Collaboration (Abdalla *et al.*, 2020c). The discovery of this extension is, in fact, the first spatially resolved object detected beyond the Milky Way's virial radius ($\sim$300 kpc) in the VHE band (see left panel in Fig. 6.17).

Interestingly, the origin of the new spectral component at an energy of 10 GeV appears to relate to the extended nature of the VHE gamma-ray emission. The extension length scale of the TeV gamma rays from Centaurus A relates to the size in excess of 2.2 kpc for the emission region. On this distance scale, the inner lobes of the jets from the AGNs are found. These lobe structures (see black radio contour line in the left panel of Fig. 6.17) have previously been studied at both X-ray, infrared and radio wavelengths. Indeed, whether the X-ray emission observed from this AGN is synchrotron or inverse Compton has previously been an issue of contention.

The detection of the VHE TeV emission indicates that HE ($> 10^{14}$ eV) electrons are present out at large kpc distances away from the central Galactic nuclei from where the jet originates. The presence of these electrons out at this distance along the jet, combined with the fast cooling times that such electrons experience in the jet's magnetic fields, indicates definitively that in situ particle acceleration takes place along the jet. This result sits contrary to the view that particle acceleration only occurs close to the Galactic nuclear region and provides further credibility to the view that the X-ray emission observed from the jet is synchrotron in origin.

With the discovery that TeV electrons are present out at kpc length scales along the jet, a considerable level of inverse Compton scattering of both the Cosmic Microwave Background (CMB) and infrared radiation fields present at this distance scale is subsequently expected. Indeed, new calculations factoring in the discovery of such HE electrons with the jet reveal a new consistent picture in which the 10 GeV to TeV energy gamma-ray emission from Centaurus A results from the inverse Compton scattering of these two radiation field components.

Injection spectrum in Galactic and extragalactic sources: The origin of the HE photons observed

The injection spectrum in *AGNs* and the modeling of their SED is usually assumed to be a spherical emission region that contains a single population of accelerated electrons with a power-law energy spectrum. Similar to the case of the *PWN* above, the power-law distribution approach does not represent the data satisfactorily and is usually replaced by a broken power law. The reasoning, in this case, is based on escape arguments: the highest-energy electrons escape the emission region, producing a change of the spectral index by one above a break energy. The differences between the two observed spectral indices before and after the break are typically close to one, favoring this argument. The electron spectrum is subsequently characterized by five parameters: minimal electron energy $\gamma_{\rm min}$ (typically, results are not sensitive to the choice of this parameter); maximum electron energy $\gamma{\rm max}$; break energy $\gamma'_{\rm b}$ and the two spectral indices, α^1 and α^2. The emission region is described further by three parameters: the radius of the region R, the mean of the magnetic field B (which is considered to have a random orientation inside the volume) and the bulk Doppler factor Γ. These are, in total, eight parameters that are fit to a typical data set consisting of a snapshot of optical, UV, X-ray and gamma-ray data. Radio data of blazars are typically some orders of magnitude higher than the prediction of this model, which is explained by a much higher volume in which radio waves are emitted compared to the region where X-rays and TeV gamma rays are produced. The success of this simple steady-state model is remarkable, as it seems to fit most of the available data sets. This success, however, might be misleading: while the model gives us a satisfactory description of *snapshots* of what might be happening in the jet, it neither accounts for variability nor explains the origin and shape of the electron spectrum used as input. With improved data quality in the TeV regime and a better sampling in time and energy-range coverage, in particular, having strictly simultaneous data, the one-zone SSC quickly reaches its limits.

One possible solution to take into account the complicated spectral and temporal behavior of AGNs is to assume that their relativistic jets are structured. In particular, Ghisellini *et al.* (2005) and Tavecchio and Ghisellini (2008) argue that there is compelling observational evidence that the jet of low-power TeV blazars could be structured (e.g. Giroletti *et al.*, 2008), with a fast (bulk Lorentz factor $\Gamma_s = 10\text{--}20$) central spine, surrounded by a slower ($\Gamma = 2\text{--}4$) layer. A structure of this type is also required to unify BL Lacs objects with the parent population of FRI radio galaxies (Chiaberge *et al.*, 2000). This scenario leads to the enhancement of the radiative efficiency of both components, since the electrons of each region can inverse Compton scatter the beamed soft photons coming from the other (Ghisellini *et al.*, 2005). An example is shown in Fig. 6.18 from Tavecchio and Ghisellini (2008) for BL Lac, a typical blazar, from which — due to the jet orientation — the observer can see the emission from the spine only, and for M87, a radio galaxy, from which the observer can see both components: the spine and the layer. Depending on the source, the synchrotron component can originate from the spine or layer and the inverse Compton component vice versa (Tavecchio and Ghisellini, 2014).

Hadronic models in Galactic and extragalactic sources as an alternative solution

In competition with the previous interpretation, hadronic models are usually invoked to explain the gamma-ray emission, therefore decoupling it from the synchrotron component. Even if, for many Galactic and extragalactic sources, a leptonic approach seems to work well based on the cooling signature or correlated variability, the possible contribution of heavy nuclei accelerated together with the electrons is unavoidable. In the case of PWNe, for instance, several authors (Bednarz and Ostrowski, 1998; Horns *et al.*, 2007) have proposed that protons can be efficiently accelerated in such winds and contribute to the total emission observed in the HE regime. Additionally, in blazars and radio galaxies, these models can alleviate some of the problems that arise from leptonic modeling. It also opens an attractive possibility of association with other messengers such as neutrinos. While earlier hadronic models struggled to explain fast

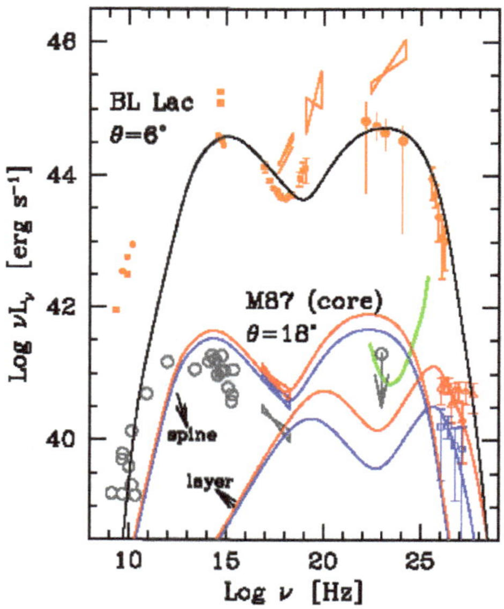

Fig. 6.18. Spine and layer jet model. SED of the core of M87 (gray open points) together with the H.E.S.S. spectra taken in 2004 (open blue squares) and 2005 (open red triangles), from Aharonian *et al.* (2006). The upper black line reports the emission from the spine that would detect an observer located at an angle of 6° from the jet axis (the corresponding emission from the layer is well below that of the spine and is not reported for the sake of simplicity). For comparison, we show the data of BL Lac (filled orange symbols from Ravasio *et al.*, 2002, TeV data from Albert *et al.*, 2007). See text and Tavecchio and Ghisellini (2008) for details.

variability and correlated behavior in X-rays and gamma rays, advances in modeling have allowed accounting for variability as well as for spectral shape variations. The parameter space spanned by the hadronic models, however, is large, giving the freedom to couple or decouple X-rays from the gamma-ray emission at the author's comfort, depending on the data set. Real-life light curves, e.g. Mrk 421, are still too complex and our understanding is too limited to conclude on the applicability of this type of model. The ultimate proof that hadronic acceleration — and, at least, part of the gamma-ray emission — has hadronic origin would come only with unambiguous detection of HE neutrino sources. Recently, a strong hint of such a source was found in TXS 0506+056 (Aartsen *et al.*, 2018), which opens a window to a new class of TeV gamma-ray sources and strengthens the efforts in hadronic modeling.

In addition to the three general trends discussed earlier, there is one important aspect that we would like to highlight related to the large luminosity and thus distances of AGNs. This aspect will be discussed in the following.

6.5.2.4. *AGNs: Lighthouses of the Universe*

The improved sensitivity, together with the lower-energy threshold of the imaging Cherenkov telescopes, has allowed the exploration of the Universe to go to larger distances. One good example is perhaps the gravitationally lensed blazar S3 0218+35 (see Ahnen *et al.*, 2016b and references therein) located at a distance of $z = 0.954$. The attraction of gravitationally lensed objects is that they provide several images of the same object and, more relevantly, they allow for multiple detections of the same flare with a fixed time offset between them due to different path lengths. This particular case is gravitationally lensed by the face-on spiral galaxy B0218+357G located at a redshift of $z = 0.685$, resulting in two images resolved in optical and X-rays with an angular separation of only 335 mas. The emission from the two images arrives with a delay of 10–12 days between the leading and trailing images. The detection of a flare in a leading image by *Fermi*-LAT (Buson and Cheung, 2014) triggered ground-based observatories to prepare a campaign to observe the trailing image. MAGIC succeeded to detect it, which not only allows us to study distant blazars more predictably, but also provides insight into properties of the gravitational lens itself by comparing the intensities between leading and trailing images. Besides this indirect technique to observe faraway objects, the extreme bright emission of these jetted galaxies permits the radiation to reach us from faraway distances after propagating throughout the extragalactic medium. To explain such bright radiation in the TeV regime in an inverse Compton interpretation, high Lorentz bulk motion factors and corresponding Doppler boosting factors in the order of 10–50 are required. Such a fast component, however, has never been observed in the X-ray synchrotron radiation, associated with the inverse Compton radiation. Moreover, high Lorentz factors should be applied to avoid large absorption due to the high density of optical photons via electron–positron pair production. This seemingly consistent scenario is challenged by the observations in radio, optical and X-rays, through which the measurement of the super-luminary motion (Ghisellini *et al.*, 1993), of actual Doppler factors can be determined. Indeed, typical super-luminal motions seen in blazars are of the order 1–5 — one order of magnitude

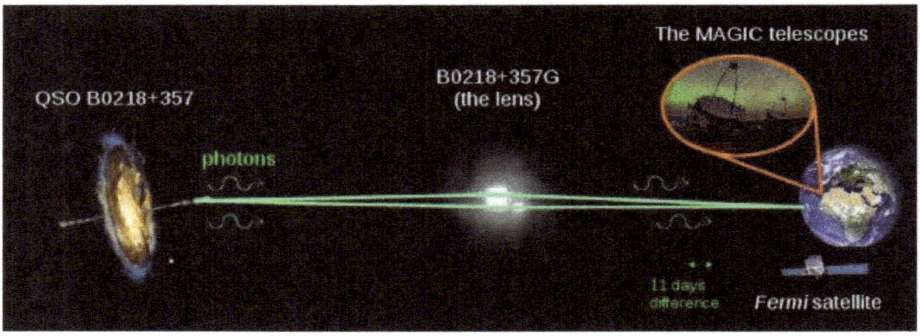

Fig. 6.19. Sketch of the detection of QSO B0218+357 ($z = 0.944$). The gamma rays from a single flare emitted at the blazar location were gravitationally lensed by the elliptical galaxy B0218+357G ($z = 0.685$) leading to two flares detected on Earth, with a delay between the two of about 11 days. The second flare was detected with the MAGIC telescopes (see Ahnen *et al.*, 2016b for details).

lower than required by the models dealing with their TeV emission. This could mean that the jet structure is more complex than the one resolved in radio to X-rays with TeV emission regions being smaller (and farther out) compared to the structures seen at lower energies. Alternatively, it might be that the models are too simplistic and that accelerated particles actually reach much higher energies in the co-moving frame than estimated. A challenge of the latter scenario is that at higher energies, the inverse Compton scattering takes place in the Klein–Nishina regime (in contrast to the Thomson regime at lower energies), in which case a super-exponential cutoff is expected at VHE gamma rays, which is not seen in the data.

A second consequence regarding the distant location of these sources concerns the measurement of the extragalactic background light (EBL) by means of evaluating the unavoidable absorption of the TeV emission when traversing through it. The level of EBL absorption probed by TeV observations of AGNs at different redshifts has provided constraints at multiple EBL wavelengths (Abdalla *et al.*, 2017b; Mazin and Raue, 2007). The resultant EBL level consistent with these limits has proven to be that obtained if one assumes no significant additional contributions to the EBL from hitherto undetected sources of the EBL (e.g. dark stars, Population-III stars and exotic decays). It should be noted, however, that the EBL constraints rely on the assumption that the intrinsic (i.e. those that are emitted locally) blazar's VHE gamma-ray spectra show neither pile-ups at the HE end of the spectrum nor sharp spectral features.

These are reasonable assumptions given the observations at lower energies, which are also dominated by non-thermal emission but do not show such features. Nevertheless, in some scenarios, pile-ups and potential sharp spectral features could be masked by the EBL, allowing for additional EBL contributors in particular. Remarkably, VHE blazar spectra were used not only to place upper bounds on the EBL density, but also to successfully determine the minimum level of the EBL in the nearby universe to explain the spectral shapes of detected blazars (see Chapter 7). Again, this method assumes a degree of smoothness of the intrinsic blazar spectra, which is strengthened by the fact that a consistent result is derived by several experiments using different data samples of different sources independently. These results underline the potential of the young field of ground-based gamma-ray astrophysics with the Cherenkov telescopes to contribute to the observational cosmology (see Chapter 7).

6.5.3. *Observations of relativistic shocks in high-density, weakly magnetized plasma*

The energetic power of relativistic shocks can also be investigated through their interaction with weakly magnetized, relatively dense media. Jets in active nuclei traveling through dense accretion disks or interacting with stars in the surroundings, or similar processes observed in Galactic-scale jets or relativistic pulsar winds in binary systems, are optimal scenarios to test the physics processes at play. The following discusses the implications of the observations of stellar-scale and Mpc-scale scenarios in gamma rays in the context of relativistic shocks in high-density plasma.

6.5.3.1. *Stellar-scale shocks: Binary systems*

Concerning Galactic binary systems, a diversity of systems have been postulated to be efficient particle accelerators: novae (Ackermann *et al.*, 2014a), transitional millisecond pulsars (Torres *et al.*, 2017), binaries composed by a pulsar and a massive companion (O-type or WR stars) (see Dubus *et al.*, 2017 and reference therein), binaries formed by two massive stars (Abramowski *et al.*, 2012d) and microquasars (Zanin *et al.*, 2016). In those systems, gamma rays are thought to arise either from the interaction of a relativistic wind or jet with the stellar wind of the companion, or from processes related to an accretion disk injecting matter onto a black hole or neutron star. To date, only gamma-ray binaries (compact objects with massive companions, either OB-type or WR star) have been firmly detected

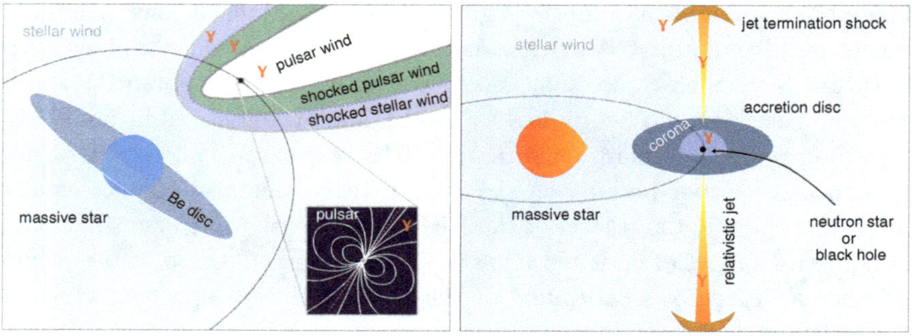

Fig. 6.20. Scenarios for gamma-ray emission from binaries. *Left*: Relativistic wind from a rotation-powered pulsar interacts with the stellar wind (and Be disc if present) of its massive companion star. The pulsar inset shows the magnetic field lines within the light cylinder. Gamma-ray emission can occur near the pulsar, within the pulsar wind, or at the shocks terminating the pulsar and stellar wind. *Right*: Compact object accretes matter from the stellar wind or Be disc. Part of the energy released in the accretion disc is used to launch a relativistic jet. Gamma-ray emission can arise from the corona of the accretion disc, within the jet, or at the termination shock of the jet with the ISM (from Dubus, 2013).

in the TeV regime, whereas the rest shows bright emission in the GeV regime that seems to quickly decay after a few GeVs. Remarkable exceptions are the massive-star binary system Eta Carinae, which is discussed in the next section, and the nova RS Ophiuchi, recently detected in the TeV regime by H.E.S.S. and MAGIC (Abdalla *et al.*, 2022; Acciari *et al.*, 2022).

A few VHE gamma-ray binaries have been firmly identified, among which LMC P3 lies on the Large Magellanic Clouds. Two more binary systems, HESS J1832-093 and Cyg X-1, have also been proposed as gamma-ray emitters, but still lack clear confirmation.

Given the different phenomenology regarding orbital modulation, flaring episodes and luminosity, a long debate has been going on in the astrophysical community as to the nature of the compact object powering the relativistic wind and the gamma-ray emission regions. The questions about the nature of the compact star, black hole with relativistic jets or neutron stars have triggered many observations (see Fig. 6.20). No clear answers have been provided yet, except for one case (PSR B1259-063, Aharonian *et al.*, 2005a) for which pulsations were detected when the compact object was located far from the companion, revealing its neutron star nature. The multi-wavelength comparison of the light curve for the different systems demands a detailed source-to-source investigation. The observed variability in all of them, however, implies a compact emission region with systems as close as $\sim 10^{12}$ cm (see Fig. 6.3). Interestingly, the measured spectra extend beyond 10 TeV without signs of cutoff or curvature (except in the case of LS 5039 in inferior conjunction where the curvature is clearly observed, (Aharonian *et al.*, 2005b)), implying an extremely efficient accelerator toward high energies in such a compact region. The gamma-ray emission is usually interpreted as inverse Compton on the photon field provided by the stellar companion by an energetic electron population powered by the compact object. Given the energy of the stellar photons (2.7 kT$\sim$ 9 eV), electrons with energies larger than $(\mathrm{m_e c^2})^2/(\mathrm{h}\nu) \sim 25\,\mathrm{GeV}$ upscatter photons in the Klein–Nishina regime in which all of the electron energy is transferred to the photon in the scattering process. In the Klein–Nishina regime, the photon energies detected in the TeV regime require the presence of electrons of similar energy, which in turn requires that they are accelerated faster than their radiative energy-loss timescale. These energy losses, given the magnetic fields in this system of fraction of Gauss, are dominated by synchrotron losses, limiting the maximum energy to a few tens of TeV (Khangulyan *et al.*, 2007). Still, these binary systems seem to be extremely efficient, with photon spectra extending beyond 10–30 TeV. The constraints derived from these systems depend strongly, though, on the location, acceleration efficiency and magnetic field of the emission region. The parameter space has been explored by several authors, delimiting the location in which the gamma-ray emission can occur (Zabalza *et al.*, 2013; Dubus *et al.*, 2008).

Modulation of the energy spectrum could arise from changes in the maximum energy electrons responsible for the radiation, changes in the dominant radiative mechanism and/or the scattering angle dependence of the inverse Compton scattering effect. Moreover, the different phenomenology of the multi-wavelength component adds additional complexity to the acceleration region(s), preventing the *standard model* from being able to explain globally the emission observed in gamma-ray binaries. It seems clear, however, that the orbital parameters play a particularly important role in shaping the gamma-ray light curve and spectral shape. Current modeling suggests that more than one electron population is likely required to explain the multi-wavelength spectral energy distribution. To add more complexity, exceptionally bright flares were discovered in LS1 +61 303 (Archambault *et al.*, 2016) on timescales of a few hours. A possible explanation implies the presence of an inhomogeneous stellar wind provided by the BO Ve companion star. The clumpy wind would fall into the shock region, reducing the size of the acceleration region and promoting a larger acceleration time. Similar to other systems (see, e.g. Abramowski *et al.*, 2015a), the upper limit derived on the magnetic field requests a large distance to the compact object (see above), pointing to different acceleration regions or more complex geometries. Indeed, LS 1+61 303 displays a super-orbital variability that has been observed in several energy bands (Li *et al.*, 2012; Ahnen *et al.*, 2016d; Paredes, 1987; Chernyakova *et al.*, 2012). This variability can be attributed to quasi-cyclical expansion and shrinking of the circumstellar disk of a Be star: when the compact object (a putative neutron star) approaches periastron, it would face a region of the disk with different density, producing a change in the strength of the interaction and emission each time. A different interpretation invokes a change of state from a propeller regime during periastron to an ejector region at apastron, also known as the flip-flop scenario. These changes of state would be driven by the interaction with the circumstellar disk and the stellar wind. All of these different clues point to different directions, providing a complex scenario to explain without a clear idea as to the nature of the compact object. Direct measurement of the mass of the compact object would firmly unveil its nature and help to explain the phenomenology of this and other types of binary systems.

Recently, a new object has been added to the collection of gamma-ray binaries, adding an extra twist to the problem: the pulsar PSR J2032+4127 associated with the extended gamma-ray source TeV J2032+4130 has been found to be embedded in a decades-long binary system, LS 2883, with the pulsar moving rapidly toward periastron. The periastron passage that occurred in 2017 was accompanied by a large increase of the gamma-ray emission (Mirzoyan and Mukherjee, 2017). If indeed the two emissions (the extended TeV J2032+4130 and the brighter point source) are related, it would represent the first time that the ablation of a PWN by a companion wind is observed, providing extraordinary information about the formation of wind nebulae and the acceleration of particles in two colliding relativistic winds.

Finally, similar processes are proposed to be at play in extragalactic jets colliding with red giants (Barkov *et al.*, 2012), as discussed in the previous section. Better-sampled light curves with higher sensitivity, however, are required.

6.5.3.2. *Relativistic shocks on large length/timescales*

Despite a significant effort by the major IACTs to detect gamma-ray emission since around 2004, there was no positive detection of any GRB

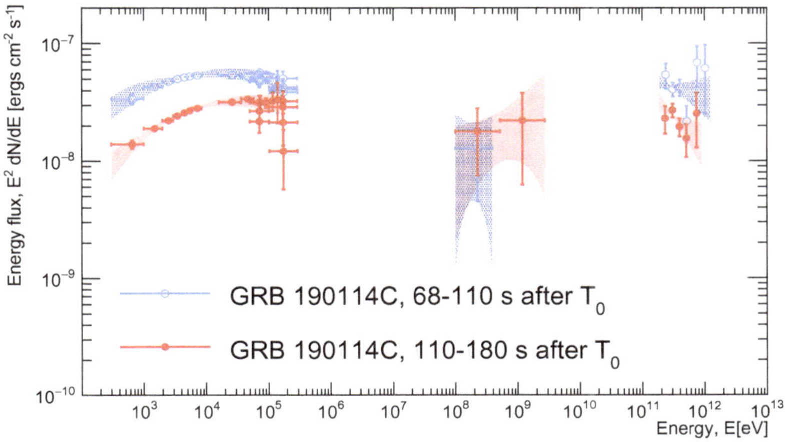

Fig. 6.21. Measurements of GRB 190114C taken from Acciari *et al.* (2019) in two different time intervals: 68-110s and 110-180s after T_0.

signal at VHE beyond a couple of photons detected by *Fermi*-LAT until 2018 (Acciari *et al.*, 2019b). A long period of non-detection had triggered speculations that GRBs do not generally emit gamma rays beyond 100 GeV, and even then only in a very few exceptional cases. In the last few years, however, this paradigm has radically changed. Both MAGIC and H.E.S.S. collaborations reported highly significant detection of currently four different GRBs: GRB 180720B (Abdalla *et al.*, 2019b), GRB 190114C (Acciari *et al.*, 2019b; Acciari *et al.*, 2019), GRB 190829A (Abdalla *et al.*, 2021b) and GRB 201216C (Blanch *et al.*, 2020). Spectral energy distributions of two GRBs in different wavelengths and time intervals are shown in Figs. 6.21 and 6.22. All of the VHE-detected GRBs belong to the class of long-duration GRBs, i.e. when the t_{90} measure of the prompt emission phase lasts longer than 2 s, and all four have been detected in the afterglow phase when HE electrons are believed to be accelerated by the external shock of the burst. Surprisingly, not only have gamma rays in these four GRBs been detected with energies up to 1 TeV, but also significant emission up to three days after the start of the burst (in the case of GRB 190829A, see Fig. 6.22). Generally, the light curves of VHE gamma-ray emission seem to follow those in X-rays, suggesting a leptonic origin of the emission. The fact that the emission is detected up to several hundreds of GeV points to a high acceleration efficiency and high transparency of the

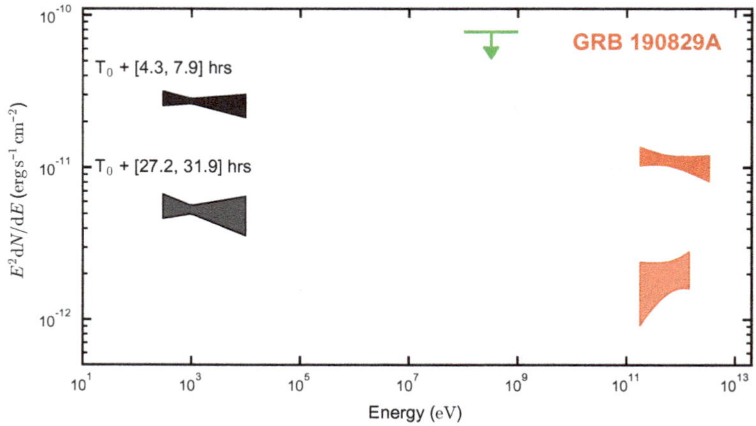

Fig. 6.22. Measurements of GRB 190829A taken from Abdalla *et al.* (2019b) in two different time intervals: 4.3–7.9 h and 27.2–31.9 h after T_0.

emission region to gamma rays, i.e. low level of IR radiation or a very-high Doppler factor inside the emission region.

There is, however, presently a controversy as to whether the detected VHE emission by H.E.S.S. and MAGIC can be unambiguously associated with inverse Compton upscattering of the synchrotron photons, or whether it is actually a continuation of the synchrotron radiation of very energetic electrons. While inverse Compton scenarios are easier to model, there remains a striking similarity between the X-ray and VHE gamma-ray spectra, where the higher energies seem to form a continuation of the lower energy spectra, supporting the same (thus synchrotron) origin of the emission. The sparsity of the available simultaneous data (Figs. 6.21 and 6.22) does not yet allow one to unambiguously distinguish between these two scenarios.

More GRB detections are expected to follow, leading to a better understanding of the origin of their VHE gamma-ray emission.

6.5.4. *Observations of non-relativistic shocks*

Typically, non-relativistic shocks are efficiently originated in ejecta moving supersonically, originated by either an explosion or the combined effects of two or more stellar winds. These phenomena result in a plasma front that creates a shock wave that propagates at velocities as high as $\sim 10^4$ km/s ($\beta\Gamma \sim 0.03$). These shock waves create several sites up and downstream in

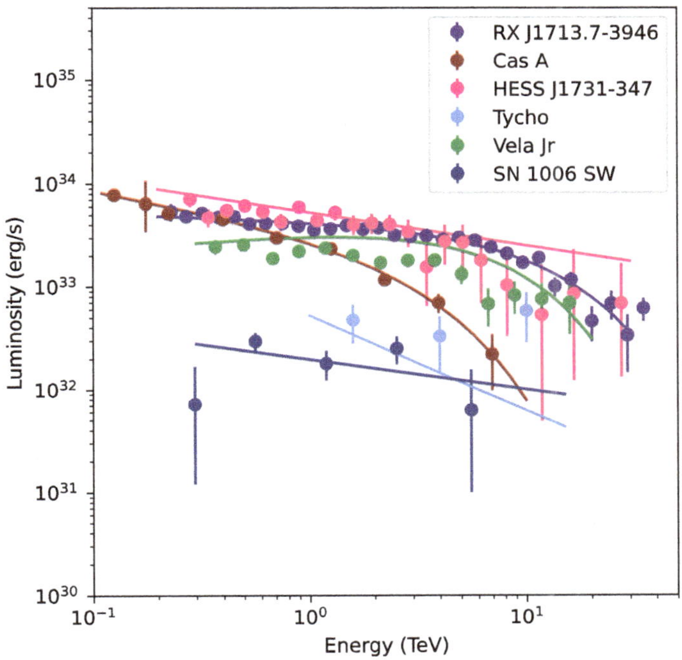

Fig. 6.23. Luminosity in different logarithmic energy bins for several young, TeV-bright SNRs.

which particles can be accelerated to very high energies. SNR explosions are of particular interest since they are believed to be the acceleration sites of Galactic CR. This assumption is based on energetic arguments (Baade and Zwicky, 1934) and is discussed in Chapter 5 Section 5.4 of this book whereas here we will discuss the astrophysical implication of the gamma-ray emission observed in SNRs. Dozens of those SNRs were detected above a few tens of GeV; the spectra of the most prominent, young SNRs are shown in Fig. 6.23 (adapted from Aharonian *et al.*, 2019). Most of them show a shell-like morphology supporting the general predictions of Fermi-I (shock) acceleration theory and thus establish the SNR shocks as effective particle accelerators. Indeed, SNRs are complex systems that also provide several regions in which particles can be accelerated. The supersonic expansion of the SN ejecta into the ambient medium gives rise to a forward shock expanding into the medium, and a reverse shock expanding back into the ejecta. In the former, the magnetic field can be amplified to a few mG, providing in principle acceleration power to accelerate electrons and protons up to few tens of TeVs. This high magnetic field results in a quick synchrotron burnoff of electrons, whereas protons can survive long enough to produce VHE gamma rays by interacting with the surrounding gas. The reverse shock instead could provide very different conditions since it propagates in a less-dense medium, allowing a wider range of magnetic field intensities able to account for a leptonic gamma-ray origin.

SNRs are luminous sources possessing luminosities in the range between 10^{32}erg/s and 10^{35}erg/s and displaying a large variation in the strengths of their magnetic field values, from a few μG for old SNRs to $\sim$ mG for young SNRs. In general, when the gamma-ray emission extends up to the GeV regime, it seems to be better described by leptonic emission, except in a few cases in which a flatter spectrum seems to point to a hadronic origin. For the latter, however, the gamma-ray emission is either very steep in the TeV regime — favoring an early cutoff due to lack of acceleration power — or indicates that particle escape is shaping the observed spectrum. For SNRs that have actually been detected above 10 TeV, the spectrum quickly turns over with increasing energies (see Fig. 6.23). The inverse Compton interpretation is often challenged when modeling the multi-wavelength spectrum. To reproduce the X-ray and gamma-ray spectrum, magnetic fields of the order of a few μG are needed, which are at odds with the values expected in fast SNR shocks. A different approach might overcome this problem if two-zone models are considered based on the forward-shock and reverse-shock regions (as discussed earlier).

Similar shocks occur when the winds of two or more young stars interact, collectively creating a front in which particles can be efficiently accelerated. This acceleration can take place in the vicinity of the stars, in the stellar clusters (Cesarsky and Montmerle, 1983) (where SNR shocks can also be energized by stellar winds (Gupta *et al.*, 2020)), or in the superbubbles — multi-parsec structures caused by the collective activity of massive OB stars around the compact stellar associations (Aharonian *et al.*, 2019; Yang *et al.*, 2018). Indeed, bright and very-extended TeV gamma-ray emission has been detected from the direction of some of the most energetic clusters of stars: Westerlund 1, Westerlund 2 and the Galactic center (Abramowski *et al.*, 2011a, 2012b, 2016). The same extended nature of the emission, however, can hamper the identification of the extent of the emission with current Cherenkov telescopes for objects whose angular extension is comparable or larger than the fields of view of the instruments. For instance, around Westerlund 2, a large $\sim 0.2^{\circ}$ TeV source was detected, which could be either interpreted as emission originating in the cluster wind or as the result of a bright pulsar inflating an energetic wind nebula in the line of sight (Abramowski *et al.*, 2011a; Mestre *et al.*, 2021). A larger structure observed in the GeV regime, called the Westerlund 2 cocoon (Yang *et al.*, 2018), results in an extended structure that is too large to be efficiently detected by H.E.S.S. Likewise, very-extended emission around Westerlund 1 of $\sim 1°$ was observed around the cluster in the TeV regime (Abramowski *et al.*, 2012b), but a clear multi-wavelength counterpart makes the unequivocal identification hard to assess. The CR radial distribution derived in both the GeV and TeV regime points to a $1/r$ propagation profile, which corresponds to a continuous operation of CR accelerators in these objects. This interpretation challenges the long-held belief of SNRs being the origin of the CRs up to the "knee" and opens a new possibility of efficient acceleration of CRs (see Chapter 5). Similar phenomenology is believed to happen in the exceptional binary system Eta Carinae. The binary stellar system of Eta Carinae is composed of a luminous blue variable and its lighter companion star of O- or B-type in a 5.5 yr orbit. Orbital modulated gamma-ray emission was detected in the GeV regime with *Fermi*-LAT (Martí-Devesa and Reimer, 2021), triggering observations with Cherenkov telescopes. H.E.S.S. detected a second component, reaching energies up to 0.7 TeV (Abdalla *et al.*, 2020a). The new narrow component probes the acceleration of particles up to a few TeV for both proton or leptonic models, opening a window for a new type of accelerator.

Regardless of whether CRs are accelerated in SNRs or other Galactic accelerators such as the previous discussion on star clusters, once they are accelerated by the Fermi acceleration process, they are subsequently injected into the Galaxy and diffuse away from their initial sources. This injected energy flux may either escape from the host galaxy or interact with it through hadronic and leptonic energy-loss interactions. In the case in which this energy flux fails to escape the host galaxy in the form of CRs (i.e. the CRs undergo significant energy losses before escaping the host galaxy), the galaxy is said to operate as a CR calorimeter. The Milky Way itself does not appear to act as a hadronic CR calorimeter, with only $\sim 10\%$ of the energy in CRs being lost within the Galaxy before they escape. Contrary to this, the Milky Way does appear to act as a leptonic CR calorimeter for sufficiently HE electrons.

Such hadronic CR energy-loss interactions occur not only in the Milky Way but also in other galaxies. In particular, starburst galaxies observed in the TeV regime (Abramowski *et al.*, 2012; Acciari *et al.*, 2009a) appear to operate much closer to the calorimetric limit for CR hadrons (Ackermann

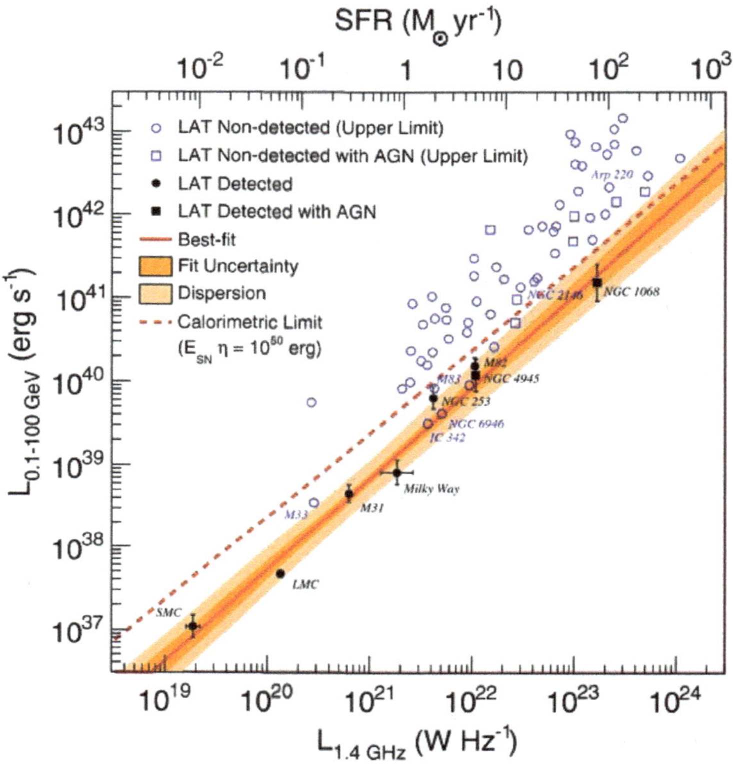

Fig. 6.24. Gamma-ray luminosity (0.1–100 GeV) vs. radio continuum luminosity at 1.4 GHz. Galaxies significantly detected by *Fermi*-LAT are indicated with filled symbols, whereas galaxies with gamma-ray flux upper limits (95% confidence level) are marked with open symbols. Galaxies hosting Swift-BAT AGNs are shown with square markers. The upper abscissa indicates SFR estimated from the RC luminosity. The red solid line and associated shaded regions show the best power-law fit to the data and associated uncertainties. The dashed line represents the expected gamma-ray luminosity in the calorimetric limit, assuming an average CR luminosity per supernova of 10^{50} erg (from Ackermann *et al.*, 2012b).

et al., 2012b). Compared with normal star-forming galaxies like the Milky Way, these starburst galaxies have much higher star-formation rates and harbor regions of very dense gas. The TeV radiation observed is believed to be related to the combined effect of the numerous galactic accelerators, providing a collectively large luminosity of accelerated CRs that illuminate the matter in the host Galaxy. The combined high and VHE spectrum, which can be reproduced with a single hard-index power-law component, also favors this interpretation. Moreover, the gamma-ray observations indicate that between 10%–100% of the energy of the CRs capable of producing hadronic interactions is channeled into pion production in starburst galaxy systems (Abdalla *et al.*, 2018l) (more details can be found in Chapter 6). A precise determination of the calorimetric fraction of such starburst galaxies, however, is hard to determine due to the considerable uncertainty in the underlying CR luminosity of these systems.

The center of our Galaxy also hosts a number of potential accelerators, and significant gamma-ray activity has indeed been detected in the TeV regime (Abramowski *et al.*, 2016; Archer *et al.*, 2014; Acciari *et al.*, 2020a). Several point sources and a remarkable extended emission have been detected from the region of the galactic nucleus (Sagittarius A*), reflecting considerable CR-production activity. A point source compatible with the central black hole has been detected, but despite the deep search, no variability has been detected so far, leaving the question of the actual size of the gamma-ray emission region still open. An extended diffuse emission associated with the molecular ridge has been also detected, providing important information about the acceleration and propagation of particles, mainly protons, illuminating the dense molecular surrounding (Abramowski *et al.*, 2016). These observations again reveal a $1/r$ profile of the CR centered on the galactic nucleus region. Furthermore, these measurements have revealed this source to be the first known Pevatron within the Galaxy. An extended explanation of the Galactic center phenomenology is also given in Chapter 5.

6.6. Discussion

HE emission processes require particular environmental conditions that can only be found in certain astrophysical sources. These involve outflows of supersonic plasmas whose properties can be studied through the interaction of this plasma with the surrounding media. The last several years' advances in the imaging Cherenkov technique have provided a plethora of new results, which, on many occasions, tend to uncover new mysteries rather than resolve existing ones. The past several years' discoveries have highlighted basic gaps in our understanding of the origin of the highest-energy photons in the Galaxy. Basic questions, such as whether the emission originates in hadronic or leptonic processes, remain unanswered. The future, however, is bright. The advent of the new generation of imaging Cherenkov telescopes, namely, the CTAO, with full coverage of the gamma-ray spectra from

20 GeV to more than 100 TeV with approximately 10-times-better sensitivity than the instruments of the current generation, aided by X-ray and radio observations from existing and planned satellite missions and complementary gamma-ray observations from water Cherenkov instruments, will be critical to elucidate questions regarding the production mechanisms, diffusion and interaction of cosmic rays, residual leptonic emission and processes of particle acceleration. This multi-wavelength, multi-messenger approach is crucial to constrain parameters such as the medium density, the non-thermal leptonic population or the hadronic component.

In the following section, we would like to mention some of the open problems that we believe will be leading the field in the coming decade.

Transients: One of the most urgent problems in the field is related to the mechanisms behind short timescale transient emission. Indeed, these are proving to be the least-understood phenomena in terms of the acceleration process, emission location and emission process. From pulsars to GRBs, the very-short variability observed poses many challenges to our current understanding. Concerning Galactic transients, *Fermi*-LAT has opened a window to a large and varied phenomenology ranging from compact objects such as pulsars, microquasars, novae and solar flares to more extended objects such as PWNe. These flaring events are usually attributed to synchrotron variation due to a change in the leptonic population and/or the magnetic field. The expected counterpart above few tens of GeV, however, remains missing. New observations with more sensitive instruments with low-threshold energy would allow testing for those mechanisms by seeing variations on the tails of the GeV emission or maybe the inverse Compton component, and also will help to constrain the maximum energy reached and therefore the location and population of injected particles. Likewise, these types of studies are of utmost relevance for a unique object in our Galaxy, the Galactic center. A very broad range of phenomenology has been observed in lower-energy bands, whereas in the VHE band the emission seems to remain stable. Whether this apparent difference in activity is due to lack of sensitivity, however, remains unclear.

Extragalactic transients are even more interesting since their observations reveal that an extremely complex range of phenomenology is at play. Indeed, the flaring mechanism is far from being well understood: are they due to the injection of a new particle component (new shock) into the jet? A restructuring of the magnetic field (magnetic reconnection)? Or something else? The connection between the black hole characteristics and the flaring events is also not understood. A new focus on coordinated multi-wavelength monitoring observations being carried out promises to provide further insights into the relationship between the variability in the temporal structure (the spectral slope of the temporal power spectrum) and the size of the black hole or the location of the emitter with respect to its position. Regardless, future in-depth observations planned with CTAO (Acharya *et al.*, 2019) will definitively reveal any transient phenomena associated with the accelerator, resulting in a clear identification of the still-mysterious origin of the VHE emission.

Acceleration processes: The imprints of the acceleration processes are usually overlooked by assuming a simple power law or broken power law to fit the gamma-ray emission. Different acceleration mechanisms, however, should be clearly reflected in the spectrum and morphology in the upcoming observational results when a large statistic in a wide energy range is expected. For instance, the role of the magnetic field structure (and magnetic reconnection) should become clear with better angular and temporal resolution and higher statistics in Galactic sources. Observations of pulsar "tails" should unveil the acceleration processes in the surroundings of these objects, whereas the spectro-morphological information in different regions on the winds should reflect the particle population for a good knowledge of the photon field in situ. Likewise, we will be able to derive the spectrum in different parts of SNR shells, which will reveal the mechanism and location of the particle accelerator. The detailed study of the cutoff shape should provide crucial information about the shape of the emission that can be directly translated to the acceleration mechanism behind it (see Section 3). The better sensitivity that new instruments will reach, together with the lower threshold, is fundamental to also understanding the acceleration processes in extragalactic sources (Romoli *et al.*, 2017). Features in the spectrum — for instance, revealing a new component — should be accessible with the new IACT facilities, as well as the detection of other source types rather than the ones we already know.

Particle propagation: Extended sources are excellent tools to study the propagation of particles in different scenarios. Already with the current imaging Cherenkov instruments, very extended TeV sources like the Galactic center ridge or the open cluster Westerlund 1, or even extragalactic sources like the very recent detection of the extension of Cen A (Abramowski *et al.*, 2016, 2012b; Abdalla *et al.*, 2020c), are successfully detected. The number of photons observed from these sources is still low, however, preventing accurate morphological and spectral studies. Moreover, above a certain size of $\sim$ 1–2°, we run into technical issues related to the camera size and the effective assessment of the background, making it very difficult to study very extended emission. This problem is/will be partially solved by using current and future Water Cherenkov instruments such as HAWC (Tepe *et al.*, 2012), LHAASO (Di Sciascio and LHAASO Collaboration, 2016) or SWGO in the future,[a] which are more sensitive to extended regions, yet lack an optimal angular resolution. CTAO, and in particular the SSTs, with camera FoV of 10°, will provide an excellent opportunity to study these types of sources with high statistics, which will allow us to understand the role of transport and cooling of particles at large timescales and their effect on the morphology and spectrum of the extended sources.

Multi-wavelength/multi-messenger: Finally, the studies performed in the last several years have demonstrated the need for a complete multi-wavelength/multi-messenger coverage in order to strongly constrain the processes producing the VHE radiation. Transients, in particular, require simultaneous observations to trace the variability features at the very same moment to be able, for instance, to palliate the poor angular resolution of Cherenkov telescopes by correlating the signal with the much better resolution available in the lower-energy bands. These observations are also crucial to understanding the mechanisms behind the emission: for instance, the observations of some BL Lac in the TeV regime suggest that Doppler factors of the order of 20–50 are required (Begelman *et al.*, 2008; Aharonian *et al.*, 2007; Albert *et al.*, 2007d); however, in observations at lower energies (in particular in radio waves), such large values are not observed. Instead, moderated values of Doppler boosting of $\sim$ 5 are found (Piner and Edwards, 2004; Giroletti *et al.*, 2004). Simultaneous, multi-wavelength campaigns are crucial to understanding such differences. Another important aspect is related to changes in the optical polarization angle (see, e.g. Hovatta *et al.*, 2016). It has been shown that TeV flares are occasionally accompanied by a significant change in optical polarization. This might be related to newly accelerated particles; the amount of polarization might be associated with an ordered magnetic field (thus, the synchrotron emission is polarized) and the rotation of the polarization angle to newly injected particles, which change the orientation of the magnetic field. The timescales of the optical

[a] www.swgo.org.

polarization changes, however, do not match the ones in gamma rays; the delay between the offset of the two phenomena is different flare-by-flare, and the behavior after the flare is different case-by-case, too. It is therefore not yet possible to draw a self-consistent picture of how or if the optical polarization is connected to the gamma-ray activity without more sensitive and simultaneous observations.

Comparison with imaging polarimeters in X-rays such as IXPE[b] is also crucial to understanding the magnetic field structure in extended Galactic sources such as PWNe and SNRs. The direct comparison between the synchrotron emission and the inverse Compton emission (if applicable) would shed light on how the magnetic field is distributed in the source region, providing important information needed to understand long-standing questions such as the equipartition departure or the amplification of magnetic fields in shocks. In fact, not only is polarimetry important, but a good sampling of the X-ray counterparts (if any) with instruments, e.g. ATHENA (Piro *et al.*, 2021), would constrain the leptonic acceleration in the source, which would directly translate into a constraint on the VHE counterpart. Ultimately, observations of other messengers, such as neutrinos, would constitute unequivocal proof of hadronic-origin TeV emission. Recently, a hint was reported on an associated neutrino excess from the direction of a flaring blazar (Aartsen *et al.*, 2018). If it was not a coincidence and the neutrino is associated with the blazar, does it mean that at least some ordinary looking blazars accelerate protons to 10^{14}–10^{18} eV, which would be required to produce a neutrino with the detected energy? Since so far we have measured only one such coincidence, we might not know the answer to this question until the next generation of neutrino telescopes starts operating, most likely in the coming 10–15 years. The detection of the first neutron-star merger event via its gravitational-wave (GW) signal, in coincidence with a short GRB, has finally validated the case for the origin of this GRB class (Abbott *et al.*, 2017a). Both the short delay ($\sim$2 s) between the GW and prompt GRB signals and the long delay time ($\sim$9 days) of the non-thermal X-ray and radio detections (Margutti *et al.*, 2017), however, indicate that there is still much to be learned about the nature of these non-thermal electromagnetic components.

Finally, there is also a large space for the unexpected, which hopefully will bring new and exciting physics to the field. Some of the open questions are regarding, for instance, the identification of the nature of FRBs by

---

[b]https://ixpe.msfc.nasa.gov/.

exploring counterparts in the TeV range, or the search for gravitational-wave counterparts in the VHE domain, or further, whether there is a new population of TeV sources waiting to be discovered with more sensitive instruments. The answers to those questions, and many more, will hopefully be unveiled in the coming years (Spitler *et al.*, 2016).

https://doi.org/10.1142/9789813275720_0007

Chapter 7

Cosmology with Very-High-Energy Gamma Rays

Elisa Pueschel[*,§] and Jonathan Biteau[†,‡,¶]

[*]Ruhr-University Universitätsstraße 150, 44801 Bochum, Germany
[†]Université Paris-Saclay, CNRS/IN2P3, IJCLab, 91405 Orsay, France
[‡]Institut universitaire de France (IUF)
[§]elisa.pueschel@desy.de
[¶]biteau@in2p3.fr

7.1. Introduction

Modern cosmology depicts the history of the Universe as starting from a hot initial phase filled with particles and antiparticles and expanding to eventually give rise to large-scale structures. While the first moments of the Universe, near the Planck scale at $t_{\rm Pl} = 5.4 \times 10^{-44}$ s, remain elusive on theoretical grounds, modern theories successfully describe the large-scale components and composition of the Universe through a succession of epochs: inflation creating the flatness and causal connection of our Universe; abrupt transitions describing the symmetry-breaking of the electroweak and strong forces; decoupling of matter and radiation, giving rise to the Cosmic Microwave Background (CMB); the Dark Ages ending with the emergence of the first astrophysical sources of reionization; and finally, the formation of structures and galaxies.

Observations constraining Big Bang Nucleosynthesis, the cosmic distance ladder, the spectrum and anisotropies of the CMB, and large-scale structures have been instrumental over the past 50 years in forming the modern vision of the Universe's history, yielding the current ΛCDMs model largely dominated by Cold Dark Matter (CDM) and dark energy (Λ). In this chapter, we discuss the contributions of gamma-ray astronomy at TeV energies to our understanding of the visible content and structure of the Universe. We start from the present epoch with the second-most-intense electromagnetic background field after the CMB: the Extragalactic Background Light (EBL). The EBL is composed of all light emitted by stars and galaxies since the beginning of reionization, including light absorbed and re-emitted by dust. As such, the EBL traces the history of radiating matter in the Universe. We then dive further into the large voids of the Universe to study the large-scale magnetic fields that should permeate them. These fields could originate from the onset of structure formation or early-phase transitions, bringing us back to the infancy of the Universe. We conclude by looking back to the elusive Planck timescale, where the standard models of cosmology and particle physics are no longer applicable. Observations with current-generation gamma-ray astronomy experiments have now started to scratch the surface of cosmology, as we will show in this chapter.

7.2. The Extragalactic Background Light

The study of the EBL is intertwined with an old scientific riddle, introduced by Digges in his English translation of Copernicus's work (1576), further discussed by astronomers such as Kepler, Huygens and Halley, to finally be formulated in its mathematical version by de Chéseaux in 1744.

Let us divide a static, homogeneous and infinite Universe in concentric shells of radius r and width dr. Each layer contains $4\pi r^2 dr \rho_*$ galaxies, with ρ_* their number density. Each galaxy emits a flux $L_*/4\pi r^2$, with L_* its individual luminosity. The flux reaching the observer from each shell is then a constant, $L_* \rho_* dr$, and the sum from multiple shells diverges when integrated over sufficiently great distances, resulting in an infinitely bright sky. This is known as the "dark night sky paradox", or Olbers's paradox, from the amateur astronomer who popularized the riddle in the 19th century (Wesson, 1991). Both de Chéseaux and Olbers proposed an absorbing medium as a solution, but Herschel and Kelvin argued that absorption would inescapably lead to re-emission at longer wavelengths. The elucidation of the problem shook the premise of an infinite and immutable Universe. Instead, light emission has a history that, from the epoch of reionization to the present epoch, is summarized in the starlight that composes the EBL.

7.2.1. *Production and evolution: Modeling the EBL*

The EBL intensity and its shape as a function of wavelength are dictated by the Cosmic Star-Formation History (CSFH). Most of the direct and reprocessed emission from stars lies in the spectral range 0.1–1000 μm: limited at its lower end by the Lyman limit (0.0912 μm) and dominated at its higher end by the CMB. About half of the starlight emitted in the ultraviolet (UV, 0.1–0.3 μm), optical (0.3–0.8 μm) and near-infrared (NIR, 0.8–3 μm) bands escapes from galaxies, while the other half is absorbed by dust and reprocessed into the far-infrared FIR band (50–1000 μm). Accretion of matter onto supermassive black holes of Active Galactic Nuclei constitutes a secondary source of UV and optical photons, about half of which are absorbed and re-radiated in the mid-infrared (MIR) band (3–50 μm) by the hot dusty torii surrounding the accretion disks. Integrating the UV-FIR emission in time over the cosmic ages results in the broadband EBL spectrum at $z = 0$, presented in Fig. 7.1. While also extending to lower and higher wavelengths, the EBL features two main components: the cosmic optical background (COB, 0.1–8 μm), composed of photons that escaped their environment without attenuation, and the cosmic infrared background (CIB, 8–1000 μm), composed of photons absorbed and re-radiated by dust. Because about half of the light emitted by the sources of the EBL escapes directly from its environment and half is reprocessed, the bolometric intensities of the COB and CIB end up being comparable, with both being compatible with 30 nW m^{-2} sr^{-1}, corresponding to a total energy budget for the EBL at the level of 6%–7% that of the CMB. Three state-of-the-art models are compared in Fig. 7.1, each representing a distinct subclass: empirical, phenomenological and semi-analytic.

Deep-field observations from UV-FIR satellites such as GALEX, Hubble, Spitzer and Herschel have been key in deriving the luminosity functions of various galaxy types, down to luminosities low enough that the integrated galaxy light converges at most wavelengths (Driver *et al.*, 2016). These luminosity functions and associated observations are an important input for EBL models.

Empirical models (e.g. Domínguez *et al.*, 2011b; Franceschini and Rodighiero, 2017) exploit galaxy counts and inferred luminosity functions to accumulate average UV-to-FIR template spectra from various subtypes of galaxies. Phenomenological models (e.g. Finke *et al.*, 2010;

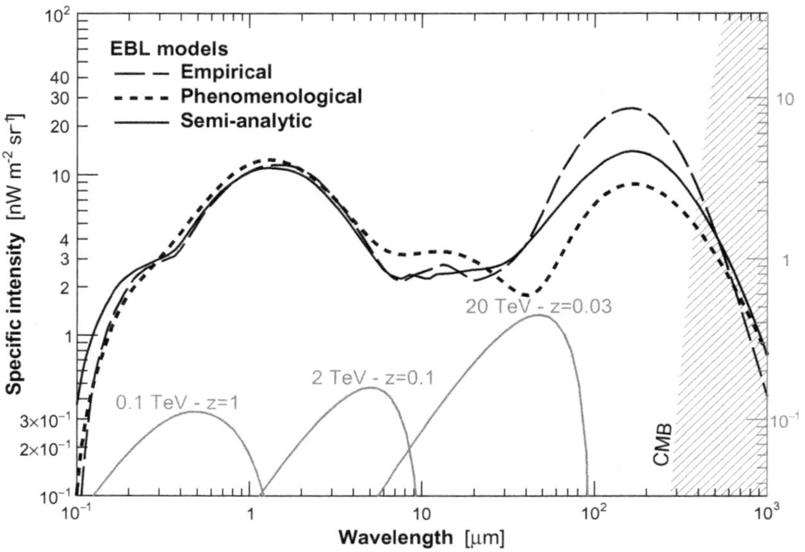

Fig. 7.1. Spectrum of the EBL at $z = 0$, following the empirical model of Domínguez *et al.* (2011b), the phenomenological model of Finke *et al.* (2010) and the semi-analytical model of Gilmore *et al.* (2012). The spectrum of the CMB, peaking around $1000\,\mathrm{nW\,m^{-2}\,sr^{-1}}$, is shown as a dashed area. Gray curves, denoted as optical depth kernels on the right-hand-side axis, display the integral over redshift and angle of interaction of the pair-production cross-section and evolution factors. The integral of the product of these functions and the EBL intensity results in the gamma-ray optical depth, provided here at three characteristic energies and redshifts.

Andrews *et al.*, 2018) additionally aim to reproduce the CSFH and evolution of metallicity. Combining constraints on the cosmic star-formation rate with the emission expected from each star, obtained, e.g. from stellar population synthesis models for a given initial mass function, yields the direct emission contributing to the COB. The CIB is modeled assuming a fraction of reprocessed light and using template-emission spectra for AGNs and dust. These templates use high-quality spectral observations of galaxies and include emission from amorphous or polycylic aromatic hydrocarbons, resulting in the bump around $\sim$15 μm, visible in Fig. 7.1. Finally, semi-analytical models (e.g. Gilmore *et al.*, 2012) additionally aim at tracking accretion throughout cosmic ages by exploiting N-body simulations of dark-matter halo mergers. Such models can include baryonic physics, in particular AGN and supernova feedback, which affects the star-formation rate of their host galaxies.

All three types of models fundamentally aim at reproducing the emission from known, observed galaxies and as such reproduce in a fairly consistent manner the integrated galaxy light inferred from galaxy counts. This consistency is illustrated by a good match of the models around the peak of the COB at 1 μm. While some uncertainties remain in the UV range (see Khaire and Srianand, 2019 and references therein), the clearest mismatch between the models in Fig. 7.1 is observed at mid- and far-infrared wavelengths, where the amount of reprocessing by dust in the environment of the sources plays a key role. It should be noted that, as current models aim at reproducing galaxy counts, they necessarily miss any unresolved population, such as primordial stars, and any truly diffuse components.

7.2.2. *VHE observables*

The very-high-energy (VHE; $E_\gamma > 100\,\mathrm{GeV}$) band opens a unique window to the density of EBL photons integrated over the line of sight, be they from known galaxies or of truly diffuse origin (see, e.g. Franceschini, 2021 for a recent review). Gamma rays from extragalactic sources can interact with EBL photons through the production of electron–positron pairs (Nikishov, 1962; Gould and Schréder, 1967a,1967b). For this process to occur, the product of the gamma-ray energy, E'_γ, and of the EBL photon energy, ϵ', as measured in the cosmological comoving frame, must satisfy a threshold condition imposed by kinematics, that is

$$E'_\gamma \epsilon' \geq \left(m_e c^2\right)^2, \tag{7.1}$$

where $m_e c^2$ is the rest energy of the electron–positron.

The gamma-ray optical depth, $\tau(E_\gamma, z_0)$, with E_γ the gamma-ray energy in the observer's frame and z_0 the redshift of the source, quantifies the number of interactions on the line of sight. As indicated in Eq. (7.2), $\tau(E_\gamma, z_0)$ is the product of the EBL photon density, $\frac{\partial n}{\partial \epsilon}$ (units of $\mathrm{cm}^{-3}\,\mathrm{eV}^{-1}$), and of the differential cross-section for pair production, $\sigma_{\gamma\gamma}$ (units of cm^2), integrated over the distance between the observer and the source, $\frac{\partial L}{\partial z}$ (units of cm), over the energy of the EBL photons, ϵ' (units of eV), and over the angle, θ, between the momenta of the gamma-ray and target-EBL photons.

$$\begin{aligned}\tau(E_\gamma, z_0) = &\int_0^{z_0} dz \frac{\partial L}{\partial z}(z) \int_0^{\infty} d\epsilon' \frac{\partial n}{\partial \epsilon'}(\epsilon', z) \\ &\int_1^{-1} d\cos\theta \frac{1-\cos\theta}{2} \sigma_{\gamma\gamma}\left(E_\gamma \times (1+z), \epsilon, \cos\theta\right).\end{aligned} \tag{7.2}$$

Using $\mu = \cos\theta$, the term $(1-\mu)/2$ in the integral, normalizes the cross-section so that if it were isotropic ($\sigma_{\gamma\gamma}$ independent from θ), one would obtain $\int_{-1}^{1} d\mu \frac{1-\mu}{2} \sigma_{\gamma\gamma} = \sigma_{\gamma\gamma}$.

The distance element in a flat ΛCDM cosmology, with H_0 the Hubble constant, Ω_Λ the dark-energy density and Ω_M the matter–energy density, is given by

$$\frac{\partial L}{\partial z} = \frac{c}{H_0} \frac{1}{1+z} \frac{1}{\sqrt{\Omega_\Lambda + \Omega_M (1+z)^3}}. \tag{7.3}$$

The Breit–Wheeler formula provides the pair-creation cross-section as

$$\sigma_{\gamma\gamma}(\beta) = \frac{3\sigma_T}{16}(1-\beta^2)\left[2\beta(\beta^2-2) + (3-\beta^4)\ln\left(\frac{1+\beta}{1-\beta}\right)\right], \tag{7.4}$$

where σ_T is the Thomson cross-section, and β is given by

$$\beta = \sqrt{1 - \frac{2(m_e c^2)^2}{E'_\gamma \epsilon'} \frac{1}{1-\cos\theta}}. \tag{7.5}$$

The attentive reader should note that $E'_\gamma = E_\gamma \times (1+z)$ and ϵ' are the energies of the gamma-ray and EBL photon interacting at redshift z. It is worth noting that the physical condition that β be a real number and that $\sigma_{\gamma\gamma} \geq 0$ imposes the threshold condition

$$E'_\gamma \epsilon' \geq \frac{2\left(m_e c^2\right)^2}{1-\cos\theta}, \tag{7.6}$$

which translates to the lowest-possible threshold for head-on collisions, i.e. Eq. (7.1) for $\theta = \pi$.

As illustrated in Fig. 7.1, the pair-production cross-section, integrated over the line of sight,[a] has a full width at half-maximum, covering almost a decade in EBL photon energy and peaking at an energy about twice that obtained from the threshold condition in Eq. (7.1). This results in gamma rays interacting preferentially with EBL photons in the wavelength range

$$\lambda_{\rm EBL} \simeq 0.5 - 5\ \mu{\rm m} \times \left(\frac{E_\gamma}{1\ {\rm TeV}}\right) \times (1+z)^2, \tag{7.7}$$

where the factor $(1+z)^2$ accounts for the redshift of the energies of the photons. Interaction with photons from the CMB, which reaches an

[a]More accurately, we refer here to the quantity called the "EBL kernel" (Biteau and Williams, 2015), properly normalized so that its unit is the inverse of that of the EBL intensity.

intensity comparable to that of the CIB around $\sim$250 μm, becomes relevant for gamma-ray energies larger than $\sim$ 50 TeV. From Eq. (7.7), one can infer that measurements of gamma-ray absorption around 20 TeV, obtained from nearby sources ($z \lesssim 0.05$), probe the CIB up to 100 μm, while distant sources ($z \gtrsim 1$) detected up to a few hundred GeV probe the UV-optical part of the COB.

The transparency of the Universe to gamma rays scales with the attenuation factor $\exp(-\tau(E_\gamma, z))$. At first order, the optical depth is a linear function of energy and of EBL photon density integrated over the line of sight, so that the transparency to gamma rays decreases with increasing energy and source distance, as illustrated in Fig. 7.2. Substructures can be identified in Fig. 7.2, with inflection points tracing the variations of the EBL spectrum as a function of wavelength. The features observed in the attenuation reflect the dependence of the EBL intensity on wavelength, with low-energy gamma rays (up to $\sim$ 1 TeV) interacting with photons from the COB, while higher-energy gamma rays interact mostly with photons from the CIB.

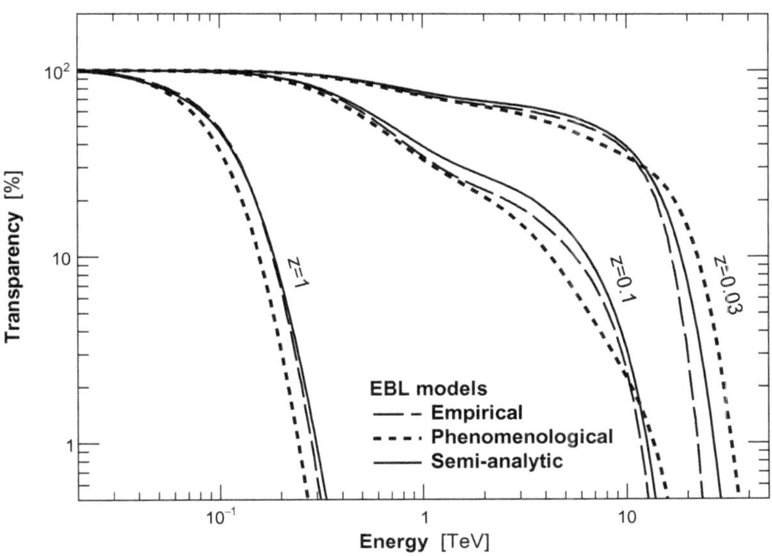

Fig. 7.2. Attenuation factor, in percent, as a function of gamma-ray energy on Earth for sources located at $z = 0.03$, $z = 0.1$ and $z = 1.0$, following the empirical model of Domínguez *et al.* (2011b), the phenomenological model of Finke *et al.* (2010), and the semi-analytical model of Gilmore *et al.* (2012).

The gamma-ray spectrum observed from an extragalactic source is the product of the intrinsic spectrum of the source and the attenuation factor, $\exp(-\tau(E_\gamma, z))$. The intrinsic spectrum is not known a priori, thus multiple features can be used to constrain or measure the EBL density: the variation of the spectral index from the unabsorbed region to the attenuated one (the $\Delta\Gamma$ method, e.g. Sanchez *et al.*, 2013); the measurement of the cosmic gamma-ray horizon (the energy for which $\tau = 1$, e.g. Domínguez *et al.*, 2013); and the reconstruction of the full spectral signature that is shown in Fig. 7.2 (Ackermann *et al.*, 2012h; Abramowski *et al.*, 2013b). The exploitation of these features and the resulting constraints are further discussed in the next section.

7.2.3. *VHE constraints*

The extragalactic gamma-ray sky at GeV and TeV energies is dominated by blazars (BL Lac objects and flat-spectrum radio quasars), which have been detected by ground-based instruments to redshifts of about $z = 1$. The energy spectra of less-distant sources can extend to high energies, in some cases to tens of TeV for sources with particularly hard intrinsic spectra (see Biteau *et al.*, 2020 for a review). Many blazars show dramatic flux variability: for example, up to 10,000 gamma rays have been collected in less than an hour during the highest states of the blazar PKS 2155-304 (Aharonian *et al.*, 2007), enabling the construction of energy spectra with small statistical uncertainties. Blazars thus appear to be ideal gamma-ray beacons to probe the EBL. Long gamma-ray bursts (GRBs), a new class of VHE-emitting sources, are also promising targets (Acciari *et al.*, 2019b; Abdalla *et al.*, 2019b, 2021b). The current generation of Imaging Atmospheric Cherenkov Telescopes (IACTs), together with *Fermi*-LAT (Ackermann *et al.*, 2012h), has mostly used blazar observations to probe the EBL spectral energy distribution (SED) (see, e.g. Desai *et al.*, 2019).

Individual measurements of sources at boundaries of the TeV-accessible redshift range ($z \sim 1$) are relevant for testing the redshift evolution of the EBL. Both MAGIC (Aleksić *et al.*, 2015b; Ahnen *et al.*, 2016b) and VERITAS (Abeysekara *et al.*, 2015) have detected flat-spectrum radio quasars at $z > 0.9$. Studies of the spectra of these sources have placed competitive constraints on the optical peak in the EBL SED. These measurements can be compared to EBL constraints obtained with nearby sources, which probe similar optical depths at the high-energy ends of their spectra. The comparison of measurements with low- and high-redshift gamma-ray sources provides an important consistency check for our understanding of the EBL evolution, provided that the EBL SED is well-constrained.

The most robust measurements use large ensembles of blazar spectra, mitigating the sensitivity to any unexpected spectral properties affecting individual sources. These include absorption of gamma rays by local radiation fields in the vicinity of the source (Reimer, 2007) and unaccounted-for variations in spectral shape over time.

While the latest EBL measurements have focused on a few well-tested methods, we pause briefly to describe alternative strategies. In one approach, intrinsic spectra in the VHE range were predicted from the blazars' multi-wavelength SEDs (from radio to *Fermi*-LAT measurements), assuming that the SEDs could be described by synchrotron self Compton models. The optical depth as a function of energy and cosmic gamma-ray horizon were calculated from the predicted and observed VHE spectra and compared against EBL model predictions (Domínguez *et al.*, 2013); the observed values were found to be consistent with model predictions.

Another method utilizes features in the optical depth as a function of energy. It is visible from Fig. 7.2 that there is a flattening in the optical depth around 1 TeV (most visible for $z = 0.1$). This flattening should propagate to a significant change (a "break") in the spectral shape around 1 TeV for a precisely measured energy spectrum. Testing for a spectral break in sources located at different redshifts thus provides an alternative method for probing the spectral shape of the EBL. Such a study was performed using *Fermi*-LAT and VHE spectra (Orr *et al.*, 2011), and the authors estimated the EBL intensity at 15 μm to be 1.36 ± 0.58 nW m^{-2} sr^{-1}, while demonstrating agreement with model predictions at other wavelengths.

The study of Abramowski *et al.* (2013b) developed the method now commonly employed by the community. While confirming earlier VHE findings that the EBL is primarily due to resolved sources (Aharonian *et al.*, 2006a), Abramowski *et al.* (2013b) established the first significant detection of the EBL imprint at VHE. More recently, H.E.S.S. made a multi-source measurement of the EBL imprint using 21 spectra from nine blazars, covering a redshift range from $z = 0.031$ to $z = 0.287$ and an energy range from 190 GeV to 19.5 TeV (Abdalla *et al.*, 2017b). Datasets were divided into low- and high-flux states, resulting in multiple spectra per source for strongly variable sources, an approach now standard to address flux-level-dependent spectral variability. The method of measuring the EBL was to fit the spectra with a log-parabolic (or power-law, in the case of no curvature) shape multiplied by an EBL attenuation factor of the form $\exp(-\tau(E_\gamma, z, \rho_i))$. Crucially, the level of the EBL intensity, denoted by the parameter ρ, was allowed to vary independently in several adjoining energy bands, indicated by the index i. The only assumption about the shape or normalization of the EBL is the absence of spectral peaks spanning a range of wavelengths smaller than the tested energy bands. The evolution of the EBL with redshift is tuned to replicate the predictions of Franceschini *et al.* (2008); this injects some dependence on a theoretical EBL model, although the impact is limited for gamma-ray sources with redshifts $z < 0.7$ (Biteau and Williams, 2015).

The combined best fit to all spectra reflects both the shape and normalization of the optical depth as a function of energy. The best fit, including EBL attenuation, was strongly favored over a fit assuming no EBL attenuation, indicating evidence of a non-zero EBL at 9.5σ. The optical depth as a function of energy was translated back into an EBL intensity as a function of wavelength. The resulting EBL SED is shown in Fig. 7.3 (open squares labeled "B"). Both statistical and systematic uncertainties are shown; the latter are dominated by uncertainties on the H.E.S.S. energy scale.

Similar approaches to measuring the EBL were adopted by the MAGIC and VERITAS collaborations (Acciari *et al.*, 2019; Abeysekara *et al.*, 2019) using their own sets of blazar spectral measurements. In the case of MAGIC, contemporaneous *Fermi*-LAT spectra were also included, providing a more constraining fit, albeit with larger systematic uncertainties. VERITAS used a different analysis approach to that of MAGIC and H.E.S.S., testing the properties of a set of generic, smoothly varying EBL SED shapes. The results of both of these studies were consistent with the H.E.S.S. results.

A different approach (but also with minimal model assumptions) was taken in a measurement of the EBL based on a large sample of archival spectral measurements (Biteau and Williams, 2015). Altogether 30 sources and 86 spectra were analyzed, bringing together spectral measurements from all of the current IACTs as well as older instruments. Additionally, information from the *Fermi*-LAT spectra was used, namely, the requirement that the VHE spectra be softer than the contemporaneous *Fermi*-LAT spectra.

The EBL signature was detected at the 11σ C.L., and the SED resolved, as shown in Fig. 7.3 (filled squares labeled "A"). This strong detection and the relatively small errors on the points describing the EBL shape made it possible to disfavor several theoretical EBL models.

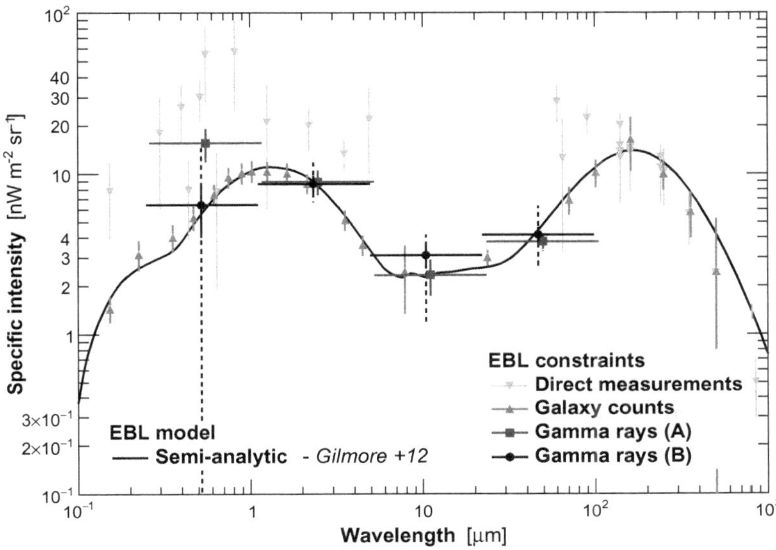

Fig. 7.3. Observational constraints on the spectrum of the EBL at $z = 0$. Grey upward-pointing triangles: lower limits obtained from galaxy counts, which could miss unresolved populations or any truly diffuse EBL component (Driver *et al.*, 2016). Light-gray downard-pointing triangles: constraints from direct measurements, possibly contaminated by foregrounds (Dwek and Krennrich, 2013). Gamma-ray measurements are taken from the works of Biteau and Williams (2015) (a) and Abdalla *et al.* (2017b) (b). Continuous line: semi-analytic model of Gilmore *et al.* (2012).

The assessment of systematic uncertainties was, however, complicated by the use of archival measurements from multiple instruments. It should be noted that instrumental effects can be more thoroughly accounted for in the analysis of lower-level data.[b]

Selected results are summarized in Fig. 7.3. For clarity, Fig. 7.3 only shows the semi-analytic model of Gilmore *et al.* (2012). While no tension with the empirical model of Domínguez *et al.* (2011b) was found in Biteau and Williams (2015), the phenomenological model of Finke *et al.* (2010) appears to be disfavored due to an underestimation of the CIB amplitude, also suggested by the integrated galaxy light measured above $100\,\mu$m (Driver *et al.*, 2016; Duivenvoorden *et al.*, 2020). The most up-to-date

[b]Besides the VHE energy scale, instrumental effects relevant to gamma-ray studies of the EBL, in particular, include energy resolution and selection effects related to the choice of spectral points included in the analysis.

empirical (Franceschini and Rodighiero, 2017; Saldana-Lopez *et al.*, 2021) and phenomenological models (Andrews *et al.*, 2018), driven by the recent gamma-ray detections and updated constraints from galaxy counts (Driver *et al.*, 2016), show an agreement with the semi-analytic model of Gilmore *et al.* (2012) displayed in Fig. 7.3 at the 10%–20% level in the far-infrared region. This level of agreement suggests that both the models and the galaxy count measurements are converging to comparable intensities, not only in the wavelength region of the COB, but also of the CIB.

We note that the current generation of IACTs has not reached the physical limit for resolving structures in the EBL spectrum imposed by the natural width of the EBL kernel. Both of the measurements shown in Fig. 7.3 are based on large data samples, i.e. bright flares, spanning years of observations. In spite of this, the EBL intensity can be extracted for only a limited range of EBL wavelengths, leaving ample room for progress with future observations.

7.2.4. *Relation to non-VHE measurements*

Our knowledge of the EBL benefits from the existence of a variety of measurement techniques beyond the VHE approach described above. Non-VHE measurements can be divided into direct measurements and constraints from galaxy counts (see Mattila and Väisänen, 2019; Driver, 2021 for reviews).

Direct measurements attempt to observe the sky (usually from space, as far from Earth as Pluto's orbit for the New Horizons mission) and subtract foreground contamination from non-EBL light (e.g. Matsuura *et al.*, 2017; Zemcov *et al.*, 2017; Lauer *et al.*, 2021, 2022). This includes a number of light sources, some of which are diffuse: stars, scattered light in Earth's atmosphere and light scattered by dust in the solar system (zodiacal light) and in the Milky Way. Due to the difficulty of adequately modeling and controlling against the foregrounds, direct measurements are generally treated as upper limits on the EBL intensity. Indeed, it has been pointed out that the spectrum of the EBL inferred from direct measurements shows similarities with the zodiacal light spectrum, suggesting an incomplete foreground subtraction (Dwek *et al.*, 2005). An alternative approach that does not rely on the subtraction of a model-dependent foreground consists in measuring the night-sky brightness in the vicinity and within the shadow of a dark, opaque nebula. This approach has been used to measure the EBL intensity in the optical range and its results are consistent with integrated galaxy light and VHE measurements, albeit with larger uncertainties (Mattila *et al.*, 2017a, 2017b).

On the other hand, robust lower limits on the EBL intensity can be derived by summing the observed number of galaxies, accounting for their brightness as a function of wavelength (Levenson and Wright, 2008; Madau and Pozzetti, 2000; Driver *et al.*, 2016). These measurements have become increasingly precise, as deep surveys have resolved fainter and fainter galaxies. One of the best examples to date lies in the 0.4–2μm range, where a precision on integrated galaxy light close to 5% has been claimed (Koushan *et al.*, 2021). Such measurements, however, are incapable of resolving diffuse emission. By comparing them to the VHE measurements, one can test for the presence of a diffuse component. It is interesting to note from Fig. 7.3 that the EBL intensity derived from VHE measurements does not differ significantly from the integrated galaxy light. This rules out a large diffuse component of the EBL — although the uncertainties on both sets of measurements still allow for some diffuse contributions at the level of a few $\text{nW m}^{-2}\,\text{sr}^{-1}$.

The spectrum and evolution of the EBL are tightly connected to the history of star formation, including that of core-collapse supernovae. The cumulative emission of neutrinos in the MeV energy range from all of these objects builds up the diffuse supernova neutrino background (Horiuchi *et al.*, 2009). A signal from the latter could start to emerge in future observations, e.g. of the Super-Kamiokande Gadolinium experiment (Lunardini, 2016), providing constraints not only on the rate of core-collapse supernovae but also on the CSFH and the initial mass function. The interplay of MeV neutrino measurements with high-precision EBL measurements provides exciting prospects for understanding the evolution of baryonic matter in the Universe.

Finally, gamma-ray constraints on the EBL, particularly in the FIR band, could help further constrain the propagation of other astrophysical particles — e.g. ultra-high-energy cosmic rays (UHECRs), which are nuclei with energies larger than 10^{18} eV ($\equiv 1$ EeV). UHE protons interact with CMB photons through the Greisen–Zatsepin–Kuzmin (GZK) effect, losing energy via pion and electron–positron pair production. Heavier nuclei, ranging from He to Fe, can also photodissociate on low-energy photon fields, i.e. release a neutron or a proton carrying a fraction of the energy of the parent nucleus. Photodissociation on FIR EBL photons is the dominant energy-loss process for low-to-intermediate-mass nuclei with a Lorentz factor $\Gamma \sim 10^9$ (Alves Batista *et al.*, 2015). The uncertainty about the intensity of the CIB, as shown in Fig. 7.1, has a non-negligible impact on the interpretation of the UHECR spectrum above a few tens of EeV. It additionally impacts the flux of cosmogenic neutrinos, i.e. those produced along the line of sight by UHECR, particularly around 10^{17} EeV.

7.2.5. *Connecting the EBL and the Hubble constant*

An enticing inversion of EBL measurements is to use the same VHE gamma-ray propagation effects to measure the Hubble constant, H_0. Proposed from the early days of gamma-ray astronomy (Salamon *et al.*, 1994; Blanch and Martinez, 2005a, 2005b, 2005c; Barrau *et al.*, 2008), this alternative method of measuring H_0 remains relevant, as tension between different measurements of H_0 persists at a significance currently estimated to be larger than 4σ (Di Valentino *et al.*, 2021). Two approaches have been employed to constrain the Universe's expansion rate with measurements of gamma-ray absorption.

The first approach, largely independent of the EBL model, consists in comparing the integrated galaxy light to the gamma-ray optical depth inferred from VHE emitters in the local Universe ($z \leq 0.2$), where evolutionary effects have a negligible impact on gamma-ray absorption. Such approaches are labeled here as "gamma-ray/local EBL". Since the optical depth is the product of the EBL intensity, pair-production cross-section and distance element, with the latter scaling as H_0^{-1}, the Hubble constant can be constrained by assuming that the EBL consists entirely of integrated galaxy light. The latter measurements are independent of the Hubble constant, but may underestimate the total EBL, particularly if truly diffuse components or unresolved populations of UV-optical photons are non-negligible.

The second approach exploits the latest models of the EBL, which aim at reproducing the galaxy counts that result in the integrated galaxy light, in addition to the CSFH and metallicity evolution for phenomenological models. Such approaches are labeled here as "gamma-ray/CSFH". For a fixed astrophysical model, the additional dependence of the EBL density on the Hubble constant, scaling as H_0^3, provides a further handle on the Universe's expansion rate. This approach is more sensitive to variations of the Hubble constant, but could also be more dependent on the underlying astrophysical model.

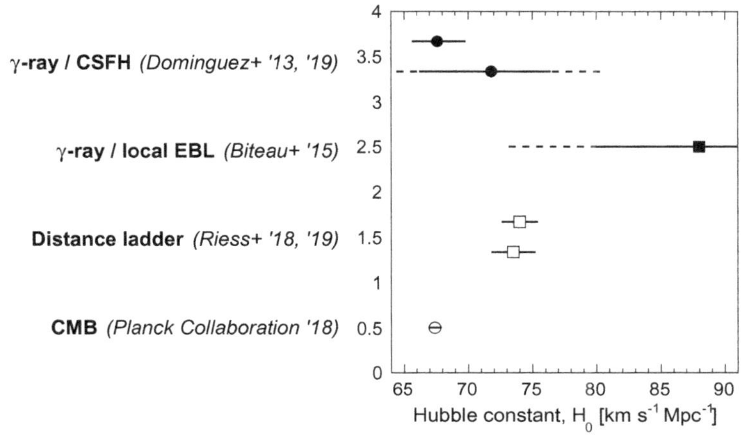

Fig. 7.4. Comparison of measurements of the Hubble constant by Planck (Aghanim *et al.*, 2020) and the Hubble Space Telescope (Riess *et al.*, 2018, 2019) with VHE gamma-ray measurements (Domínguez and Prada, 2013; Biteau and Williams, 2015; Domínguez *et al.*, 2019).

The first measurements exploiting the local (Biteau and Williams, 2015) and evolutionary techniques (Domínguez and Prada, 2013) were dominated by systematic uncertainties about the integrated galaxy light and EBL model, respectively, as well as those about the gamma-ray energy scale. As shown in Fig. 7.4, the systematic uncertainties about H_0 are on the order of 10–15 $\mathrm{km\,s^{-1}\,Mpc^{-1}}$, preventing a useful contribution to the debate between early-type and late-type measurements of H_0, which differ by about 5 $\mathrm{km\,s^{-1}\,Mpc^{-1}}$. Recently, Domínguez *et al.* (2019) further developed the evolutionary approach and tested both the phenomenological model of Finke *et al.* (2010) and the empirical model of Domínguez *et al.* (2011b) against the cosmic gamma-ray horizon inferred from VHE archival data and *Fermi*-LAT data. The authors claimed a measurement with an accuracy of 3 $\mathrm{km\,s^{-1}\,Mpc^{-1}}$ when fixing Ω_M to its reference value and with an accuracy of 6 $\mathrm{km\,s^{-1}\,Mpc^{-1}}$ when leaving Ω_M free (the former is reported in Fig. 7.4).

Independent analyses are critically needed to assess and compare the potential of such gamma-ray constraints. If confirmed, current and future gamma-ray observations could provide relevant inputs to the debate on the current cosmological paradigm. The number of extragalactic gamma-ray sources detectable by the CTAO and the ability to measure their redshifts will be key in the development of this scientific avenue.

7.2.6. *Outlook*

The EBL encodes information about the CSFH and accretion of matter over cosmic ages. Ground-based gamma-ray astronomy is now mature enough to detect the imprint of the EBL in the spectra of gamma-ray blazars and measure the EBL spectrum at $z = 0$ from optical to MIR wavelengths.

Nonetheless, wide areas of research on the EBL remain open. Tighter constraints on the UV part of the EBL spectrum, probed by distant sources beyond $z = 1$, will provide insight into the role of the AGN in ionizing the Universe (Andrews *et al.*, 2018). The signature of polycyclic aromatic and amorphous hydrocarbons, responsible for the bump around 15 μm in the models shown in Fig. 7.1, remains underconstrained at multi-TeV energies. Theoretical and observational uncertainties in the FIR band, which could be solved by high-quality measurements beyond 10 TeV, still hinder a full grasp of UHECR propagation. Exciting prospects can also be found in the detection of the spectral signature of the first stars, microquasars or outflows that may have been responsible for the reionization of the Universe. The signature is expected to be visible as a broad feature above 1 μm in the core EBL wavelength range probed by TeV instruments (Gilmore, 2012). Finally, the wavelength range close to the peak of the COB offers interesting prospects for constraining the amount of light emitted at the outskirts or even outside of galaxies by stars in the intra-halo, intra-group and intra-cluster media (Mattila and Väisänen, 2019).

Besides providing a wide wavelength coverage of the EBL thanks to its order-of-magnitude improvement in sensitivity with respect to current-generation instruments, CTAO will also enable the study of the evolution of the EBL by reconstructing the absorption in successive layers of redshift (Abdalla *et al.*, 2021b). The expected accuracy on the EBL reconstruction is 2–3 times better for CTAO than for current-generation instruments, and it is anticipated that EBL constraints will be derived with gamma-ray sources located at redshifts as distant as $z = 2$. This will enable ground-based gamma-ray astronomy to place meaningful constraints on the CSFH, as recently done by Abdollahi *et al.* (2018). The low-energy threshold of CTAO will further help in disentangling the intrinsic spectral curvature due to internal absorption processes, energy cutoffs related to particle acceleration and escape, and extrinsic spectral features due to the propagation of gamma rays on cosmological scales. Finally, it is expected that CTAO will set robust constraints on the Hubble constant, testing the current cosmological paradigm.

7.3. The Intergalactic Magnetic Field

Understanding the generation and evolution of large-scale magnetic fields in the Universe is a key topic in cosmology, and one for which persistent questions remain. Magnetic fields permeate the Universe from scales as small as stars, to galaxies and galaxy clusters, to the filaments separating the voids in the large-scale structure, and at the largest scales, to the voids themselves.

Two possibilities have been proposed for the origin of cosmological magnetic fields: primordial or astrophysical origins. In the former case, a weak magnetic field is generated in phase transitions in the early Universe. This "seed" field is then amplified in astrophysical systems to the strong magnetic fields ($\sim \mu$G or greater) observed in galaxies and galaxy clusters. In the latter case, the magnetic fields are generated and amplified within galaxies and then ejected beyond the immediate environment of the astrophysical accelerator into the intergalactic medium. The timing of the magnetic field generation is notably different in the two cases: shortly after the Big Bang, or after the onset of galaxy formation. This difference propagates to differences in the magnetic field strength, correlation length (distance over which the field behaves coherently) and the evolution of the magnetic energy density with redshift in the different generation scenarios.

It is worth briefly discussing the underlying mechanisms for the generation and amplification of magnetic fields in plasmas. A magnetic field in a plasma can be spontaneously generated via the Biermann battery process. In a plasma with a temperature and density gradient, the lighter, free electrons drift down the pressure gradient faster than protons. For misaligned temperature and density gradients, this creates an electromotive force that generates a magnetic field. Once created, a weak magnetic field can be amplified via dynamo mechanics. The movement and rotation of plasma in an astrophysical system stretches and compresses magnetic field lines, amplifying the field as kinetic energy is converted into magnetic energy.

Extracting clues about magnetic field generation from observations of magnetic fields in galaxies and filaments is extremely difficult due to the uncertainties associated with the dynamics of the amplification process. On the other hand, the magnetic field in the voids of the large-scale structure — the Inter-Galactic Magnetic Field (IGMF) — reflects the conditions of its generation for primordial production scenarios. In the case that the magnetic fields in the voids are not primordial, but due to plasma outflows from astrophysical systems, they would provide information about the efficiency of magnetic energy transfer into the voids. Consequently, regardless of the generation scenario, magnetic fields in voids are of particular interest (Neronov and Semikoz, 2009; Durrer and Neronov, 2013).

Measuring the IGMF poses experimental challenges, primarily because it is expected to be extremely weak. While some experimental and theoretical bounds exist, the allowed range of strength and correlation length is distressingly broad, as shown in Fig. 7.5 (Alves Batista and Saveliev, 2021). The lowest guaranteed theoretical bound on the IGMF exploits the Harrison effect (Harrison, 1970). Magnetic fields are generated during the radiation-dominated era due to the difference in angular momentum of electron–photon and ion gases as rotating protogalaxies expand. The minimum magnetic field strength generated by this process gives a lower

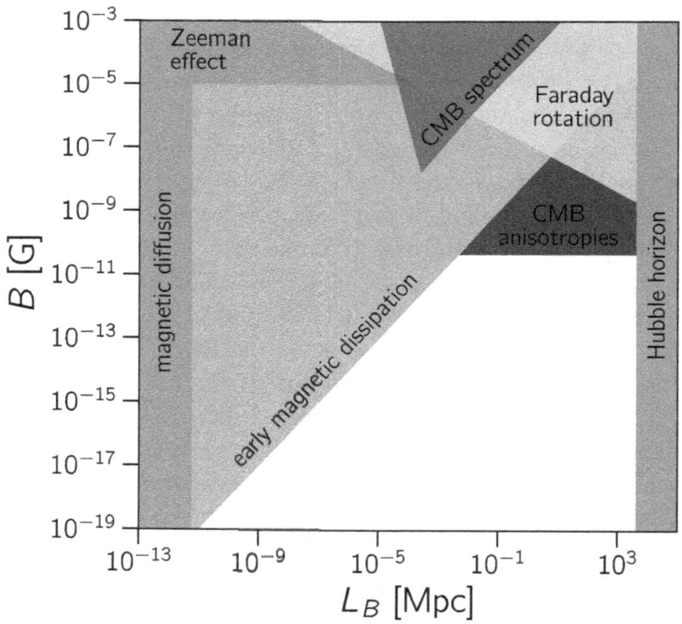

Fig. 7.5. The allowed region of IGMF strength and correlation length. Bounds from non-VHE measurements (discussed in the text) are shown by the shaded regions. Credit: Rafael Alves Batista, adapted from Alves Batista and Saveliev (2021).

bound on the IGMF strength in voids of 10^{-29} G for a correlation length of 10 Mpc. The gamma-ray measurements described in the following set more aggressive lower limits on the IGMF strength.

7.3.1. *Field generation*

7.3.1.1. *Early universe*

Magnetic fields can be generated in the early Universe due to the charge separation that occurs during phase transitions. The Hubble horizon at the time of field generation determines the possible range of correlation lengths for the case of magnetic fields generated after inflation. An IGMF generated during a QCD phase transition, for example, is expected to have a longer correlation length than one generated during an electroweak phase transition, although it must be noted that there is a large overlap region in the strength and correlation-length predictions in the two generation scenarios (see, e.g. Neronov and Semikoz, 2009). The correlation length of inflation-generated magnetic fields, on the other hand, is not limited by the Hubble horizon and can reach the current horizon scale (Kahniashvili *et al.*, 2016).

7.3.1.2. *Astrophysical*

Turbulent plasma movement occurs in a number of astrophysical systems and could lead to the injection of magnetic fields into the intergalactic medium. Star-formation activity and supernovae in starburst galaxies are candidates for IGMF generation. Starburst activity of early forming dwarf galaxies at a time when the Universe was smaller and more densely packed could also plausibly magnetize the intergalactic medium. Alternatively, at later stages in the Universe's evolution, the giant lobes of radio galaxies and the jets of AGNs could expel magnetized plasma; indeed, much more energy is available from these sources than from the outflows of starburst galaxies.

The expected correlation length for an astrophysically generated IGMF is shorter than for a primordial field due to the later generation. The field strength is expected to be high in the vicinity of the generating sources, but to decay steeply toward zero in voids. The predicted strength in the voids is of the order 10^{-10} G or less, several orders of magnitude less than the strength in sheets and filaments (Neronov and Semikoz, 2009; Vazza *et al.*, 2017).

7.3.1.3. *Beyond strength and length*

In addition to measuring the strength and correlation length of the IGMF, constraining its helicity is of particular interest, both for understanding the conditions at generation and for modeling its evolution (Vachaspati, 2021). The diffusion of magnetic energy from small scales to large scales depends on the helicity of the field, with a slower dissipation time for helical fields (Neronov and Semikoz, 2009).

Lastly, measuring the redshift evolution of the field strength and energy density is important for tracing the generation of the field. An astrophysically generated IGMF is expected to have a lower energy density at high redshifts than in the local Universe.

7.3.2. *VHE observables*

We have already discussed the impact of the EBL on VHE photons traveling from distant blazars. Neglected in that discussion was the further trajectory of the electron–positron pairs produced in interactions with EBL photons. As charged particles, they are deflected by the IGMF prior to interacting with the CMB. The electrons and positrons upscatter the CMB photons to higher energies via inverse Compton scattering. For sufficiently High-Energy (HE) particles, the process repeats and an electromagnetic cascade develops. A useful estimation of the energies involved is given in Neronov and Semikoz (2009):

$$E_\gamma = \frac{4}{3}(1+z_{\gamma\gamma})^{-1}\epsilon_{\rm cmb}\frac{E_{\rm e}^2}{m_{\rm e}^2c^4} \simeq 0.32\,{\rm TeV}\left[\frac{E_{\gamma 0}}{20\,{\rm TeV}}\right]^2, \tag{7.8}$$

for which E_γ is the energy of the observed photon on Earth, $E_{\gamma 0}$ is the energy of the primary photon at the source, $m_{\rm e}$ and $E_{\rm e}$ are the mass and energy at pair production of the electron or positron, respectively, and $\epsilon_{\rm cmb} = 0.6\,{\rm meV}(1+z_{\gamma\gamma})$ is the CMB photon energy. A ~ 10 TeV primary photon is thus reprocessed to $\sim$100 GeV, not far above the energy threshold of current-generation IACTs.

Deflection of the electrons and positrons by the IGMF broadens the cascade in space and delays the cascade arrival time at the observer by increasing the path length from the pair-production site to the observer. Both of these effects make it possible to probe the IGMF strength along the line of sight with observations of the VHE gamma-ray emission of distant blazars.

The narrowly beamed ensemble of relativistic electron–positron pairs constitutes a plasma and can develop electromagnetic and electrostatic plasma beam instabilities. Whether these instabilities develop faster than the timescale for inverse Compton scattering, and how efficiently they dissipate energy into the intergalactic medium, is important for understanding whether inverse Compton scattering is the dominant cooling process and whether an electromagnetic cascade can indeed develop (Broderick *et al.*, 2012). Some simulations indicate that beam cooling from plasma instabilities is not expected to be a large effect in the case of cascades from TeV blazars (Sironi and Giannios, 2014; Rafighi *et al.*, 2017). Other results indicate, however, that the beam cooling from plasma instabilities could proceed on a similar timescale to inverse Compton cooling, reducing the expected cascade emission (Vafin *et al.*, 2018). The theoretical debate on this topic thus awaits conclusive answers. The reader should note that the results that follow assume that inverse Compton scattering is the dominant cooling process.

7.3.2.1. *Spectral signature*

Due to the reprocessing of emission from higher to lower energy, the shape of the GeV–TeV spectrum is sensitive to the IGMF strength. Some fraction of electron–positron pairs are deflected out of the observer's FoV for a non-zero IGMF, reducing the number of observed GeV gamma rays and the relative ratio of GeV to TeV emission. By comparing the observed source spectrum with predicted spectra with and without considering an IGMF, it is possible to set lower limits on the IGMF strength. It is necessary, however, to make some assumptions about the shape of the intrinsic source spectrum.

Blazar flux variability is an important confounding factor in such studies. Unless the GeV and TeV observations are made simultaneously, it is impossible to differentiate an IGMF-induced effect from sampling different flux states with the GeV and TeV observations.

7.3.2.2. *Pair halos and magnetically broadened cascades*

A second approach for measuring the IGMF is to search for a low-energy angular extension produced by angular broadening of the cascade emission. As distant sources, even the closest blazars are expected to be point-like within the angular resolution of the current instruments. Depending on the strength of the IGMF, however, two types of angular extensions can occur, referred to here as pair halo and magnetically broadened cascades. The key point is that some fraction of electrons (positrons) isotropize in the magnetic field and gyrate around their point of production. The stronger the IGMF, the higher the energy of the charged particles that are isotropized.

A pair halo refers to the case where a strong IGMF (10^{-12}–10^{-9} G) isotropizes a large fraction of high-energy charged particles in the cascade, causing them to accumulate around the source. The reprocessed emission arriving at the observer is consequently extremely time-delayed on the order of hundreds of years (Eungwanichayapant and Aharonian, 2009).

In the case of magnetically broadened cascades, a weaker IGMF (10^{-16}–10^{-12} G, still strong enough to produce an extension beyond the point spread function of current instruments) isotropizes only lower-energy charged particles, while much of the cascade travels to the observer with time delays of the order of years with respect to the primary emission.

Predictions for angular extensions can be arrived at through analytic approximations (Neronov and Semikoz, 2009; Dermer *et al.*, 2011). An approximation of the deflection angle of an inverse Compton gamma ray illustrates the important factors (Neronov, 2007):

$$\theta_{ext} \simeq \frac{0.7^\circ}{\tau(E_{\gamma_0}, z)} \left[\frac{E_\gamma}{6\ \mathrm{TeV}}\right]^{-1} \frac{B}{10^{-13}\ \mathrm{G}}. \tag{7.9}$$

As expected, the extension scales with the IGMF strength B and inversely with the energy of the upscattered photon E_γ. It also depends on the optical depth for the primary photon (the equation holds for $\tau > 1$). Consequently, the more distant the source and the higher the energy of the primary photon, the smaller the predicted extension. This only holds to a point; if the mean free path of the primary gamma ray has the same scale as the distance to the observer, the cascade will not have developed enough to be observable by the time it reaches the observer. Consequently, sources at $z \simeq 0.1$–0.2 are particularly suitable for IGMF studies in the energy range of interest for IACTs.

7.3.3. *Simulating cascades*

In addition to analytical approximations, there is an abundance of numerical simulations that treat the cascade process and magnetic field structure with varying levels of detail (Eungwanichayapant and Aharonian, 2009; Taylor *et al.*, 2011; Kachelriess *et al.*, 2012; Arlen *et al.*, 2014; Abramowski *et al.*, 2014d; Finke *et al.*, 2015; Alves Batista *et al.*, 2016; Archambault *et al.*, 2017b; Fitoussi *et al.*, 2017). Numerical simulations, although often computationally intensive, provide more accurate predictions of the cascade spectrum and angular extension as well as the ratio of cascade to total emission (cascade fraction). They also allow a detailed treatment of the primary spectrum of the source, most commonly taken as a power law with an exponential cutoff at HE in the literature. The impact on the cascade emission of factors, such as the Doppler factor of the blazar jet, the viewing angle of observer with respect to the jet and the choice of EBL model, can also be studied.

7.3.4. *VHE constraints*

Similarly to EBL studies, hard-spectrum blazars — particularly extreme TeV blazars — are good candidates for IGMF studies (Biteau *et al.*, 2020). As discussed previously, the primary emission is reprocessed to much lower energies through cascade development. The cascade emission coming from sources with soft spectral indices and low-energy (sub-TeV) spectral cutoffs falls below the GeV–TeV energy range. Extreme blazars thus represent the most viable candidates for setting VHE constraints on the IGMF.

The constraints presented below can be best read as order-of-magnitude estimates. The uncertainties on predictions of the cascade emission are still large and have a strong impact on the derived constraints.

7.3.4.1. *Constraints from spectral measurements*

Many attempts have been made to set lower limits on the IGMF strength based on the GeV–TeV broadband spectrum. We only discuss selected results here. All studies nominally assumed a correlation length $\lambda_B \geq 1$Mpc.

An early study used archival TeV spectra of three hard-spectrum blazars, together with simultaneous *Fermi*-LAT observations, to produce broadband spectra (Taylor *et al.*, 2011). These were then fit with numerical predictions for the spectral shape with and without the IGMF. Example spectra are shown in Fig. 7.6. As shown in the figure, the predicted spectrum for IGMF strength $B = 0$ overshoots the observed *Fermi*-LAT flux. Assuming the GeV deficit is due to the cascade emission being deflected out of the FoV of the observer, B must be at least $\sim 10^{-15}$ G. Taking the pessimistic assumption that the source was only active during the observation time (three years at the time the study was made) and the

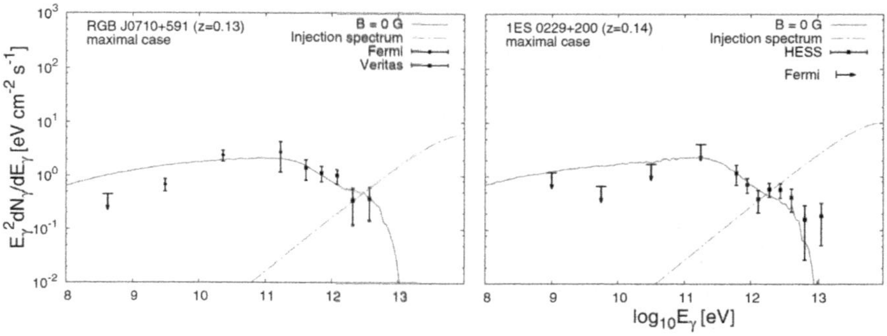

Fig. 7.6. The HE/VHE spectra for two gamma-ray blazars, shown with the predicted cascade component in the absence of the IGMF (solid line). The predicted GeV component of the spectrum is higher than the observed flux. Upper bounds denoted by arrows: 99% C.L. Dashed line: injection spectrum prior to the cascade development. Extracted from Taylor *et al.* (2011).

GeV deficit is due to the time delay of the cascade emission with respect to the primary emission, a lower limit of $B \gtrsim 10^{-17}$ G was obtained.

A later study used the four blazar data sets of Taylor *et al.* (2011) as well as four additional hard-spectrum blazars (Arlen *et al.*, 2014) to create a scan of the possible parameters (jet opening angle, Doppler factor, etc.) within reasonable limits. Accounting for these uncertainties, the authors concluded that the broadband spectra are compatible with $B = 0$. Some tension was observed in the case of one source (1ES 0229+200); it was noted, however, that this could be resolved by shifting the assumed EBL shape within experimental uncertainties.

An overlapping set of sources, covering observations of five hard-spectrum blazars, and including several more years of *Fermi*-LAT data than previous studies, has also been considered (Finke *et al.*, 2015). While an analytical approach to the cascade predictions was taken, the predicted cascade spectra were checked against simulation. An IGMF strength of less than 10^{-19} G was ruled out at $\geq 5\sigma$ confidence.

7.3.4.2. *Constraints from searches for angular extension*

All of the currently operating IACT collaborations have made searches for angular emission around blazars. Searches for extended emission have the advantage that they can be carried out by a single instrument. The sensitivity to extended cascade emission is driven by two factors: the energy threshold and the angular resolution of current-generation IACTs. The former determines the total measurable flux of cascade emission; the latter is directly correlated to the smallest-detectable extension and the weakest-detectable IGMF. A standard approach has been to compare the angular profile observed in the data to the prediction for a simulated point source, where the extension is entirely due to the instrument point spread function.

An early attempt was made by MAGIC to constrain the IGMF with a search for extended emission around the nearby blazars Markarian 501 and Markarian 421 (Aleksić *et al.*, 2010c). In lieu of the detection of an IGMF signature, placing constraints is complicated by the close proximity of the sources, their extreme variability and the presence of source-intrinsic spectral cutoffs.

H.E.S.S. studied three hard-spectrum blazars located at redshifts $z = 0.140$–0.186 using numerical simulations to predict the cascade flux and angular profile (Abramowski *et al.*, 2014d). An example of the angular profile in data and simulation is shown in Fig. 7.7. Extended emission was ruled out at the level of a few percent of the Crab Nebula flux for the most constraining source. Additionally, H.E.S.S. set limits on the cascade fraction, which were translated into a limit on the IGMF strength. The cascade simulation predicts the cascade fraction and angular extension as a function of IGMF strength for an assumed intrinsic spectrum. Figure 7.7 shows the evolution of the predicted cascade fraction as a function of IGMF strength, compared to upper limits on the cascade fraction extracted from data. The latter are produced by finding the maximum cascade fraction that could result in a non-detection of an angular extension. Uncertainties due to assumptions about the intrinsic spectra of the sources and the EBL model used in the cascade simulation were considered. The impact of these uncertainties is significant, but it was still possible to rule out a range of IGMF strengths around 10^{-16} G to 10^{-15} G at the 99% C.L., assuming a correlation length of 1 Mpc.

VERITAS produced a complementary result to the H.E.S.S. measurement and arrived at similar conclusions (Archambault *et al.*, 2017b). Observations of seven hard-spectrum blazars were utilized, covering a range of redshifts from $z = 0.031$ to $z \sim 0.5$. As with the H.E.S.S. measurement, the angularly extended flux was constrained to be less than a few percent of the Crab Nebula flux. One source (1ES 1218+304) was found to be the most useful object for IGMF studies due to its brightness, redshift and hard intrinsic spectrum with no identified cutoff. It was thus the

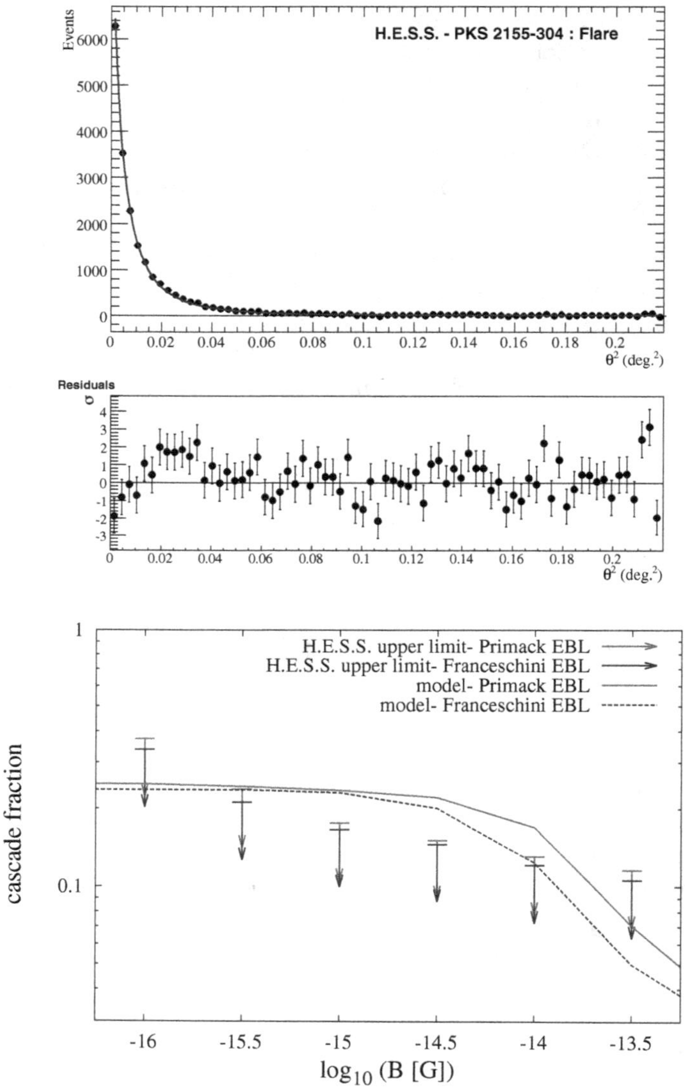

Fig. 7.7. *Top*: Angular profile of a flaring blazar observed by H.E.S.S. (shown as data points) whose angular extension is expected to be only due to the instrument point spread function. It is in good agreement with a simulated point source (solid line), as illustrated by the residuals to the best fit shown in the middle panel. *Bottom*: Upper limits on the observed cascade fraction are compared against predictions as a function of IGMF strength. Extracted from Abramowski *et al.* (2014d).

only source considered for extracting IGMF limits. Limits on the cascade fraction as a function of IGMF strength were set and used to constrain the IGMF strength. The choice of sources makes the H.E.S.S. and VERITAS measurements overlapping but complementary, with VERITAS ruling out a set of slightly stronger IGMF strengths between 10^{-15} G and 10^{-14} G at the 95% C.L., again assuming a correlation length of 1 Mpc.

The definitive study, at the time of writing, is a search that combined angular and spectral information from *Fermi*-LAT data and archival IACT spectra (Ackermann *et al.*, 2018), with the spectral information proving key to the extraction of IGMF constraints. *Fermi*-LAT observations are particularly useful as they can probe broad extensions on degree scales, which would result from cascades of primary gamma-rays at energies covered by IACTs. A set of six hard-spectrum blazars with low detected variability and spectral points extending to an optical depth $\tau = 2$ were studied individually, as well as jointly in a stacked analysis. The time period for which the sources were assumed to be active was varied from 10 to 10^7 years, the former corresponding to the observation period and the latter to the typical AGN activity timescale, and the IGMF constraints were reevaluated. Even in the case of the most conservative assumption of the blazars only being active for ten years, the extracted limits were significantly more constraining than the other measurements discussed here. A lower limit of $B \gtrsim 10^{-16}$ G was set at the 95% C.L., assuming a correlation length greater than 10 kpc. Combined with the smaller H.E.S.S. and VERITAS exclusion regions, the *Fermi*-LAT results substantially reduce the allowed region for a weak IGMF.

7.3.5. *Relation to non-VHE measurements*

As is shown in Fig. 7.5, upper limits on the IGMF strength are also available based on non-VHE measurements (Heiles and Troland, 2004). Magnetic fields encountered en route to the observer lead to Zeeman splitting of the 21-cm hydrogen absorption line in quasar spectra. The primary expected contributor to this is the Galactic magnetic field, thus the IGMF must clearly be less than the total magnetic field derived from Zeeman splitting (Neronov and Semikoz, 2009). Another constraint comes from polarized emission by quasars, which undergo Faraday rotation. The polarization angle of the emission changes with wavelength, with the ratio of the change in polarization to the change in wavelength varying directly with the magnetic field strength along the line of sight (Blasi *et al.*, 1999).

Lastly, tangled primordial magnetic fields are expected to produce anisotropies in the CMB, the non-observation of which can be used to constrain the field strength (Ade *et al.*, 2015, 2016; Jedamzik and Saveliev, 2019). Further constraints can be derived from Big Bang nucleosynthesis, namely, that a large primordial magnetic field would drive an overly fast expansion of the Universe, slowing heavy-element formation and leading to a greater fraction of hydrogen to heavy elements than is observed (Grasso and Rubinstein, 2001).

Upper limits on the IGMF correlation length are set under the assumption that the correlation length must be less than the size of the Universe, while lower limits on the correlation length can be set based on the predicted decay time of the IGMF due to magnetic diffusion.

In addition to its interest as a cosmological observable, the IGMF strength and structure are experimentally relevant for a number of measurements. UHECRs are bent by the IGMF, although the effect of magnetic fields in voids is expected to be subdominant compared to deflections by the stronger magnetic fields in filaments, galaxy clusters and in the Milky Way. As discussed in Chapter 8, dedicated to fundamental physics, the IGMF is also important for studies of axion-like particles, hypothetical particles which couple to the magnetic field, converting to and from VHE photons.

7.3.6. *Outlook*

Magnetism on cosmological scales remains a mystery. As discussed in the introduction, lower limits on the IGMF strength based on the Harrison effect are nearly 15 orders of magnitude too low to explain the missing cascade component at low gamma-ray energies. Is the IGMF strength larger than constrained by H.E.S.S. and VERITAS? Which magnetic field generation scenarios are the most viable and what would be the resulting strength, correlation length and helicity of the IGMF in voids today? Could plasma instabilities play a significant role in cooling electron–positron beams that develop over dozens of Mpc?

More precise non-VHE measurements and convincing theoretical arguments may reduce the allowed region for the IGMF strength and correlation length, and further study of collective plasma effects may clarify predictions for the cascade development. Observational searches with IACTs, however, have already shown the possibility of probing an uncharted territory in the parameter space of the IGMF. These searches can be expected to expand beyond blazar-based studies in the coming years, as a catalog of TeV-detected GRBs is assembled, enabling a parallel set of searches Associations between neutrinos and blazars, discussed in the chapter dedicated to multi-messenger astronomy, also present a potentially promising route to constrain the IGMF, using the time delay between the neutrino measurement and gamma-ray flare detection.

Regardless of the source class considered in IGMF searches, a low-energy threshold, sensitivity from tenths to tens of TeV to constrain both the primary and secondary components and an angular resolution at the few arcminute level constitute the key characteristics of an instrument that will be able to either measure for the first time or strongly constrain the large-scale magnetism of the Universe. With these capabilities, CTAO will probe IGMF strengths up to 0.3 pG (Abdalla *et al.*, 2021b). Future VHE gamma-ray observations will undoubtedly improve our understanding of the origin of cosmic magnetism.

7.4. Lorentz Invariance Violation

We now turn to a topic focused on the fundamental physical laws of the Universe rather than its evolution. Lorentz invariance and the constancy of the speed of light c for all inertial observers underpin the current understanding of relativity and quantum mechanics. A number of theories that attempt to unite gravity with the other fundamental forces and incorporate it into a quantum mechanical framework predict violation of Lorentz invariance at or below the Planck scale, $E_{\rm Pl} \approx 1.22 \times 10^{19}$ GeV (e.g. Rovelli, 2008). This prediction can be tested with gamma-ray astronomy (see, e.g. Amelino-Camelia and Smolin, 2009; Ellis *et al.*, 2008; Amelino-Camelia, 2013 and references therein). Such theories are most simply expressed in terms of effective field theories, with two consequences. First, general and special relativity are recovered at leading order of the Lagrangian, but higher-order terms, scaling with powers of the energy, appear in the Lagrangian. Second, the effective field theory breaks down at a specific energy, presumably below the Planck scale. These consequences dictate the search strategy and energy scale of interest in the measurements described in the following. Observations can be used to constrain particular effective field theories (see, e.g. Kislat and Krawczynski, 2015, where constraints are placed on a specific model, the Standard Model Extension), but the approach taken in most studies has been simpler: to set a lower limit on the energy scale at which a modification to Lorentz invariance can occur.

7.4.1. *Theoretical framework*

A natural consequence of quantum gravity is that the uncertainty principle makes the vacuum richer ("fuzzy" or "foamy"). This introduces a small refractive index for particles traveling through the vacuum, resulting in an energy-dependent dispersion.

Rather than the classic relation $E^2 = p^2c^2 + m^2c^4$, the energy in effective field theories incorporating Lorentz invariance violation (LIV) can be generically expanded as

$$E^2 = p^2c^2 + m^2c^4 \pm E^2\left(\frac{E}{\xi_n E_{\rm LIV}}\right)^n, \tag{7.10}$$

where c is the standard speed of light, $E_{\rm LIV}$ sets the energy scale of the modifications and the coefficients ξ_n give the relative size of the terms in the expansion (note that ξ_n can be very large/infinite, rendering terms negligible/zero). The additional terms can be positive or negative, corresponding to a superluminal or subluminal velocity, respectively. The latter is generally considered more theoretically probable (Jacob and Piran, 2008). It should be noted that odd-powered terms violate CPT invariance (charge conjugation/parity/time, where parity refers to a spatial inversion) and thus theories that respect CPT will only contain even-powered terms. While the standard model of particle physics respects CPT, it can, however, be violated in models incorporating LIV (Amelino-Camelia, 2013).

7.4.2. *VHE observables*

Gamma rays traveling over large distances will accumulate effects of LIV, and the agreement between their expected and observed propagation provides a probe of modifications at a given energy scale. This is of particular interest, as it is an energy regime that cannot be tested with the particle physics experiments upon which we base our understanding of quantum mechanics and special relativity.

7.4.2.1. *Modification of the pair-production threshold*

Returning to the earlier discussion of the EBL, the energy threshold for pair production of gamma rays interacting with EBL photons is modified by LIV terms (Kifune, 1999; Jacob and Piran, 2008). Assuming energy/momentum conservation and considering all particle types to be equally affected by LIV, this modifies Eq. (7.1) for the subluminal case to

$$E'_\gamma \epsilon'_{\rm EBL} \geq \left(m_{\rm e}c^2\right)^2 + \frac{1-2^{-n}}{4}\left(\frac{E'_\gamma}{\xi_n E_{\rm LIV}}\right)^n E'^2_\gamma. \qquad (7.11)$$

It can be seen from Eq. (7.11) that LIV raises the energy threshold for pair production for a subluminal modification of the dispersion relation. As shown in Fig. 7.8, a subluminal LIV decreases the opacity of the Universe to gamma rays by effectively depleting the target photon field interacting with gamma rays above 10 TeV. Assuming that the pair-production cross-section retains its functional dependence on energy under LIV, the loss and recovery of transparency are nearly symmetric around a critical energy (Jacob and Piran, 2008), as shown in Fig. 7.8. As the effect is most noticeable at the highest energies in a gamma-ray spectrum, tests of

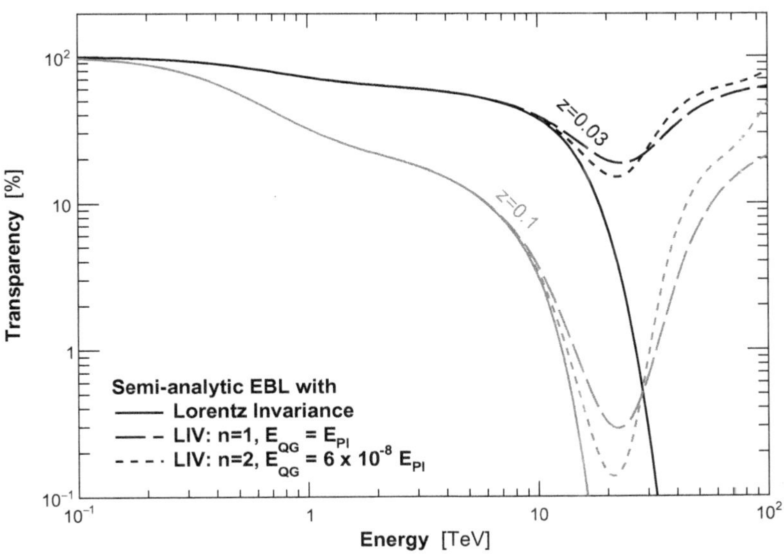

Fig. 7.8. Attenuation factor, in percent, as a function of gamma-ray energy on Earth, for sources located at $z = 0.03$ and $z = 0.1$, following the semi-analytical EBL model of Gilmore *et al.* (2012) and the LIV formalism of Jacob and Piran (2008). Both linear ($n = 1$) and quadratic ($n = 2$) modifications of the pair-creation threshold are shown at the Planck scale for $n = 1$, and at an ad-hoc energy scale of $6 \times 10^{-8}\,E_{\rm Pl}$ for $n = 2$, resulting in a similar effect.

LIV-induced threshold modifications are best performed using observations of blazars with hard photon spectra extending to tens of TeV.

7.4.2.2. *Energy-dependent time dispersion*

Another approach to searching for LIV, which requires no assumptions about kinematics, is to search for a correlation between the energy and arrival time of gamma rays. In this case, the limiting assumption is that all of the gamma rays were produced at the same time at the source (we will discuss the uncertainties associated with this assumption in the following). For an individual photon, a subluminal modification to the velocity is parametrized as

$$v = c \times (1 - (E/\chi_n E_{\rm LIV})^n). \tag{7.12}$$

The parameter χ_n is the equivalent ξ_n in Eq. (7.11) within a numerical factor. A more direct link between Eqs. (7.11), Eq. (7.12) can be found by defining the velocity following the formalism of Hamiltonian mechanics, that is, as $v = \partial E/\partial p$.

Measuring the arrival time of gamma rays of different energies coming from the same emitter enables the extraction of $E_{\rm LIV}$. The limitation of this method is obvious: how does the observer know that the gamma rays were generated in the source at the same time? Such searches focus on bursts of gamma rays, whether from GRBs, AGN flares or the well-defined flux peaks in pulsar light curves, where a sharp increase in flux occurs across the gamma-ray energy range. For single sources, this method remains vulnerable to intrinsic energy dependence of emission at the source (for instance, due to the presence of different emission regions), which could either increase or decrease the observed time dispersion.

A robust detection of an LIV effect would consequently rely on the combination of observations from multiple sources, and indeed multiple types of sources, located at different redshifts. The time difference due to an LIV should increase as a function of redshift as the effects accumulate over the photon's path length. Detection of an energy-dependent dispersion without such a trend would point to intrinsic source effects.

In terms of source selection, the properties of GRBs, AGNs and pulsars offer different advantages and disadvantages. The ideal candidates for LIV studies are distant (LIV effects accumulate over distance), rapidly variable (producing well-defined bursts of gamma rays), and detected to high energies (as the modifications to the propagation scale with energy).

AGNs and GRBs are attractive candidates for the first and third properties, located at cosmological distances and with energy spectra that are observed to extend to tens of TeV for AGNs and beyond 1 TeV for GRBs. The flux variability of AGNs is, however, not optimal, as it is unpredictable and often slow (timescales of days to years). GRBs behave more predictably, with a rapid burst of gamma rays followed by a steep decay in the gamma-ray flux, but there is variation between bursts as well as uncertainty in the underlying structure of the flux as a function of time. Pulsars are excellent candidates from the perspective of flux variability. The peaks of their light curves are sharp and predictable (enabling improvements in statistical uncertainties of LIV tests by simply observing the sources for longer), and changes in their behavior, due to spin-down, for instance, are precisely predicted. For energy range and distance, though, pulsars are less attractive candidates, located within our galaxy and emitting gamma rays to ~1 TeV rather than tens of TeV. Only four pulsars are detected at VHE; of these, the Crab pulsar is detected to the highest energies, up to ~1.5 TeV (Ansoldi *et al.*, 2016).

7.4.3. *VHE constraints*

7.4.3.1. *Search for the modification of pair-production threshold*

The large blazar spectral sample of Biteau and Williams (2015) facilitated the extraction of limits on $E_{\rm LIV}$ while leaving the EBL SED free within observational constraints. A range of possible $E_{\rm LIV}$ was scanned under the assumption of subluminal LIVs, with a recalculation of the optical depth with the modified pair-production threshold of Eq. (7.11). No significant evidence of LIV was observed. Consequently, a lower limit of $E_{\rm LIV} > 0.6 \times E_{\rm Pl}$ was set at the 99% C.L. This can be translated into a limit of $4 \times 10^{-8} \times E_{\rm Pl}$ for a second-order modification (see Eq. (32) in Biteau and Williams, 2015).

Similarly, Lang *et al.* (2019) evaluated three distinct EBL models and selected the most promising spectra for detecting LIV effects from a sample of 111 archival AGN energy spectra. The final sample included 18 spectra from six blazars, dominated by bright and only moderately distant ($z < 0.2$) objects. The extracted limits on $E_{\rm LIV}$ for subluminal LIV are well above the Planck energy for the linear ($n = 1$) term of Eq. (7.10), with $E_{\rm LIV} < 10 \times E_{\rm Pl}$ excluded at the 95% C.L. For the quadratic term, $E_{\rm LIV} < 2 \times 10^{-7} \times E_{\rm Pl}$ is excluded.

As for studies of the EBL, analyses exploiting archival data are not devoid of possible biases. In particular, for LIV, the treatment of upper limits at the high-energy end of the gamma-ray spectra is expected to impact the constraining power on the LIV energy scale, which could contribute to the differences in the results from the two aforementioned studies.

7.4.3.2. *Search for energy-dependent time dispersion*

In 2006, H.E.S.S. observed a massive flare from PKS 2155-304, peaking above ten times the flux of the Crab Nebula. The source displayed significant variability on the timescale of minutes. A time-dependent analysis was performed wherein the structure of the intrinsic flux variability was assumed to be given by the observed flux variability of the lowest-energy gamma rays, between 0.25 TeV and 0.28 TeV. No distortion of the light curve stage at higher energies was allowed — only an overall shift in arrival time.

No evidence for an energy-dependent time dispersion was observed, and lower limits were set on E_{LIV}: on the linear term at $0.2 \times E_{\mathrm{Pl}}$ and on the quadratic term at $5 \times 10^{-9} \times E_{\mathrm{Pl}}$, both at the 95% C.L. While these limits were calculated for the subluminal case, the superluminal case gives similar, slightly weaker limits (Abramowski *et al.*, 2011c).

A similar study was undertaken by MAGIC for two flares of the blazar Mrk 501, yielding similar limits for the quadratic term to the H.E.S.S. result, and order-of-magnitude weaker limits on the linear term (Albert *et al.*, 2008c; Martínez and Errando, 2009).

Turning to pulsars, the MAGIC collaboration produced LIV constraints based on 12 years of observations of the Crab pulsar (Ahnen *et al.*, 2017b). As a pulsar, its flux shows a predictable time structure, in the case of the Crab characterized by two sharp peaks of differing intensity, and a low state or non-detected state in between the peaks, as shown in Fig. 7.9. While the variability behavior of the source is controlled, the distance to the Crab pulsar is, to date, only measured to 25% accuracy (Kaplan *et al.*, 2008), which propagates to an uncertainty on the derived LIV results. Both a simple comparison of the peak positions in time for gamma rays in different energy ranges and a full likelihood analysis were performed, yielding consistent results. In the absence of an LIV signal, lower limits were calculated at the 95% C.L. for E_{LIV}, corresponding to $0.05 \times E_{\mathrm{Pl}}$ for the linear term and $5 \times 10^{-9} \times E_{\mathrm{Pl}}$ for the quadratic term, assuming

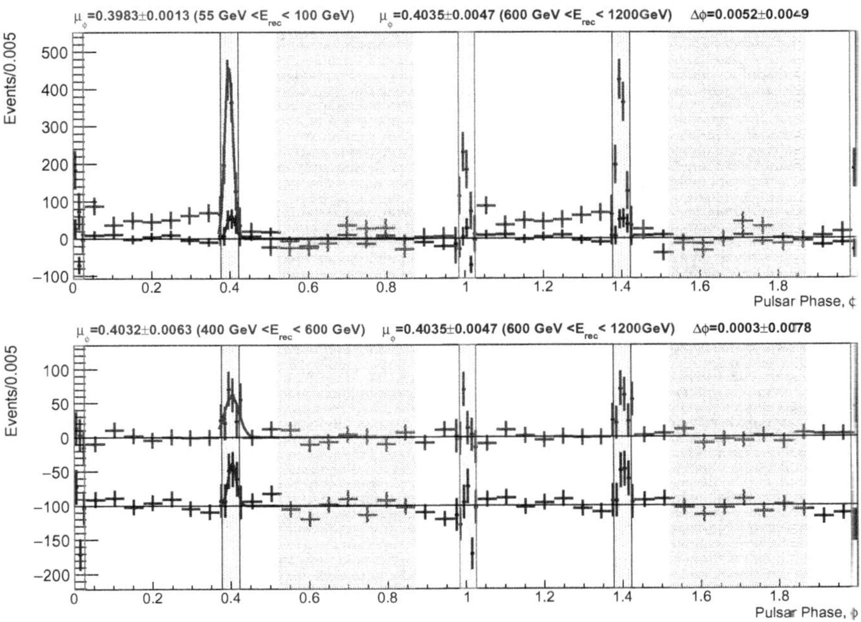

Fig. 7.9. The Crab pulsar folded light curve as measured by MAGIC. Two periods are shown, for high and low gamma-ray energies. The shaded areas define the pulse and inter-pulse regions. Extracted from Ahnen *et al.* (2017b).
Note that the values of μ and $\Delta\phi$ indicated in the figure have been obtained with the less accurate *peak comparison method* described in the above-mentioned paper, which also provides more accurate values in its Table 3.

subluminal LIV. As for the results of Abramowski *et al.* (2011c), the limits for the superluminal case are slightly weaker.

With the detection of the first GRBs at VHE, a new source class is available for time-dispersion measurements. In particular, the MAGIC-detected GRB190114C had the large redshift ($z \sim 0.4$) and high-energy photons (largest-detected photon energy was $\sim$2 TeV) necessary to produce competitive LIV constraints (Acciari *et al.*, 2020). The dominant uncertainty in this measurement is the shape of the intrinsic emission as a function of time. The authors considered a minimal model, which assumes zero emission before the multi-wavelength GRB alert was issued, triggering MAGIC observations, as well as a theoretically motivated model in which the peak emission occurs before the beginning of MAGIC observations. Neither model of the intrinsic emission provided evidence for LIV, and

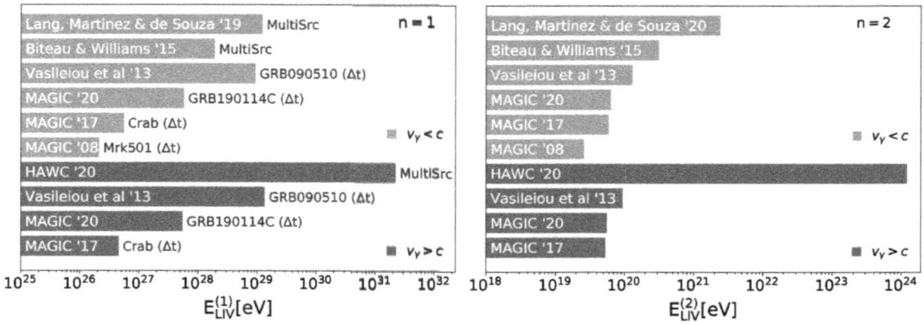

Fig. 7.10. Summary of the lower limits on the energy scale of an LIV, E_{LIV}. *Left*: Limits on the linear term. *Right*: Limits on the quadratic term. Light gray bars assume a subluminal modification while dark gray bars assume a superluminal modification. Credit: Humberto Martínez-Huerta, adapted from Martínez-Huerta *et al.* (2020).

the extracted 95% C.L. limits were comparable, with $0.5 \times E_{\mathrm{Pl}}$ for the subluminal linear term, assuming the physically motivated emission model, and $5 \times 10^{-9} \times E_{\mathrm{Pl}}$ for the subluminal quadratic term. Such limits suggest a sensitivity similar to that of AGN time-dispersion measurements; the lower energy of gamma rays from the GRB are compensated by the greater redshift of the source.

A summary of the limits discussed here, including the non-IACT measurements discussed in the next section, is given in Fig. 7.10. Further discussion of current limits on LIV can be found in Martínez-Huerta *et al.* (2020).

7.4.4. *Relation to non-IACT measurements*

The most constraining limits on the subluminal linear term for LIV come from *Fermi*-LAT. These constraints are based on observations of GRBs, which — within the *Fermi*-LAT-sensitive energy range — are ideal candidates, with sharp flux variability and broadband emission and occurring at cosmological distances. A large sample of GRBs has been measured, although several of the canonical limits are from individual (Abdo *et al.*, 2009a) or small samples of GRBs (Vasileiou *et al.*, 2013). GRBs, however, exhibit intrinsic energy-dependent time lags that must be taken into account (Ackermann *et al.*, 2013). Lower limits at the 95% C.L. on E_{LIV} exceed the Planck scale by a factor of 7.6 for the linear term, going down to a factor of two when accounting for systematic uncertainties related to emission processes at the source. For the quadratic term, the limits are comparable to many of the VHE limits, at $1 \times 10^{-8} \times E_{\mathrm{Pl}}$ (Vasileiou *et al.*, 2013).

The HAWC observatory has strongly constrained superluminal LIV by searching for high-energy spectral cutoffs induced by photon decay to an electron–positron pair or to multiple photons (Albert *et al.*, 2020b). This search utilized four Galactic high-energy emitters, including the Crab Nebula, and placed 95% C.L. lower limits on $E_{\rm LIV}$ three orders of magnitude above the Planck energy for the linear term and $7 \times 10^{-6} \times E_{\rm Pl}$ for the quadratic term.

Upper limits on the flux of photons at ultra-high energies ($E > 10^{18}$ eV) might also provide stringent constraints on LIV, provided that the sources are identified and a significant fraction of protons is present at the highest energies. Such a probe could reach the Planck scale even for second-order modification of the pair-creation threshold (Gelmini *et al.*, 2008; Guedes Lang *et al.*, 2018). LIV could also impact the spectrum of UHECRs by increasing the threshold for photo-production of pions (see Section 3.5 in Amelino-Camelia, 2013, for an excellent review of the debates on this matter). The constraints on the composition and sources of UHECRs expected from the upgrade of the Pierre Auger Observatory will help in determining the magnitude of the constraints on LIVs from UHECR propagation.

7.4.5. *Outlook*

The violation of Lorentz invariance at energies beyond the reach of man-made accelerators can be studied with gamma rays at GeV–TeV energies. Two paths have been followed in the VHE regime, both based on phenomenological modifications of the dispersion relation of photons, resulting in high-energy correction terms.

Assuming that speed can be computed from Hamiltonian mechanics within a quantum-gravity framework, such modifications result in an energy-dependent speed of light whose effect, accumulated over cosmological scales, could be observed as an energy-dependent time lag between successive gamma-ray bands. The lack of such lags has already allowed ground-based instruments to place limits on the LIV energy scale that, for a linear modification, approach the Planck scale. The timing capabilities of CTAO will be more than three orders of magnitude better than *Fermi*-LAT in the 30–100-GeV band (Funk and Hinton, 2013). This will facilitate the detection of numerous highly variable extragalactic sources (GRBs, blazars), constituting a gold mine for LIV studies.

An alternative approach, based on four-momentum conservation, suggests that LIV corrections should result in a modification of the pair-creation threshold, directly affecting the transparency of the Universe beyond tens of TeV. In contrast to timing approaches, the limits for a linear modification have been quoted an order of magnitude above the Planck scale. Limits on the subluminal quadratic modification, however, which are particularly motivated by the theoretical perspective, remain well below the Planck energy on the order of $10^{-7} \times E_{\text{Pl}}$. Thanks to its enhanced sensitivity at the highest energies, CTAO will be able to probe LIV predictions in both nearby (Fairbairn *et al.*, 2014) and more distant objects (Tavecchio and Bonnoli, 2016), either improving existing constraints or finding evidence for LIV (Abdalla *et al.*, 2021b).

The two approaches, based on timing, on the one hand, and spectral reconstruction, on the other, are complementary in their assumptions: the first assumes an energy-dependent modification in propagation time, the second, a modification to the kinematics of pair production. The discovery of LIV in one channel does not demand it in the other. The surprise with CTAO may come from studies of the second-order modification, which affects gamma-ray observables at an LIV energy scale beyond 10^{20} eV.

7.5. Status and Prospects of TeV Gamma-Ray Cosmology

In this chapter, we have attempted to demonstrate the relevance of VHE gamma-ray observations for understanding cosmology. We have examined timescales ranging from the Planck time to the current day, considering Lorentz invariance violation and cosmological magnetic and photon fields. The current generation of IACTs has made progress in measuring or constraining all of these cosmological observables. We summarize by pointing out that gamma-ray-based measurements are beginning to probe EBL wavelengths, where theoretical models developed prior to the gamma-ray detections disagree, are rapidly narrowing the allowed range of intergalactic magnetic field strength and correlation length, and are pushing the lower limit on the energy scale at which Lorentz invariance violation may occur beyond the Planck scale. We have also discussed the relevance of these measurements for other fields of astrophysics and astroparticle physics, particularly the spectral energy distribution of the diffuse supernova neutrino background and the propagation of UHECRs. We have noted the extent to which the gamma-ray measurements are limited by astrophysical uncertainties, namely, those associated with the production of VHE gamma rays in astrophysical sources. While current-generation IACTs have not made their final statements on the topics discussed above, we can look to CTAO to substantially improve our knowledge and possibly make groundbreaking discoveries. The precise energy and angular resolution of CTAO, coupled with its high-flux sensitivity, will allow it to quickly supersede existing instruments, dramatically improving our understanding of the light content of the Universe, cosmic magnetism and the structure of space–time.

https://doi.org/10.1142/9789813275720_0008

Chapter 8

Dark Matter and Fundamental Physics

Michele Doro[*,§], Miguel A. Sánchez-Conde[†,¶] and M. Hütten[‡,||]

[*]*Università degli studi di Padova, Padova, Italy*
[†]*Universidad Autónoma de Madrid, Madrid, Spain*
[‡]*University of Tokyo, Tokyo, Japan*
[§]*michele.doro@unipd.it*
[¶]*miguel.sanchezconde@uam.es*
[||]*huetten@icrr.u-tokyo.ac.jp*

8.1. Introduction

Gamma rays make up the highest energy end of the electromagnetic spectrum and are generated when interactions involve highly energetic leptons or hadrons. These accelerated particles reaching us from outer space are collectively called cosmic rays (CRs) and have been the subject of intense studies for nearly a century. Gamma radiation is created in diverse non-thermal environments in the Universe. Today, CR studies are performed with several tens of different instruments of different classes, including Imaging Atmospheric Cherenkov Telescopes (IACTs). The astrophysical reach of gamma-ray astronomy is huge, as largely proven in the past decade of scientific production and comprehensively resumed in this book. Aside from non-thermal astrophysical environments, gamma rays are excellent probes to search for new physics. Some of these exotic and exciting scenarios are the subject of this contribution.

All current IACTs have invested a great deal of time and resources in searching for these signatures over the past two decades, e.g. through a large fraction of their observation time and precious assets for the other astrophysical target observations described in this book. In this chapter, we focus on a selection of topics in fundamental physics, such as the indirect search for Weakly Interacting Massive Particles (WIMPs; Section 8.2) and the search for Axion-like Particles (ALPs; Section 8.3). We follow with less debated, yet simultaneously interesting studies on the search for primordial black holes (PBHs; Section 8.4), tau-neutrinos (Section 8.6) and magnetic monopoles (Section 8.5). Other topics in Fundamental Physics, such as Lorentz Invariance tests, are reported elsewhere in this book (see Chapter 7). We will close this chapter with a summary of the material discussed and final remarks.

We structured this chapter particularly to provide a self-contained (yet minimal) theoretical framework, as well as a complete report of all IACTs' published contributions to the topics under consideration. Our aim is to provide a reference review and capture a photograph of the effort of the current generation of IACTs and the challenges undertaken in preparation of the next-generation instrument to come: the Cherenkov Telescope Array Observatory (CTAO).

8.2. Massive Particle Dark Matter

For several decades now, the term "dark matter" has not only been known and used by all physicists worldwide, but also by most people — scientists or not — especially as it has secured a foothold in pop culture.[a] The rising awareness and popularity of the term are due to the innumerable studies that have written about the evidence of dark matter (DM), the so-far inconclusive effort to identify its nature (but knowing plenty now about what DM *is not*), and the theoretical mapping of potential candidates. There are also countless reviews of this knowledge. It is therefore a challenging task to summarize even a fraction of what we have learned so far about DM. For the purpose of this book, we will review the search for massive DM candidates — also called WIMPs — performed with IACTs. Within this scope, we will limit ourselves to the basic information that will be needed for such a review. In particular, we will limit ourselves to *particle* DM as opposed to theories of modified gravity (as e.g. by Milgrom, 1983) that would have no evident signatures for IACTs. Beyond that, we refer the interested reader to the literature: we recommend Bertone and Hooper (2016) for a profound, detailed account of the history of DM in which some myths are dismounted, such as that affirming that the term itself was coined by Zwicky (1933). Other comprehensive reports are Bertone *et al.* (2005b), Feng and Kumar (2008) and many more. Fine lecture notes

[a] "I don't want to be human. I want to see gamma rays, I want to hear X-rays, and I want to smell dark matter..." — Cylon Model Number One in the fiction TV series "Battlestar Galactica".

taken from schools are e.g. Profumo (2013); Hooper and Witte (2017) and Lisanti (2017). Dedicated textbooks are e.g. Hooper (2006); Sanders (2014); Borkowski *et al.* (2017) and Profumo (2017).

8.2.1. *Basics of gamma-ray searches for WIMP dark matter*

To date, we still experience DM only through its gravitational manifestation. We do not know either its microscopic character or its composition and coupling to ordinary matter, or if it interacts with the non-dark sector whatsoever. We do know, however, that DM does not interact electromagnetically, and any cross-section scattering must be very weak. Therefore, having excluded neutrinos to make up the DM, the nature of this particle — or several particles — must go beyond the ensemble of particles of the standard model (SM). Such strange particles would have permeated the whole Universe from its beginning and shaped its evolution and structure formation due to their dominance. In fact, today we have a remarkably accurate knowledge about these structures and the distribution of DM throughout the cosmos. We've measured that DM is clustered in "halos" at very diverse cosmic scales, in turn hosting astrophysical systems from the scale of dwarf galaxies to that of galaxy clusters. We know that — despite being elusive, weakly interacting and very much stable — there is room for many theoretical models predicting interactions that could be observable with different classes of instruments, from X-ray to gamma-ray telescopes.

Tracking the particle dark-matter evolution: One of the most compelling pieces of evidence of DM is its imprint on the spatial anisotropy spectrum of the cosmic microwave background (CMB). This diffuse radiation encodes the acoustic oscillations of the Universe matter and radiation fields at the epoch of matter recombination, when the Universe was about 380,000 years old. The power of a specific scale of anisotropy sensibly depends on the composition of matter species. For example, a particle such as DM exhibits a different anisotropy spectrum than a particle such as a quark that additionally couples to photons. Today's reference for the measurement of the CMB anisotropy is obtained with the Planck satellite detector. By decoding the matter composition at the recombination epoch and extrapolating it to current times, this measurement yields values of

$$\Omega_{\rm DM}\, h^2 = \frac{\rho_{\rm DM}}{\rho_{\rm crit}}\, h^2 = 0.1198 \pm 0.0012 \quad \text{and} \quad \Omega_{\rm SM}\, \mathrm{h}^2 = 0.02233 \pm 0.00015, \tag{8.1}$$

where $\rho_{\rm crit} = 3\,H_0^2/8\,\pi\,G = 10^{-6}$ GeV cm^{-3}, $h = 10^{-2}H_0$ km s^{-1}Mpc^{-1} = 0.6737 ± 0.0054 is the reduced Hubble parameter at redshift zero, and G is the gravitational constant (Aghanim *et al.*, 2020). That is, roughly 26% of the current Universe's energy density is in the form of DM, and DM makes up 84% of the total mass of the Universe.

Being over-abundant with respect to ordinary matter and having left thermal equilibrium before baryons, DM would then drive and shape galaxy formation: after their decoupling, baryons would fall into already existing and growing DM overdensities. These overdensities would gather and merge, slowly giving rise to the large-scale filamentary structure that is supported by current observations. Since the observed matter distribution is very clumpy, there are strong requirements on any DM candidate made of elementary particles: generically, particle DM candidates are classified into hot, warm and cold DM (HDM, WDM and CDM, respectively) according to whether they were highly relativistic (hot) or non-relativistic (cold) at the time they decoupled in the early Universe by thermal freeze-out or any other production mechanism.[b] Due to their high velocity, HDM particles inherit a large free-streaming length which would lead to extremely diluted DM clustering, in contradiction with observations. This was clarified only thanks to N-body cosmological simulations (Gershtein and Zeldovich, 1966; White *et al.*, 1983), which, for example, ruled out cosmogenic neutrinos as DM candidates. In turn, CDM provides at large a better agreement with observations.

It is important to track the evolution of DM in the early Universe in order to assess its observability now. The evolution of a comoving particle field density n_χ in a Universe with Hubble parameter H is governed by the law

$$\frac{dn_\chi}{dt} + 3\,H\,n_\chi = -\langle\sigma v\rangle\left(n_\chi^2 - n_{\chi,\rm eq}^2\right), \tag{8.2}$$

where $\langle\sigma v\rangle$ is the self-annihilation cross-section between two particles in the field averaged over their relative velocity, and the density at equilibrium has reached $n_{\chi,\rm eq}$ when particles are in thermal and chemical equilibrium with the environment (Dodelson, 2003). The Hubble parameter, expressing the expansion rate of the Universe, evolves with the effective field composition as well as the equilibrium temperature T. As a consequence, the number density of a given particle species evolves with the effective Universe

[b]HDM must consist of very light, $\lesssim$eV particles while CDM can be at any mass scale, depending on the production mechanism (Primack and Gross, 2000).

composition and temperature as well. Specifically, when the temperature decreases to values below a certain particle mass $T \lesssim m_\chi$, the comoving density of that particle starts decreasing exponentially as[c]

$$n_{\chi,\mathrm{eq}} = g_\chi \left(\frac{m_\chi T}{2\pi}\right)^{3/2} e^{-m_\chi/T}. \tag{8.3}$$

This trend stops when the particle self-annihilation becomes inefficient; that is, when the Hubble expansion rate grows larger than the particle annihilation rate. This event is called a "freeze-out" of the particle density, which subsequently stays constant in a comoving volume. For particle masses at the GeV scale, the freeze-out happens at a temperature T_F roughly 20 to 30 times smaller than the particle mass

$$T_\mathrm{F} \sim \frac{m_\chi}{25},$$

which means that these particles are at that moment already non-relativistic, e.g. cold. Considering such frozen-out particle species χ as DM candidates, a curious fact emerges: if one evolved the DM density from freeze-out to today, considering plausible values for the total number of degrees of freedom g^* according to the known particles in the Universe, one obtains for $m_\chi \sim$ GeV

$$\rho_\chi h^2 \simeq 0.12\,\rho_\mathrm{crit} \left(\frac{80}{g^*}\right)^{1/2} \left(\frac{m_\chi}{25\,T_\mathrm{F}}\right) \left(\frac{2.2\times 10^{-26}\ \mathrm{cm^3s^{-1}}}{\langle\sigma v\rangle}\right), \tag{8.4}$$

which demonstrates that if that particle χ made up all of the DM content of the Universe found according to Eq. (8.1),

$$\Omega_\mathrm{DM} = \Omega_\chi = \frac{\rho_\chi}{\rho_\mathrm{crit}} = 0.12,$$

then a velocity-averaged annihilation cross-section at a freeze-out of

$$\langle\sigma v\rangle \approx 2.2\times 10^{-26}\,\mathrm{cm^3s^{-1}} \tag{8.5}$$

is required. Such a value is remarkably close to the leading order scattering cross-section at the weak scale, $\langle\sigma v\rangle \sim \hbar^2/c \times \alpha_W^2/m_W^2$. This means that in a DM scenario where DM is made up of a yet unknown particle species of mass at the weak scale, a weak-scale freeze-out annihilation cross-section would naturally be required. This fact is commonly dubbed as the "WIMP miracle".

[c]In Eq. (8.3), g_χ denotes the degrees of freedom of the particle species χ. In the following, we presume Majorana fermions (Spin 1/2 and $\bar\chi = \chi$), such that $g_\chi = 2$.

This picture is only one possible realization. There are theories of valid particle DM alternatives that were not in thermal equilibrium in the Universe, or interacted in more complicated ways with the existing ambient fields. Yet, in addition to being one of the most studied and preferred DM particle models, the WIMP scenario provides a very generic framework, and several of the search methods developed for this scenario can be easily reformulated for other cases. As such, we limit ourselves to this case hereafter.

Dark matter particle candidates: The appeal of such generic WIMPs as DM candidates was increased with the advent of the theory of Supersymmetry (SUSY). Originally proposed to solve open questions unrelated to DM, such as the electroweak hierarchy problem or the unification of interactions in particle physics, the SUSY theory soon recognized that supersymmetric particles could also account for WIMP DM (e.g. Pagels and Primack, 1982).[d] SUSY depicts a symmetry of SM particles to a new sector populated with particles of similar quantum numbers but spin changed by $\pm 1/2$, thus swapping fermions with bosons and expanding the number of elementary particles. Why supersymmetric particles have never been observed, however, remains an unanswered question. Clearly, a quantum rule must protect the couplings between the SM and SUSY sectors: the R-parity — which, combining particle spin, baryon number and lepton number, provides SM particles with the parity number $R = 1$ and SUSY particles with $R = -1$. The conservation of this symmetry not only shields the SM from the SUSY sector, it also automatically makes the lightest SUSY particle stable. Here is the most studied WIMP candidate, the Lightest Supersymmetric Particle (LSP). There is not one single LSP candidate. Yet the one that emerged the most as the best candidate for WIMP DM is the lightest neutralino, the mass eigenstate of the superpartners of the photon, the W,Z particle and the two Higgs fields. In principle, if the R-parity was slightly broken, an LSP could have a finite lifetime, as long as it is much larger than the age of the Universe or in agreement with lower limits for specific models (e.g. Buchmuller *et al.*, 2007, for gravitino DM). This also holds for any other mechanism to provide stability to a heavy particle. Therefore, it is well worth it to also watch out for DM decays, as long instruments are able to probe them for such

[d]See Bertone and Hooper (2016) for a full historic account.

sufficiently large lifetimes. As we will see later, we can indeed probe lifetimes larger than these lower bounds.

Gravitational anomalies and the dark-matter distribution in astrophysical targets: The term DM was probably first used by H. Poincaré in 1906, albeit with the more classical meaning of "non-emitting light" ("matière obscure" in Poincare, 1906).[e] In fact, at the turn of the last century, astronomers started to investigate the distribution of visible objects in the sky and their motion, trying to understand from first principles why matter was clustered in a specific form. Early studies by Lord Kelvin, using thermodynamics arguments, and Poincaré, using the virial equilibrium hypothesis, applied to the Milky Way (MW), started to show that a significant fraction of mass was not attributable to stars but needed to be explained by different "dark bodies" such as meteors or gas clouds. Evidence of missing mass also started to pile up at a much larger scale than that of galaxy clusters. First Hubble and Humason (1931) and then Zwicky (1933), in the first decades of the last century, found a large scatter in the velocities of galaxies within galaxy clusters, up to 2,000 km/s, to be difficult to explain with only the visible matter in such clusters. Indeed, Zwicky used the virial theorem on a sample of a thousand galaxies in the Coma cluster and arrived at the conclusion that 500 times more mass was present in the cluster than in the one inferred from the visible counterpart.[f] This is referred to as the mass-to-light ratio — that is, the ratio of the total estimated mass with respect to the mass expected from the luminous matter alone. Thus, this mass-to-light ratio is a measure of how much an astrophysical object is DM-dominated. Similar gravitational evidence continued accumulating by the middle of the 20th century, but the "missing mass" problem started to become more intriguing as soon as standard explanations started to be ruled out: X-ray observation limited the amount of hot gas in galaxies to below 2% (Meekins *et al.*, 1971) and stellar nucleosynthesis could not generate enough helium to accommodate the observed abundance (Burbidge *et al.*, 1957). The advent of precise spectrographs in the 1970s allowed the measurement of the velocity of gas and stars within galaxies, which in turn allowed the determination of the mass of individual galaxies. In particular, in spiral

[e]See Bertone and Hooper (2016).

[f]This value was based on the estimate of the Hubble parameter H_0 at the time, overestimated by a factor of $\sim$10. With the present estimation, the imbalance noted by Zwicky (1933) would reduce down to a factor of eight.

galaxies, stars and gases move approximately in circular orbits around the galaxys' center. Newtonian dynamics infers that an object placed at distance r from a barycenter should display a circular velocity

$$v_{\rm c}(r) = \sqrt{\frac{GM(<r)}{r}}\,. \tag{8.6}$$

That is, objects well outside the size R of a mass distribution, where the total mass $M(<R)$ no longer grows, should have circular velocities following simply $v_{\rm c}(r) \sim \sqrt{1/r}$. In the 1970s, however, Bosma (1978), Rubin *et al.* (1978) and Rubin *et al.* (1980) used samples of tens to dozens of spiral galaxies and found unambiguous evidence of flat circular ($v_{\rm c}(r) \sim k$) velocities of gas regions well outside the size R determined from the optical extension of the galaxies. These observations allowed for one to robustly infer for the first time that the distribution of the missing mass extends several factors beyond the visible mass in individual galaxies, confirming previous findings by Einasto *et al.* (1974) and early findings for the M31 galaxy (Babcock, 1939). Spectrography is still the best technique nowadays to infer the DM content in galaxies, and is being widely used with instruments such as the Dark Energy Spectroscopic Instrument (DESI) (Aghamousa *et al.*, 2016).

Driven by insights from N-body cosmological simulations, the DM density profiles of cosmic DM overdensities can be commonly parametrized with the Zhao–Hernquist expression (Hernquist, 1990; Zhao, 1996):

$$\rho(r) = \frac{\rho_0}{\left(\frac{r}{r_s}\right)^{\gamma}\left[1+\left(\frac{r}{r_s}\right)^{\alpha}\right]^{(\beta-\gamma)/\alpha}}, \tag{8.7}$$

where r is the distance from the DM halo center, r_s and ρ_0 are the scale radius and characteristic DM density, respectively, and α, β, γ are free parameters (Kuhlen *et al.*, 2012). A specific case of this profile, widely used in the literature, is the parameterization Domanda a Michele (NFW) parametrization (Navarro *et al.*, 1996) with fixed $\alpha = 1$, $\beta = 3$ and $\gamma = 1$. A profile very similar to the NFW parametrization, but describing the innermost halo slopes in simulations slightly better with one additional parameter α (Navarro *et al.*, 2004; Springel *et al.*, 2008b; Wang *et al.*, 2020), is the Einasto profile

$$\rho(r) = \rho_0 \exp\left\{\frac{-2}{\alpha}\left[\left(\frac{r}{r_{\rm s}}\right)^{\alpha} - 1\right]\right\}. \tag{8.8}$$

A different analytic description was found for some dwarf spiral galaxies to better fit the inferred DM profile — namely, the Burkert profile (Burkert,

1996)

$$\rho(r) = \frac{\rho_0}{\left(1 + \dfrac{r}{r_s}\right)\left(1 + \dfrac{r^2}{r_s^2}\right)}, \tag{8.9}$$

which shows a density plateau in the inner halo center.

Gamma-ray signatures of dark matter: Besides trying to produce DM in collider events and observe their annihilation or decay into more known particles, DM can be observed through nuclear or atomic recoils in noble gas or semiconductors. This method is usually called the "direct detection" technique, and there is ample literature about it, e.g. Schumann (2019). A third avenue is to explore DM signatures "indirectly" by observing VHE gamma rays generated in relic annihilations or decays of DM throughout the Cosmos.

High-energy (HE) photons are found primarily as a result of neutral pion decay, where the pions are created after the fragmentation of quarks expected to be produced in DM annihilations or decays. Such an origin of gamma radiation results in a continuous spectrum (Fig. 8.1). As such, the latter may be indistinguishable from other astrophysical processes in which pion decays are found. If the DM is cold, e.g. non-relativistic, however, an abrupt spectral cutoff is expected at the DM mass for annihilating DM, or at its half for decaying DM. In more sophisticated scenarios, enhanced or even

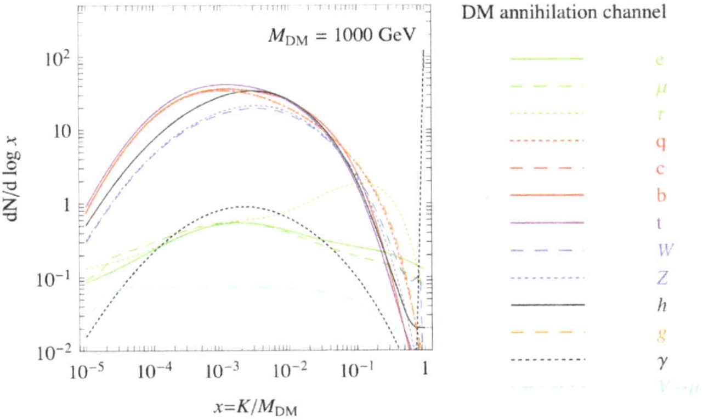

Fig. 8.1. Primary DM gamma-ray spectra for various annihilation models. For DM decays, the same spectra are expected, yet with the energy scale shifted to half these values, as the energy is equally distributed in two decayed particles. Extracted from Fig 3 of Cirelli *et al.* (2011) with the permission of the authors.

line-like emission may occur at the termination energy set by the particle mass (Bergstrom and Kaplan, 1994; Bringmann *et al.*, 2008). Because of the peculiarity of searches for line-like signals with IACTs, these will be discussed separately. In any case, all of these cutoff features make the resulting spectra from DM annihilation or decay very distinct from those of other astrophysical phenomena. Another consequence arises from the energy scale of the annihilation products set by the mass of the CDM particle: should the latter be of the order of GeV to TeV, gamma rays would be generated at these same energies, and ground-based VHE gamma-ray detectors like IACTs would be perfectly suited for such indirect searches for DM. Indeed, at WIMP masses above several TeV, IACTs are currently the most sensitive instruments to search for such particles. Finally, there are several other avenues for indirect DM searches, e.g. looking for other types of DM annihilation or decay products, namely, charged CRs or neutrinos. While these searches are not in the scope of this review, we will mention on page 326 the possibility of using IACTs to study the CR electron and positron fluxes to search for DM signatures.

Although DM possesses a nearly unique or very characteristic gamma-ray spectrum, the number of expected annihilation or decay events — and therefore the detectability at Earth — ultimately depends on the integrated density, squared in the annihilating case, of DM along the line of sight (*l.o.s.*). Because of that, overdense DM regions are clearly preferred, such as those at the center of galaxies or clusters of galaxies. The computation of the double differential gamma-ray DM flux at Earth can therefore be written as[g]

$$\begin{aligned}
\text{Annihilation}: \quad & \frac{d\Phi}{dE\,d\Omega} = \frac{1}{4\pi}\frac{\langle\sigma v\rangle}{2m_{\rm DM}^2}\frac{dN_\gamma}{dE}\frac{dJ_{\rm ann}}{d\Omega}, \quad \text{with} \\
& \frac{dJ_{\rm ann}}{d\Omega} = \int_{\rm l.o.s.} dl\,\rho^2(l,\Omega). \\
\text{Decay}: \quad & \frac{d\Phi}{dE\,d\Omega} = \frac{1}{4\pi}\frac{1}{\tau_{\rm DM}\,m_{\rm DM}}\frac{dN_\gamma}{dE}\frac{dJ_{\rm dec}}{d\Omega}, \quad \text{with} \\
& \frac{dJ_{\rm dec}}{d\Omega} = \int_{\rm l.o.s.} dl\,\rho(l,\Omega).
\end{aligned} \tag{8.10}$$

Here, $m_{\rm DM}$ is the DM mass; $\tau_{\rm DM}$, the DM lifetime; $\langle\sigma v\rangle$, the velocity-averaged annihilation cross-section; dN_γ/dE, the average number

[g]The annihilation formula is valid for self-conjugate DM particles ($\bar{\chi} = \chi$), otherwise there is an additional factor 1/2.

of photons per unit of energy per reaction; $J_{\rm dec}$, the "astrophysical factor" (Bergstrom *et al.*, 1998) for decaying DM; and $J_{\rm ann}$, the astrophysical factor for the annihilating case. Note that, in order to derive the expected DM-induced flux from a given target, one needs to integrate along the line of sight and the aperture $\Delta\Omega$. Note also that Eq. (8.10) is factorized so that one factor entails information about the microscopic DM properties (mass, cross-section or lifetime, average photon emission) while the astrophysical factor encloses separately all the information on the DM spatial distribution.[h] In most cases, for collisionless DM, the latter is almost independent of the specific microscopic nature of DM and thus the astrophysical factor is very general.[i] Yet, other cases are also possible, such as self-interacting DM, in which the exact DM density profile will depend on the DM particle's characteristics (Kahlhoefer *et al.*, 2019). In all cases, the expected gamma-ray spectra should look the same across different astrophysical targets. Therefore, observing two independent targets with identical spectra would constitute a strong case in favor of the DM interpretation.

The density integral in Eq. (8.10) is very sensitive to the density scaling in the halo center, in particular for integrating over the square-density in case of annihilation. Therefore, if we are interested in maximizing the DM-induced gamma-ray signals, the telescope should be pointed to those astrophysical environments exhibiting the highest DM concentrations at their centers. Also, since fluxes decrease as the inverse of the distance is squared, the closer the targets, the better.[j] Fortunately, a plethora of objects are viable candidates for DM searches with IACTs: the closest and probably brightest in terms of DM-induced flux is the center of the MW. Further out, dwarf spheroidal galaxies (dSphs) are optimal targets as well, especially those nearby and with large mass-to-light ratios. Much further out, galaxy clusters host a huge DM content, which could provide a significant emission at Earth as well. Less favored or more debated objects may also be interesting, such as Galactic globular clusters; their origin is still

[h]Strictly speaking, the factorization only holds for integration over a small redshift span, $\Delta z/z \ll 1$.

[i]The mass of the smallest predicted CDM halos, however, is ultimately set by the properties of the DM particle itself; this fact may have important implications in computations of the DM-induced gamma-ray flux involving halo substructure, these will be commented later on in this section.

[j]The inverse square-distance scaling is contained but hidden in Eq. (8.10) by the factor $1/l^2$, canceling out the l^2 in the volume element.

a matter of debate and there is a possibility they could retain significant DM density spikes at their very centers (Brown *et al.*, 2018). IACT DM searches in all of these objects will be presented in the following subsection.

8.2.2. *Review of WIMP DM searches with IACTs*

The search for WIMP DM with IACTs began in tandem with the birth of this class of instruments, and strongly evolved in expectations, methods and dedications over the years. In the following subsection, we gather all of these observations as a summary of what has been done so far in order to guide our experience in the future.

Classical targets were first considered: The central region of the MW, the MW dwarf spheroidal galaxies (dSphs) and galaxy clusters. The interest expanded to less-established targets such as IMBHs and dark satellites. In general, searches have been conducted looking for DM-induced continuum emission and line-like emission, and for both decaying and annihilating DM. Not only gamma-ray signatures, but also antiparticle signatures, which may originate from DM annihilation or decay, have been searched for — demonstrating the capability of IACTs as antiparticle detectors from the ground. It is interesting to see how early efforts invested relatively low observation times of the order of a few hours, while later observation campaigns show more than a hundred hours devoted to single targets. Considering that the total available observation time in a year of a single IACT experiment is about 1000 h, it is obvious that a large time investment has been made, with more than 1500 h of observation time devoted exclusively to dSphs as of today.

In Table 8.1, we have collected, to the best of our knowledge, all of the DM search campaigns carried out by IACTs. The table, an update version of the one in Doro (2014) (now including results up to 2021), gathers the numerous targets scrutinized as of today. The content is organized by target class and year of observation as well as the duration of observation. In the following, we discuss in detail all of these observations according to target class.

Observations of the multi-wavelength center and halo: The Galactic Center (GC) is a prime target for DM searches with IACTs. It is best observed from the Southern Hemisphere, where the H.E.S.S. telescopes are located. This guarantees the lowest energy threshold and longer observability. In 2006, H.E.S.S. published the observation of a signal from

Table 8.1. Summary of the campaigns carried out by IACTs to indirectly search for DM with gamma rays.

Target	Year	Time (h)	IACT	Limit	Ref.
		The Milky Way central region & halo			
MW center	2004	(48.7)	H.E.S.S.	Ann.	Aharonian *et al.* (2006e)
MW inner halo	2004–2008	(112)	H.E.S.S.	Ann.	Abramowski *et al.* (2011b)
	2010	9.1		Ann.	Abramowski *et al.* (2015)
	2004–2014	254		Ann.	Abdalla *et al.* (2016b)
	2014–2020	546	H.E.S.S.†	Ann.	Montanari *et al.* (2021)
MW outer halo	2018	10	MAGIC	Decay	Ninci *et al.* (2019)
		Dwarf Satellite Galaxies			
Draco	2003	7.4	Whipple	Ann.	Wood *et al.* (2008)
	2007	7.8	MAGIC‡	Ann.	Albert *et al.* (2008d)
	2007	(18.4)	VERITAS	Ann.	Acciari *et al.* (2010)
	2007–2013	(49.8)		Ann.	Archambault *et al.* (2017)
	2007–2018	114		—	Kelley-Hoskins (2018)
	2018	52.6	MAGIC	Ann.	Acciari *et al.* (2022)
Ursa Minor	2003	7.9	Whipple	Ann.	Wood *et al.* (2008)
	2007	(18.9)	VERITAS	Ann.	Acciari *et al.* (2010)
	2007–2013	(60.4)		Ann.	Archambault *et al.* (2017)
	2007–2018	162		—	Kelley-Hoskins (2018)
Sagittarius	2006	(11.0)	H.E.S.S.	Ann.	Aharonian *et al.* (2008b)
	2006–2012	90		Ann.	Abramowski *et al.* (2014c)
	2006–2012	(85.5)		Ann.	Abdalla *et al.* (2018h)
Canis Major	2006	9.6	H.E.S.S.	Ann.	Aharonian *et al.* (2009e)
Willman 1	2007–2008	13.7	VERITAS	Ann.	Acciari *et al.* (2010)
		(13.6)		Ann.	Archambault *et al.* (2017)
	2008	15.5	MAGIC‡	Ann.	Aliu *et al.* (2009b)
Sculptor	2008	(11.8)	H.E.S.S.	Ann.	Abramowski, A. and others (2011)
				Ann.	Abdalla *et al.* (2018h)
	2008–2009	12.5		Ann.	Abramowski *et al.* (2014c)
Carina	2008–2009	(14.8)	H.E.S.S.	Ann.	Abramowski, A. and others (2011)
	2008–2009	(12.7)		Ann.	Abramowski *et al.* (2014c)
	2008–2010	22.9		Ann.	Abdalla *et al.* (2018h)
Segue 1	2008–2009	29.4	MAGIC‡	Ann.	Aleksić *et al.* (2011d)
	2010–2011	(47.8)	VERITAS	A.+D.	Aliu *et al.* (2012)
	2010–2013	(92.0)		Ann.	Archambault *et al.* (2017)
	2010–2013	157.9	MAGIC	A.+D.	Aleksić *et al.* (2014b)
				Ann.	Ahnen *et al.* (2016c)
	2010–2018	184	VERITAS	—	Kelley-Hoskins (2018)
Boötes 1	2009	14.3	VERITAS	Ann.	Acciari *et al.* (2010)
		(14.0)		Ann.	Archambault *et al.* (2017)
Coma Berenices	2010–2013	(8.6)	H.E.S.S.	Ann.	Abramowski *et al.* (2014c)
	2010–2013	10.9		Ann.	Abdalla *et al.* (2018h)

(*Continued*)

Table 8.1. (*Continued*)

Target	Year	Time (h)	IACT	Limit	Ref.
	<2018	37	VERITAS	—	Kelley-Hoskins (2018)
	2018	50.2	MAGIC	Ann.	Acciari *et al.* (2022)
Fornax	2010	6.0	H.E.S.S.	Ann.	Abramowski *et al.* (2014c)
				Ann.	Abdalla *et al.* (2018h)
Ursa Major II	2014–2016	94.8	MAGIC	Ann.	Ahnen *et al.* (2018a)
	<2018	181	VERITAS	—	Kelley-Hoskins (2018)
Triangulum II*	2014–2016	62.4	MAGIC	Ann.	Acciari *et al.* (2020)
	<2018	15	VERITAS	—	Kelley-Hoskins (2018)
Segue II	<2018	19	VERITAS	—	Kelley-Hoskins (2018)
Canes Ven I	<2018	14	VERITAS	—	Kelley-Hoskins (2018)
Canes Ven II	<2018	14	VERITAS	—	Kelley-Hoskins (2018)
Hercules	<2018	13	VERITAS	—	Kelley-Hoskins (2018)
Sextans	<2018	13	VERITAS	—	Kelley-Hoskins (2018)
Draco II	<2018	10	VERITAS	—	Kelley-Hoskins (2018)
Leo I	<2018	7	VERITAS	—	Kelley-Hoskins (2018)
Leo II	<2018	16	VERITAS	—	Kelley-Hoskins (2018)
Leo IV	<2018	3	VERITAS	—	Kelley-Hoskins (2018)
Leo V	<2018	3	VERITAS	—	Kelley-Hoskins (2018)
Reticulum II	2017–2018	18.3	H.E.S.S.†	Ann.	Abdalla *et al.* (2020d)
Tucana II	2017–2018	16.4	H.E.S.S.†	Ann.	Abdalla *et al.* (2020d)
Tucana III*	2017–2018	23.6	H.E.S.S.†	Ann.	Abdalla *et al.* (2020d)
Tucana IV*	2017–2018	12.4	H.E.S.S.†	Ann.	Abdalla *et al.* (2020d)
Grus II*	2018	11.3	H.E.S.S.†	Ann.	Abdalla *et al.* (2020d)
			Dark satellites		
1FGL J2347.3+0710	2010	8.3	MAGIC	—	Nieto *et al.* (2011b)
1FGL J0338.8+1313	2010–2011	10.7	MAGIC	—	Nieto *et al.* (2011b)
2FGL J0545.6+6018	2013–2015	8.5	VERITAS	Ann.	Nieto *et al.* (2015)
2FGL J1115.0–0701	2013–2015	13.8	VERITAS	Ann.	Nieto *et al.* (2015)
H3FHL J0929.2–4110	2018–2019	7.8	H.E.S.S.†	Ann.	Abdalla *et al.* (2021a)
3FHL J1915.2–1323	2018–2019	3.0	H.E.S.S.†	Ann.	Abdalla *et al.* (2021a)
3FHL J2030.2–5037	2018–2019	8.8	H.E.S.S.†	Ann.	Abdalla *et al.* (2021a)
3FHL J2104.5+2117	2018–2019	5.5	H.E.S.S.†	Ann.	Abdalla *et al.* (2021a)
			Intermediate mass black holes		
Galactic Plane Survey	2004–2007	400	H.E.S.S.	Ann.	Aharonian *et al.* (2008)
	2005–2006	25	MAGIC‡	Ann.	Doro *et al.* (2007)

(*Continued*)

Table 8.1. (*Continued*)

Target	Year	Time (h)	IACT	Limit	Ref.
		Globular clusters			
M15	2002	0.2	Whipple	Ann.	Wood *et al.* (2008)
	2006–2007	15.2	H.E.S.S.	Ann.	Abramowski *et al.* (2011)
NGC 6388	2008–2009	27.2	H.E.S.S.	Ann.	Abramowski *et al.* (2011)
		Other galaxies			
M33	2002–2004	7.9	Whipple	Ann.	Wood *et al.* (2008)
M32	2004	6.9	Whipple	Ann.	Wood *et al.* (2008)
WLM	2018	18.2	H.E.S.S.†	Ann.	Abdalla *et al.* (2021)
		Galaxy clusters			
Abell 2029	2003–2004	6.1	Whipple	—	Perkins *et al.* (2006)
Perseus (Abell 426)	2004–2005	13.5	Whipple	—	Perkins *et al.* (2006)
	2008	24.4	MAGIC‡	Ann.	Aleksić *et al.* (2010d)
	2009–2017	202.2	MAGIC	Decay	Acciari *et al.* (2018a)
Fornax (Abell S0373)	2005	14.5	H.E.S.S.	Ann.	Abramowski *et al.* (2012)
Coma (Abell 1656)	2008	18.6	VERITAS	Ann.	Arlen *et al.* (2012)
		Line searches			
MW inner halo	2004 – 2008	(112)	H.E.S.S.	Ann.	Abramowski *et al.* (2013c)
	2014	15.2	H.E.S.S.†	Ann.	Abdalla *et al.* (2016a)
	2004–2014	(254)	H.E.S.S.	Ann.	Abdalla *et al.* (2018g)
	2013–2019	204	MAGIC	Ann.	Inada *et al.* (2021)
Segue 1 dSph	2010–2013	(157.9)	MAGIC	Ann.	Aleksić *et al.* (2014b)
Five dSph galaxies	2006–2012	(137.1)	H.E.S.S.	Ann.	Abdalla *et al.* (2018h)
Five dSph galaxies	2007–2013	(229.8)	VERITAS	Ann.	Archambault *et al.* (2017)
WLM	2018	(18.2)	H.E.S.S.†	Ann.	Abdalla *et al.* (2021)
		Charged particles			
All-electron	2004–2007	239	H.E.S.S.	—	Aharonian *et al.* (2008c, 2009c)
	2009–2012	296	VERITAS	—	Archer *et al.* (2018)
	2009–2010	14	MAGIC	—	Borla Tridon *et al.* (2011)
Moon shadow	2010–2011	20	MAGIC	—	Colin *et al.* (2011)
	2014	1.2	VERITAS	—	Bird *et al.* (2016b)

We report observations from the Whipple, H.E.S.S., MAGIC and VERITAS experiments, sorted by astrophysical target class. The first column is the name of the target; the second and third columns refer to the year and duration (in hours) of observations, respectively; and the fifth column describes the type of limit, either annihilation or decay. Unconfirmed dSph candidates are marked with an asterisk. H.E.S.S. data, including the H.E.S.S. Phase II telescope, are marked with a †, while MAGIC data from monoscopic observations with MAGIC-I are marked with a ‡. Times in parentheses denote data which also function as a subset of other analyses. Additionally, we include yet unpublished data presented at various conferences.

the GC above 160 GeV and its interpretation in terms of DM after an observation campaign of about 50 h (Aharonian *et al.*, 2006e). In that work, DM density profiles as obtained from N-body cosmological simulations of MW-size halos were adopted, namely, the NFW (Eq. 8.7 with $\alpha, \beta, \gamma = 1, 3, 1$) and Einasto (Eq. 8.8). Large differences between the two DM profile parametrizations occur only at the very center of the halo, where the NFW profile is more strongly peaked, while already at distances from the GC larger than $\sim$10 pc the density descriptions become very similar when adopting the same profile normalization at the Solar position. This was the reason for H.E.S.S. to focus on an analysis region at a projected Galactocentric distance of 45–150 pc that corresponds to an angular distance of 0.3–1.0° from the GC. Furthermore, the contamination from known and expected gamma-ray astrophysical sources at this distance from the GC is largely reduced with respect to the very central region. Limits on the annihilation cross-section were first presented in Abramowski *et al.* (2011b) using 112 h of data and long remained as the most constraining upper limits on DM annihilations from IACTs. More recently, Abdalla *et al.* (2016b) gathered 10 years of H.E.S.S. datataking involving a total of 254 h of high-quality data, which resulted in the DM exclusion limits shown in Fig. 8.2 for the $b\bar{b}$ and $\tau^+\tau^-$ pure annihilation cases. The limits were obtained after an elaborated selection of both the signal and background regions in the proximity of the GC that removes the Galactic plane and known TeV emitters. It can be seen that, while the limits for $b\bar{b}$ approach the thermal relic cross-section value at a few TeV, the limits for $\tau^+\tau^-$ actually already skim the canonical parameter space. These limits represent the strongest ones obtained by IACTs at present, recently even extended by the announcement of the publication of a data set comprising almost 550 h of data (Montanari *et al.*, 2021). Additionally, the MAGIC (Albert *et al.*, 2006) and VERITAS telescopes (Beilicke *et al.*, 2012) observed the GC over the years, yet with a more modest performance in comparison to H.E.S.S. due to their location in the Northern Hemisphere (whereas H.E.S.S. is located in the Southern Hemisphere). Because of that, neither MAGIC nor VERITAS was able to draw conclusions on DM physics.

It must be noted that all mentioned DM limits are only valid as long as a cuspy profile of the distribution of DM toward the MW center is

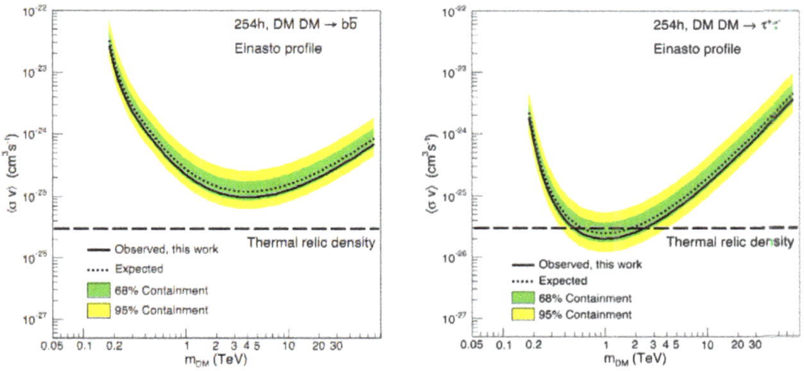

Fig. 8.2. Constraints on the velocity-weighted annihilation cross-section derived from 10 years of GC observations by H.E.S.S. for the $b\bar{b}$ (left panel) and into $\tau^+\tau^-$ (right panel). The observed upper limit is plotted as a solid black line. Expected limits are derived from blank-field observations at high Galactic latitudes. The mean expected limit (black dashed line) is plotted together with the 65% (dark gray) and 95% (light gray) confidence bands. The natural scale for the annihilation cross-section is given as long-dashed black line. Figures from Abdalla *et al.* (2016b), courtesy of the H.E.S.S. collaboration.

assumed, as is the case for both the NFW and Einasto profiles. It is still equally plausible, however, that the actual DM profile toward the GC is more cored and closer to the Burkert profile, Eq. (8.9). This would strongly reduce the DM sensitivity, not only because of a much smaller J-factor, but also because most DM analyses of IACT data rely on the existence of a DM density contrast between a signal and background region. Such cored profiles can appear, e.g., when baryonic feedback occurs in connection with strong stellar activity that sweeps the matter outward and dilutes the DM concentration in the innermost regions of the Galaxy (Di Cintio *et al.*, 2013; Tollet *et al.*, 2016; Read *et al.*, 2019; Lazar *et al.*, 2020). Additionally, self-interacting DM is expected to form pronounced halo cores (e.g. Rocha *et al.*, 2013; Tulin and Yu, 2018). Due to the crowded line of sight toward the GC and the complex dynamics in the Galactic bulge, an empirical direct determination of the inner GC DM density profile from, e.g., kinematics data of different tracers in the area is very difficult to obtain. Therefore, the inner profile is yet poorly constrained (Benito *et al.*, 2019, 2021). For these reasons, despite being far more constraining than others, GC limits need to be treated with special caution. In 2015, the H.E.S.S. collaboration published a study constraining DM annihilation from a DM core around the GC based on 9 h of dedicated observations with a background subtraction region $\gtrsim 5°$ away from the GC (Abramowski *et al.*, 2015). A similar methodology addressing DM decay in the Galactic halo has been pursued by the MAGIC collaboration (Ninci *et al.*, 2019). In the conclusion of this section, we will discuss the significant improvement that the CTAO will bring even for this pessimistic scenario of a cored DM profile in the GC. Note that the caveat of potential DM cores is also present in the case of dSphs — yet, it is considered much less severe compared to the GC case: first, because the inner dSph DM profiles are better constrained, and second, because the dSph galaxies are expected to show, at least for standard CDM, cuspier inner profiles due to their comparatively low baryonic content.

Observation of dSphs. DSphs are small galaxies with masses of the order 10^7–$10^9 M_\odot$ that are gravitationally bound to their host galaxy (Strigari *et al.*, 2008). In the MW, they are located within the MW virial radius at typical heliocentric distances of 20–250 kpc.[k] They are considered to be in internal dynamical equilibrium and exhibit mass-to-light ratios of the order of 100–1000, meaning that they are highly DM-dominated systems. In addition, the great interest in these objects for indirect DM searches stems from the fact that dSphs display very reduced stellar activity in the past few gigayears (Gyrs). This results in an unlikely emission of HE photons of an astrophysical origin, that is, photons coming from the depletion of interstellar gas and sources of HE CRs. Thus, any putative gamma-ray signal in these objects would be, most likely, of exotic origin. As a result, dSphs are considered one of the cleanest targets for indirect DM searches. Furthermore, the low stellar activity may have allowed an efficient DM contraction, with reduced matter redistribution to, e.g., supernova explosions and shock propagation (Read *et al.*, 2019; Lazar *et al.*, 2020). Because of this, such compact DM overdensities result in promising conditions for annihilating DM searches according to Eq. (8.10), yet less for decays (Geringer-Sameth *et al.*, 2015; Bonnivard *et al.*, 2015a). Estimates of the total number of MW dwarf satellites in a ΛCDM universe predict the existence of hundreds of these objects within the MW virial radius (Tollerud *et al.*, 2008; Hargis *et al.*, 2014). Discrepancies between observations and expectations from N-body cosmological simulations were reported in the past (also called the missing satellite problem, discussed in, e.g., Moore

[k]See Strigari (2018) for an excellent overview on these targets.

et al., 1999; Klypin et al., 1999; Strigari et al., 2007; Hargis et al., 2014). However, not only is our census of dSphs known to be incomplete, but more importantly, the discrepancy may be related to the fact that a large fraction of dSphs, especially at the smallest scales, are expected to contain little or no baryonic material in the form of stars and gases. Therefore, they would basically remain as dark DM overdensities (Sawala et al., 2015; Bullock and Boylan-Kolchin, 2017). An additional potential issue with these targets is the fact that ΛCDM predicts a larger number of massive dSphs than what is actually observed. In other words, these massive galaxies should necessarily contain visible counterparts that are nevertheless unseen. This is the "too big to fail" problem (Read et al., 2006; Boylan-Kolchin et al., 2011). Possible arguments to mitigate this issue include a more accurate description of subhalo disruption and mass loss, and of the role and evolution of the baryonic contents in these objects (Helmi et al., 2012; Brooks and Zolotov, 2014; Tomozeiu et al., 2016; Garrison-Kimmel et al., 2019).

DSphs are commonly classified in "classical" and "ultra-faint" galaxies, depending on whether they were discovered before or after the Sloan Digital Sky Survey (SDSS; York et al., 2000), which roughly translates into a brightness cut. Yet, this distinction shadows more subtle differences: ultra-faint dSphs possess more compact sizes and less visible stars, and are typically more DM-dominated systems than the classical ones, with mass-to-light ratios sometimes exceeding 1000 $M_\odot/L_\odot$ (Bonnivard et al., 2015a). In addition, observation (Hayashi et al., 2020) and numerical simulations (e.g. Read et al., 2019; Lazar et al., 2020) suggest that the inner cusp of the DM profile increases with decreasing baryonic activity, as expected from classical dwarf systems down to ultra-faint ones. In addition to the nine classical dSphs known, nearly 50 ultra-faint dSph candidates have been discovered in photometric surveys since then, mostly with SDSS itself and, most recently, with Dark Energy Survey (DES) data (Drlica-Wagner et al., 2015, 2018). From this number, about 40 have already been confirmed to be dSphs by means of follow-up spectroscopic measurements. For each dSph, the J-factor is computed via the Jeans hydrostatic equilibrium equation from the measured positions and velocities of member stars. To solve the equation, one needs to input a model for the star phase-space and another for the DM density profile, whose normalization is left as a free parameter.[1] This technique relies on several assumptions of the geometry of

[1] All of the details of these calculations are reviewed, e.g., by Strigari (2013a).

the DM subhalo hosting the dSph: triaxiality, anisotropy and co-rotation all constitute relevant ansatzes. Another factor is the misreconstruction of interloper stars, especially that of binary systems, whose intrinsic high velocities strongly affect the computation of velocity distribution and, ultimately, of the dSph J-factor. The main limiting factor, however, is the very small number of stars, down to a few tens, observed in the dimmer ultra-faint dSphs compared to the thousands of stars observed in the larger dSphs. The interplay between these ansatzes and limitations are discussed by, e.g., Bonnivard *et al.* (2015b, 2016); Chiappo *et al.* (2017); Klop *et al.* (2017); Ichikawa *et al.* (2017, 2018), and Ando *et al.* (2020).

Since 2004, these targets have been extensively studied by IACTs in the context of DM searches, as shown in Table 8.1. Many targets have been sighted with exposures ranging from a few hours to more than 150 h. Over almost 20 years we have observed the classical dSphs Canis Major, Carina, Draco, Fornax, Leo I-II, Sextans, Sculptor and Ursa Minor; the tidally disrupted Sagittarius; and the ultra-faint systems Boötes, Canes Venatici I-II, Coma Berenices, Draco 2, Grus-II, Leo IV-V, Hercules, Reticulum II, Segue 1-2, Triangulum II, Tucana II-III-IV, Ursa Major 2 and Willman 1.[m] We already noted that in early times, only a few hours were devoted to single targets. Part of the reason is that, back then, the prospects of detection were in general more optimistic than today (Bergstrom and Hooper, 2006). Additionally, the large uncertainties related to the estimation of the DM content in dSph galaxies warranted a diversification strategy, such that a larger pool of different dSphs was targeted. As a matter of fact, the exclusion DM limits have significantly improved since the first publications for the classical dSphs to the latest ones on both classical and ultra-faint dwarf systems. This was possible thanks to extended observation times and improved analysis techniques, as reviewed in detail by Rico (2020). Aleksic *et al.* (2012) were the first to present a dedicated maximum likelihood approach for DM analyses with IACTs, which allows one to account for both the spectral and morphological shape of putative DM emission, significantly enhancing the sensitivity to DM. This work was also the first to combine data from different sources and instruments. The same approach was consequently applied to analyses of the deep observations performed with the MAGIC stereo system (Aleksić *et al.*, 2014b; Ahnen *et al.*, 2016c, 2018a; Acciari *et al.*, 2020). As used in Ahnen *et al.* (2018a) and Acciari *et al.* (2020), the likelihood binned in energy

[m]Note that not all of these ultra-faint systems are already confirmed dSphs.

is built as follows:

$$\mathcal{L}(\langle\sigma v\rangle;\nu_i|\mathcal{D}_i) = \mathcal{L}\left(\langle\sigma v\rangle; s_{ij}, J, \tau_i | N_{ij}^{\mathrm{ON}}, N_{ij}^{\mathrm{OFF}}\right)$$
$$= \prod_{j=1}^{\mathrm{bins}} \left[\frac{\left(s_{ij}(\langle\sigma v\rangle + b_{ij})\right)^{N_{ij}^{\mathrm{ON}}}}{N_{ij}^{\mathrm{ON}}!} e^{-((s_{ij}(\langle\sigma v\rangle + b_{ij})))} \times \frac{(\tau_i b_{ij})^{N_{ij}^{\mathrm{OFF}}}}{N_{ij}^{\mathrm{OFF}}!} e^{-(\tau_i\, b_{ij})} \right]$$
$$\times \prod_{j=1}^{\mathrm{bins}} \left[\mathcal{G}(\tau_i|\tau_{\mathrm{obs},i}, \sigma_{\tau_{\mathrm{obs},i}})\right] \times \mathcal{G}(J|J_{\mathrm{obs}}, \sigma_J), \tag{8.11}$$

where the likelihood is the product of likelihoods over the number of N_{bins} energy intervals that have $\langle\sigma v\rangle$ as parameter of interest, the astrophysical J-factor, the normalization between signal and control regions τ_i and the estimated number of signal and background events s_{ij}, b_{ij} as nuisance parameters. The number of measured events in the signal and background extraction regions are N_{ij}^{ON} and N_{ij}^{OFF}, respectively; the terms $\mathcal{G}(J)$ and $\mathcal{G}(\tau_i)$ are the Gaussian likelihoods for the J-factor and the ratio of the signal to background-region exposure. The estimated signal s_{ij} depends on the parameter of interest $\langle\sigma v\rangle$ as

$$s_{ij}(\langle\sigma v\rangle) = T_j^{\mathrm{obs}} \int_{\mathrm{bin},j} dE'_{\mathrm{bin},j} \int dE\, \frac{d\phi(\langle\sigma v\rangle)}{dE}\, A_{\mathrm{eff}}(E)\, G(E'|E)\,, \tag{8.12}$$

where T_j^{obs} is the observation time, $\phi(\langle\sigma v\rangle)$, the intrinsic flux, $A_{\mathrm{eff}}(E)$, the effective area and $G(E'|E)$, the migration matrix between true and reconstructed energy.

Marginalizing the instrument response functions and the J-factor uncertainties in a full likelihood approach not only allows one to stack results of the same instrument, but also, as mentioned earlier, to combine limits from different instruments. In Ahnen *et al.* (2016c), the unbinned approach from Aleksic *et al.* (2012) was used to combine for the first time IACT data from MAGIC on Segue 1 with multi-dSph data from *Fermi*-LAT. These limits are shown in Fig. 8.3 and still provide the most constraining results for indirect DM searches at dSphs with IACTs.[n] A similar unbinned likelihood analysis method based on "event weighting" has been developed by the VERITAS collaboration using individual event

[n]Note, however, that these limits could be deteriorated by some yet unaccounted-for uncertainty on the Segue I DM content and its corresponding J-factor (Bonnivard *et al.*, 2016).

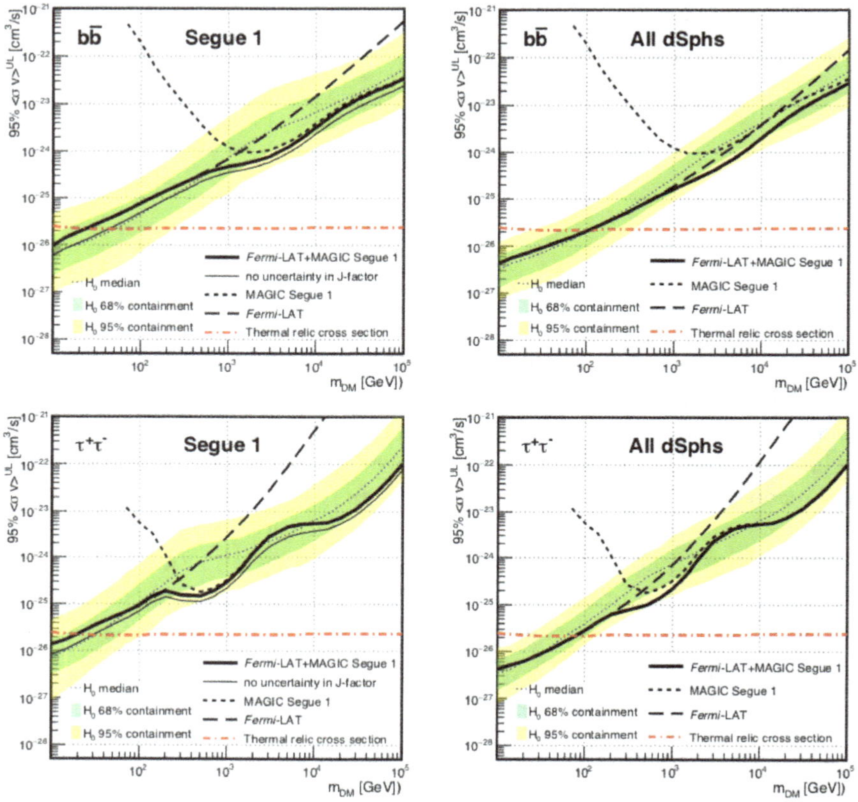

Fig. 8.3. 95% C.L. upper limits on the thermally averaged cross-section for DM particles annihilating into $b\bar{b}$ (top row) and $\tau^+\tau^-$ (bottom row). Thick solid lines show the limits obtained by combining Segue 1 observations from *Fermi*-LAT and MAGIC (left column), and the observations of 15 dSphs with the *Fermi*-LAT and Segue 1 with MAGIC (right column). Dashed lines show the observed individual MAGIC (short dashes) and *Fermi*-LAT (long dashes) limits. The thin dotted line, dark and light gray bands show, respectively, the median and the symmetrical, two-sided 68% and 95% containment bands for the distribution of limits under the null hypothesis. The red dashed-dotted line shows the thermal relic cross-section from Steigman *et al.* (2012). Image courtesy of the MAGIC collaboration and reproduced from Ahnen *et al.* (2016c).

information to improve the sensitivity from a joint analysis of five individual dSph targets into a single limit (Archambault *et al.*, 2017). Additionally, the H.E.S.S. collaboration had adopted a simplified version of Eq. (8.11) to combine data from five dSphs, to search both for continuum (Abramowski *et al.*, 2014c) and line (Abdalla *et al.*, 2018h) emission. Currently, a collaborative effort is ongoing among the H.E.S.S., MAGIC, VERITAS, *Fermi*-LAT and HAWC collaborations to provide combined DM limits using a vast amount of data collected from dSphs so far (Armand *et al.*, 2021). This study shall include more than 500 h of IACT data plus 10 yr of *Fermi*-LAT integrated time on 20 dSphs as well as 1038 days of HAWC exposure on 12 dSphs.

In summary, limits from dSphs now reach values of cross-section of the order of 10^{-24} cm^3 s^{-1} and are perceived by the community as the most robust obtained with IACTs. In the outlook section, we will briefly discuss the expectation for the CTAO with regard to dSphs.

Observation of dark satellites. In the standard cosmological framework, small dense DM structures form first in the early Universe and later merge to form larger halos. A natural consequence of this scenario of hierarchical bottom-up structure formation is that it predicts abundant substructures, or "subhaloes", inside larger halos like our own Galaxy. The most massive of these subhaloes in our own Galaxy host the known dwarf satellite galaxies of the Milky Way, while smaller subhaloes with masses below $\sim 10^7$ $M_\odot$ may harbor no stars or gases at all and thus may remain completely dark. These less-massive DM subhaloes, with no astrophysical counterparts at other wavelengths, can represent excellent targets for indirect DM searches as well (Strigari, 2013b); they may be located close to us and thus ease the detection of gamma radiation from DM annihilation or decay, while background processes are absent by definition. Indeed, N-body simulations of Milky-Way-sized halos predict the existence of thousands of such subhaloes (Diemand *et al.*, 2008; Springel *et al.*, 2008a; Stadel *et al.*, 2009) and some are expected to yield large DM annihilation fluxes at Earth given their typical distances and masses (Zechlin *et al.*, 2012; Ackermann *et al.*, 2012g; Schoonenberg *et al.*, 2016; Hütten *et al.*, 2016; Coronado-Blázquez *et al.*, 2019a).

Since we do not know a priori the exact location of these "dark subhaloes", current IACTs are not particularly suited to discover them given their narrow FoV of a few degrees wide. This is in contrast with *Fermi*-LAT, which continually surveys the entire gamma-ray sky and thus naturally emerges as the ideal instrument to search for dark subhalo candidates in the GeV energy range. Recently, the HAWC observatory, with a broad FoV as well, came into operation and enabled such searches also in the TeV regime. In most cases, the dark satellites' search strategy is based on looking for them among the pool of unidentified gamma-ray sources (unIDs) in current point-source catalogs. The key point is to identify those unIDs that exhibit spectral or spatial properties compatible with those expected from DM annihilation. Indeed, it is intriguing that about one-third of the sources in all available gamma-ray catalogs are still unidentified and some of them may be dark satellites just awaiting a proper classification. In the absence of an unequivocal DM signal, this strategy allows one to set competitive DM constraints by comparing the number of DM subhalo candidates among the list of unIDs to that expected from a combination of N-body simulation results and flux instrumental sensitivity (Buckley and Hooper, 2010; Belikov *et al.*, 2012; Mirabal *et al.*, 2012; Zechlin *et al.*, 2012; Zechlin and Horns, 2012; Ackermann *et al.*, 2012g; Berlin and Hooper, 2014; Bertoni *et al.*, 2015, 2016; Schoonenberg *et al.*, 2016; Mirabal *et al.*, 2016; Calore *et al.*, 2017; Hooper and Witte, 2017; Abeysekara *et al.*, 2019; Calore *et al.*, 2019; Coronado-Blázquez *et al.*, 2019a, 2019b; Coronado-Blázquez and Sánchez-Conde, 2020). Traditionally, however, these DM constraints have been subject to large uncertainties given the lack of a precise characterization of the DM subhalo population, such as their structural properties and radial distribution in the Galaxy. A statistical approach on the computation of the number of expected dark subhaloes and their prospects of detection with the CTAO were reported by Coronado-Blázquez *et al.* (2021).

While current IACTs are not the best-suited instruments to discover dark satellites, they have been used to scrutinize in further detail some of the best DM subhalo candidates among the pool of unIDs identified in the aforementioned works, given their superior sensitivity at TeV energies and angular resolution. In Nieto *et al.* (2011a, 2011b), the MAGIC collaboration reported on the observation of two objects in 2010 and 2011, 1FGL J2347.3+0710 and 1FGL J0338.8+1313, respectively, among the list of 630 unIDs present in the first *Fermi*-LAT catalog of point sources (Abdo *et al.*, 2010, 1FGL). No VHE gamma rays were detected after an exposure of around 10 h on each source. These two objects were selected as the most promising out of 10 after having survived several selection criteria that were based on the expected DM signal properties. In particular, they selected only non-variable sources located at high ($|b| > 10°$) Galactic latitudes to avoid contamination from Galactic objects like pulsars, and whose spectra are well described by a power law in the *Fermi*-LAT energy band and are particularly "hard" (spectral index $\Gamma > -2$). They also performed a careful study to search for counterparts of all unIDs at other (X-ray, optical, radio) wavelengths. Later, the VERITAS collaboration reported on the observation of two other unIDs in a DM context, 2FGL J0545.6+6018 and 2FGL J1115.0-0701, with again around 10 h exposure on each target (Geringer-Sameth *et al.*, 2013; Nieto *et al.*, 2015). These objects were identified as the best candidates among the 576 unIDs in the second *Fermi*-LAT catalog of point sources (2FGL; Nolan *et al.*, 2012). As summarized by Nieto *et al.* (2015), VERITAS adopted the same selection criteria proposed by MAGIC in Nieto *et al.* (2011a, 2011b), the only difference with respect to MAGIC being the VERITAS peculiarities in terms of observation site and instrumental sensitivity. Once again, the null detection of VHE gamma rays from the location of these two unIDs only led to set upper limits to the flux from these objects. In Abdalla *et al.* (2021a),[o] the H.E.S.S. collaboration recently reported the follow-up observation of dark satellite candidates among unIDs in the third catalog of hard *Fermi*-LAT sources (3FHL; Ajello *et al.*, 2017). Applying similar selection criteria as by the MAGIC and VERITAS works to 177 unIDs in the 3FHL, they obtained six subhalo candidates out of which four were observed for between three and nine hours (see Table 8.1). And among these four, the source 3FHL J2104.5.2117 was recently associated with high probability to an AGN in the 4FGL catalog (Abdollahi *et al.*, 2020). No signal was detected from any of the four targets. As no robust assumption on the J-factor of the putative DM subhaloes can be made for this source class, the H.E.S.S. collaboration adopted a reverse conclusion: on the assumption of a thermal DM annihilation cross-section in the order of 8.5, upper limits on the dark satellite J-factor were derived. This is illustrated in Fig. 8.4(left).

Yet, we note that although no detection was obtained in either MAGIC, VERITAS or H.E.S.S. observations of unIDs, the resulting flux or J-factor upper limits in all cases may still be useful to test the DM subhalo hypothesis and to shed further light on the nature of these sources. So far, no IACT dark satellite observations have yet been performed of any of the unIDs in the most recent *Fermi*-LAT 4FGL catalog (Abdollahi *et al.*, 2020) containing more than 1000 unIDs. In addition, Coronado-Blázquez *et al.* (2019b, 2019a) have recently highlighted potential DM subhalo candidates among the 2FHL (Ackermann *et al.*, 2016a), 3FGL (Acero *et al.*, 2015) and 3FHL catalogs, which have also not yet been observed by IACTs.

Finally, thanks to a significantly larger FoV and foreseen large-scale survey observations, dark satellite discovery searches may also be feasible with

[o]Note that the recent publication by Abdalla *et al.* (2021a) is the only IACT study on this topic so far published in a peer-review journal.

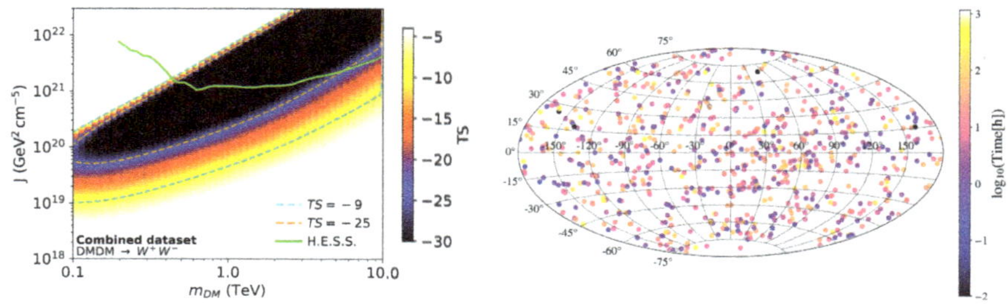

Fig. 8.4. The two approaches to search for dark satellites with IACTs. The first is by following up observations by suitable unidentified gamma-ray sources detected by *Fermi*-LAT, which is the most viable strategy with current IACTs: the left panel, taken from Abdalla *et al.* (2021a), shows the upper limit on the sum of the J-factors in a combined analysis of three unIDs observed by H.E.S.S. under the assumption of a thermal DM annihilation cross-section (solid light gray line). Under the assumption that the signals detected by *Fermi*-LAT in fact stem from DM subhaloes and annihilation into W^+W^- bosons (likelihood contours in the background), the H.E.S.S. null results significantly restrict the possible J-factor range. Second, with the future CTAO, the serendipitous discovery in any observation's FoV also becomes a feasible search strategy: the right figure shows a random realization of sky coverage with the CTAO within the first decade, extrapolated from MAGIC stereo observations. Figure from Coronado-Blázquez *et al.* (2021); see page 332 Section 8.2.3 for further discussion.

the next-generation IACT: the CTAO. This is illustrated in Fig. 8.4 (right) and further discussed on page 332 Section 8.2.3 in the outlook section.

Observation of intermediate mass black holes. There are theories that predict the existence of black hole (BHs) of masses between 10^2–10^6 $M_\odot$, sometimes referred to as intermediate mass black holes (IMBHs); they could form as remnants of the collapse of Population III stars, resulting in $10^2\,M_\odot$-mass BHs (Madau and Rees, 2001; Heger *et al.*, 2003, scenario A) or from the direct collapse of primordial huge gas overdensities in early forming haloes, resulting in $10^6\,M_\odot$-mass BHs (scenario B Koushiappas *et al.*, 2004). Many of them could reside in the MW halo. The average number of IMBHs in the MW was estimated by numerical simulations to be of the order of 100 (scenario A) to 100 (scenario B) (Bertone *et al.*, 2005a). As a result of the increased gravitational potential due to infalling baryons on a central accreting system, the DM could have readjusted and shrunk, giving rise to the formation of "mini-spikes" (Bertone *et al.*, 2005a, 2009). On the other hand, mini-spikes are rapidly disrupted as a result of dynamical processes like BH formation or merging events. The interesting fact is that, starting from a typical NFW distribution for the DM, the adiabatic growth of the spike leads to a final DM density profile even cuspier than NFW, with a central slope of index $-7/3$ as opposed to -1 for NFW (Bertone *et al.*, 2009). For scenario B, which foresees more luminous objects than scenario A, the corresponding gamma-ray luminosity would be of the order of the gamma-ray luminosity of the entire MW halo, which made IMBHs very interesting targets for DM searches (Bertone *et al.*, 2005a). Similar to the dark satellites' case discussed above, it is possible that some IMBHs are hidden in unidentified EGRET or *Fermi*-LAT point-like sources. Before the advent of *Fermi*-LAT, and after having performed a selection of the best EGRET unidentified sources, MAGIC observed the brightest of them in 2006 for 25 h, with no detection (Doro *et al.*, 2007). The source was later on associated with a bright pulsar using *Fermi*-LAT data. Also, by using 400 h of data collected from 2004 to 2007 in the region between -30 and $+60$ degrees in Galactic longitude, and between -3 and $+3$ degrees in Galactic latitude, H.E.S.S. could exclude the IMBH formation scenario B at a 90% confidence level for DM particles with annihilation cross-section values $\langle \sigma v \rangle$ above 10^{-28} cm^3 s^{-1}and masses between 800 GeV and 10 TeV (Aharonian *et al.*, 2008).

Observation of globular clusters. Globular clusters are star compounds that share properties with dSphs. They are in general more compact and with narrower metallicity distributions than dSphs. There is an ongoing debate on how these clusters formed, considering that it is possible to find globular clusters already 1 Gyr after the Big Bang. Despite the fact that they are generally thought to contain low amounts of DM, the debate on whether or not they reside in a DM halo is still undecided (Mashchenko and Sills, 2005a, 2005b; Lane *et al.*, 2010; Bradford *et al.*, 2011; Ibata *et al.*, 2013; Den Brok *et al.*, 2014; Baumgardt, 2017; Creasey *et al.*, 2019). From the cosmological point of view, there is no doubt that they were formed in DM overdensities, just as the dSphs — however, there is no strong observational evidence for the presence of a DM halo as of today. One possibility is, in fact, that the luminous part of a globular cluster is just located in the inner region of a flat DM overdensity, which would not significantly modulate the stellar velocities for different radii (Ardi and Baumgardt, 2020). The interest in globular clusters from the DM search perspective has recently been revived by the observational evidence of a $\sim$30 GeV DM signature in the gamma-ray spectra of 47 Tuc (Brown *et al.*, 2018) and Omega Cen (Brown *et al.*, 2019b), though a pure DM interpretation of this observational evidence is still a matter of debate (Bartels and Edwards, 2019; Brown *et al.*, 2019a).

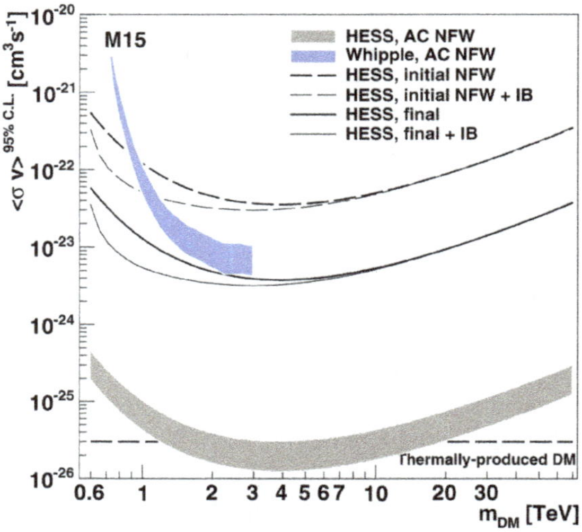

Fig. 8.5. 95% C.L. upper limits on the DM annihilation cross-section by H.E.S.S. (Abramowski *et al.*, 2011) and Whipple (Wood *et al.*, 2008) observations of the globular cluster M15. Limits were calculated for Whipple (blue band) and H.E.S.S. (brown band) assuming an NFW density profile, adiabatically contracted due to the presence of an IMBH. Additionally, taking into account kinetic heating by stars results in the final profile assumed by H.E.S.S. (solid lines). The dashed lines denote an initial DM halo model assuming $M_{\rm vir} = 10^7\,M_\odot$. The final-state gamma-ray spectrum is modeled in this work according to Bergstrom *et al.* (1998). "IB" refers to accounting for internal Bremsstrahlung effects. Figure from Abramowski *et al.* (2011).

With regards to the observation of globular clusters with IACTs, M15 was observed from 2002 to 2004 by Whipple (Wood *et al.*, 2008) and later re-observed together with NGC 6388 by H.E.S.S. in 2006–2009 (Abramowski *et al.*, 2011), shown in Fig. 8.5. MAGIC also invested more than 150 h on M15, with which only upper limits on the gamma-ray emission were obtained and discussed in the context of astrophysics (Acciari *et al.*, 2019). (The interpretation in terms of DM is currently in preparation.) The best exclusion curves for annihilating DM come from the observation of NGC 6388 by H.E.S.S., and are of the order of 10^{-24} – 10^{-25} cm^3 s^{-1}. It must be noted, however, that these results rely on strong assumptions about the amount of DM in this particular object.

Observation of galaxy clusters. Within the standard ΛCDM scenario, galaxy clusters, with masses around 10^{14}–$10^{15}\,M_\odot$, are the largest gravitationally bound objects in the Universe and the most recent structures to form (Voit, 2005). They are relevant for what concerns indirect DM searches (e.g. Jeltema *et al.*, 2009; Colafrancesco *et al.*, 2011; Sanchez-Conde *et al.*, 2011; Pinzke *et al.*, 2011; Gao *et al.*, 2012), because DM is the dominant mass component within this type of system, accounting for up to 80% of the total virial mass.[p] Galaxy clusters, however, are extragalactic objects, many of them located at cosmological distances, and therefore any radiative signal from DM annihilation or decay would be significantly diluted or attenuated by the Extragalactic Background Light (EBL). Given their large halo masses, the presence of a large amount of substructures in clusters is expected to be particularly important for DM searches, as they intrinsically boost the total J-factor for annihilating DM by a factor of 30–50 (Sanchez-Conde *et al.*, 2011; Sánchez-Conde and Prada, 2014; Zavala and Afshordi, 2016; Moliné *et al.*, 2017; Hiroshima *et al.*, 2018). Because of this, galaxy clusters are targets competitive to dSphs for indirect DM searches in terms of expected J-factors (Sanchez-Conde *et al.*, 2011). Although this "subhalo boost" is not present in a decaying DM scenario, for which the "clumpiness" of DM is not relevant, galaxy clusters are even more competitive compared to dSphs in the case of decaying DM; this is to be understood by the huge total mass budget of a cluster integrated within a small opening angle. Note also that, as in other targets, in galaxy clusters, any potential DM-induced gamma-ray emission is not only expected as prompt emission, but also as secondary emission mostly generated by synchrotron and inverse Compton processes from HE electrons and positrons produced in DM interactions (Colafrancesco *et al.*, 2006).

Clusters of galaxies also host strong astrophysical gamma-ray emitters in many cases (Blasi and Colafrancesco, 1999). Thus, any careful search for DM signatures in galaxy clusters must disentangle these from more "conventional" astrophysical contributions. For instance, AGNs sitting in individual cluster galaxies or a sizeable interaction of CRs with the ICM may potentially yield gamma-ray fluxes larger than the one expected from DM interactions. Fortunately, a different spatial morphology of the gamma-ray signal is expected in each case: while the signal from DM is more extended, especially in the case of decay, the signal from CR interactions and radiative energy losses is more compact, and individual galaxies are point-like. This can be used as a powerful tool for discriminating between the different components, as discussed by Doro *et al.* (2013).

[p]The remainder component is dominated by intracluster medium (ICM) gas.

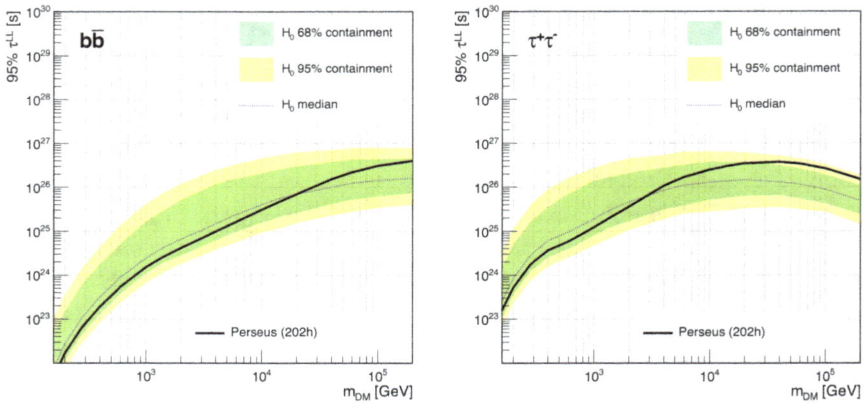

Fig. 8.6. 95% C.L. lower limits on the DM decay lifetime $\tau^{\rm LL}$ (solid black lines) as obtained from the 202 h observation of the Perseus cluster of galaxies with the MAGIC telescopes for the $b\bar{b}$ (left panel) and $\tau^+\tau^-$ (right) decay channels. The expected limits (dashed lines) and the two-sided 68% and 95% containment bands are also shown. These are the most constraining decay limits obtained so far above few hundred GeV. From Acciari *et al.* (2018a).

To date, the deepest scrutiny of a galaxy cluster in gamma rays is the one performed with the MAGIC telescopes for the Perseus cluster (Acciari *et al.*, 2018a). MAGIC observed this object for more than 400 h over several years. This deep observation of the Perseus cluster allowed to set the strongest constraints to date on decay DM using a subset of 202 h of data (Acciari *et al.*, 2018a). Such constraints are reported in Fig. 8.6 for two pure decay channels that roughly bracket the minimum and maximum number of photons expected from decay, e.g. by decays into purely $b\bar{b}$ quarks or $\tau^+\tau^-$ leptons. These MAGIC limits are of the order of 10^{26} s, that is, more constraining than any other limit set to decaying DM in the DM mass range above few hundred GeV. At lower energies, stronger constraints are found using *Fermi*-LAT data of the diffuse emission (Ackermann *et al.*, 2012).

For DM annihilation, the strongest DM constraints from clusters as of today were set by H.E.S.S. by the observation of the Fornax cluster (Abramowski *et al.*, 2012), which is one of the clusters with the highest expected DM annihilation signal. Apart from Perseus and Fornax, the Coma cluster was also observed for 18.6 h by the VERITAS telescopes (Arlen *et al.*, 2012), however yielding less stringent constraints on DM annihilation.

Constraints on line-like spectral signatures: Self-annihilation or decay of DM particles may create unique line-like gamma-ray spectral features not expected at TeV energies from any other astrophysical process (Bertone *et al.*, 2005a). In particular, narrow spectral lines can be expected from events such as $\chi\chi \to \gamma\gamma$ and $\chi\chi \to X\gamma$, where $X = Z^0, H, \ldots$ are found at loop or enhanced level (Bergstrom and Kaplan, 1994; Jungman and Kamionkowski, 1995; Bergstrom and Ullio, 1997; Arina *et al.*, 2010; Rinchiuso *et al.*, 2018; Beneke *et al.*, 2018). Furthermore, in case the annihilation is mediated by a heavy particle, peculiar features such as "box-like" features in the gamma-ray spectral energy distribution may occur (Ibarra *et al.*, 2012). In addition, emission from virtual internal states can happen for specific particle models via Bremsstrahlung, providing a pronounced component peaked toward the spectral cutoff at the highest energies (Bringmann *et al.*, 2008; Bringmann, 2009; Bringmann and Calore, 2014).

Such a signal would be readily distinguishable from astrophysical HE gamma-ray sources that are expected to only produce continuous spectra, and may constitute a robust "smoking-gun" identification of DM. Moreover, for WIMP masses above several hundreds of GeV and above the energy reach of space-based detectors, Earth-bound gamma-ray detectors are the only instruments to search for such features. Additionally, for IACTs, line-like features are even easier to search for than continuous spectra; the large residual hadronic is a generic challenge for IACTs, yet line-like gamma-ray spectra are very distinct from the background spectral energy distribution and, thus, a powerful discrimination between background and signal is possible in terms of their spectral properties. Nevertheless, specific analysis algorithms must be developed in order to look for such signatures in IACT data. In particular, IACTs — like any other instrument — are not only affected by a limited energy resolution (in this case of the order of 10%), but may also show specific systematic uncertainties at the energy scale as a result of using the atmosphere as a calorimeter.

Using data collected with H.E.S.S., upper limits on line-like emission have been obtained in the energy range between 100 GeV and 70 TeV from the central part of the MW halo (Abdalla *et al.*, 2016a, 2018g). For a DM particle mass of 1 TeV, limits on the velocity-averaged DM annihilation cross-section $\langle\sigma v\rangle$ reach the level of 10^{-27} cm^3 s^{-1} when adopting an Einasto DM density profile for the Galaxy. Using combined dSph observations, limits were derived between 300 GeV and 20 TeV by H.E.S.S. (Abdalla *et al.*, 2018h) and between 100 GeV and 10 TeV by VERITAS (Archambault *et al.*, 2017). MAGIC obtained additional limits

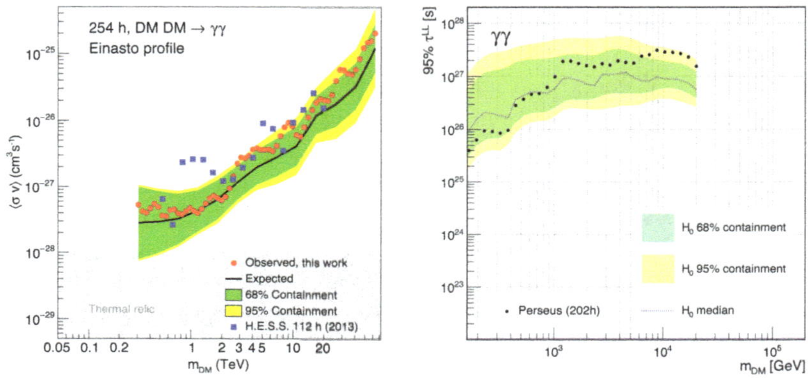

Fig. 8.7. Constraints on the velocity-averaged annihilation cross-section $\langle \sigma v \rangle$ for the prompt annihilation into two photons derived from H.E.S.S. observations of the inner GC region (left), and of the lifetime from prompt DM decay into two photons derived from MAGIC observations of the Perseus galaxy cluster (right). From Abdalla *et al.* (2018g) and Acciari *et al.* (2018a), respectively.

after a 158 h observation of the Segue 1 dSph (Aleksić *et al.*, 2014b), and further, presented an analysis of the GC region (Inada *et al.*, 2021). Line emission can likewise be emitted in DM decay processes, and this was scrutinized by Acciari *et al.* (2018a) in the analysis of 202 h of data toward the Perseus galaxy cluster. The current best limits on line-like emission from DM annihilation (left) and decay (right) at TeV energies are shown in Fig. 8.7. Note that these limits correspond to a pure annihilation or decay process into two photons. For this process only constituting a small fraction among other annihilation or decay channels, as expected for a loop-suppressed process, the total constrained annihilation cross-section may be much larger (and correspondingly, the lifetime may be shorter) than suggested by the shown data. In Fig. 8.7 (left), this is considered by the gray-shaded region of an expected signal for thermal WIMPs downshifted to $\langle \sigma v \rangle \lesssim 10^{-28}$ cm^3 s^{-1}, already implying annihilations into two photons happening with a branching ratio smaller than 10^{-2}.

Searches with charged particles: In the annihilation or decay process of DM particles, hadrons and leptons are abundantly produced (see, e.g. Donato *et al.*, 2004). Below these, antiparticles would constitute a species that could be easier discriminated against the astrophysical hadronic background, such as the antiparticles produced after the spallation of CRs in the interstellar medium. This possibility attracted great interest a decade ago after the accumulation of evidence of a positron excess from satellite-borne particle detector experiments such as PAMELA (Adriani *et al.*, 2009), *Fermi*-LAT (Ackermann *et al.*, 2012d) and subsequently the AMS collaboration (Aguilar *et al.*, 2013). As no antiproton excess was found simultaneously, an explanation for this positron excess that generated particular interest in the community was to invoke the existence of "leptophilic" DM, e.g. DM particles that prefer leptonic annihilations or decays over hadronic channels. In addition, the balloon-borne ATIC reported an excess in the electron+positron spectrum around 500 GeV (Chang *et al.*, 2008). The sensitivity of these balloon or space-borne measurements, however, was limited to energies below a few TeV. In turn, IACTs are sensitive to Cherenkov light generated in atmospheric showers initiated by sufficiently energetic photons or charged particles. In this sense, they are also suitable detectors for cosmic electrons, positrons and antiprotons beyond TeV energies.

Leptonic showers from cosmic electrons or positrons are currently indistinguishable from gamma-ray induced showers. Yet, in contrast to gamma-ray induced showers, which are always aligned to a very localized cosmic gamma-ray source, cosmic leptons reach the Earth from perfectly isotropical directions. As such, in a data sample without a gamma-ray source in the FoV, electrons or positrons can be discriminated from the residual hadronic background in IACT data by comparing the whole data sample with corresponding Monte Carlo (MC) simulations of hadron events, via the definition of discriminator parameters specifically designed for that purpose. Consequently, the systematic uncertainties need to be carefully controlled; even a small mismatch between MC and real data would result in a bias in the estimated electron flux. This search was pioneered by H.E.S.S. back in 2008, which used 239 h of data, extending the measured electron+positron spectrum deeper into the TeV range (Aharonian *et al.*, 2008c). They found evidence for a steepening in the all-electron spectrum above 600 GeV, while they excluded the spectral peak previously claimed by ATIC (Aharonian *et al.*, 2009c). An analogous measurement by VERITAS confirmed a spectral break in the energy spectrum at around 700 GeV (Archer *et al.*, 2018), as did a preliminary measurement presented by MAGIC (Borla Tridon *et al.*, 2011). The DAMPE satellite detector published an all electron+positron spectrum up to 4.6 TeV, confirming a spectral break in the all-electron spectrum at around 900 GeV (Ambrosi *et al.*, 2017), as did the CALET experiment measuring the spectrum to 4.8 TeV on the International Space Station (Adriani *et al.*, 2018).

Unfortunately, the IACT detection technique does not directly allow us to discriminate electrons from positrons or investigate the excess of positrons at TeV energies. The Earth's magnetic field, however, is separating electrons from positrons, and this separation could be observable where the isotropic flux is shadowed by the Earth (as exploited by the *Fermi*-LAT measurement; Ackermann *et al.*, 2012d) or the Moon. In fact, the feasibility of disentangling electrons and positrons by observing a CR "Moon shadow" with IACTs has already been investigated, and some data have been collected with the MAGIC and VERITAS telescopes despite the big observational challenges (Colin *et al.*, 2011; Aharonian *et al.*, 2009c; Bird *et al.*, 2016a).

8.2.3. *Conclusions and outlook*

The length of Table 8.1 substantiates the huge effort in the pursuit of DM made by IACTs over the past decade. Not only have target classes been diversified, but also novel analyses and algorithms have been developed specifically for such searches. In Fig. 8.8, we report some of the most important limits produced by IACTs so far. It is vitally important that we comprehensively discuss this effort.

On the left panel of Fig. 8.8 we report some of the most relevant limits obtained on annihilating DM for the $b\bar{b}$ channel. We show the H.E.S.S. observation of the GC halo adopting an NFW profile as a solid line (Abdalla *et al.*, 2016b), the MAGIC stereo observations of the Segue 1 dSph as a

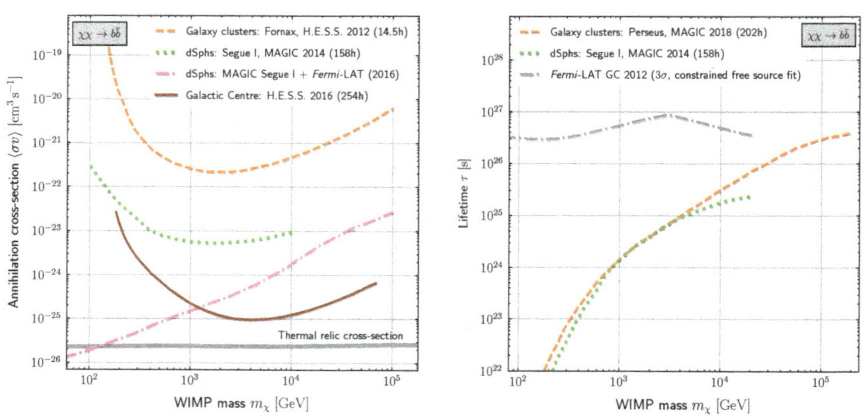

Fig. 8.8. Selection of some of the most representative WIMP limits by IACTs for the $b\bar{b}$ annihilation channel (left) and the $b\bar{b}$ decay channel (right).

dotted line (Aleksić *et al.*, 2014b), the *Fermi*-LAT combined limits from the observation of 15 dSphs with 6 yr of data together with MAGIC Segue I as a dot-dashed line (Ahnen *et al.*, 2016c) and the 14.5 h limits obtained with H.E.S.S. observation of the Fornax galaxy cluster as a dashed line (Abramowski *et al.*, 2012). This figure is emblematic. We see that only searches in an object with a large DM concentration and close distance such as the GC (J-factor of the order of 10^{21} Gev^2 cm^{-5}) have a chance to skim the thermal relic value at masses at about 1–10 TeV. The second best limits, although possibly less subject to model uncertainties, are those obtained with Segue 1, a very DM-dominated dSph (J-factor of the order of 10^{19} Gev^2 cm^{-5}). Galaxy clusters provide weaker constraints, albeit the corresponding limits were obtained with a much more reduced exposure than the others shown in the same figure. On the other hand, at lower masses, *Fermi*-LAT dominates the sensitivity.

This is not necessarily the end of the game. For instance, there are several mechanisms that can provide a cross-section larger than the thermal one, such as the Sommerfeld effects (Nagayama *et al.*, 2021), which are activated resonantly for some DM masses due to the particles' extremely non-relativistic velocities today. These resonant phenomena predict that there are some values of mass for which the cross-section may be much larger than predicted by the thermal limit scale, still consistent with today's observed DM density. Such resonances, however, can appear at virtually any mass. Therefore, indirect detection of DM could be around the corner for such models. Furthermore, it is important to remember that, should more than one DM particle exist, the thermal value would need to be re-discussed. In any case, the IACTs' DM annihilation limits obtained so far are the strongest for DM masses in the TeV scale.

On the right panel of Fig. 8.8 we report some of the most relevant limits obtained on decaying DM with IACTs. We report the results from 202 h observation of the Perseus galaxy cluster (dashed line) and those obtained with 158 h exposure on the Segue 1 dSph galaxy (dotted line) with MAGIC (Aleksić *et al.*, 2011d, 2010d). We also show, for comparison, the *Fermi*-LAT decay limits from the GC (Ackermann *et al.*, 2012). One can see that for energies larger than 10 TeV, lower limits at the order of 10^{26} s are obtained with IACTs, and they are competitive with those obtained at lower energies with *Fermi*-LAT. Here, there is no reference value to be reached, in contrast to the annihilation case — in principle, any value above the exclusion curve is equally probable. We are not in a position to assess the likelihood of a model.

Outlook: IACTs have certainly produced unique information about DM candidates at the TeV mass scale, with a sensitivity unmatched by any other technique or experiment. Yet, these results should be put into context with the other gamma-ray detectors in the field, operating or planned. Throughout this chapter, we already presented results from *Fermi*-LAT, a pair-production detector hosted aboard a satellite, sensitive to gamma rays from hundreds of MeV to hundreds of GeV. Data from *Fermi*-LAT dominate the DM limits shown in Fig. 8.8 at the lowest DM masses. Towards the HE end of the IACT sensitivity range, we find a different kind of detector class: that of the shower-front detectors that sample the very particles in the extended atmospheric showers. Instruments of this class have been the decommissioned ARGO-YB and Milagro experiments, now followed by the HAWC and LHAASO (He *et al.*, 2019) detectors. All of these were or are hosted in the Northern Hemisphere. A project for a Southern Hemisphere shower-front detector goes under the name of SWGO (Barres de Almeida, 2021). The sensitivity of these instruments for DM particles is competitive with current IACTs, as shown in Fig. 8.9 for LHAASO (dashed line). At lower energies, direct detection experiments hardly reach the sensitivity needed to test DM candidates with masses of few hundred GeV, and particle accelerators similarly cannot access the energy scale of multi-TeV particles. Therefore, IACTs are crucial to probe the parameter space of heavy WIMP candidates between a few TeV and the unitary upper bound for WIMP masses expected at around a few hundreds of TeV (Griest and Kamionkowski, 1990).

A huge leap will be made with the CTAO (Acharya *et al.*, 2013), which will supersede the IACTs we have discussed in this chapter: H.E.S.S., MAGIC and VERITAS, in the coming decade. The CTAO will not only have an improved sensitivity, but also both a better energy and angular resolution compared to current IACTs. A better sensitivity will obviously boost the performance obtained for the same observation time. In practice, this will significantly increase our chances for detection and, in the absence of it, translate into more constraining DM limits. The improved energy and angular resolutions will allow us to provide more robust constraints: the peculiarity of DM spectra can be assessed only if the energy resolution is sufficient enough to not wash out their unique properties. An improved angular resolution will allow for a better investigation of the DM signal region, especially in those cases for which abundant and complicated astrophysical emission is expected to be found, such as the GC or galaxy clusters, or in cases where the spatial extension itself can be a DM smoking

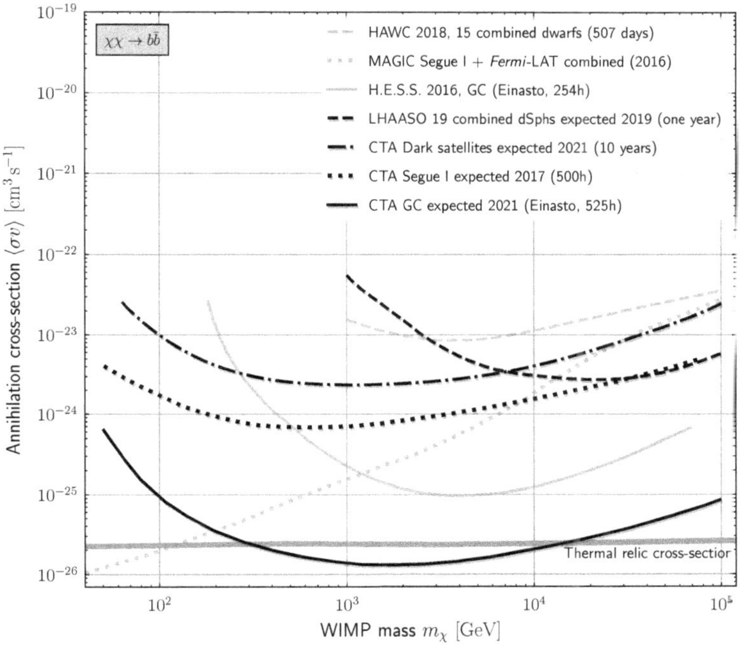

Fig. 8.9. Selection of some of the most representative sensitivity prospects for WIMP searches with future gamma-ray detectors, namely, CTAO (Acharya *et al.*, 2019, 2021; Coronado-Blázquez *et al.*, 2021). In shaded colors, a selection of current limits by IACTs (as shown in Fig. 8.8) and HAWC (Albert *et al.*, 2017) are overlaid.

gun, as expected in dark satellites (Coronado-Blázquez *et al.*, 2019b). The envisioned CTAO strategy for DM searches is discussed in Acharya *et al.* (2019): a deep survey of the GC region, a dedicated pointing on the best available dSph at the time, and deep surveys of both the Perseus galaxy cluster and the Large Magellanic Cloud are all parts of the CTAO Key Science Programs for the first decade.

For the GC region, a deep scan over 525 h will be performed in the first 3 years of operation, primarily with CTAO South, covering a circular area of ten degrees in diameter around the GC. Additionally, an extended survey with over 300 more hours on a large area north or south of the GC region is envisaged (Acharya *et al.*, 2019). Acharya *et al.* (2021) studied in detail how this wealth of data can be used to probe WIMP candidates with annihilation cross-sections below the thermal relic value. Figure 8.9 displays how this expectation compares to current limits on DM annihilation in the

TeV range, with a sensitivity exceeding the best current limit by H.E.S.S. by one order of magnitude (solid line). Notably, Acharya *et al.* (2021) show that this sensitivity is even guaranteed when accounting for the uncertainty of the instrumental background and astrophysical emissions presented in the data. Furthermore, they found that, even for a cored inner Galactic DM density profile, a sensitivity closely reaching the thermal value can be achieved thanks to the large covered area and a template-fitting analysis approach. As a result, the CTAO may robustly allow for the discovery of TeV DM at the canonically expected thermal cross-section. In parallel to the observation of the GC region, 300 h of the most promising dSph target are guaranteed in the first 3 years of operation of the CTAO. In addition to the interest that such deep observation itself warrants to search for DM in this type of object, there is one more reason that makes it particularly useful. Indeed, in the case of a putative detection of a DM signal in the GC region, this additional time on a dSph may allow us to further scrutinize such a discovery in a DM context by means of other types of targets, and to test or exclude possible alternative explanations. In Fig. 8.9, DM prospects are shown assuming the Segue I dSph with its currently best-estimated J-factor and for a total of 500 h exposure time (dotted line). Independently of the success of these initial DM searches with the CTAO, 700 more hours will be dedicated in the first decade of CTAO operations to hunt for DM either at the GC or at the position of the best dSph target (Acharya *et al.*, 2019).

The GC and dSphs are not the only viable targets for the CTAO. Brun *et al.* (2011), Hütten *et al.* (2016) and Coronado-Blázquez *et al.* (2021) also discussed the CTAO's capabilities to search for dark satellites. In contrast to current IACTs, the CTAO may also have a high chance of realistically discovering these objects. This is not only due to the improved sensitivity, but also to its larger field of view and planned dedicated large-area Galactic-plane and extragalactic surveys (Dubus *et al.*, 2013). While Brun *et al.* (2011) and Hütten *et al.* (2016) only investigated the individual surveys for such discovery potential, Coronado-Blázquez *et al.* (2021) recently scrutinized the chances for a serendipitous discovery of a dark satellite in all CTAO data taken within the first decade of CTAO operations. As shown in Fig. 8.9 (dot-dashed line), this strategy could yield a sensitivity competitive to those obtained by a dedicated deep CTAO observation of the most promising dSph galaxy to date.

Can we expect novelties beyond the advent of the CTAO? Certainly new dSphs have been discovered in recent years thanks to the Dark Energy

Survey (Abbott *et al.*, 2018), and opportunities to identify promising targets for indirect DM searches will be greatly increased with the Very Rubin Observatory in the near future (Drlica-Wagner *et al.*, 2019). It's possible that in a decade from now, all or the majority of MW dSph satellites will be detected (Hargis *et al.*, 2014). Improved spectroscopic capabilities, like DESI (Aghamousa *et al.*, 2016) or the Prime Focus Spectrograph recently put in operation at the Subaru telescope (Tamura *et al.*, 2018) will also allow for more precise member star velocity estimations and, thus, for a more accurate DM modeling. Data from the GAIA satellite will also enable a better removal of interloper stars (Helmi *et al.*, 2018). Altogether, this information should lead to a greater accuracy of the determination of dSph J-factors. It is also still possible to find a dSph exhibiting a higher J-factor than any other dSph currently known. It was shown in Doro *et al.* (2017) that current IACTs may reach the thermal cross-section value if they were able to observe a dSph with a J-factor five times the one of Segue 1 for 500 h. Another possibility is that this J-factor was reached by a dark satellite serendipitously detected, both massive and close enough to the Earth.[q] Additionally, multi-wavelength approaches complementing IACT DM searches from the radio band to MeV gamma rays are identified as very useful more and more. Data from other wavelengths may either set strong constraints on heavy DM candidates on their own (Regis *et al.*, 2021) or allow us to improve the VHE gamma-ray analysis by providing additional information (Di Mauro and Winkler, 2021).

All in all, the next decade will give rise to great expectations for the discovery of TeV WIMP DM or in challenging the paradigm of massive-particle DM. Even if particle DM may be first discovered in colliders or direct detection experiments, an accurate DM identification and characterization will only be possible by combining signatures from different detectors and methods (Cahill-Rowley *et al.*, 2015). A signature in telescopes is furthermore necessary to establish the definitive link between a potential DM signal at the lab and the long-sought DM in the Universe. To this end, IACTs that study the properties and budgets of massive-particle DM in astrophysical reservoirs play a crucial role, today already and all the more in the future.

[q]In this case, however, it may be difficult to assess the underlying DM particle physics model until we have some knowledge of its exact distance and mass by other means.

Verification

(1) Why is the expected gamma-ray spectrum of DM annihilation *exactly* cut off at the rest mass of the DM particle (at half the rest mass for decay)?
(2) Why does one not search for WIMP relic annihilations or decays of local DM particles in a detector?
(3) According to the canonic scenario of thermal DM freeze-out in the early Universe, would there be more or less DM left over today if the annihilation cross-section was larger than $\langle \sigma v \rangle \sim 10^{-26}$ cm^3 s^{-1}?
(4) After first results of IACT DM searches were published with a few hours of data, why were improved limits from the same instruments not published before hundreds of more hours had been accumulated?

8.3. Axion-like Particles

As of today, the Peccei–Quinn (PQ) mechanism, proposed in the late 1970s (Peccei and Quinn, 1977), represents the most convincing solution to the strong Charge-Parity (CP) problem in Quantum Chromodynamics (QCD), e.g. the absence of CP violation in strong interactions. It was soon realized that a byproduct of the proposed PQ mechanism is the existence of a pseudo-Nambu-Goldstone boson called "axion" (Weinberg, 1978; Wilczek, 1978).[r] Fascinatingly, these QCD axions may account for a fraction or the whole (cold) DM content in the Universe and, indeed, constitute one of the most valid alternatives to WIMPs at present. Contrary to WIMPs, however, the axion would be very light, as its mass has already been experimentally constrained to be $\ll$eV. In fact, axions represent the most famous exponent of the whole class of weakly interacting slim particles (WISPs) (see, e.g. Arias *et al.*, 2012 for a recent review).

One interesting property of these axions, particularly relevant for the purposes of this book, is a predicted interaction with photons in the presence of external magnetic or electric fields described by the following Lagrangian (Dicus *et al.*, 1978; Sikivie, 1983):

$$\mathcal{L}_{a\gamma} = -\frac{1}{4}\, g_{a\gamma}\, F_{\mu\nu}\tilde{F}^{\mu\nu} a = g_{a\gamma}\, \mathbf{E}\cdot\mathbf{B}\, a, \tag{8.13}$$

where a is the axion field, $g_{a\gamma}$, the coupling strength, F, the electromagnetic field-strength tensor, $\tilde{F}$, its dual, $\mathbf{E}$, the electric field, and $\mathbf{B}$, the magnetic

[r]The name "axion" is typically attributed to Frank Wilczek, who supposedly took it from a laundry detergent alluding to the fact that the new particle would "remove a stain".

field. For the QCD axion, the mass is proportional to the coupling strength — yet there are other, more general states arising from extensions of the SM, for which this relation does not hold. These are frequently known as ALPs because they still share many similarities with the QCD axion (Chikashige *et al.*, 1978; Langacker *et al.*, 1986; Wilczek, 1982; Witten, 1984; Conlon, 2006). In string theory, for example, it is possible to properly formulate the entire PQ mechanism, which leads us to predict the existence of many of these "generalized" axions or ALPs, the QCD axion being just a particular case (Witten, 1984; Conlon, 2006; Arvanitaki *et al.*, 2010; Cicoli *et al.*, 2012).

The interaction with photons described by Eq. (8.13) represents the main vehicle used to experimentally search for axions and ALPs up to the present (Zioutas *et al.*, 2005; Duffy *et al.*, 2006; Andriamonje *et al.*, 2007; Wagner *et al.*, 2010; Anastassopoulos *et al.*, 2017; Irastorza and Redondo, 2018). This interaction results in conversions or "oscillations" of photons into ALPs and vice versa (see Fig. 8.10). This same property of axions and ALPs can also be used to look for them by means of astrophysical observations, whose main observable is precisely photons. Indeed, should ALPs and photons couple to each other, significant distortions in the spectra of astrophysical sources may be induced that could be measured, e.g. in the gamma-ray energy range with current or future IACTs.[s] Thus, photon-ALP conversions may potentially have very important implications for gamma-ray astrophysics and, excitingly, ALPs offer the possibility to search for new physics with gamma rays. Even in the absence of ALP-induced spectral features in gamma-ray data, a null result might be used to provide valuable insight, e.g. competitive experimental constraints, on the precise ALP nature.

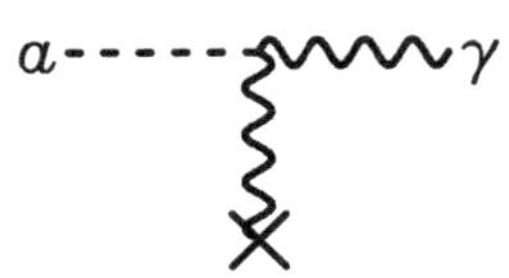

Fig. 8.10. Feynman diagram showing the interaction of photons and ALPs in the presence of an external electric or magnetic field. This interaction obeys the Lagrangian in Eq. (8.13).

[s]Note that we only refer to ALPs here; the search for QCD axions is in principle not possible in gamma rays given the insufficiently large couplings expected at the axion masses that can be probed by gamma-ray telescopes (see, however, Farina *et al.*, 2017 for "boosted" photon/axion couplings). This will be further discussed later in the section.

ALP-induced imprints in the spectra of gamma-ray sources are expected to happen around and above a particular critical energy for conversion that, in convenient units for gamma-ray experiments, can be expressed as (Hooper and Serpico, 2007)

$$E_{\rm crit} \equiv \frac{m_a^2}{2\ g_{a\gamma}\ B} \sim 2.5\ {\rm GeV}\ \frac{m_{\mu{\rm eV}}^2}{g_{11}\ B_{\rm G}}, \tag{8.14}$$

where the sub-indices refer to dimensionless quantities: $m_{\mu eV} \equiv m_a/\mu$eV, $g_{11} \equiv g_{a\gamma}/10^{-11}$ GeV and $B_{\rm G} \equiv B/$Gauss; m is an effective ALP mass $m^2 \equiv |m_a^2 - \omega_{\rm pl}^2|$, with m_a the ALP mass and $\omega_{\rm pl} = 0.37 \times 10^{-4}\mu{\rm eV}\sqrt{n_e/{\rm cm}^{-3}}$ the plasma frequency of the medium under consideration (n_e being its electron density). Recent results from the CAST experiment (Anastassopoulos *et al.*, 2017) provide an upper limit of $g_{11} \leq 6.6$ for ALP masses $m_a \leq 0.02$ eV. Currently, this bound represents the most general and stringent limit in this ALP mass range.

Just around this $E_{\rm crit}$, an oscillatory spectral pattern is expected whose exact properties will depend on both the morphology and orientation of the magnetic field along the direction of propagation of the ALP. The mixing between photons and ALPs becomes maximal above $E_{\rm crit}$ and independent of the energy, the probability for conversion being just proportional to $(g_{a\gamma} \times B)^2$. This is also called a strong mixing regime. A modification of the observed photon flux is then expected within this energy interval, because photons convert into ALPs and vice versa. The mixing remains maximum up to an energy $E_{\rm max} \sim 2.2 \times 10^6$ GeV $\rm g_{11}$ $\rm B_{\mu G}$, where birefringence effects start to heavily suppress the conversions. We note that other effects might significantly alter this general picture. For instance, at TeV energies, Kartavtsev *et al.* (2017) recently noticed that photon–photon refraction is expected to diminish the expected photon–ALP maximal mixing, this attenuation being more and more efficient as the energy increases.

The probability for photon–ALP conversion is proportional to the modulus of the magnetic field multiplied by the physical size, s, where this field is confined. More precisely, it is (Hooper and Serpico, 2007; Sánchez-Conde *et al.*, 2009)

$$P_{a\gamma} = \frac{1}{1 + (E_{\rm crit}/E_\gamma)^2}\ \sin^2\left[\frac{B\ s\ g_{a\gamma}}{2}\sqrt{1 + \left(\frac{E_{\rm crit}}{E_\gamma}\right)^2}\right]. \tag{8.15}$$

This means that an efficient conversion can happen not only in relatively small regions of the Universe, exhibiting very strong magnetic fields (e.g. AGNs, neutron stars), but also along large physical volumes filled with weak magnetic fields (e.g. the intergalactic medium between galaxies, the intracluster medium in galaxy clusters or the interstellar medium in a galaxy like our own). Very interestingly, this particular "property" is very closely connected to the "Hillas criterion", which is required to accelerate HE CRs in astrophysical objects (see Fig. 8.11) in such a way that this criterion can be actually used as a good proxy for the identification of the best astrophysical targets for ALP searches.

The precise effect of photon–ALP conversions on gamma-ray spectra can be very diverse — as diverse as are the possible astrophysical scenarios that potentially allow for efficient conversions; thus, a detailed, dedicated analysis must be done for each particular physical case. At first sight, it might seem that photon–ALP conversions would always lead to an

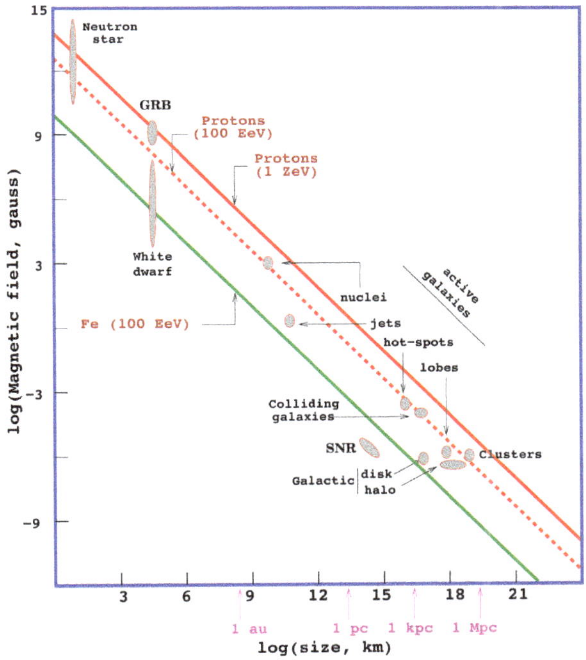

Fig. 8.11. The Hillas criterion establishes a relation between physical sizes and magnetic fields in astrophysical objects/environments so that an efficient acceleration of ultra-high-energy CRs is possible. This correspondence can also be used as an excellent proxy to understand where in the Universe an efficient conversion between photons and axions/ALPs can be expected. Figure taken from Hooper and Serpico (2007).

attenuation of the gamma-ray flux at Earth, as a fraction of the emitted photons would convert into ALPs and therefore never reach the Earth. Yet, reconversions of ALPs into gamma-ray photons along the line of sight are possible, which would alter the level of attenuated flux. An additional ingredient needed for an accurate prediction is the precise role of the Extragalactic Background Light (EBL). This EBL — the result of the light emitted by galaxies over the entire cosmic history — permeates the whole Universe and induces an attenuation of the gamma-ray flux by pair production, e.g. $\gamma\gamma^{EBL} \rightarrow e^- e^+$. The attenuation is maximal for a background photon energy $\epsilon \sim$ (500 GeV/E) eV, which means that infrared/optical background photons are the ones responsible for the attenuation in the energy range specific to IACTs. The inclusion of the EBL in the computations of photon–ALP conversion leads to very interesting predictions. Indeed, for distant VHE sources, ALP conversions in the intergalactic medium can lead to either an attenuation or an enhancement of the observed flux at Earth, depending on the exact configuration of the magnetic fields, energies considered and source distance. An enhancement is possible because ALPs may travel unimpeded through the EBL and, only at later times before reaching the observer, some of them would convert back into gamma rays, for instance, in the magnetic field of our galaxy. Intriguingly, this might potentially allow us to observe gamma-ray emitters located at distances larger than those expected from conventional EBL models. Additionally, it is interesting to note that photon–ALP conversion might also happen at the sources themselves, in this way providing an alternative scenario to avoid absorption of gamma rays in the dense ultraviolet radiation field of AGNs (see e.g. Tavecchio *et al.*, 2012).

Note that none of the aforementioned arguments explains the reason to perform the astrophysical ALP search, particularly in the gamma-ray energy band vs. at any other photon wavelengths. Indeed, the latter is certainly possible and has already been done in some works (see e.g. Wouters and Brun, 2013 and Berg *et al.*, 2017, for X-ray astrophysical searches). When typical values of the astrophysical magnetic field strengths are considered and the CAST upper bound is taken into account, photon–ALP conversions are expected to show up in the ~GeV-TeV range, provided that the ALP mass is ultra-light, e.g. $neV \leq m_a \leq \mu eV$; see Eq. (8.14).

Current status of ALP searches with IACTs. From a historical perspective, the event that triggered a rapidly increasing interest for ALPs within the gamma-ray astrophysics community was, very likely, the claim by the PVLAS collaboration back in 2006 of the observation of optical rotation generated in a vacuum by a magnetic field, possibly associated to a neutral, light boson (Zavattini *et al.*, 2006). Certainly, should such a light boson exist, it might have important astrophysical and cosmological consequences that could be tested (Frieman *et al.*, 1995; Sikivie, 2008; Mirizzi *et al.*, 2008; Visinelli and Gondolo, 2010; Wantz and Shellard, 2010; Archidiacono *et al.*, 2015; De Angelis *et al.*, 2008). The PVLAS claim was excluded soon after an upgraded setup (Zavattini *et al.*, 2008); the whole effort aimed at looking for ALP-induced features in the spectra of gamma-ray sources, however, had already been boosted. The idea of using astrophysical observations to set constraints on ALP parameters was certainly not new and had been already exploited well before 2006 (e.g. Grifols *et al.* 1996; Brockway *et al.* 1996; Csáki *et al.* 2002, 2003; Mirizzi *et al.* 2005); yet the mentioned PVLAS claim first and, later, some astrophysical gamma-ray hints of photon–ALP conversions, as described in the following, generated a momentum in the community that remains today.

As said previously, efficient photon–ALP conversions may happen in those astrophysical environments that exhibit the right combination of magnetic field strength and size of the magnetized region. Indeed, prospect studies have been published that predict the size and shape of the expected ALP-induced spectral features due to conversions in or around the gamma-ray emitters themselves (Hooper and Serpico, 2007; Hochmuth and Sigl, 2007), within galaxy clusters (Horns *et al.*, 2012), the intergalactic medium along cosmological distances (De Angelis *et al.*, 2007; Mirizzi *et al.*, 2008; Mirizzi and Montanino, 2009), our own galaxy (Simet *et al.*, 2008) or a combination of these scenarios (Sánchez-Conde *et al.*, 2009). With these predictions at hand, different approaches to look for ALPs with IACTs are possible; all of them, however, will most likely be based on the search and analysis of a systematic residual after applying the best-fit conventional model to the available gamma-ray data. Only if we found significant departures with respect to this best-fit model, and after having discarded other possible conventional sources of error for such residuals, we might seriously consider it a signal of exotic physics. On the contrary, the absence of any significant deviations with respect to the conventional baseline model can be used to set constraints on ALP parameters in such a way that all those combinations of ALP parameters that should have yielded a measurable effect in the data can now be excluded.

Through the years, two competing but complementary data analysis strategies slowly arose as a consequence of both the predicted features and the matureness of the field: one aimed at statistically detecting the predicted spectral oscillations or "irregularities" around the critical energy for photon–ALP conversion, the other focused on the strong mixing regime, where the ALP-induced effect on the gamma-ray spectra is expected to be maximal. In most cases, the involved gamma-ray analyses were used to set limits on the ALP parameter space in the absence of any clear ALP-induced spectral trace in the data.[t] The obtained limits are nicely complementary to those achieved in the lab by other means, both in terms of the masses and of the couplings being tested. This is illustrated in Fig. 8.12, which shows the current summary of limits obtained in the region of the ALP parameter space relevant to IACTs. In particular, the H.E.S.S. collaboration searched for ALP-induced irregularities in the spectrum of PKS 2155-304, an AGN located at $z = 0.116$ (Abramowski *et al.*, 2013a). The authors of this work considered photon–ALP conversions happening both in the intergalactic medium between us and the source, and those in the intracluster magnetic field where the AGN resides. Since the magnetic field strength in each of these astrophysical scenarios is different by orders

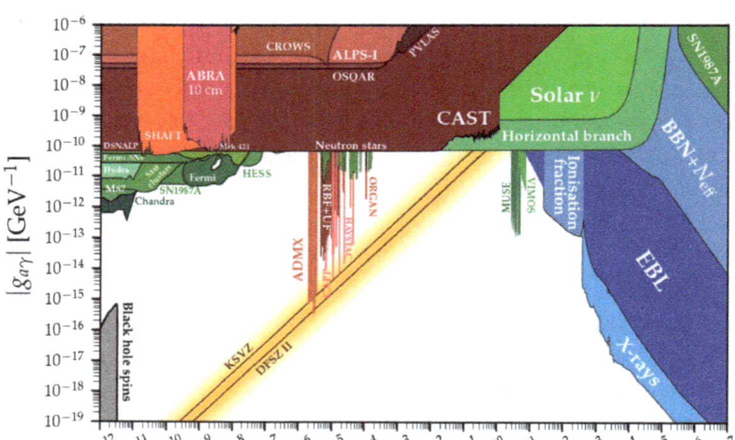

Fig. 8.12. Summary of current constraints to the ALP–photon coupling. Results obtained by H.E.S.S. are displayed in the upper left regime of the displayed parameter space for ALP masses around 0.1 μeV. Figure taken from https://cajohare.github.io/AxionLimits/ (version as of June 11, 2021).

[t]See, however, the claims made and briefly described later this same section.

of magnitude (nG vs. μG, respectively), according to Eq. (8.14), the data analysis will lead to two different sets of limits in the ALP parameter space. This can be seen in Fig. 8.12. The most competitive upper limits are obtained for the case of the intracluster medium and are as good as $g_{11} < 2.1$ for $15 < m_{neV} < 60$ (at 95% C.L.).

Perhaps surprisingly, the H.E.S.S. collaboration is thus far the only imaging Cherenkov experiment that has published ALP limits from their proprietary data, namely, on PKS 2155-304 (Abramowski *et al.*, 2013a). In addition, using public H.E.S.S. data, Liang *et al.* (2019) have set limits on the ALP parameter space from a compilation of ten bright Galactic sources. Similarly, Guo *et al.* (2021) have combined public data from H.E.S.S. and *Fermi*-LAT on PKS 2155-304 and another extragalactic object (PG 1553+113) for a study of ALP implications.[u] Tavecchio *et al.* (2012) and Galanti *et al.* (2020) have studied an ample sample of blazar spectra at VHE energies seen by *Fermi*-LAT, H.E.S.S., MAGIC and VERITAS, as did similarly Xia *et al.* (2019) with combined public data from *Fermi*-LAT, H.E.S.S., MAGIC, and VERITAS from supernova remnants to search for irregularities in the gamma-ray spectra. Although not strictly in the VHE band, other limits exist that were obtained with gamma-ray data as well, namely, those using *Fermi*-LAT observations at lower but partially overlapping energies than those covered by IACTs. In particular, a search for spectral oscillations in the spectrum of NGC 1275, an AGN located at the center of the Perseus galaxy cluster, was used to set the most competitive ALP constraints derived from gamma rays as of today (Ajello, 2016). These limits exclude couplings $g_{11} > 0.5$ at ALP masses $0.5 \lesssim m_a \lesssim 5\,\text{neV}$ (95% C.L.) under the default (conservative) scenario considered by the authors to describe the configuration of the involved magnetic fields: the one in the galaxy cluster hosting the AGN and the one in our galaxy. Additional works to set constraints on the ALP parameter space using *Fermi*-LAT data are the ones by Xia *et al.* (2018) and Zhang *et al.* (2018). In the former, the authors used three supernova remnants as targets and reported suspicious oscillatory behavior of their spectra at small ($\sim 4.2\sigma$) significance. The uncertainties in the measurement do not allow us

[u]A similar work by Malyshev *et al.* (2018) focuses on the AGN labeled as NGC 1275 in the Perseus galaxy cluster by a joint analysis of *Fermi*-LAT and public MAGIC data. The latter result suggests that those results in Ajello (2016) can be improved by roughly an order of magnitude under an optimistic, yet uncertain, configuration of the magnetic field. Yet, we note that this work has not been accepted as a refereed article as of today.

to extract robust conclusions and, instead, the authors speculated that this result could also be caused by different parts of the remnant contributing to the observed emission. In Zhang *et al.* (2018), the authors focused on PKS 2155-304 and set competitive limits with respect to those in Ajello (2016). Yet, their results are largely affected by important uncertainties in the modeling of the involved magnetic fields and, thus, the derived results should be taken with some caution. More recently, Li *et al.* (2021b) and Li *et al.* (2021a) searched for ALP–photon oscillations in the gamma-ray spectra of blazars to set competitive ALP constraints as well, in the former work using *Fermi*-LAT and ARGO-YB data of Mrk 421 and in the latter combining *Fermi*-LAT and MAGIC observations of both Mrk 421 and PG 1553+113. Their constraints turned out to be nicely complementary to those in Ajello (2016) in the ALP mass range between a few μeV to a few neV (95% C.L.).

In addition to setting meaningful limits on the ALP parameter space, some hints of the presence of ALPs were claimed through the years using gamma-ray data. As recent as 2007, ALPs with masses $\ll 10^{-10}$ eV were invoked to explain alleged anomalies in the flux of distant AGNs as measured by IACTs (De Angelis *et al.*, 2007). At that time, the Universe seemed less opaque to gamma rays than predicted by the latest available EBL models (e.g. Aharonian *et al.*, 2006a; Albert *et al.*, 2008f; Stecker and Scully, 2009; Helgason and Kashlinsky, 2012). This transparency hint within the ALP and VHE communities, pointing to ultra-light ALPs as a possible (exotic) solution to the transparency problem, was reinforced a few years later by a statistical analysis of 50 VHE sources (Horns and Meyer, 2012). This work found an apparent suppression of the pair-production process at a 4.2σ level; in other words, there was a systematic overcorrection of the intrinsic VHE spectra of cosmological sources by EBL absorption, especially at the highest attenuations. The same study showed that ultra-light ALPs could represent a viable explanation for such an anomaly as well. In a subsequent work (Meyer *et al.*, 2013), the same authors calculated the ALP parameter values that were needed to account for the apparent VHE flux anomalies measured by IACTs, e.g. they shaped a specific contour in the ALP parameter space for the mentioned "ALP transparency hint". Despite these early and subsequent works supporting both an overly low level of EBL density and an ALP explanation for it (e.g. De Angelis *et al.*, 2009, 2011), this hint of the existence of ALPs has not survived over time. Nowadays, it is generally accepted that state-of-the-art EBL models are in good agreement with gamma-ray data, e.g. that the measured transparency of the Universe to VHE gamma rays is as expected (Belikov *et al.*, 2011; Ackermann *et al.*, 2012h; Abramowski *et al.*, 2013b; Sanchez *et al.*, 2013; Biteau and Williams, 2015; Domínguez and Ajello, 2015) and, consequently, that there is no need to invoke the existence of ALPs in this context. In parallel, the transparency hint region itself in the ALP parameter space has also considerably shrunk since it was first proposed as a consequence of more recent ALP searches, which have, as already described above, imposed tight and robust constraints in this same region.

Other potential issues with conventional physics that might point to ALPs, however, were reported in the VHE regime. To cite some, the hard spectra exhibited by some AGNs seemed difficult to explain by conventional emission mechanisms (De Angelis *et al.*, 2011); the intrinsic spectrum of a few AGNs showed marginal deviation at the highest measured energies with respect to the expected power-law behavior (the "pile-up problem"; Domínguez *et al.*, 2011a); the existence of spectral hardening precisely at those energies where the EBL absorption is essential as well as its significant (12σ) unexpected dependence with redshift (Rubtsov and Troitsky, 2014); the existence of spectral breaks at a few GeV (Mena and Razzaque, 2013); the energy spectrum of some Galactic pulsars (Majumdar *et al.*, 2018), etc. These apparent puzzles — although inconclusive and possible to explain within the boundaries of conventional VHE physics, e.g. particles accelerated at relativistic shocks (Stecker *et al.*, 2007), internal photon–photon absorption (Aharonian *et al.*, 2008a), secondary gamma rays produced along the line of sight by CR proton interaction with background photons (Essey and Kusenko, 2010; Essey *et al.*, 2011; Essey and Kusenko, 2012; Aharonian *et al.*, 2013) — gradually stacked up to offer the ultra-light ALP as a single, viable and very appealing explanation able to simultaneously resolve all of them. As a result, the momentum to perform the ALP quest in gamma rays persisted over the years and, indeed, continues intensively at present.

The future of ALP searches with IACTs: There is definitely room for further work on ALP searches in gamma rays, even with current IACTs. As described above, up to now, only the H.E.S.S. collaboration has published data to search for ALPs and, in the absence of a signal, to set constraints on the ALP parameter space. Yet, both MAGIC and VERITAS are equally suited to perform similar studies in the Northern Hemisphere. Even more, archival data taken by these telescopes over the last years could be easily used as well to search for ALP-induced spectral features. A good example is the deep 400 h observation of the Perseus galaxy cluster recently accomplished by the MAGIC telescopes. The initial goal of this observational campaign — the longest ever performed by an IACT for a single object — was to look for the long-awaited first detection of CR-induced gamma-ray emission from clusters (Ahnen *et al.*, 2016a) and, more recently, to set limits on DM decay (Acciari *et al.*, 2018a). Since these observations include NGC 1275 and IC 310 in the FoV, two AGNs sitting in the Perseus cluster and observed at a very high significance with MAGIC, the data possess all that it is necessary to search for spectral irregularities in a similar way to what was done by the *Fermi*-LAT collaboration in Ajello (2016). We note that this LAT work, which also adopted Perseus as a target, provided the best ALP limits derived from gamma-ray observations so far. Another potential line of action to be further exploited by current IACTs can be provided by the measurement of new photons exhibiting very high opacities.[v] A statistical analysis combining this hypothetical new data with what has been already observed, in a similar way to that done by Horns *et al.* (2012), Meyer *et al.* (2013) and Biteau and Williams (2015), could provide further insight on the role of ALPs on photon propagation through cosmological distances.

Beyond current IACTs, the CTAO, with its ~10 times better sensitivity with respect to current IACTs, should make it possible to test the photon–ALP conversion scenario at an unprecedented sensitivity in the energy range between a few tens of GeV up to a few dozen TeV (Sanchez-Conde *et al.*, 2013). Indeed, a 300-h CTAO observation of the Perseus galaxy cluster is foreseen during the first five years of CTAO operations (Acharya *et al.*, 2019), which could be used to improve upon existing limits over a wider ALP mass range. A dedicated sensitivity estimate for gamma rays from NGC 1275 inside the Perseus cluster shows that the CTAO could reach couplings as low as $g_{11} \gtrsim 0.1$ for masses around 10 neV (Abdalla *et al.*, 2021b). In this way, the CTAO may become the most sensitive instrument to ALPs at these masses, largely exceeding laboratory-based dedicated experiments such as CAST (Zioutas *et al.*, 2005) or ALPS II (Bähre *et al.*, 2013) and even competing with the future IAXO (Irastorza *et al.*, 2013) at masses of $5\,\text{neV} \lesssim m_a \lesssim 100\,\text{neV}$. These prospects were based on the search for ALP-induced spectral irregularities around the

[v]These may be possible to obtain from both new discoveries of sources located at cosmological distances and/or for already known sources observed at even higher energies.

critical energy for conversion. In addition to the already discussed overlap and complementarity with *Fermi*-LAT, other next-generation instruments will complement the CTAO, particularly in its low energy end as well, such as LHAASO (Bai *et al.*, 2019; Long *et al.*, 2021), the future eASTROGAM (De Angelis *et al.*, 2017) or AMEGO (Hartmann, 2018) missions. By design, these next-generation experiments are expected to possess superb spectral capabilities in the search for ALP-induced spectral features. Indeed, preliminary results focusing on photon–ALP conversions in the Perseus galaxy cluster (Caputo *et al.*, 2018) show that these instruments could be sensitive to couplings $g_{11} > 0.3$ in the approximated ALP mass range $0.5 \lesssim m_a \lesssim 500\,\text{neV}$. Preliminary prospect work that focused on using the spectral hardening of distant gamma-ray sources instead was also performed (Sanchez-Conde *et al.*, 2013; Doro *et al.*, 2013). This was explored in further detail for the CTAOs by Meyer and Conrad (2014) by adopting a scenario consisting of four bright gamma-ray sources combined under the same consistent likelihood analysis. From this work, a significant detection of an ALP signal could be possible for ALP masses below 100 neV and couplings $g_{11} > 2$. At the highest energies covered by CTAOs, not only CTAOs but also HAWC (Abeysekara *et al.*, 2017b) and HiSCORE (Tluczykont *et al.*, 2014), working in a complementary (even higher) energy range, will be much more sensitive to HE photons from distant gamma-ray emitters compared to current IACTs. This increased sensitivity will allow us to test at a higher confidence if the once-claimed higher transparency of the Universe to gamma rays than expected from current EBL models is on solid ground or not and, if so, whether ALPs could still represent a viable explanation. These and other instruments covering the VHE range up to tens to hundreds of TeV could also be used to search for spectral oscillations (Galanti and Roncadelli, 2018).

This whole picture of the near future of ALP searches with planned gamma-ray and other instruments is summarized in Fig. 8.13. In order to put this particular effort into the more general ALP search context, as done for Fig. 8.12, the figure also shows a broader energy range than the one that can be actually probed by gamma rays. Fascinatingly, gamma rays may have the potential to look for ALPs in a region of the parameter space where they could constitute a portion or all of the DM content in the Universe, which is difficult to explore with any other current or planned detection technique or instruments. We refer the interested reader to the recently published review by Batković *et al.* (2021) for a further reading of the methods and technical challenges of ALP searches with IACTs.

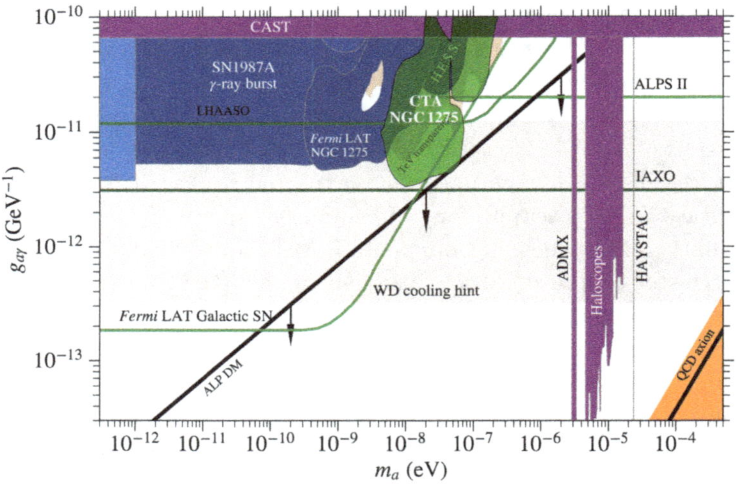

Fig. 8.13. Future ALP search prospects for different experiments and astrophysical targets (depicted as horizontal lines and by the CTAO excluding region). The CTAO should be able to provide very competitive limits around a few tens of neV, skimming the part of the parameter space where ALPs could constitute the entirety of the DM in the Universe (region below the diagonal line). Figure taken from Abdalla *et al.* (2021b).

Verification

(1) If all of the DM in the Universe was made out of ALPs, would this scenario be in agreement with all gravitational evidence for DM (CMB, large-scale structure, mass distribution in galaxies and clusters)?
(2) Why is an enhancement of the gamma-ray flux at Earth possible for cosmological sources if photons oscillate into ALPs?
(3) Could there also be a way to probe the ALP DM particles present in the Universe?

8.4. Primordial Black Holes

The search for PBHs traces back to several decades ago (Zel'dovich and Novikov, 1967; Hawking, 1971). PBHs are different from astrophysical BHs. The latter form by gravitational collapse of dying heavy stars, thus retrieving masses close to that of the progenitors. PBHs instead formed in the radiation-dominated era via different mechanisms that bound their mass scale to the moment of generation. Because of their origin, they are not contributing to the count of baryonic matter. In half a century, numerous theoretical studies and experimental constraints have accumulated in search of evidence for PBHs, especially as DM candidates (see, e.g. Carr *et al.*, 2016; Sasaki *et al.*, 2018; Green and Kavanagh, 2021 for recent reviews). There are several theoretically proposed formation mechanisms that allow the mass to range from the Planck mass scale to that of the supermassive BHs (Carr, 2005; Carr *et al.*, 2010). An exact mapping of the PBH formation time and evolution combines constraints from cosmology, thermodynamics, general relativity and quantum mechanisms, at the extremes of their validity. The search for PBHs gained significant momentum after the discovery of the GW150914 and merger event of two stellar mass BHs by the Ligo-Virgo Collaboration (LVC; Abbott *et al.*, 2016c). Since then, a large number of further-confirmed gravitational wave (GW) events from merging BHs has been reported (Bailes *et al.*, 2021). It was soon realized that BHs compatible with such merger events could constitute a large fraction or even the totality of DM (see Section 8.2.1) and be of primordial nature, thus opening new, very intriguing possibilities (Bird *et al.*, 2016b). Clearly, a discovery of PBHs would constitute a major breakthrough, shedding light onto the early Universe, quantum mechanics mechanisms, particle physics, and BH thermodynamics.

Most of the fate of a PBH relates to the time of formation, which foremost determines its mass. By relating the Universe density at time t with the critical density to generate a BH, one finds a relation between the PBH mass and the time of formation (Hawking, 1971):

$$M(t) \sim \frac{c^3\, t}{G} \sim 10^{38}\, \frac{t}{1s}\, [g] \;=\; 0.5 \times 10^5\, \frac{t}{1s}\, [M_\odot], \qquad (8.16)$$

where t is the Universe age at the time of creation, M the PBH mass, c the speed of light and G the Cavendish constant. Different mechanisms allow the generation of PBHs. One general possibility is the collapse of regions with higher density than the surroundings. This could be the case during sudden pressure reduction, as possibly happened during the quark-hadron era, or during dust-like phases or slow reheating after inflation. Another possibility is phase transitions — for example, connected to bubble collisions, the collapse of cosmic strings or domain walls. It is beyond the purpose of this text to properly summarize all theories and thus the reader is referred to specific reviews, such as Carr (2005) and Khlopov (2010).

The mass of the PBH has important consequences in the PBH evolution and affects several observable quantities. If the PBHs were formed at a precise moment, their mass distribution could be extremely narrow;

however, this may very well not be the case: if PBHs were generated through scale-invariant mechanisms, like generic quantum fluctuations of the space time, or during a wide time elapse, or have been subject to significant further mass accretion, they could have a wider mass distribution (Carr *et al.*, 2010). The number density of PBHs is primarily constrained to be smaller than the DM at current times: $\Omega_{\rm PBH} < \Omega_{\rm DM} = 0.25$. Their initial number, however, could have been larger but constantly reducing over time due to the Hawking mechanism (Hawking, 1974). According to Hawking, BHs radiate over time, thus reducing their entropy and increasing their temperature, with faster pace as time goes by, eventually reaching a final stage of complete disruption called "evaporation". The mass-loss rate during the BH evolution in quantum decay scenarios is governed by

$$\frac{dM_{\rm BH}}{dt} = -\frac{\alpha(M_{\rm BH})}{M_{\rm BH}^2}, \tag{8.17}$$

where $\alpha(M)$ is a parameter that depends on the number of degrees of freedom (DoF) available at any given time of the evaporation process. Clearly, the larger the temperature and the larger the number of DoF, the faster the pace of mass loss and the shorter the rate to evaporation. In particular, when the temperature exceeds the chromodynamics scale of $200 - 300$ MeV, quarks and gluons start to be directly radiated, which subsequently fragment and generate hadrons, photons and leptons, which in turn produce a strong non-thermal emission. Above this value, considering the number of DoF for quarks and gluons is larger than that for leptons and photons, the former dominate the observed gamma-ray spectrum (see, for example, Fig. 1 of Halzen *et al.* 1991). While below the chromodynamics scale, the number of DoF is more accurately known, one should note that there are also models like that of Hagedorn (1968) that foresee a strong multiplication of the number of DoF during the final stages of evaporation, which would imply an even faster final stage evolution before the evaporation, with complete disruption in nanoseconds to microseconds. It should be noted that the Hawking mechanism in its simpler description may not be accurate. It was found recently, for example, that small-mass BHs may become unstable due to quantum-gravitational effects earlier than predicted, with relevant consequences on observables like galaxy clustering (Raccanelli *et al.*, 2018). A PBH with mass M has a temperature (Halzen *et al.*, 1991)

$$T_{\rm BH}(M) = \frac{\hbar\, c^3}{8\pi\, G\, k_{\rm b}} \frac{1}{M} \sim 100 \left(\frac{10^{15}\,{\rm g}}{M}\right) \,[\rm MeV], \tag{8.18}$$

which increases over time as long as mass is lost, so that the life expectation for a BH, also called "evaporation time", is

$$\tau_{\rm BH}(M) = \frac{G^2 M^3}{\hbar\, c^4} \sim 10^{10} \left(\frac{M}{10^{15}\,{\rm g}}\right)^3 \; [{\rm yr}], \tag{8.19}$$

where $\hbar$ is the reduced Planck constant and $k_{\rm b}$ the Boltzmann constant. The temperature can be connected to the lifetime through the following formula (MacGibbon *et al.*, 2015):

$$T_{\rm BH}(\tau) \approx \left[4.8 \times 10^{11} \frac{1\,{\rm s}}{\tau_{\rm BH}(M)}\right]^{1/3} \; [{\rm GeV}]. \tag{8.20}$$

The reason to use a reference mass of $M_\dagger = 10^{15}$g ($\sim 10^{-18}$M$_\odot$) is that Eq. (8.19) shows that all PBHs roughly lighter than this value would have already evaporated at present time. PBHs at about M$_\dagger$ would instead be evaporating now, possibly associated with cataclysmic events such as long-duration GRBs. Heavier PBHs would be only mildly emitting radiation and almost stable, constituting a sort of non-baryonic collisionless cold fluid, and thus a possible contributor to the DM (Feng, 2013; Barack *et al.*, 2019). Evaporated, evaporating and non-evaporating PBHs imply therefore rather different observables and means for investigation, and quite different chances for detection. A collection of constraints can be found online.[w]

For already-evaporated PBHs, the limits are obtained with cosmological probes such as the CMB or theoretical arguments. In the range $< 10^{17}$ M$_\odot$, evaporated PBHs would have filled the Universe with particles and radiation that would have affected baryogenesis, for example, through the injection of a substantial amount of antiparticles, or through interaction with photons of the CMB. In the range 10^{-17} M$_\odot$ instead, PBHs would have strongly radiated MeV photons in their final phase, as shown in Eq. 8.16. This radiation would have filled the Universe, a prediction that is not observed on top of a standard astrophysical MeV diffuse background, thus providing very robust constraints as well (Laha *et al.*, 2020). PBHs with masses around M$_\dagger$ would instead be evaporating at present time. We could observe such events — for example, as gamma- or X-ray bursts or even through emission lines such as those observed from the GC. IACTs are sensitive in the TeV range and would therefore be valid

[w]https://github.com/bradkav/PBHbounds.

probes for PBHs evaporating now, as we will discuss in more depth in the following section. Finally, for larger initial masses, PBHs would appear as massive non-baryonic objects. As such, they would naturally act like weakly interacting massive objects–a standard requisite for a DM particle theory. For such objects, we could expect electromagnetic or hadronic signatures aside from the gravitational ones. The detection of GWs from black holes with an estimated mass of a few tens of GeV (Abbott *et al.*, 2016b, Abbott *et al.*, 2016a, Abbott *et al.*, 2016c) revived the interest of PBHs as DM candidates of similar mass (Barack *et al.*, 2019; Bird *et al.*, 2016b).

Search for PBHs with IACTs: How well do IACTs fit into this wide world of PBH observables? IACTs have a narrow FoV, which makes any attempt of detection of diffuse emission difficult to attain. This is probably the main avenue for IACTs to search for signatures of ongoing PBH disruptions. As discussed, in the final phase, evaporating PBHs would deliver an increasing emission, culminating with a disruption and burst of very-high-energy (GeV to TeV) radiation, lasting however for a short time, from fractions of seconds (Hagedorn, 1968) to several seconds (see Fig. 3 of MacGibbon *et al.*, 2015). The expectation for the gamma-ray spectrum is a superposition of two components: a primary component comes directly from Hawking radiation and is thus peaked at around the PBH mass, and a secondary component comes from the decay of hadrons produced in the fragmentation of primary quarks and gluons and peaks at somewhat lower energies. Due to the available number of DoF, a large fraction of Hawking radiation from evaporating PBHs would be in the form of quark and gluon jets. The secondary components would provide a photon contribution larger than the primary one.

Despite presumable intense radiation during evaporation, such disrupting events would not be easy to catch with IACTs; their short duration would in fact reduce the number of external triggers such as those provided by the Fermi GMB or LAT instruments, due to the small sensitivity for seconds-long PBH evaporation. Evaporating PBHs could alternatively appear serendipitously as extremely ephemeral but bright emission anywhere in the FoV of IACTs during regular observations. Such events would likely not be automatically recognized by the standard trigger and data reconstruction mechanisms. A specific search for evaporating PBHs must be developed on IACT data.

Results: The idea of using Cherenkov light to detect PBHs was first developed by Porter and Weekes (1978, 1979) and further implemented with the realization of a specific trigger system called SGARFACE (Krennrich *et al.*, 2000; LeBohec *et al.*, 2005). The SGARFACE detector, mounted on the single-dish Whipple telescope, was a specific trigger based on analog pixel summation, and sensitive to gamma-ray "glows" developing in the atmosphere after short and intense bursts of gamma rays above 100 MeV. The Cherenkov light at the ground from sub-GeV showers would be extremely spread due to the increased Coulomb scattering of the atmospheric shower particles, much more than for TeV showers, almost impeding a directional reconstruction. Individually, they would not be detectable, but in case of a strong burst, the sum of all of these glows was simulated and shown to be detectable with SGARFACE. The motivation for SGARFACE was two-fold. First, it was to constrain the short timescale bursts from PBHs in the sub-microsecond to 100-μs scale, as predicted by Hagedorn (1968). But second and foremost, it was to be used to look for other short timescale bursts in the sub-GeV to GeV regime, including counterparts to giant radio pulses from pulsars. The intensity peaks were searched in windows of 60, 180, 540, 1620, 4860 ns and 15 μs, with a digitization bin of 20 ns. The predicted sensitivity of SGARFACE ranges from 0.8 to 50 ph cm^{-2}s^{-1} for primaries above 200 MeV and pulse width of 60 ns and 15 μs, respectively, and for a burst modeled as a power law with spectral index of -2.5. Results obtained with this instrument were presented by Schroedter *et al.* (2009) using 2.2 million events taken over six years of operation. That study also presents in detail the instrument, reconstruction and subsequent analysis. The sensitivity limits were computed for a PBH of mass 10^{13} g, for different burst durations, and ranged between 1 and 10^3 bursts per cubic parsec per year for the duration from ns to μs if assuming the Hagedorn mechanism. Larger duration bursts, of the order of seconds, were also constrained using the Whipple telescope standard trigger (Connaughton *et al.*, 1998; Linton *et al.*, 2006), providing additional results on PBH evaporation from IACTs. In particular, the latest result with Whipple (Linton *et al.*, 2006) made use of about 2200 h of high-quality data taken between 1998 and 2003. The search was made over time windows of one, three and five seconds. A peak in emission was determined when a certain number of consecutive events were recorded within a certain window of time. It is clear that the larger the window, the larger the probability of contamination from background events. The background contamination was attenuated by scrambling all the time of events. Statistically generated bursts surviving the search criteria on the scramble sample were thus removed. The probability of observing a burst event of intensity b in a time window Δt at an angular direction θ and an astrophysical distance r can be written as (Linton *et al.*, 2006)

$$n_{\rm BH}(b, \Delta t) = \rho_{\rm BH}\, \tau \int_{\Delta\Omega} d\omega \int r^2\, P(b, N_{\rm d}(r, \theta, \Delta t))\, dr\,, \tag{8.21}$$

where $\rho_{\rm BH}$ is the BH numerical volume density, τ, the lifetime, ω, the solid angle of the signal search and $P(b, N_{\rm d}(r, \theta, \Delta t)$, the Poisson distribution of a BH-emitting $N_{\rm d}$ gamma ray and producing an event of intensity b where the number of emitted gamma rays is

$$N_{\rm d}(r, \theta, \Delta t) = \frac{A(\theta)}{4\pi\, r^2} \int_{\Delta t} dt \int \frac{d^2N(E,t)}{dE\, dt}\, A_0(E)\, dE, \tag{8.22}$$

where $A(\theta)$ is the effective area. These formulas can be used to provide upper limits on the PBH density using the number of detected bursts, once the initial gamma-ray spectrum is specified. Of course, this depends on the exact knowledge of the quantum process at work during the evaporation phase, as well as with BH properties (e.g. spin) and its surroundings (e.g. if a chromosphere is formed). This search provided the first constraints to the PBH density at 1.08×10^6 PBH pc^{-3} yr^{-1} (see Fig. 8.14):

$$\frac{dN_\gamma}{dE} = 9 \times 10^{35} \times \begin{cases} \left(\dfrac{T_\tau}{1\,\rm GeV}\right)^{-3/2} \left(\dfrac{E}{1\,\rm GeV}\right)^{-3/2} {\rm GeV}^{-1} & \text{for } E < k_{\rm b}\, T_\tau, \\ \left(\dfrac{E}{1\,\rm GeV}\right)^{-3} {\rm GeV}^{-1} & \text{for } E > k_{\rm b}\, T_\tau, \end{cases} \tag{8.23}$$

where $T_\tau = 7.8\, k_{\rm b}^{-1} (\tau/1\,{\rm s})^{-1/3}$ [TeV] is the PBH temperature with a lifetime of τ.

In the thesis of Cassanyes (2015), the potential for MAGIC to search for PBHs was investigated for the first time. Cassanyes assumed 1/10,000 h of high-quality data, and photons from 158 GeV to 10 TeV. He found that the largest sensitivity comes from integration times of about several minutes — a value larger than the few seconds found by VERITAS and H.E.S.S. This is probably explained by a different emission model than Eq. 8.23. Within this window, upper limits of the order of $\sim 10^4$ pc^{-3} yr^{-1} can be achieved. MAGIC has not published any results on actual data with such a technique, however.

Recently, there has been increased activity in the field to analyze accumulated existing IACT data for PBH signatures: Tavernier *et al.* (2019, 2021) are so far analyzing almost 5000 hours of H.E.S.S. data in

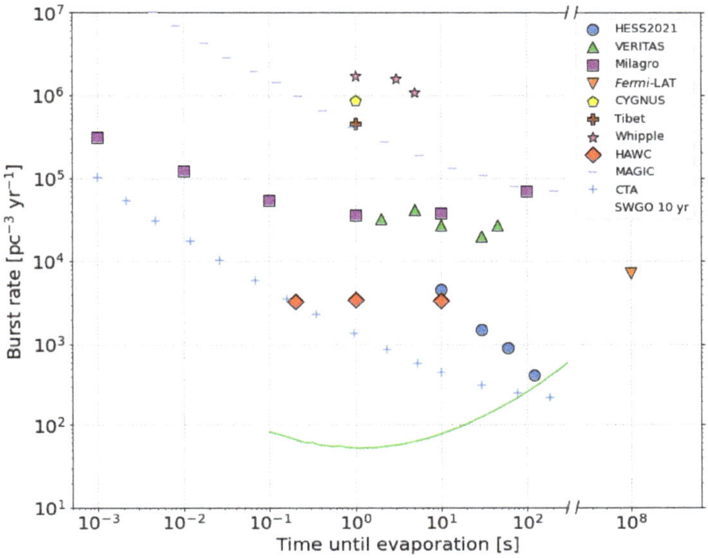

Fig. 8.14. A collection of limits on the PBH volume density as a function of the time to evaporation for PBHs of mass of the order of 10^{15}g. The image is an update of Fig. 4 of López-Coto *et al.* (2021) after the inclusion of recent H.E.S.S. data from Tavernier *et al.* (2021) and the predictions for MAGIC and CTAO by Cassanyes (2015).

search of PBH events, reaching upper limits on the evaporation rate to < 527 pc^{-3}yr^{-1} (see Fig. 8.14). Additionally, the VERITAS collaboration is pursuing such reanalysis of their data (Pfrang *et al.*, 2021b) and also looking for short-term optical fluctuations by PBH microlensing events (Pfrang *et al.*, 2021a).

Outlook: Are the current limits interesting? These limits are several orders of magnitude away from the limits obtained with, e.g. the EGB, in the range of PBH evaporation at present times. The theoretical mapping of PBH formation and evolution, however, still allows for a wide range of possibilities to make these limits significantly stronger. There are at least two direct ways to improve the limits. First, the mass loss of the PBH is governed by the number of DoF. As such, if one relaxes the assumption that only Standard Model DoF are to be used, boosted emission can occur as a result of the multiplication of DoF during the final stage of evaporation (Hagedorn, 1968). Secondly, those limits assume a homogeneous PBH distribution in the Universe. Clustering in galaxies is plausible (Young and Byrnes, 2020), however, which would drastically change these limits. Finally, one should not forget that PBHs may have a narrow mass range and therefore a limit in one mass region does not necessarily affect limits in other parts of the mass spectrum.

There are instruments sensitive to PBHs in the TeV range,[x] specifically the wide FoV air shower detectors such as Milagro (Abdo *et al.*, 2015) and HAWC (Albert *et al.*, 2020a) provided results with 4.6 and 2.6 years of data, respectively. Such a large observation time compared to IACTs is justified by the fact that Milagro observed particles within the showers and thus had a daily duty cycle. The optimal search time for Milagro was between 1 and 100 s. At smaller intervals, the sensitivity was low; at larger intervals, the background contamination became dominant. Limits were obtained of the order of $\sim 10^5$ pc^{-3}yr^{-1}. In the same work by Abdo *et al.* (2015), an extrapolation for the performance of HAWC was attempted, resulting again with limits in the ballpark of $\sim 10^4$ pc^{-3} yr^{-1}. More recently, López-Coto *et al.* (2021) investigated the prospects for this search with the planned instrument SWGO, demonstrating how this instrument detector can improve by more than one order of magnitude of the limits from HAWC as shown in Fig. 8.14. An endeavor to compute the prospects for the CTAO was made by Cassanyes (2015). Cassanyes used the same method to estimate the performance of MAGIC with the CTAO's figure of merit. He found that the sensitive PBH volume of the CTAO would be a factor of about 790 larger than in MAGIC: a factor of 25 due to the larger FoV, and a factor of $10^{3/2}$ because the 10-times better sensitivity allows for a $\sqrt{10}$ farther distance. With this performance, the CTAO could set constraints of the order of $\sim 10^2$ pc^{-3} yr^{-1}. The advantage is that no specific time is required for IACTs: PBHs could appear serendipitously as bright and short bursts, lasting from sub-seconds to seconds, anywhere in the FoV. Detecting these events requires one to build a specific offline reconstruction pipeline. As discussed by Doro *et al.* (2017), considering that the CTAO will surely strongly compress and reduce the global information from its hundreds of cameras in order to save disk space, these pipelines should be developed and investigated as soon as possible in order to avoid losing important data for PBH searches.

[x]Note that there are also efforts to search for PBHs at lower, MeV gamma-ray energies (Ray *et al.*, 2021; LeyVa *et al.*, 2021; Ghosh *et al.*, 2021).

Verification

(1) Can the imprint of non-baryonic matter in the CMB anisotropies be caused by PBHs?
(2) What gamma-ray instruments are best suited to search for PBH evaporation?
(3) What gamma-ray instruments are best suited to search for evaporated PBHs?
(4) Why is a tailored analysis needed by IACTs in the search for PBHs?
(5) Why may the CTAO be less competitive than other gamma-ray instruments in the same energy band?

8.5. Magnetic Monopoles

In the classical Maxwellian formulation of electromagnetism, if a magnetic charge g and a magnetic current ρ_m are introduced, the Maxwell equations would become symmetric if one swaps electric and magnetic fields, and the electric and magnetic constants as follows:

$$\vec{\nabla} \cdot \epsilon\vec{E} = \rho_e \qquad \vec{\nabla} \times \vec{E} = -\vec{j}_m - \frac{\partial \vec{B}}{\partial t}, \tag{8.24}$$

$$\vec{\nabla} \cdot \vec{B} = \rho_m \qquad \vec{\nabla} \times \vec{B}/\mu = \vec{j}_e + \frac{\partial \epsilon\vec{E}}{\partial t}. \tag{8.25}$$

This symmetry, albeit tantalizing, is valid only in the trivial case in which neither an electric charge nor an electric current is present. In any other case, it is incompatible with the concept of vector potential $\vec{A}$, either because the divergence theorem states that the monopole field would be null, or because any magnetic field line is closed.

An elegant solution to this conundrum was found almost a century ago when Dirac (1931) introduced the concept of the (now-called) "Dirac string", a hypothetical dimensionless entity that can be visualized as a solenoid and shifts the magnetic field phase by multiples of 2π. This kind of a phase-shift is similar to the observed Aharonov–Bohm solenoid effect (Aharonov and Bohm, 1959), a quantum-mechanical effect in which the electromagnetic field of a charged particle undergoes a phase-shift even in regions of null electric and magnetic fields such as within a long solenoid. The Dirac string theory embodies the "classical" Magnetic Monopole (MM) and it is still suggestive because it naturally predicts the quantization of the electric charge (the so-called "Dirac quantization rule"). This rule affirms that the product of magnetic g and electric e charge is quantized as an integer multiple of 2π, as a consequence of the 2π phase shift induced by the Dirac string:

$$e\,g = 2\pi\hbar\,n, \tag{8.26}$$

with $g = g_D$ defined as the Dirac elementary magnetic charge, or the unit "Dirac charge". It naturally motivates the hitherto unexplained quantization of the electric charge. By assuming a Dirac string size such as that of the electron, it is possible to estimate that the Dirac string would have a mass of 2.4 GeV. The signatures of the Dirac string would have been seen at accelerators and the missing detection provides strong motivation for rejecting this elegant solution (Patrizii and Spurio, 2015). Dirac himself said of MM, "One would be surprised if Nature had made no use of it."

A change of paradigm happened a few decades later, when 't Hooft (1974) and Polyakov (1974) realized that MMs were not only a feature arising from the asymmetry in the Maxwell equations, but also naturally predicted in Grand Unified Theories (GUTs). Unlike the Dirac MM, GUT MMs would be extremely heavy, because they are directly related to the mass of the mediator of the unified interaction, in the range $10^{16}/10^{17}$ GeV (Preskill, 1984), or even above, in GUTs, including gravity (Giacomelli and Patrizii, 2001). Considering that they arise from a complete particle physics theory, GUT MMs have the advantage of allowing a more complete theoretical mapping, with predictable properties such as internal structure and interactions with other fields. They are not structureless like the Dirac MM, but instead a huge conglomerate of particle fields arranged in layers, formed by a core of gauge bosons of the unified interaction, surrounded by shells of electroweak bosons and confined by a fermion-antifermion condensate region (Patrizii and Spurio, 2015).

At distances larger than a few fm, such an MM would behave like a pointlike magnetic charge generating a radial magnetic field. These objects are also called "topological solitons" and they could even show excited states in which they acquire a net electrical charge, going under the more general name of dyon (Schwinger, 1969). These MMs would be stable relics of events that happened in the early Universe, such as the freeze-out of some conditions or a symmetry breaking. Their decay would thereafter be forbidden due to the conservation of the magnetic charge. A further scenario is that of intermediate mass MM, generated at later cosmic times, during other phase transitions (Polyakov, 1974), allowing for masses of the order of $10^7/10^{13}$ GeV (Patrizii and Spurio, 2015). Depending on their mass and number density, MMs could actually constitute all or a fraction of the total DM of the Universe (Khoze and Ro, 2014). For further readings see, e.g. Shnir (2005).

The Universe expansion must have slowed down the initial velocity of MMs down to their non-relativistic speed. Dynamo effects in magnetic fields of galaxy clusters, AGNs or even possible magnetars, however, could have re-accelerated a fraction of MMs: such an acceleration process could be, in fact, highly efficient. An MM traversing a magnetic field B with coherence length L gains $K = g\,B\,L$ in kinetic energy (Wick *et al.*, 2003). For values of $B = 3\ \mu$G and $L = 1$ kpc, like those typical of galaxy clusters, the gain corresponds to 1.8×10^{11} GeV. This mechanism could even allow the thesis that MMs could be associated with ultra-high-energy (EeV) CRs (UHECRs) (Kephart and Weiler, 1996). The acceleration of MMs happens at the expense of local magnetic fields. Out of this, Parker (1970) obtained a conservative limit on the MM density requiring that this "draw" of magnetic energy was sufficiently slow so as to not disrupt galactic magnetic fields completely. This is the so-called Parker bound. Galactic objects such as the Sun could accumulate MMs in their core. The enhanced overdensity could ignite MM-catalized nucleon decay. Because of the presence of GUT gauge bosons in their core, interactions such as $p + \mathrm{MM} \rightarrow e^+ + \mathrm{MM} + \pi_0$ may have ample cross-sections if the interaction is independent from the grand unification gause boson mass (Callan-Rubakov process; Callan, 1982; Rubakov, 1982), generating huge pion fluxes and providing strong photons or neutrino bursts after the pion decays (Tanabashi *et al.*, 2018). Not long ago, Super Kamiokande provided very strong constraints on MMs (Ueno *et al.*, 2012). The Callan–Rubakov process is efficient only for GUT MMs of moderate speed $\beta < 0.3$.

In the passage through matter, an MM with $\beta > 0.1$ would behave like an equivalent huge electric charge, with strong ionizing power. Dyons, being electrically charged, would also directly generate Cherenkov light. The number of photons N per unit length x and wavelength λ for charged particles is computed by Tamm and Frank (1937) in air as follows:

$$\frac{d^2N}{dx\,d\lambda}(\mathrm{MM}) \simeq \frac{n^2}{4\,\alpha^2}\,\frac{d^2N}{dx\,d\lambda}(e,\mu), \tag{8.27}$$

and computed for MM using Eq. 8.26. In Eq. 8.27, n is the air refraction index and α is the electromagnetic fine structure constant. In air, $n^2/(4\,\alpha^2) \simeq 4{,}700$ and therefore the number of photons produced by an MM is significantly larger than that by an electron or muon, and it corresponds to an energy loss of about 8 GeV g^{-1} cm^{-2} compared to 2 MeV g^{-1} cm^{-2} for the muons (Tompkins, 1965; Cecchini *et al.*, 2016). Such a strong interaction with matter opens up interesting possibilities for the detection of MMs — for example, by generating local damages on high-Z material slabs when traversed by MMs that are inspected offline. For masses above 10^3 GeV, the MACRO (Ambrosio *et al.*, 2002) and MoEDAL (Patrizii and Spurio, 2015) instruments are successful examples of highly sensitive nuclear track detectors.

MMs could be also identified by their passage through the Earth's atmosphere. In this occurrence, they will strongly radiate Cherenkov light, either directly, if electrically charged, or through the secondary charged particle generated through ionization, as well as through luminescence (Obertacke Pollmann, 2017). Even a globally neutral MM, due to its charged outer layer, would excite atmospheric nuclei throughout the full length of the atmosphere. Such an effect does not happen for CRs that are absorbed in the high atmosphere. The search for such a signature was investigated with the Pierre Auger Observatory (PAO). Aab *et al.* (2016) reported strong constraints searching for fluorescent light induced by MMs. The IceCube collaboration (Abbasi *et al.*, 2013; Aartsen *et al.*, 2014, 2016b), improving on the results obtained with AMANDA (Achterberg *et al.*, 2010), performed extensive searches for non-relativistic and relativistic MMs. GUT MMs could catalyze proto decays in the surrounding Antarctic ice via the Callan–Rubakov effect. This would generate relativistic charged particles emitting Cherenkov light with a characteristic hit pattern in the detector. No such pattern was observed, allowing us to limit the flux of non-relativistic GUT MMs to three orders of magnitude below the Parker bound. A collection of the limits discussed above on the MM flux is shown in Fig. 8.16.

Magnetic monopole searches with IACTs: The search for MM signatures was so far only investigated with a simplified approach by Spengler (2009). Spengler (2009) provided a preliminary discussion of the sensitivity to MMs by studying dedicated MC simulations, optimization of the selection cuts, and real data selection.

An MMs entering the Earth's atmosphere will start generating Cherenkov light in the high atmosphere, at 80 km altitude or above. This emission will start within a narrow Cherenkov cone angle of 0.1°, increasing to angles of 1.2° when the MM will hit the ground. Differently than CRs, the

Cherenkov emission will last throughout the full length of the MM track. The Cherenkov emission depends on the MM Lorentz factor γ, which is related to the local air refraction index n by the relation

$$\gamma > \frac{1}{\sqrt{2\,(n-1)}}. \tag{8.28}$$

For the atmosphere close to the ground, $n-1 = 10^{-4}$. Therefore, a minimum boost of $\gamma > 70$ is required to have Cherenkov emission throughout the full length of the track. This condition in turn allows for significant effectiveness in the separation from the background. This is illustrated in Fig. 8.15: the MM would be observed as a double-spot system in the IACT cameras — a first due to the very high-altitude emission, and a second due to the low-altitude emission, while the emission from the central part of the track would not be geometrically focused. A second simplification was applied in order to guarantee minimal energy loss of the MM during the track. Spengler (2009) computed a total energy loss of 22.5 TeV throughout the atmosphere. This required therefore that $\gamma > 22.5\ \mathrm{TeV}/\mathrm{m_{MM}c^2} \sim 10^5$ or, in other words, that the analysis is limited to ultra-relativistic MMs. We will discuss the effect of these limitations further in the following.

The images formed by MMs in the IACT cameras will be mostly composed of small clusters of very illuminated pixels for the high-altitude signal, or very bright fraction or rings, similarly to the muon case, for low-altitude originating emission. An example of a simulated event is shown in the right panel of Fig. 8.15. Considering the large Cherenkov photon yield, thousands of times larger than that of a muon, all illuminated pixels will show extremely high signals, often saturating the reach of the camera PMTs. This further characteristic was used by Spengler (2009) as a main selection cut: an event was to show a large fraction of saturated pixels in at least two out of four telescopes. Further cuts, also optimized through MC, were applied to exclude all of the remaining background. For example, less than 200 pixels were to trigger in each camera in order to remove HE hadronic showers surviving previous cuts. The selection cuts were optimized on a sample of about 10^6 MC events from 0 to 60° zenith angle, and applied to five years of (selected) H.E.S.S. data, for a total of 2 400 h. No MM candidate event was found, and from this, limits of $< 4.5 \times 10^{-14}$ $\mathrm{cm^{-2}s^{-1}sr^{-1}}$ were computed and shown in Fig. 8.16. These values are extremely poorly constraining — less than the conservative Parker bound — and orders of magnitude less constraining than PAO or IceCube limits. This is due to the larger aperture and duty cycle of these instruments. In any

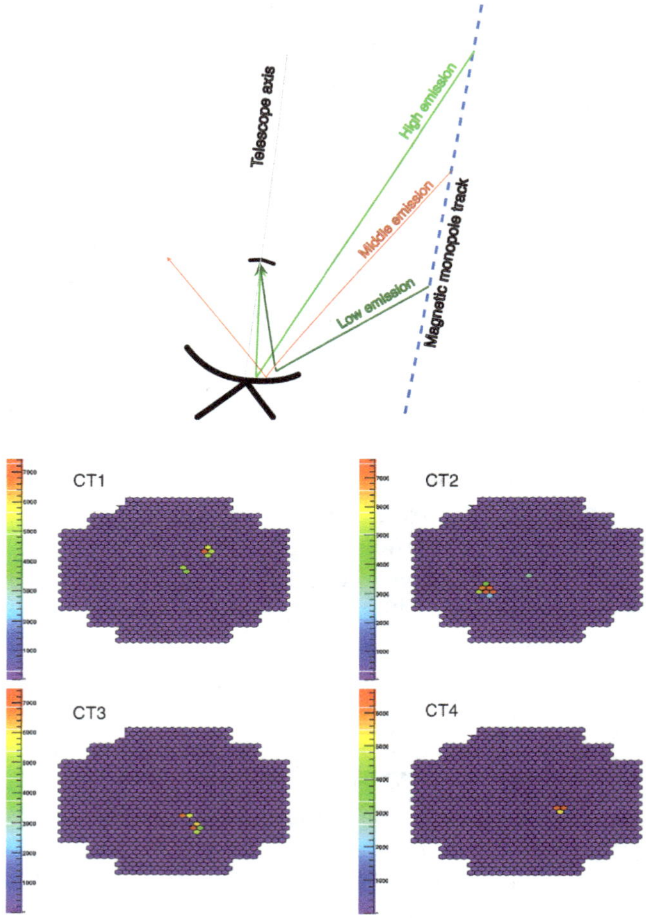

Fig. 8.15. *Left*: Emission scheme from an ultra-relativistic MM emitting Cherenkov radiation throughout the full length of the atmosphere. *Right*: A simulated MM event on H.E.S.S. cameras. Courtesy of Spengler (2009).

case, these limits should be taken as conservative limits in a very simplified scenario. Further exploration is required.

Outlook: Improving on this work would require us to reassess the MC simulations by allowing lower-speed connected effects, such as a more significant energy loss during the MM tracking, the appearance of secondary hadronic sub-showers which would alter the signatures in the IACT camera

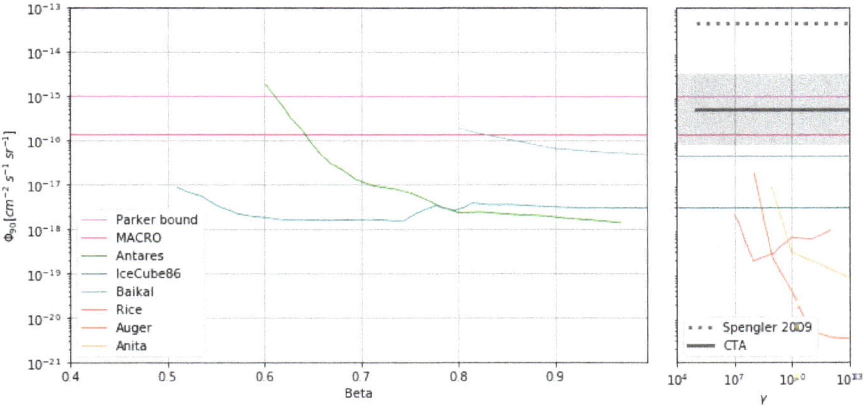

Fig. 8.16. Current limits on the flux of MM as a function of the speed of MM at the detector. Upper limits obtained with the H.E.S.S. instrument are shown as a gray dashed line in the inset to the right (Spengler, 2009). Preliminary predictions for the CTAO are also shown as a thick gray line (Doro *et al.*, 2017).

and therefore the simplified picture of background/signal separation. Furthermore, Spengler (2009) did not use the time of arrival of photons to separate candidate events from background ones, which requires an isochronous paraboloid such as that of the CTAO's Large-Size Telescopes (LSTs). The CTAO also features a wider FoV that allows us to increase the acceptance to MM events, allowing for those with farther impact parameters. Finally, we should mention that by pointing the telescope to very high zenith angles (low altitude above the horizon), one could in principle modify the geometry of the event and largely increase the effective area. MMs' investigation with the CTAO would also profit by a tenfold increase in the effective area, and result in the improvement in energy resolution, angular resolution and off-axis performance. A back-of-the-envelope estimation for the CTAOs performance was discussed in Doro *et al.* (2013) — however, a full quantification of the performance is yet to be done. CTAO also features a more flexible software inter-trigger system, which could allow for dedicated search campaigns. Still, MM events could survive trigger criteria and be hidden in current IACT data archives. In MAGIC, the number of background events stored on disks is roughly 1.2×10^9 events/year for two telescopes (Doro *et al.*, 2017),[y] thus providing

[y]For CTAOs, this number could be a factor of 50 larger in principle.

a rich search sample. Nonetheless, specific online or offline fast selection criteria must be envisaged to avoid rejecting possible triggers from MMs and thus increasing the chances of detection or the constraining power of any consequent limit. If detected, MMs would possibly provide a clear signature in an IACT. Despite the smaller strengths of IACTs compared to other techniques, it is important to note that this search would be achievable at no cost of dedicated telescope pointings nor specifically allocated observation time, but making use of archival data and tailored analyses.

Verification

- Q1. Why does an MM leave a different imprint in the cameras of IACTs with respect to a primary γ shower?
- Q2. What is the time imprint of a shower initiated by an MM?
- Q3. What is the main reason why IACTs are mostly sensitive to ultra-relativistic MMs?
- Q4. If you were to optimize your search analysis on MMs, on what would you base your classification algorithm?

8.6. Tau Neutrinos

The nature of the astrophysical emission of neutrinos has been one of the hottest topics in astrophysics for decades. The interest has been boosted recently by the measurement of an astrophysical diffuse flux of HE neutrinos by IceCube (Aartsen *et al.*, 2014b) and the significant spatial and temporal coincidence observed between a 290 TeV neutrino event by IceCube and a flaring blazar by *Fermi*-LAT and MAGIC (Aartsen *et al.*, 2018). Several unknowns are yet to be understood: What is the full spectrum and composition of the diffuse neutrino flux? What astrophysical targets are neutrino emitters and where does the neutrino emission take place? Are neutrinos produced via leptonic or hadronic processes? Are neutrinos also the result of interaction of HE CRs with local ambient fields? These and others questions are effectively tackled by large-scale ground-based instruments such as the aforementioned IceCube, Antares (Adrian-Martinez *et al.*, 2011), the planned KM3Net (Adrián-Martínez *et al.*, 2016) at TeV to PeV energies the PAO (Aab *et al.*, 2015), the ASHRA (Asaoka and Sasaki, 2013) or the future GRAND (Alvarez-Muñiz *et al.*, 2018) and POEMMA observatory (Olinto *et al.*, 2020) at PeV energies and above. Additionally, though not primarily designed for neutrino detection, IACTs are instruments capable of not only detecting electromagnetic counterparts to neutrino astrophysical targets, but also to possibly directly detect signatures of neutrino events crossing the Earth's atmosphere or neutrinos emerging from the Earth's crust in a technique called the "Earth-skimming" detection technique (Fargion, 1999).

Extraterrestrial neutrinos are generated (see, e.g. Becker, 2008; Baret and Van Elewyck, 2011 for recent reviews) through general production mechanisms such as nuclear decay (proton–proton) or photopion (proton–photon) interactions eventually generating neutrinos through reactions such as

$$\begin{aligned} &p\,p \to \pi^+\,\pi^-\,\pi^0, \text{ followed by:} \\ &\quad \pi^+ \to \mu^+\,\nu_\mu \to e^+\,\nu_e\,\overline{\nu_\mu}\,\nu_\mu \quad \text{or} \\ &\quad \pi^- \to \mu^-\,\overline{\nu_\mu} \to e^-\,\overline{\nu_e}\,\overline{\nu_\mu}\,\nu_\mu, \end{aligned} \tag{8.29}$$

and

$$\begin{aligned} &p\,\gamma \to \Delta^+, \text{ followed by:} \\ &\quad \Delta^+ \to p\,\pi^0\ (2/3) \quad \text{or} \\ &\quad \Delta^+ \to n\,\pi^+\ (1/3). \end{aligned} \tag{8.30}$$

Therefore, depending on the ambient matter and radiation fields, the above processes can give rise to neutrino emissions in a wide range of energies and fluxes (Ackermann *et al.*, 2019). At MeV energies, a cosmic neutrino background is expected from solar MeV neutrinos generated within our Sun as well as from farther stars. Above 0.1 GeV, the most abundant population is that of atmospheric neutrinos, generated by the interaction of CRs with the Earth's atmosphere. Atmospheric neutrinos constitute the by-far larger background in large-scale neutrino detectors. In this energy range one can also find the rarer interesting extraterrestrial neutrinos produced, e.g. in an AGN or GRB and possibly in all other non-thermal emission regions in which HE CRs are efficiently generated. Finally, above $5 \cdot 10^{19}$ eV, neutrinos are found after Δ resonances (Eq. 8.30) of UHECRs with the cosmic microwave background (Berezinsky and Zatsepin, 1969) (cosmogenic neutrinos). Therefore, the phenomenology of neutrino production is wide, and often accompanied by HE CR reservoirs.

During hadronic processes in Galactic and extragalactic environments, the expected neutrino generation rate ratio is $(\nu_e, \nu_\mu, \nu_\tau) = (1, 2, 0)$, due to the fact that positive, negative and neutral pions are produced equally during proton interactions. Therefore, ν_τs are not expected to be produced significantly through these mechanisms at the source. During propagation, however, neutrino families mix due to the non-vanishing mass eigenstates.

Already for distances larger than a few astronomical units, the mixing is total and a uniform neutrino family mix is expected at Earth: $(\nu_e, \nu_\mu, \nu_\tau) = (1, 1, 1)$ (for the computation of mixing, see Becker, 2008). It is important to recall that the family mixing depends on terms proportional to the difference of the squared masses of two neutrino flavors. Therefore, the detection of ν_τs is very relevant not only to test astrophysical CR processes and unveil new emission mechanisms at the source, but also to test neutrino physics models.

By ice or water detectors with a huge instrumented volume (such as IceCube, Antares or KM3Net), TeV–PeV neutrinos are observed through the Cherenkov light generated in particle showers produced by the neutrino interactions in the detector (Aartsen *et al.*, 2016a). Above the PeV range, the ice or water are increasingly opaque to the crossing of electron or muon neutrinos, and only ν_τs can be detected from below. Above PeV energies, the technique to detect Earth-skimming cosmic ν_τs comes into play; when neutrinos hit the Earth at its rim and cross a certain amount of the Earth's crust, they have a non-negligible chance of converting to a lepton which may subsequently emerge into the air and induce an upward-directed atmospheric shower. The advantage of this approach is very similar to seeking signatures from cosmic primaries interacting with the atmosphere: one uses the Earth itself as a suitable large interaction volume, and only needs to place a detector in or close to the induced secondary particle showers. Also, while ν_τs are the only candidates for the Earth-skimming technique, they indirectly probe the whole neutrino family of astrophysical origin due to neutrino mixing. The Earth-skimming approach is pursued by the PAO (Fargion, 2002; Fargion *et al.*, 2004), and detectors like ASHRA and GRAND are primarily designed for it. Finally, this technique can also be adopted by IACTs detecting the Cherenkov light of the ultra-relativistic particles in the neutrino-induced atmospheric showers. The setup for detecting Earth-skimming ν_τs with IACTs is illustrated in Fig. 8.17.

Neutrino interactions with matter (N) happen with charged current (CC) and neutral current (NC) mechanisms:

$$\text{CC:} \quad N + \nu_l \to \ldots \to X + l, \quad \text{where } l = e, \mu, \tau, \tag{8.31}$$

$$\text{NC:} \quad N + \nu_l \quad \to \quad X + \nu_l, \quad \text{where } l = e, \mu, \tau, \tag{8.32}$$

where X is a generic product that leads to a hadronic cascade. Of the leptons found in the reaction products, electrons are rapidly absorbed in matter and the chance of exiting the Earth's crust are low. When muons

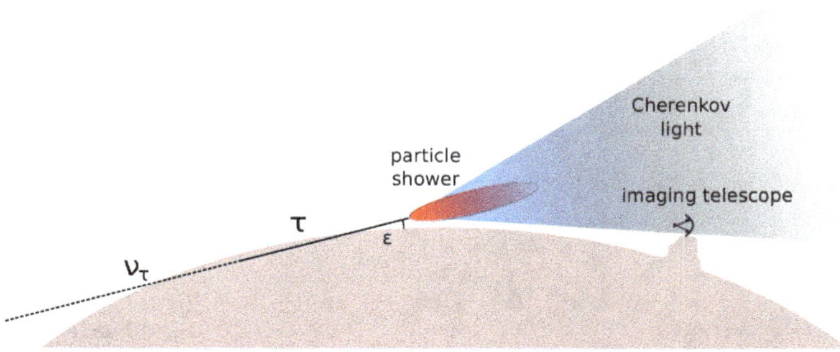

Fig. 8.17. Artistic sketch of the Earth-skimming ν_τ-technique, with the basic scheme of detection for IACTs. The Cherenkov light cone is produced by charged shower particles emerging from the direction of the ν_τ. Courtesy of N. Otte, reproduced from Otte (2019).

are created in these interactions, they also lose most of their energy inside the Earth, but with longer radiation length than electrons and occasionally pass through the crust and decay in the atmosphere. Due to its rather large lifetime, however, a muon emerging from the Earth would rarely decay into a particle shower within the telescope FoV and only generate direct Cherenkov light, which would mostly go undetected. The situation drastically changes for tau neutrinos. A PeV tau particle created after a CC ν_τ interaction in the crust can travel through a rather large amount of matter with a significant chance to decay shortly after leaving the crust at the right distance from the telescopes and generate extensive air showers.[z] Table 8.2 (extracted from Fargion, 2002) summarizes the main tau decay modes: a large cross-section is found in hadronic-induced showers (65%), electromagnetic showers (18%) or muons (17%). The first two components are detectable if generated at the right distance range and within the telescopes' FoV.

Tau-neutrino searches with IACTs: For IACTs, the Earth-skimming ν_τ observation technique requires a suitable oreography around the telescopes: the telescopes must be pointed at the ν_τ emission point, sufficiently distant, and corresponding to the right amount of crossed matter within the Earth. This is, for example, doable if a mountain with the right thickness is located

[z] At the highest $\sim$ EeV energies, the $\nu_\tau \to \tau \to \nu_\tau \to \tau \to \ldots$, recreation chain increases the chance of an HE tau leaving the crust and, consequently, the telescopes' acceptance (Alvarez-Muñiz *et al.*, 2018).

Table 8.2. Decay channels for the τ lepton. Probability and type of atmospheric showers are also shown ("e.m." for electromagnetic; "had" for hadronic). The table is reproduced from Fargion (2002).

Decay	Secondaries	Probability	Air-shower
$\tau \rightarrow \mu^- \bar{\nu}_\mu \nu_\tau$	μ^-	17.4%	–
$\tau \rightarrow e^- \bar{\nu}_e \nu_\tau$	e^-	17.8%	Pure e.m.
$\tau \rightarrow \pi^- \nu_\tau$	π^-	11.8%	Pure had
$\tau \rightarrow \pi^- \pi^- \pi^+ \nu_\tau$	$2\pi^-,\pi^+$	10%	Pure had
$\tau \rightarrow \pi^- \pi^0 \nu_\tau$	π^-, $\pi^0 \rightarrow 2\gamma$	25.8%	Mixed had/e.m.
$\tau \rightarrow \pi^- 2\pi^0 \nu_\tau$	π^-, $2\pi^0 \rightarrow 4\gamma$	10.79%	Mixed had/e.m.
$\tau \rightarrow \pi^- 3\pi^0 \nu_\tau$	π^-, $3\pi^0 \rightarrow 6\gamma$	1.23%	Mixed had/e.m.
$\tau \rightarrow \pi^- \pi^+ \pi^- \pi^0 \nu_\tau$	$2\pi^-,\pi^+,\pi^0 \rightarrow 2\gamma$	5.18%	Mixed had/e.m.

in the telescopes' vicinity, or if the telescopes can be pointed downward toward a direction in which the amount of crossed matter is optimal. The MAGIC telescopes are located at the Roque de los Muchachos Observatory in the Canary Islands. The Roque mountain can be pointed out by MAGIC. It is only a few hundreds of meters away, however, and therefore a particle shower would be initiated through or behind the telescopes. On the other hand, from the MAGIC location, there is a window of observability toward the Atlantic Ocean in the Northern direction. If the telescopes are pointed at around 1.5 degrees below the horizon, a window of 5×80 deg^2 is accessible with an extraction point at the ocean at 165 km, allowing for sufficient space for the shower development. Gaug *et al.* (2007) were among the first to investigate this technique, and performed analytical computations to find the optimal observation strategy and effective area. They also performed the first order-of-magnitude event rate estimation. At that time, however, a full simulation chain of neutrino interaction and shower development was not available. A complete MC chain of simulation was provided by Gora and Bernardini (2016) by adapting the ANIS code (Gora *et al.*, 2007; Gazizov and Kowalski, 2005), as well as with publicly available oreography maps. On this basis, Gora *et al.* (2017b) computed a precise sensitivity and angular acceptance for MAGIC.

Results: MAGIC published the results from the observation of 30 h of data in the direction of the ocean (Ahnen *et al.*, 2018b), summarizing the previous findings by Gora *et al.* (2017a, 2017b). The observation was performed by pointing at an optimized direction toward the Ocean during several nights in which high cirrus clouds were present. Such a condition normally does not allow for standard astronomical observation — however, it guarantees good visibility of the ocean. The most complex part of this technique is the discrimination of the signal parameter space from the background, as well as the corresponding MC simulations. The central analysis steps and results are briefly summarized in the following. First of all, the background was estimated by pointing the telescope for 5.5 h slightly above the ocean, at 2.5° above the horizon. This allows us to gather a significant number of background events as expected in the later below-horizon observation. A combination of the Hillas size and length image parameter dubbed Y was found to clearly separate signal and background, as shown in Fig. 8.18. The figure also shows the Y parameter for ν_τ of energy 1, 10, 46 PeV. Muon bundles are the dominant background contribution whereas contributions from CR electrons can be neglected; at horizontal directions the Cherenkov light from electromagnetic showers is strongly attenuated, even more than for proton primaries of similar energy, and below the detection threshold. To build the MC simulations, aside from the ANIS-modified version (Gora *et al.*, 2007; Gazizov and Kowalski, 2005), CORSIKA (Heck *et al.*, 1998, (v6.99)) and PYTHIA (Sjostrand *et al.*, 2006, (v6.4))) were also used. No event was found, after selection cuts, in 30 h of

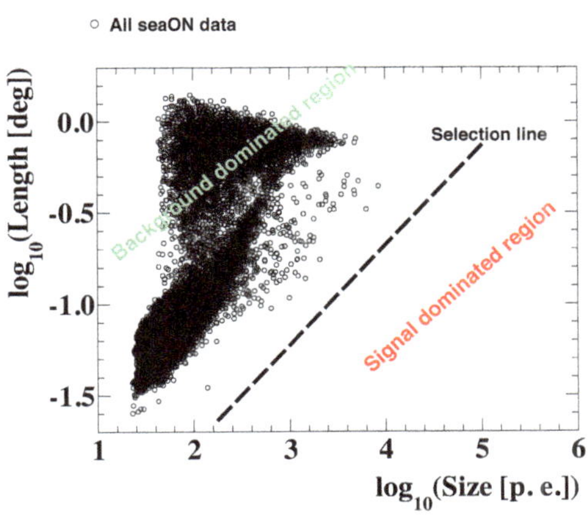

Fig. 8.18. Left: Distribution of events in the space length-size of the image parameters. Background events and signal events are clearly separated by the cut selection Y. Right: Distribution of events after the selection cuts based on the same Y parameter (see text) in search for tau-neutrino signatures with MAGIC. *seaON*, *seaOFF* and signal MC events represent the data taken pointing at the signal region (*seaON*), at the background control region (*seaOFF*) and the MC-simulated tau neutrinos, respectively. In the region $log10(Y) > 2.35$, no neutrino candidates survive. The selection criterion keeps about 40% of MC signal events. Courtesy of MAGIC Collaboration (Ahnen *et al.*, 2018b).

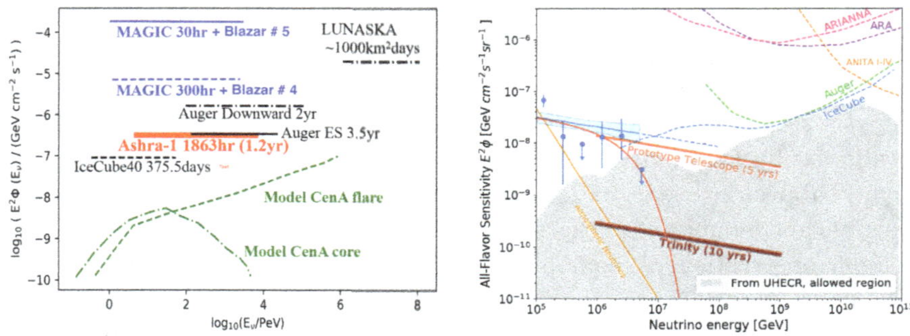

Fig. 8.19. Left: 90% C.L. upper limit on the point-source tau-neutrino flux obtained with the MAGIC telescopes with 30 h of observation (solid upper horizontal line) assuming a specific flux level from flaring blazars, as well as extrapolated to 300 h (dashed upper horizontal line) (Ahnen *et al.*, 2018b, Fig. 3). Right: Comparison of sensitivity for VHE- and UHE-neutrino experiments (Fig. 5 of Brown *et al.*, 2021).

observation. The null detection was translated into an upper limit for ν_τ emission after a thorough estimation of the acceptance for these particles, which is not straightforward considering the geometry of the observation, the presence of mixed water and rocks in the ν_τ path, the inclusion of atmospheric transparency and the fact that multiple atmospheric showers can occur. The results are reported in Fig. 8.19 considering the diffuse neutrino flux as well as different neutrino-emission models for AGNs. The latter were built based on models of real blazar flaring events. In particular, Blazar #4 and Blazar #5 of Fig. 8.19 represent predictions for PKS 2155-304 in high-state and 3C 279, respectively (Becker, 2008; Atoyan and Dermer, 2001). The point-source sensitivity corresponding to 30 h of observation (Blazar #4) are poorly constraining, of the order of 2×10^{-4} GeV cm^{-2} s^{-1}. Only if 300 h are considered, and a more optimistic flux model is adopted (Blazar #5), can the sensitivity become of the order of the one obtained with downward (all-flavor) neutrinos by PAO (Aab *et al.*, 2015). This is still far away from the sensitivity obtained with IceCube (Aartsen *et al.*, 2016b). Yet, the sensitive region is different, being that MAGIC is indeed sensitive at somewhat higher energies than IceCube but at smaller energies than PAO, thus allowing for searches for a wider class of astrophysical emitters.

Similar results were obtained with the ASHRA-1 Cherenkov telescopes at Mauna Loa (Ogawa, 2019). This telescope, composed of electrostatic lenses with an optical system to generate convergent beams, observes ν_τ skimming through the Mauna Kea mountain. It performed a campaign of

1863 h over 1.2 years that led to limits on the diffuse flux more constraining than those obtained with MAGIC (see Fig. 8.19), further demonstrating the validity of the technique.

Outlook: Though these results may not look impressive, they demonstrate that IACTs do have a potential for ν_τ searches, and that the data reconstruction is well known. ASHRA-1 will continue gathering hundreds of observational hours each year. Improving on this result is possible with MAGIC considering that approximately 100 h per year are affected by the presence of cirrus clouds, and the telescope can be pointed at the ocean. It is currently a non-standard operation, however, and it will require time to integrate more than 300 h. In the future, the CTAO could be a better-suited instrument. The CTAO has a wider FoV, especially at high energies where the SSTs operate, thus allowing for an increase in the acceptance which can be roughly estimated at a factor of four. This will be accompanied by increased sensitivity. In addition, the larger number of telescopes may also allow us to investigate less stringent selection cuts. For the Northern Hemisphere, the CTAO will be located where MAGIC telescopes are. In this sense, the CTAO telescopes could adopt the same "ocean window" used by MAGIC. If this was combined with clever scheduling (e.g. automatic pointing to the ocean whenever clouds are present, as well as an automatic scheduling of flaring objects that pass behind the ocean window at a certain time of the night), this could improve the performance by at least one order of magnitude or more with respect to MAGIC, in case bright transients events occur. In the Southern Hemisphere's CTAO site, the oreography does not allow for such kind of observations.

Recently, an ad-hoc IACT array was proposed to overcome the limits to this technique present for MAGIC and the CTAO: the rather small FoV and a non-dedicated observation program. The name of this project is Trinity (Otte, 2019) and it is composed of 18 units organized in three arrays of six units each. A single telescope covers a $5 \times 60^\circ$ FoV so that a single array covers the whole circumference. Each array is placed on mountaintops to cover the whole horizon. Due to this much wider FoV and dedicated observation mode, up to 1,200 h/year would be available to this technique. With such a performance, sensitivities of the order of 10^{-10} GeV cm^{-2} s^{-1} sr^{-1} (see Fig. 8.19, right panel) could be reached, with tens to hundreds of diffuse neutrinos detectable in 10 years of operation. If approved, such an instrument would provide very important results in a gap region between IceCube and radio-neutrino instruments such as GRAND or POEMMA.

Verification

- Q1. Why are IACTs sensitive to Earth-skimming ν_τs and not to other neutrino families?
- Q2. Why is the expected background in IACT observation of Earth-skimming tau neutrinos expected to be small?
- Q3. Why is the sensitivity of IACT observation of Earth-skimming tau neutrinos concentrated in the region of PeV to EeV?
- Q4. What are the main limiting factors for the observation of Earth-skimming tau neutrinos for MAGIC and CTAO?

8.7. Summary and Conclusions

Fundamental physics is probably a slippery term, because every piece of evidence found in scientific pursuits is fundamental to the increase of knowledge. Yet, there is no single word that can properly encompass the searches for massive-particle DM, axion-like particles, primordial black holes, tau neutrinos and magnetic monopoles that have been summarized in this chapter. In most cases, we do not possess a full theoretical mapping, therefore, we often cannot sharpen our instruments to aim for detection, but merely to set more constraints. Nevertheless, we believe that placing many independent and complementary constraints for exotic phenomena (and, in some cases, IACTs provide the strongest ones) is how we build trust in successful theories, and how we rule out alternative explanations or extended models. It is a crucial part of physical observation to continuously challenge the current worldview and to test predictions of novel phenomena. This is why we have devoted thousands of hours of IACT observations and invested tenfold hours of data analysis and investigation in the search for the "unknown". All in all, astrophysical gamma rays are wonderful probes: they not only trace back to the emitting source, but reveal to us nuclear and sub-nuclear interactions that have taken place billions of billions of kilometers away — that is, tens or hundreds of millions of years back in time — and at energy scales hardly accessible in the laboratory. For DM, if the mass of its constituent elusive particle is above several hundreds of GeV, gamma rays may represent one of the few probes (perhaps the only one) to unveil its true nature beyond gravitational pull. We may be hunting far from the right spot, yet constraints set by IACTs on the nature of DM are the strongest in the TeV regime as of today, and they will likely stay so for many years to come. Current limits from the Galactic center halo by H.E.S.S. have reached the thermal cross-section value of $3 \times 10^{-26}\,\mathrm{cm^3 s^{-1}}$, a reference value for WIMP DM models. These limits could suffer, however, from significant uncertainties regarding the precise amount and distribution of DM in the Galactic Center region. This kind of uncertainty is less severe in the case of using Milky Way dwarf satellite galaxies. Therefore, H.E.S.S., MAGIC and VERITAS have also extensively searched — not in vain — for DM in the best candidates among them, although without detecting a signal so far.

While on searches for WIMP DM IACTs are bannermen, their role in the search for axion-like particles, primordial black holes and magnetic monopoles is more that of pawns compared to other instruments. Nevertheless, the potential of these searches is large, and it is important to recall that these are made on archival data. There is no need of new specific observations, but rather of careful and complex analyses. A similar situation occurs for tau neutrinos. Even if the sensitivity of IACTs is not competitive to instruments entirely designed for the detection of astrophysical neutrinos, it has been shown that IACTs may play a significant role, especially if specific conditions are met at the site.

The upcoming CTAO will provide a performance boost in all of these fields. We might finally unequivocally exclude — or confirm — the hypothesis of massive-particle DM, and also search for axion-like particles, primordial black holes, magnetic monopoles and tau neutrinos with, in all cases, several factors of higher sensitivity. Yet, in order to adequately profit from this boost, some prerequisites have to be guaranteed: first, that the theoretical and analysis experience gained from the pioneering studies here presented should not be lost, and we hope to have contributed to that. Second, and more importantly: the IACT community should be aware of the relevance of these pursuits for fundamental physics, and take that as an obligation to devote time and resources to them, a fact that often clashes with the need of longer observations for other astrophysical endeavors. Last, but by no means least: we should guarantee that the CTAO — or any future IACT experiment — and its corresponding data-processing pipelines are put in a position to potentially catch fundamental physics events. As examples, through this chapter we have shown how peculiar MM signatures are expected to be, or how serendipitous by primordial black hole events may show up, lasting just a few seconds anywhere in the FoV. We should therefore make an extra effort to guarantee that at no place we simplify our systems so as to be sensitive only to "conventional" gamma-ray signals and not drop information to catch fundamental physics events. For example, by selecting only the brightest pixels in the camera and throwing away all others from the data acquisition and analysis, we may lose all means to identify a rare event occurring in the camera (Doro *et al.*, 2017). It is important to revive in the community the discussion on the best strategy to search for fundamental physics with the CTAO — one that will also be in a good balance with the rest of physics cases. Let nature provide the rest in our hope to ever catch signatures of new physics with IACT data.

Acknowledgments

The authors would like to warmly thank the following for useful discussions to this chapter: E. Bernardini, A. De Angelis, M. Gaug, D. Gora, D. Krennrich, M. Meyer, N. Otte, L. Patrizii, A. Raccanelli, J. Rico, U. Schwanke and G. Spengler.

https://doi.org/10.1142/9789813275720_0009

Chapter 9

Multi-Messenger Astronomy

Marcos Santander[*,§], Konstancja Satalecka[†,¶] and Fabian Schüssler[‡,‖]

[*] *University of Alabama Tuscaloosa, USA*
[†] *Deutsches Elektronen-Synchrotron Zeuthen, Germany*
[‡] *IRFU/CEA Paris-Saclay Saclay, France*
[§] *jmsantander@ua.edu*
[¶] *konstancja.satalecka@desy.de*
[‖] *fabian.schussler@cea.fr*

9.1. Introduction

For centuries, the study of the Universe has largely relied on observations made at optical wavelengths. During the 20th century, astronomy was revolutionized by new observational methods that expanded our view of the Universe to span the entire electromagnetic spectrum, from radio wavelengths to Very-High-Energy (VHE) gamma rays.

A similar revolution may await by expanding our view yet again, this time not just to other electromagnetic wavelengths but by combining these observations with the detection of other types of radiation and energetic particles arriving from the cosmos. The other cosmic "messengers" besides photons are cosmic rays (CRs), neutrinos, and Gravitational Waves (GWs).

The discovery of high-energy (HE) astrophysical neutrinos (Abbasi *et al.*, 2013) and gravitational waves (Abbott *et al.*, 2016c) has created new opportunities for the study of extreme astrophysical phenomena. Multi-messenger approaches that combine these detection channels with VHE photons can help in resolving the possible neutrino, GW, or CR sources as well as provide new insights into the emission mechanisms and source properties.

HE neutrinos and VHE gamma rays are expected to originate in the hadronic interactions of CRs with ambient material or radiation fields while being accelerated, or during propagation. Therefore, the identification of gamma rays in coincidence with high-energy neutrinos may help solve the mystery of the origin of CRs, a long-standing issue in high-energy astrophysics (Drury, 2012).

GW emitters may also produce VHE gamma rays. In this case, a merger of a compact-object system may trigger a short Gamma-Ray Burst (GRB). Since their high-energy emission has been observed to extend to multi-gigaelectron-volts (GeV) energies, flux extrapolations to the VHE range indicate that they may be detectable by Imaging Atmospheric Cherenkov Telescopes (IACTs), if the burst progenitor is bright and nearby. The observation of VHE gamma rays from a GRB would be an important step in understanding these energetic events, and a joint detection with gravitational waves would provide additional information about the merger.

Neutrinos can also be produced in compact object collapses or mergers (see, e.g. Bartos, 2017). Megaelectron-volt (MeV) neutrino emission is expected to occur in core-collapsed stars. HE neutrinos can be produced in non-thermal processes together with gamma rays, e.g. if a relativistic jet is formed after the merger. In the case of optically thick ejecta, the gamma-ray emission might be suppressed, leading to a "choked-jet" scenario where only gravitational waves and neutrinos escape.

In this chapter, we discuss the scientific motivation, current status, and recent successes of IACT follow-up observations of HE neutrino and gravitational-wave events.

9.2. Neutrinos

As you learned in Chapter 7, due to absorption in the Extragalactic Background Light (EBL), gamma rays have a limited range (see Fig. 9.1). Currently, the farthest detected very-HE gamma-ray source is S3 0218+35, at a redshift of about 1 (Ahnen *et al.*, 2016b), observed up to an energy of about 175 GeV. Similarly, CRs interact with the Cosmic Microwave Background (CMB), leading to an attenuation of the CR flux above $\sim 10^{20}$ eV. The CR horizon at these energies is limited to $\sim$20 Mpc (i.e. nearby galaxy clusters). On the other hand, neutrinos can propagate unperturbed for distances of several tens of gigaparsecs and penetrate the deep history of the Universe, as well as very dense environments such as neutron stars.

There are many sources of neutrinos in the Universe (see Fig. 9.2), starting from the cosmological neutrino background (CνB) — a relic from

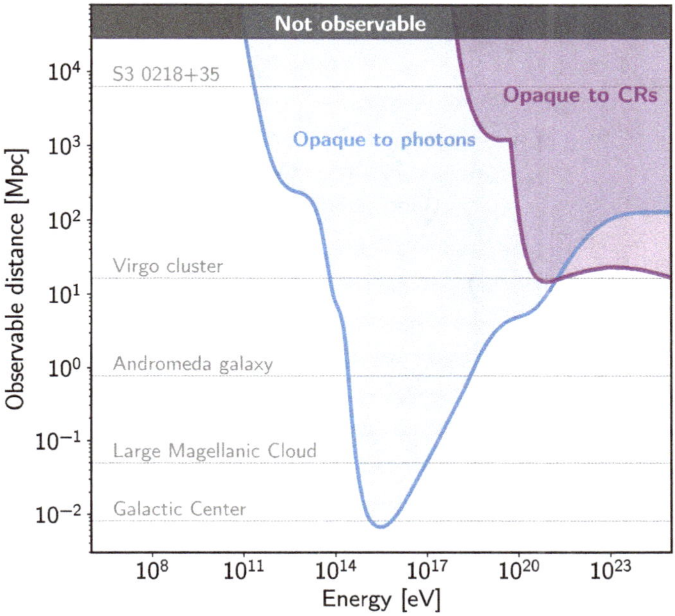

Fig. 9.1. Propagation distance of cosmic photons and protons. Photons (blue) are attenuated by interaction with the EBL (mostly CMB and InfraRed (IR) background radiation). CR protons (purple) mainly interact with the CMB. Distances to several astrophysical objects are shown for reference, including the most distant VHE source detected to date: the quasar S3 0218+35. Based on a plot by P. Gorham, updated with information from Protheroe and Biermann (1996), Kalashev *et al.* (2002), Murase (2009), and Heiter *et al.* (2018).

the Big Bang predicted by the Standard Cosmological Model but not yet measured. At energies between a few eV to GeV, the main sources are stars, including our Sun and supernovae. In fact, the Sun (Davis *et al.*, 1968; Bahcall *et al.*, 1968) and SN 1987A (Hirata *et al.*, 1987; Bionta *et al.*, 1987) were the first multi-messenger sources ever discovered, albeit at MeV energies. Finally, at the highest energies (>GeV), we observe atmospheric neutrinos produced in the interactions of CRs with our atmosphere. They are the main background for the detection of a HE astrophysical neutrino flux from potential sources such as Galactic supernova remnants, Active Galactic Nuclei (AGNs), star-forming galaxies, and GRBs, among others.

HE (>GeV) neutrino emission from astrophysical sources is usually attributed to hadronic interactions of CRs, i.e. HE protons or nuclei, with ambient material (*pp* interactions, Kelner *et al.*, 2006) or radiation fields

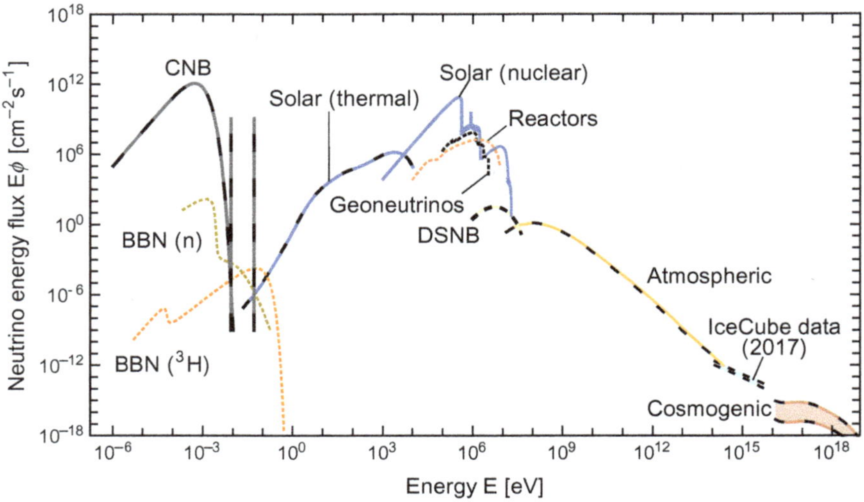

Fig. 9.2. Neutrino fluxes at Earth (in $E_\nu\phi(E_\nu)$ representation) due to different astrophysical and telluric sources. Fluxes are summed over neutrino flavors, with solid (dashed/dotted) lines representing neutrinos (antineutrinos). From Vitagliano *et al.* (2019).

($p\gamma$ interactions, Kelner and Aharonian, 2008). In these collisions, neutrinos and gamma rays are produced in the decay of pions, via

$$\begin{aligned} p + p/\gamma \rightarrow X + \pi^0 &\rightarrow \gamma\gamma \\ X + \pi^+ &\rightarrow \mu^+ + \nu_\mu \\ &\rightarrow \mu^+ \rightarrow e^+ + \nu_e + \bar{\nu_\mu} \end{aligned} \tag{9.1}$$

The $\nu_e : \nu_\mu : \nu_\tau$ neutrino flavor ratio of 1:2:0 at emission oscillates to near equipartition (1:1:1) as neutrinos propagate over astrophysical distances. The spectral production rates of neutrinos and gamma rays are related by

$$\frac{1}{3}\sum_\nu E_\nu^2 \frac{dN_\nu}{dE_\nu} \approx \frac{K_\pi}{4} E_\gamma^2 \frac{dN_\gamma}{dE_\gamma}\bigg|_{E_\gamma = 2E_\nu} \tag{9.2}$$

where $K_\pi = 1(2)$ for $p\gamma$ (pp) interactions and ν denotes the neutrino flavor. Given their deep interconnection, the joint detection of gamma rays and neutrinos from an astrophysical source would not only help resolve the cosmic-ray origin question, but also provide insights into the environments in which these hadronic processes take place.

9.2.1. *Neutrino detectors*

Neutrino astronomy is a challenging endeavor, as these particles carry no electric charge and interact only via the weak force, making them difficult to detect. Charged-current (CC) interactions of neutrinos and nucleons are mediated by a $W^{\pm}$ boson. These interactions result in the emission of a charged lepton of the same flavor as the neutrino and a hadronic cascade due to the momentum transferred to the nucleon. Neutral-current (NC) interactions are mediated by a Z^0 boson. They result in the scattering of the neutrino, also accompanied by a hadronic cascade. If the interactions take place in a refractive, transparent medium such as water or ice, the charged leptons and charged particles in the hadronic cascades may be detected via the Cherenkov radiation they emit (see Chapter 1).

It was soon realized that the low neutrino fluxes expected from astrophysical sources in the $\gg$GeV range would require the construction of colossal detectors, with masses in the gigatonne scale. To solve this issue, Markov (1960) proposed instrumenting a natural medium, such as deep sea water or glacial ice, with light sensors so that neutrinos could be detected using the Cherenkov technique. Ice and water allow the Cherenkov light to travel with minimal absorption and scattering. Deep sea water has a shorter absorption length than deep Antarctic ice, but it has the advantage of much smaller scattering. This means that in water, the direction reconstruction is potentially more accurate, while in ice the Cherenkov light can be detected from farther away. Current observatories such as ANTARES (Ageron *et al.*, 2011), Baikal (Aynutdinov *et al.*, 2005), and IceCube (Aartsen *et al.*, 2017f) are a realization of this concept. IceCube, located near the geographic South Pole, is the largest and most sensitive Cherenkov neutrino observatory currently in operation in the ~100 GeV to ~EeV energy range, with a km^3-scale volume of glacial ice instrumented by 5160 PhotoMultiplier Tubes (PMTs) at depths from 1,450 m to 2,450 m below the ice surface. A schematic view of the IceCube detector is shown in Fig. 9.3.

9.2.2. *Astrophysical neutrinos*

Astrophysical neutrinos in the teraelectron-volt (TeV)–petaelectron-volt (PeV) range were first detected by IceCube in 2013 (Abbasi *et al.*, 2013). The latest measurements of the astrophysical muon neutrino flux in the energy range between 119 TeV and 4.8 PeV yield a spectrum compatible with a power law of the form $\Phi(E) = \Phi_0(E/E_0)^{-\Gamma}$, with a

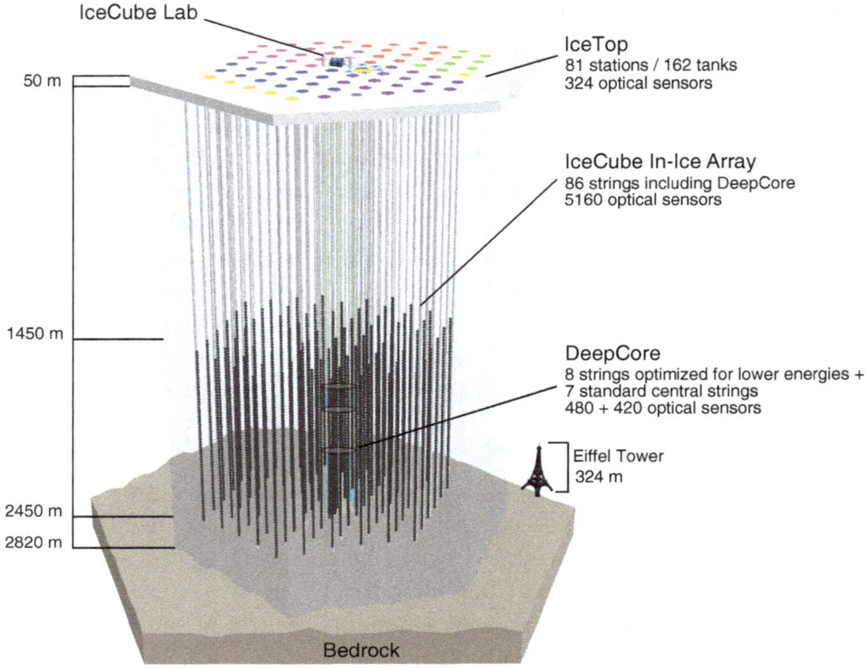

Fig. 9.3. The IceCube detector with its sub-array DeepCore (marked in green), and the cosmic-ray air shower array IceTop. Figure adapted from Aartsen *et al.* (2017f).

flux normalization $\Phi_0 = (1.01\pm^{0.26}_{0.23}) \times 10^{-18}$ GeV^{-1} cm^{-2} s^{-1} sr^{-1} and a spectral index $\Gamma = 2.19 \pm 0.10$ for a normalization energy $E_0 = 100$ TeV (Aartsen *et al.*, 2016). A measurement with a lower-energy threshold of 60 TeV sensitive to all (anti-)neutrino flavors yields a softer spectrum, with $\Phi(E) = (6.37\pm^{1.47}_{1.62}) \times 10^{-18}(E/100\text{ TeV})^{-2.87}$ GeV^{-1} cm^{-2} s^{-1} sr^{-1} (Abbasi *et al.*, 2021).

The origin of the neutrino flux is yet to be determined, although a large number of source models have been proposed (see Ahlers and Halzen, 2018 and references therein for a recent review). Potential sources may be grouped into two classes, depending on the type of hadronic interaction that gives rise to the neutrino emission.

- In *pp* sources, a large concentration of target material that CRs can transverse and interact with is required, coupled with a high CR production rate and magnetic fields that are able to contain the CRs. Supernova remnants interacting with molecular clouds in our galaxy, or "starburst" galaxies with high star-formation rates (and therefore high supernova rates), are potential pp sources. A diffuse flux of pp neutrinos is also expected at some level from the Galactic plane, as Galactic CRs interact with gas clouds during propagation. In pp scenarios, the spectrum of secondary neutrinos and gamma rays depends chiefly on the characteristics of the injected CR spectrum at the source.

- In $p\gamma$ sources, neutrinos arise from photohadronic interactions of CRs with a dense photon field. Such conditions exist near the central engine of GRBs or AGNs–therefore, we may expect this type of emission from blazars, radio galaxies and flat-spectrum radio quasars. In $p\gamma$ scenarios, the neutrino spectrum depends on the spectrum of injected CRs but also on the characteristics of the photon field.

No significant correlation has been found so far between known astrophysical sources such as blazars (Aartsen *et al.*, 2017e), GRBs (Aartsen *et al.*, 2017c), or the Galactic plane (Aartsen *et al.*, 2017b) and the positions and arrival times of neutrino events. Upper limits have been set on the contribution from these source classes to the astrophysical diffuse flux. No strong evidence of neutrino self-clustering has been found either on long (Aartsen *et al.*, 2020b) or short timescales (Aartsen *et al.*, 2015, 2017d). In the 10-year IceCube exposure map, the most significant spot ($\sim$4.2σ pre-trial) is NGC 1068, a starburst galaxy (Aartsen *et al.*, 2020b). The telescope MAGIC observed this object but did not detect it and set stringent limits on its gamma-ray and ν emission mechanism (Acciari *et al.*, 2019).

The isotropic distribution of the neutrino events favors an extragalactic origin, and while a Galactic origin cannot be excluded, its contribution must be subdominant. If the sources of neutrinos are indeed extragalactic, the TeV–PeV gamma-ray flux expected to accompany the neutrino emission would be strongly attenuated by its interaction with the extragalactic background light through the production of e^-–e^+ pairs, which in turn generate lower-energy gamma rays through electromagnetic cascading down to the MeV–GeV energy range, or lower. The connection between the astrophysical neutrino flux and the Isotropic Gamma-Ray Background (IGRB) measured by *Fermi*-LAT between 100 MeV and 820 GeV (Ackermann *et al.*, 2015a) has been explored (Tamborra *et al.*, 2014; Murase *et al.*, 2016; Bechtol *et al.*, 2017) to set constraints on the origin of both emissions, as the IGRB represents an upper bound on the hadronic gamma-ray emission that could be associated with the neutrino flux.

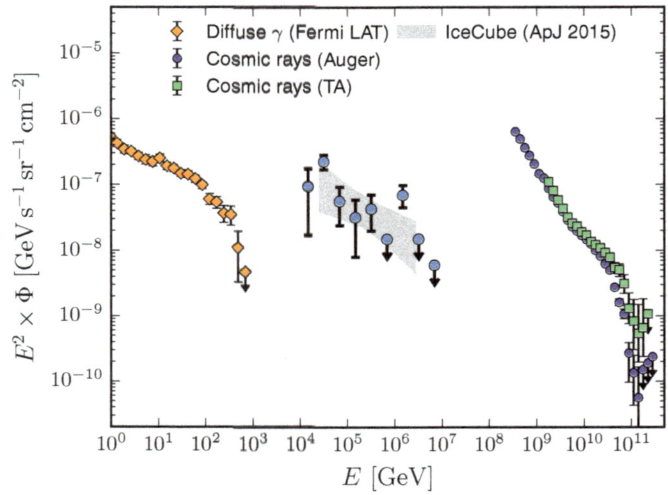

Fig. 9.4. A comparison between the astrophysical neutrino, UHECR, and gamma-ray spectra (adapted from Mohrmann, 2015). See Ahlers and Halzen (2018) for a detailed discussion on the connection between the three fluxes.

Neutrinos may also be connected with Ultra-High-Energy Cosmic Ray (UHECR) sources. Assuming that UHECR sources follow the star-formation rate and that the UHECRs are dominated by protons, Waxman and Bahcall arrived at a diffuse neutrino flux limit (per flavor) of $\sim 3 \times 10^{-8}$ GeV cm^{-2} s^{-1} sr^{-1}, near the IceCube neutrino flux level, for a $\sim$E^{-2} CR spectrum (Waxman and Bahcall, 1999; Bahcall and Waxman, 2001). This may imply that UHECRs and IceCube neutrinos share a common origin. A comparison between the astrophysical neutrino, IGRB, and UHECR fluxes is shown in Fig. 9.4 and illustrates the potential connections between the three channels. For a more detailed discussion on CRs and their sources, we refer the reader to Chapter 4.

9.2.3. *Neutrino follow-up observations*

Contrary to pointed instruments such as IACTs, neutrino telescopes constantly monitor large parts of the sky. Therefore, very early on, an idea occurred to use them as triggering devices for electromagnetic observations when an interesting neutrino event (or increase in event rate) is observed (Bernardini, 2005; Ackermann *et al.*, 2007). For a detailed description of the currently operating neutrino alert programs, see Aartsen *et al.* (2017) and Adrián-Martínez *et al.* (2016). Due to higher sensitivity of photon telescopes, they should be able to identify a potential source of neutrinos with higher precision than a neutrino detector alone.

Several neutrino follow-up programs involving IceCube and all operating IACTs are currently in operation (see Santander *et al.*, 2017; Satalecka *et al.*, 2021 and references therein). Follow-up observations with IACTs are typically limited to muon neutrino events, as their angular resolution of ~1° is comparable to the Field of View (FoV) of current IACT cameras (~3.5°–5°). Figure 9.5 depicts a sky map with all astrophysical neutrino candidate event directions followed up so far by the telescopes H.E.S.S., MAGIC, and VERITAS. Until September 2017, no VHE gamma-ray source had been detected within the error region of the IceCube events.

On September 22, 2017, IceCube detected an event (IceCube-170922A) with about 50% probability of being an astrophysical neutrino. Its best-fit reconstructed direction was within ~0.1° from the sky position of the BL Lac object TXS 0506 + 056.[a] A rapid dissemination of the event direction led to extensive follow-up observations. A high level of activity

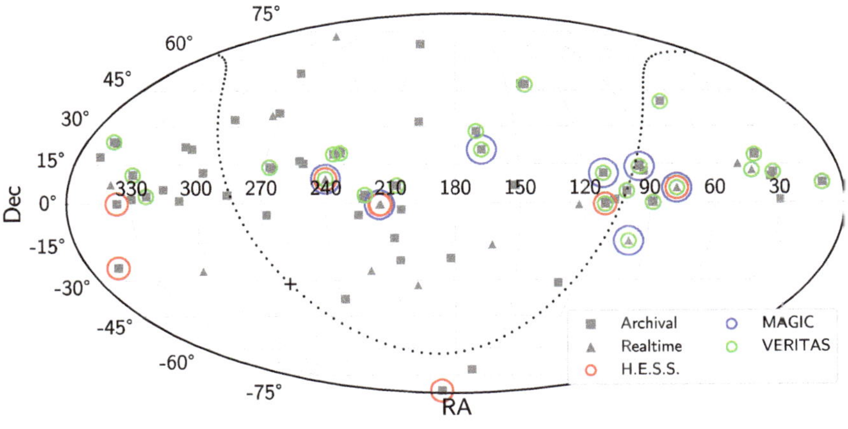

Fig. 9.5. Positions of HE astrophysical neutrino candidates observed by the IceCube observatory as reported in publications ("Archival" events, gray squares) or through realtime alerts ("Realtime" events, gray triangles) in equatorial coordinates. The color circles indicate the IACT arrays that have observed the neutrino positions as part of their neutrino follow-up program. The different latitudes of the IACTs dictate the declination range they can cover in follow-up observations. All three IACT facilities can observe moderate declinations near the celestial equator, where the IceCube effective area for the HE neutrino is maximal. See Santander *et al.* (2017) for details.

[a]The IceCube Collaboration, GRB Coordinates Network, Circular Service, No. 21916 (2017).

of TXS 0506+056 was revealed in all wavelengths, most notably in GeV gamma rays monitored by the *Fermi*-LAT (Tanaka, 2017), and VHE instruments. The source, at a redshift of $z = 0.3365 \pm 0.0010$ (Paiano *et al.*, 2018), was detected for the first time by the MAGIC telescope (Mirzoyan, 2017), and later in subsequent observations by VERITAS (Abeysekara *et al.*, 2018).

A dedicated study of the past inverse Compton observations of this object revealed another highly significant period of neutrino activity (September 2014 to March 2015), establishing it as a first potential source of HE cosmic neutrinos (Aartsen *et al.*, 2018).

Thanks to the precise determination of the direction of IceCube-170922A, the chance probability for a HE neutrino to be detected in coincidence with a flaring blazar from *Fermi*-LAT catalogs was estimated to be at the 3σ level (Aartsen *et al.*, 2018). Although the significance of the TXS 0506+056 and IceCube-170922A association is too low to claim a neutrino source discovery, it could represent the first of many such correlated detections, which would offer a unique opportunity to study the interplay between energetic photons, neutrinos, and CRs.

IceCube continues to send realtime ν alerts. The Multi-Wavelength (MW) follow-up observations sometimes reveal sources that lie within their localization error and could potentially emit neutrinos but are too faint to be detected by the current generation of IACTs. Most of them are blazars, e.g. PKS 1502+106, at redshift $z = 1.8$, proposed as a counterpart to IceCube-190730A (Rodrigues *et al.*, 2021), and 4FGL J0658.6+0636 consistent with an IceCube HE ν IceCube-201114C (De Menezes *et al.*, 2021).

A systematic search for optical counterparts to HE neutrinos with the Zwicky Transient Facility (ZTF) suggests an association of a radio-emitting Tidal Disruption Event (TDE) (AT2019dsg) with a HE neutrino IceCube-191001A (Stein *et al.*, 2021). The electromagnetic observations can be explained through a multi-zone model, with radio analysis revealing a central engine, embedded in a UV photosphere, that powers an extended synchrotron-emitting outflow. This provides an ideal site for PeV neutrino production. The association suggests that TDEs, even those with mildly relativistic outflows, contribute to the cosmic neutrino flux. No gamma-ray emission was detected for AT2019dsg using the *Fermi*-LAT telescope, but IACTs may be sensitive to the expected gamma-ray signal if the emission came from a *pp* interaction. Conversely, in case of p-γ-induced emission, the high optical depth of the UV photosphere would absorb any gamma rays. VHE gamma-ray observations of TDEs have never been conducted on a large systematic scale (the only known case is a jetted TDE Swift J1644+57 observed by MAGIC (Aleksić *et al.*, 2013) and VERITAS (Aliu *et al.*, 2011b)). Therefore, IACTs can provide very important inputs and help to distinguish between emission scenarios in these types of sources.

9.3. Gravitational Waves

In the framework of General Relativity, space–time can be considered a fabric that bends or curves under the gravitational influence of masses. In the absence of gravity (and thus mass), space–time is flat and its curvature g can be characterized by the following four-dimensional Minkowski metric $\eta_{\mu\nu}$:

$$\eta = \begin{bmatrix} -1 & 0 & 0 & 0 \\ 0 & 1 & 0 & 0 \\ 0 & 0 & 1 & 0 \\ 0 & 0 & 0 & 1 \end{bmatrix} \tag{9.3}$$

A weak gravitational field can be considered a small "perturbation" to this flat space, the curvature then becomes

$$g_{\mu\nu} = \eta_{\mu\nu} + h_{\mu\nu} \quad \text{with} \quad |h_{\mu\nu}| \ll 1 \tag{9.4}$$

The relationship between the metric tensor g and the energy–momentum tensor T is defined by the Einstein field equation:

$$R_{\mu\nu} - \frac{1}{2}Rg_{\mu\nu} = 8\pi GT^{\mu\nu} \tag{9.5}$$

The energy–momentum tensor T is a measure of the presence of matter and energy in the local space. $R_{\mu\nu}$ is the Ricci tensor and R is the scalar curvature. Given that the gravitational perturbation can be assumed to be small, one can expand the Einstein field equation in h. After linearization, one can find a relation between $h_{\mu\nu}$ and the energy–momentum tensor T (see Chakrabarty, 1999 for a detailed derivation):

$$\Box\bar{h}_{\mu\nu} = -16\pi GT^{\mu\nu} \tag{9.6}$$

In the limit of flat space–time as specified initially, where the perturbation is small and no significant mass is present, $T^{\mu\nu} \approx 0$. The relation mentioned above therefore becomes a wave equation, as follows:

$$\Box\bar{h}_{\mu\nu} = 0 \tag{9.7}$$

Gravitational waves are thus created by moving masses, very similar to electromagnetic waves being created by moving charges. On the other hand, and in contrast to electromagnetism, mass does not come in different "charges". A mass dipole is, for example, always the center of mass of the system, which is a constant by conservation of momentum. Therefore, gravitational waves can only be produced in the asymmetric accelerations of masses. At the lowest order, these are quadrupole moments found, for example, in in-spiralling binary systems or asymmetric mass ejections during supernova explosions. These violent phenomena are predicted to show emission across the full electromagnetic spectrum, motivating the search for counterparts (see Section 9.3.3). A prominent example for these scenarios is the mergers of binary neutron star systems that have been predicted to be linked to gamma-ray bursts and kilonovae (Metzger, 2017).

9.3.1. *Detection of gravitational waves*

Gravitational waves are modulations of space–time and can be detected as minute distance changes. The waves are polarized, i.e. they compress space in one direction while stretching it in the perpendicular direction. The scale of these changes is tiny: even very massive systems like two orbiting black holes that are about to merge into one will only produce displacements on the order of 10^{-18} meters, about 1,000 times smaller than the diameter of a proton, when the waves arrive at the surface of the Earth.

Current state-of-the-art gravitational-wave detectors are L-shaped and measure the relative lengths of the arms using interferometry. There are three such interferometers in operation at the moment: two in the United States, collectively called Laser Interferometer Gravitational-wave Observatory (LIGO, Abbott *et al.*, 2009), and the Virgo interferometer in Italy (Acernese *et al.*, 2015).

Detection of consistent signals by multiple interferometers is necessary to confidently detect and locate the sources of gravitational waves. Multiple detectors allow us to suppress candidate signals that are caused by local noise sources. As gravitational waves travel at finite speed (the speed of light), the arrival times of the waves at the different detectors will be typically separated by about 10–30 ms. Using these time differences for triangulation and combining them with the known sensitivity pattern of the individual detectors allows us to localize the origin of the waves. Depending on the signal strength, the associated uncertainties range from tens to several hundred square degrees. As an example, see Fig. 9.6, which shows the localization of the GW, gamma-ray, and optical signals for the

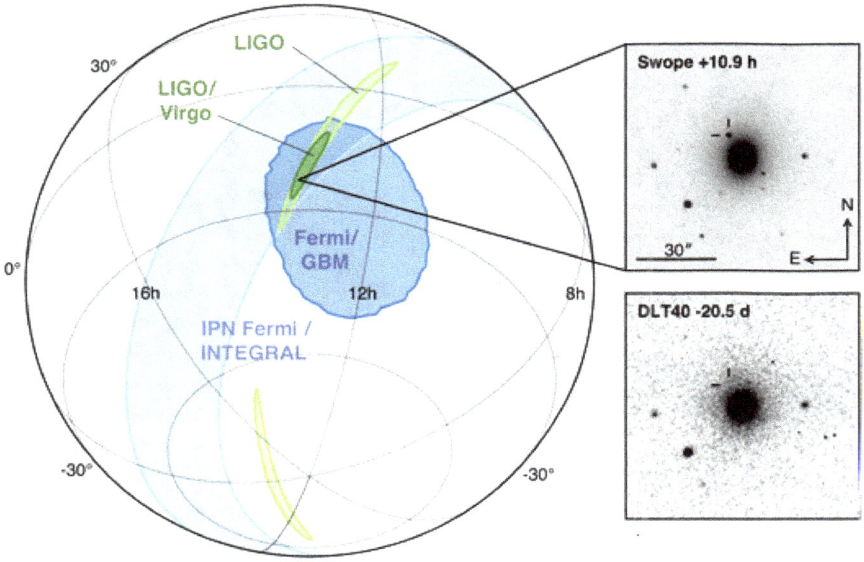

Fig. 9.6. Localization of the gravitational-wave, gamma-ray, and optical signals for the GW-170817 event. Left panel: orthographic projection of the 90% credible regions from LIGO (190 deg^2; light green), the initial LIGO–Virgo localization (31 deg^2; dark green), interplanetary network (IPN) triangulation from the time delay between Fermi and INTEGRAL (light blue), and Fermi–GBM (dark blue). Inset: location of the apparent host galaxy NGC 4993 in the Swope optical discovery image at 10.9 h after the merger (top right) and the DLT40 pre-discovery image from 20.5 days prior to merger (bottom right). The reticle marks the position of the transient in both images. Figure taken from Abdalla *et al.* (2017c).

GW-170817 event. Note how the inclusion of the Virgo interferometer helps reduce the localization error from hundreds to few tens of square degrees.

9.3.2. *First observations of gravitational waves*

After several decades of continuous upgrades and improvements of the interferometers, the era of astronomy and astrophysics with gravitational waves finally began on September 14, 2015 at 09:50:45 UTC. By then, the two interferometers of the Advanced LIGO observatory had reached the necessary sensitivity to record the first direct signal of a gravitational wave (Abbott *et al.*, 2016c). The recorded signal matched expectations derived from general relativity of the in-spiral and merger of a Binary Black

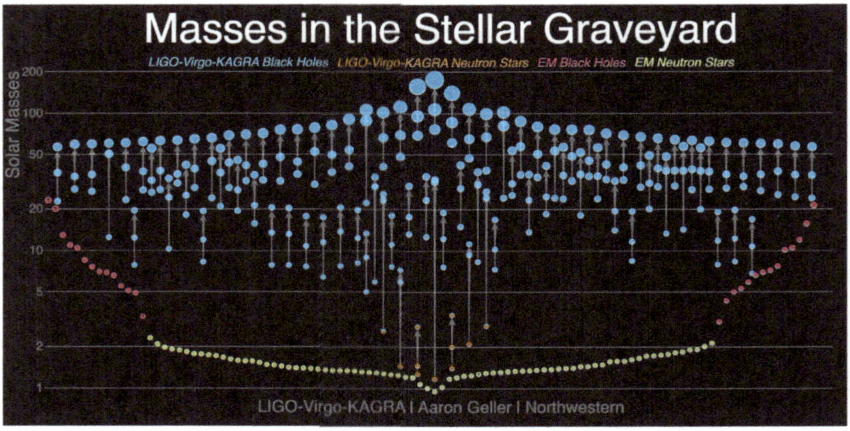

Fig. 9.7. Masses of stellar remnants as measured by gravitational-wave observations of merger events (blue and orange markers) in comparison with previous electromagnetic observations (yellow and red markers). Image credits: LIGO-Virgo-KAGRA/Aaron Geller/Northwestern University.

Hole (BBH) system. The coalescence was reconstructed at a luminosity distance of 410^{+160}_{-180} Mpc. In the source frame, the initial black hole masses were 36^{+5}_{-4} $M_{\odot}$ and 29^{+4}_{-4} $M_{\odot}$. This first detection of a binary black hole (BBH) merger clearly demonstrated the existence of binary stellar-mass black hole systems and opened the window to gravitational-wave multi-messenger astronomy.

After this extremely exciting start, both detectors operated from September 2015 until January 2016 (run O1), allowing for the detection of a second confirmed BBH merger (GW151226 Abbott *et al.*, 2016b) as well as another, very likely event (LVT151012 Abbott *et al.*, 2016a). After further upgrades, LIGO resumed data-taking in November 2016. It was joined by the Virgo interferometer in August 2017 about one month before the end of this second physics run (O2). During this run, three additional BBH mergers (GW170104, GW170608, and GW170814) and the first binary neutron star merger (GW170817) were detected. With ever-increasing sensitivity, a third observation campaign (O3) ran from April 2019 to March 2020. An overview of the events observed through March 2020 is given in Fig. 9.7, although the high detection rate (currently $\gtrsim$1 per week) continues to add events to this picture.

9.3.3. *Gravitational-wave follow-up observations*

Concurrently with the GW observations, efforts are being made on a global scale to identify electromagnetic radiation (as well as HE neutrinos) associated with these events. All IACTs actively participate in this endeavor. The origin of the detected GW is being determined at the first order by triangulation of the signals recorded by the different interferometers. Using data from two detectors, the 90% uncertainty regions spanned up to several thousand square degrees during Observation-1 (O1). At least three GW observatories were deemed necessary to enable localizations better than hundred square degrees. These conditions were fulfilled starting only in August 2017, when the Advanced Virgo observatory joined the two LIGO interferometers during run Observation-2 (O2). The large uncertainty areas, combined with delays in the alert emission induced by the reconstruction and validation of the GW data, are therefore the main difficulties for follow-up observations of GWs with pointed instruments like IACTs.

Since the beginning of the data-taking of the gravitational-wave interferometers in their advanced configuration at the end of 2015, all IACT collaborations were part of the electromagnetic follow-up group. During O1 and O2, this group comprised about 70 observatories around the world searching for electromagnetic signals related to GWs. During this early phase of GW searches (i.e. during the LIGO/Virgo runs O1 + O2), IACT follow-up started with individual, exploratory observations of BBH mergers (e.g. observations of the binary black hole merger GW151226 Abbott *et al.* (2016b), by MAGIC De Lotto *et al.* (2016), and the VERITAS observations[b] of GW170104 Abbott *et al.*, 2017b). During the run O2, these observations evolved into complete scans of the uncertainty regions (e.g. the first GW event detected by all three interferometers: the binary black hole merger GW170814 Abbott *et al.* (2017c) followed up by H.E.S.S.[c]). Aiming at optimizing the follow-up of GWs and other transients with large localization uncertainties, the H.E.S.S. collaboration implemented a complex and fully automatic framework to select the events to observe. The framework uses additional information, such as catalogs of nearby galaxies, to produce an optimal follow-up schedule (Ashkar *et al.*, 2021).

[b]https://gcn.gsfc.nasa.gov/gcn3/21153.gcn3.
[c]https://gcn.gsfc.nasa.gov/gcn3/21673.gcn3.

These preparations came to fruition during the rapid and deep follow-up of GW170817, the first-ever detected merging of two neutron stars.

9.3.3.1. *Follow-up of the first binary neutron star merger: GW170817*

On August 18, 2017, for the first time LIGO and Virgo detected gravitational waves from a pair of in-spiralling neutron stars (GW170817 Abbott *et al.* (2017d)). Mergers of binary neutron stars had been proposed as candidates for the origin of short GRBs and associated to kilonova events (see Metzger (2017) for a recent review). The announcement of the event triggered the most extensive follow-up campaign mentioned above involving 70 observatories around the world, in space, and across the entire electromagnetic spectrum (see Figs. 9.8 and 9.9). Electromagnetic emission from the collision was observed in multiple wavelength bands, and GW170817 is therefore the first cosmic event from which both gravitational waves and light have been observed. The analysis of the obtained wealth of data allowed scientists to confirm the connection between NS–NS mergers (where NS stands for neutron star) and GRBs and kilonovae.

A kilonova is a type of a transient event with a luminosity of 1%–10% that of a classic supernova. They were postulated as optical counterparts to short GRBs. This afterglow emission is powered by the radioactive decay

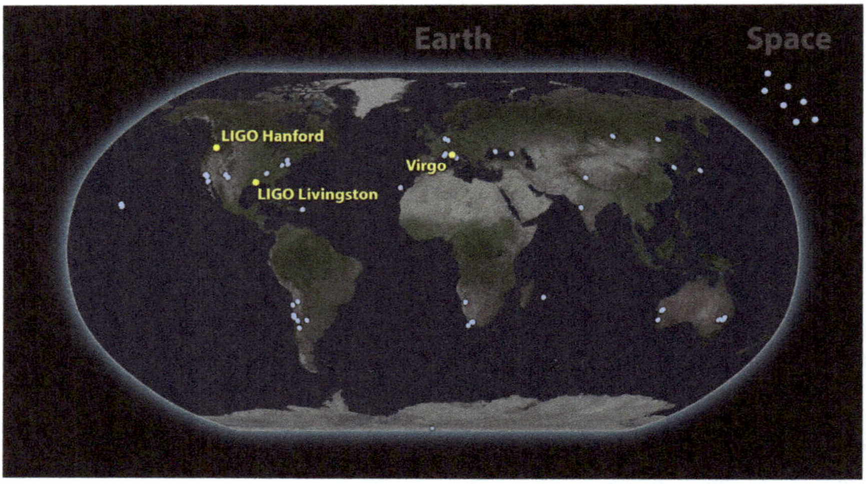

Fig. 9.8. Observatories participating in the follow-up observations of the first binary neutron star merger event, GW170817. Image credits: LIGO–Virgo.

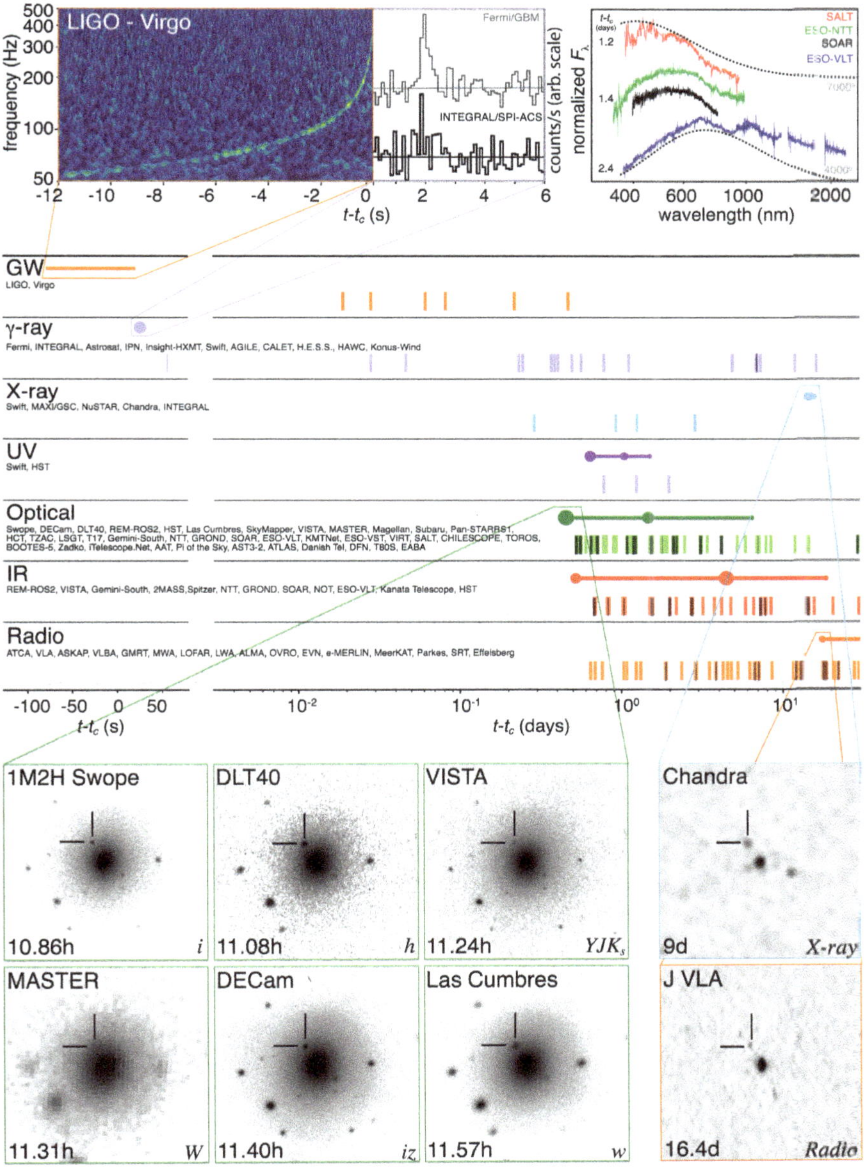

LIGO - Virgo
frequency (Hz)
Fermi/GBM
INTEGRAL/SPI-ACS
counts/s (arb. scale)
t-t_c (s)
normalized F_λ
SALT
ESO-NTT
SOAR
ESO-VLT
wavelength (nm)
GW
LIGO, Virgo
γ-ray
Fermi, INTEGRAL, Astrosat, IPN, Insight-HXMT, Swift, AGILE, CALET, H.E.S.S., HAWC, Konus-Wind
X-ray
Swift, MAXI/GSC, NuSTAR, Chandra, INTEGRAL
UV
Swift, HST
Optical
Swope, DECam, DLT40, REM-ROS2, HST, Las Cumbres, SkyMapper, VISTA, MASTER, Magellan, Subaru, Pan-STARRS1, HCT, TZAC, LSGT, T17, Gemini-South, NTT, GROND, SOAR, ESO-VLT, KMTNet, ESO-VST, VIRT, SALT, CHILESCOPE, TOROS, BOOTES-5, Zadko, iTelescope.Net, AAT, Pi of the Sky, AST3-2, ATLAS, Danish Tel, DFN, T80S, EABA
IR
REM-ROS2, VISTA, Gemini-South, 2MASS, Spitzer, NTT, GROND, SOAR, NOT, ESO-VLT, Kanata Telescope, HST
Radio
ATCA, VLA, ASKAP, VLBA, GMRT, MWA, LOFAR, LWA, ALMA, OVRO, EVN, e-MERLIN, MeerKAT, Parkes, SRT, Effelsberg
t-t_c (s)
t-t_c (days)
1M2H Swope
10.86h
i
DLT40
11.08h
h
VISTA
11.24h
YJK_s
Chandra
9d
X-ray
MASTER
11.31h
W
DECam
11.40h
iz
Las Cumbres
11.57h
w
J VLA
16.4d
Radio

Fig. 9.9.

Fig. 9.9. Timeline of the discovery of GW170817, GRB 170817A, SSS17a/AT 2017gfo, and the follow-up observations are shown by messenger and wavelength relative to the time tc of the gravitational-wave event. Two types of information are shown for each band/messenger. First, the shaded dashes represent the times when information was reported in a GCN Circular. The names of the relevant instruments, facilities, or observing teams are collected at the beginning of the row. Second, representative observations in each band are shown as solid circles with their areas approximately scaled by brightness; the solid lines indicate when the source was detectable by at least one telescope. Magnification insets give a picture of the first detections in the gravitational-wave, gamma-ray, optical, X-ray, and radio bands. Figure taken from Abdalla *et al.* (2017c).

of heavy nuclei produced via the so-called r-process. The detection of this long-sought emission during the follow-up of GW170817 confirmed this connection. This kind of merger event is likely responsible for the abundance of elements heavier than iron observed in the Milky Way (Kasen *et al.*, 2017).

GRBs as a source class (both short and long in duration) have been the target of intense IACT follow-up campaigns for over two decades (e.g. Connaughton *et al.*, 1997). These efforts have recently been rewarded with the detection of VHE emission from two long GRBs: the afterglow detection of GRB 180720B by H.E.S.S. (Abdalla *et al.*, 2018g), and a strong detection of GRB 190114C by MAGIC (Acciari *et al.*, 2019b). A similar VHE detection of a GRB associated with a GW event would allow for detailed studies of the underlying particle acceleration mechanism and thus provide significant insight into the NS–NS/GRB mechanisms. Whether short GRBs associated with kilonovae are VHE emitters is therefore a question that is of significant interest for the current generation of IACTs, as well as the next-generation instruments like CTAO.

Combining information from all three gravitational-wave interferometers, the GW170817 event could be localized within a few hours to a region spanning $\sim$31 deg^2 at a luminosity distance of 40^{+8}_{-14} Mpc, a region compatible with the localization of the short GRB (GRB 170817A Abbott *et al.* (2017a)) detected by *Fermi*–GBM and INTEGRAL $\sim$1.7 s after the GW event.

Due to the localization of the event to the Southern sky, the H.E.S.S. observatory was the only IACT able to participate in the follow-up campaign. Observations started on August 17, 2017 at 17:59 UTC, only about 5 min after the publication of the localization of the event by LIGO/Virgo (Abdalla *et al.*, 2017c). The observations thus began only 5.3 h after the detection of the gravitational-wave event. The uncertainty region of about $\sim$31 deg^2 was larger than the FoV of the H.E.S.S. telescopes — therefore, a dedicated pointing strategy had to be devised. The employed algorithm used the full 3D information provided by the GW detection and combined it with the GLADE catalog of galaxies to select the most promising regions to host the binary neutron system and thus the merger event. The first region selected and observed by H.E.S.S. contained the electromagnetic counterpart to the gravitational-wave event that has been identified in the optical domain several hours after the H.E.S.S. observations (cf. Fig. 9.10). As a result, H.E.S.S. was the first ground-based pointed instrument to obtain data on this object, which turned out to be the afterglow of GRB 170817A and the first confirmed kilonova. The transient source has been named SSS17a by the discovery team, the One-Meter Two-Hemisphere (1M2H) Collaboration using the 1m Swope telescope at Las Campanas Observatory in Chile, and was later given the IAU identification of AT 2017gfo.

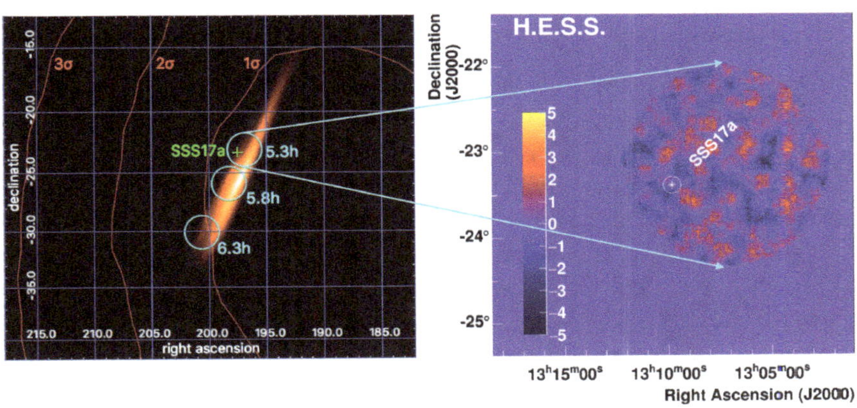

Fig. 9.10. Left: Pointing directions of the first night of H.E.S.S. follow-up observations of GW170817 starting 2017-08-17 17:59 UTC. The circles illustrate the H.E.S.S. FoV with a radius of 1.5° and the shown times are the start times of each observation with respect to GW170817. The localization map of GW170817 derived from the LIGO and Virgo observations is shown as shaded background, and the thin lines denote the uncertainty contours of GRB170817A derived from Fermi–GBM observations. Right: Map of significances of the gamma-ray emission in the region around SSS17a obtained during the first observation of GW170817. The circle has a diameter of 0.1°, corresponding to the H.E.S.S. point spread function. Image credits: H.E.S.S. Collaboration (Abdalla *et al.*, 2017c).

A subsequent monitoring campaign of the kilonova with the H.E.S.S. IACTs extended over several days, therefore covering timescales from 0.22 to 5.2 days after the merger and an energy range between 270 GeV and 8.55 TeV. No significant gamma-ray emission has been found within this time interval. The derived upper limits on the very-HE gamma-ray flux for the first time constrain non-thermal, HE emission following the merger of a confirmed binary neutron star system (Abdalla *et al.*, 2017c).

Interestingly, no HE neutrinos were observed from this event, although some models predict fluxes detectable with current instruments. This might be due to an off-axis orientation of the GRB jet (Kimura *et al.*, 2017).

Observations of the remnant of GW170817 started again with both H.E.S.S. and MAGIC between 124 and 272 days after the Binary Neutron star (BNS) merger, covering the period with the maximum X-ray and radio emission. Using 54 h (10 h) of observations, H.E.S.S. and MAGIC, respectively, could derive limits on the long-term afterglow emission and the magnetic field of the merger remnant (Abdalla *et al.*, 2020b).

9.4. IACTs in the Era of Time-Domain Multi-Messenger Astrophysics

During the follow-up observations of the first HE neutrinos and the first gravitational waves, IACTs have shown their high potential and full readiness for the era of time-domain multi-messenger astrophysics. Their comparably large FoV combined with a rapid and largely automated reaction to external alerts make them an important part of the ever-growing multi-wavelength community searching for electromagnetic counterparts of these new messengers from the violent HE universe. These observations started around 2012 as a small niche compared to more well-established observations of identified Galactic and extragalactic sources. After only a few years — and thanks to the first successful detection of a multi-messenger source such as GW170817/GRB170817A and the VHE observation of TXS 0506+056 as a potential counterpart to the neutrino event IceCube-170922A — they now play an ever increasing role in the sciencific exploitation of IACTs.

Neutrino and gravitational-wave IACT follow-ups are expected to continue expanding in the coming decade thanks to the construction of the proposed neutrino observatories such as IceCube-Gen2 (Aartsen *et al.*, 2014a) and KM3NeT (Adrián-Martínez *et al.*, 2016), and the operation of the gravitational-wave observatories advanced LIGO (aLIGO), Virgo, and KAGRA (which started observations in early 2020, Akutsu *et al.* (2018)).

9.5. Exercises

9.5.1. *Exercise 1*

The IceCube Collaboration sends an alert every time a HE muon track, with high probability of originating from an astrophysical neutrino, is detected. Recently, one of them, IC-170922A, was found in close proximity to a gamma-ray blazar TXS 0506 + 056, one of 1145 gamma-ray emitting blazars present in the 3rd *Fermi*–LAT catalog (3FGL, (Acero *et al.*, 2015)). Calculate what is the probability to find any blazar from the 3FGL catalog inside an IceCube alert error circle for:

- Extremely high energy (EHE) alerts: through-going muon tracks, with median angular resolution of 0.2°.
- High energy starting events (HESE) alerts: muon tracks which start inside the detector, with median angular resolution of 1.0°.
- High energy cascade event with median angular resolution of 15°.

IceCube issued 10 alerts before IC-170922A, how does this fact affect the significance of the spatial correlation of IC-170922A and TXS 0506+056? Calculate the trial-corrected significance assuming that IC-170922A is an EHE event.

Sample Solution

First calculate what is the probability of any random source to be inside an IC alert error circle: $P_{\text{Spatial}} = \Delta\Omega_{\text{IC}}/4\pi$, where 4π is the whole sky solid angle and $\Delta\Omega_{\text{IC}}$ is the IceCube alert error circle. Example EHE event: $\Delta\Omega_{\text{EHE}} = (0.2)^2\pi = 0.13$ deg^2, $4\pi = 41{,}000$ deg^2, $P_{\text{Spatial,EHE}} = \Delta\Omega_{\text{EHE}}/4\pi = 3.2 \times 10^{-6}$.

Now we need to include the fact that there are 1145 blazars in Fermi LAT third source catalog (3FGL catalog) and calculate a probability that at least one source, out of a catalog of N sources, is in the neutrino error ellipse: $P_N = 1 - (1 - P_{\text{Spatial}})^N$. Continuing with the EHE example: $P_{1145,\text{EHE}} = 1 - (1 - P_{\text{Spatial,EHE}})^{1145} = 3.6 \times 10^{-3}$.

Correcting for 10 trials: $P_{\text{Final}} = 1 - (1 - P_N)^N_{\text{trials}}$, for EHE track we get: $P_{\text{Final,EHE}} = 0.036$, which corresponds to a significance of $\sim 2\sigma$.

9.5.2. *Exercise 2*

Blazars are the dominant source class in the extragalactic gamma-ray sky. As such, they have been postulated as potential sources of HE neutrinos. In 2009, a strong gamma-ray flare was observed from the blazar Markarian 501 by the MAGIC and VERITAS telescopes (Abdo *et al.*, 2011b; Abbott *et al.*, 2016c). The VHE gamma-ray spectrum during the flaring period is best described by a power law of the form $\Phi(E) = \Phi_0(E/E_0)^{-\Gamma}$, with a flux normalization of $\Phi_0 = (4.17 \pm 0.24) \times 10^{-11}$ ph cm^{-2} TeV^{-1} s^{-1} and spectral index of $\Gamma = 2.26 \pm 0.06$ at a normalization energy of 1 TeV. While the HE gamma-ray emission from blazars is believed to be dominated by leptonic processes, a hadronic component cannot be ruled out. Under the very optimistic assumptions that the VHE gamma-ray flux is fully hadronic, and that the ratio between the neutrinos and gamma-ray fluxes at Earth from Markarian 421 is 1:1, how many neutrinos would IceCube observe if the flare lasted one week? What is the approximate significance of the neutrino excess assuming that IceCube detects 10^5 atmospheric events per year in the Northern sky and that the angular resolution of the events is $\sim 0.7^\circ$. Explore how the results depend on the spectral index of the neutrinos, the length of the flaring period, and the angular resolution of the events.

Hint: The IceCube effective area can be obtained from Aartsen *et al.* (2017a) for the declination of Markarian 501 and folded with the neutrino spectrum to obtain the number of observed events. The quoted atmospheric background level can be assumed isotropic over the Northern sky.

9.5.3. *Exercise 3*

Searches for temporal coincidences between different astrophysical messengers only make sense if their propagation over potentially cosmological distances does not introduce delays in the arrival time at the surface of the Earth. Imagine a hypothetical astrophysical source that emits a short (delta-function-like) burst of energy equally and synchronously in all known cosmic messengers: photons, neutrinos, gravitational waves, and CRs. At particle energies of 1 TeV and assuming a neutrino mass of 0.1 eV, calculate the arrival time difference at Earth between photons, neutrinos, and gravitational waves for source distances of 40, 100, 500 Mpc.

Charged CRs are deflected by Galactic and extragalactic magnetic fields. The additional path length leads to an arrival time delay τ which can be approximated by (Bhattacharjee, 2000):

$$\tau \approx 1.5 \times 10^{-3} q^2 \left[\frac{E}{10^{20}\ \mathrm{eV}}\right]^{-2} \left[\frac{d}{10\ \mathrm{Mpc}}\right]^{2} \left[\frac{L_c}{1\ \mathrm{Mpc}}\right] \left[\frac{B}{10^{-9}\ \mathrm{G}}\right]^{2} \mathrm{yr} \quad (9.8)$$

Assuming a coherence length of the magnetic field between the source and Earth of $L_c = 1$ Mpc and an average field strength of 10^{-9} G, use this formula to estimate the delay of 1 TeV and 10^{20} eV protons (i.e. charge $q = 1$) from the same sources (i.e. $d = 40$, 100, 500 Mpc).

https://doi.org/10.1142/9789813275720_0010

Chapter 10

Conclusions Before the Advent of CTAO

Felix Aharonian
Max-Planck-Institut fur Kernphysik
Heidelberg, Germany
felix.aharonian@mpi-hd.mpg.de

Extraterrestrial gamma rays are messengers carrying information about the non-thermal Universe. Several factors, namely the vast energy range spanning over at least six decades from MeV to PeV photons; the effective production in both electromagnetic and hadronic processes; the (almost) free propagation over cosmological distances unaffected by the intergalactic radiation and magnetic fields; and, finally, the proven effective detection methods with the space- and ground-based facilities all make gamma rays unique among other cosmic messengers. Historically being a part of traditional cosmic-ray studies, gamma-ray astronomy is now considered one of the most advanced branches of astroparticle physics. Yet, it remains a discipline in its own right, addressing an impressive range of high-energy phenomena on all astronomical and cosmological scales.

The past of gamma-ray astronomy has been ambiguous and controversial. The term "astronomy" itself was generally used with reservations. But the great (to some extent, unexpected) discoveries in the 2000s in the high-energy (MeV/GeV) and very-high-energy (TeV) bands dramatically changed the status of the field. Thanks to the intensive sky exploration using satellite-borne and ground-based gamma-ray detectors, the recent impressive advances elevated the field to a nominal observational discipline. On the other hand, these high-quality observations stimulated comprehensive theoretical and numerical studies of non-thermal phenomena in their most extreme manifestations. While the current gamma-ray detection techniques are based on technological advances achieved in particle physics experiments, large-scale astronomical facilities and space research, the theoretical models broadly use modern computational tools with the involvement of magneto hydro dynamic (MHD) simulations, methods of nonlinear and plasma physics, dark-matter-related theoretical premises, *etc.*

Thousands of high-energy and hundreds of very-high-energy gamma-ray emitters have been discovered by satellite-borne, i.e. *Fermi*-LAT and *AGILE*, and ground-based stereoscopic systems of Imaging Atmospheric Cherenkov Telescopes (IACTs), i.e. HEGRA, H.E.S.S., MAGIC, VERITAS. These sources represent more than a dozen galactic and extragalactic source populations implying that powerful particle accelerators ("Cosmic Ray Factories") are distributed throughout the Universe in an impressively wide diversity of forms and scales, from compact relativistic objects in our Galaxy (neutron stars and stellar-mass black holes) to giant extragalactic structures like hundred-kpc radiogalaxy jets to cosmological formations like galaxy clusters.

These days, we are witnessing a new revolution in gamma-ray astronomy, this time in the PeV range. The highly anticipated first reports from the HAWC and Tibet experiments that state gamma-ray spectra from multiple gamma-ray sources could extend far beyond 10 TeV were confirmed by the recently completed LHAASO experiment. Moreover, it was found that the gamma-ray spectra of two famous objects of essentially different nature — the Crab Nebula and the Cygnus star-forming region — continue beyond 1 PeV. Most likely, if in the Crab Nebula the gamma rays are produced by ultrarelativistic electrons, the TeV–PeV emission of the second source has a hadronic origin. This shows the richness of gamma-ray astronomy concerning the diversity of the radiation processes and the astronomical source populations with far richer implications.

In the first case, we deal with the first detection of an *extreme particle accelerator* where the particles are boosted at an incredible rate and efficiency — close to the theoretical margin allowed by the classical electrodynamics and ideal MHD. The gamma-ray observations of the Crab Nebula allow insight into the physics of these unique cosmic laboratories, which, despite different astronomical scales and environmental conditions, could be realized in a variety of objects: Pulsar Wind Nebulae (PWNe), binary pulsars, gamma-ray bursts (GRBs), the relativistic jets emerging from supermassive black holes and the powerhouses of AGNs. Furthermore, the exploration of the physics of *extreme accelerators* based on high-quality gamma-ray data opens a path toward *indirect* finding of sources of highest energy particles, $E \sim 10^{20}$ eV — the great mystery of extragalactic cosmic rays.

The importance of the second discovery is linked to the origin of Galactic cosmic rays, namely, the PeVatrons. The current paradigm of Galactic cosmic rays assumes that SNe — or the result of these gigantic explosions, SuperNova Remnants (SNRs) — are the major contributors to the fluxes of protons and nuclei measured in the Earth's neighborhood up to the "knee" around 1 PeV. The discovery of GeV/TeV gamma-ray emission from a dozen young SNRs, as well as Giant Molecular Clouds (GMCs) located in the vicinity of middle-aged SNRs, is one of the remarkable achievements of gamma-ray astronomy, confirming the early predictions on the significant role of SNRs in the production of Galactic cosmic rays. On the other hand, the lack of reliable detection of gamma rays well beyond 10 TeV raised doubt in the cosmic-ray community regarding the ability of SNRs to accelerate protons to PeV energies. It is premature to draw a verdict, however, given the limited sensitivity of the current IACTs arrays in the key energy regime (for these measurements) of ultra-high-energy gamma rays, $E_\gamma \geq 100$ TeV. Meanwhile, the recent reports of giant gamma-ray structures around a few young stellar clusters (YSCs), particularly around the Cygnus OB2 cluster, the energy spectrum of which continues up to 1 PeV, demonstrate the potential of these objects to act as PeVatrons.

The discovery of gamma rays from YSCs and several other galactic source populations demands a revision or at least modification of the standard Galactic cosmic-ray paradigm by considering new-type objects, in addition to SNRs, as potential suppliers of Galactic cosmic rays. The supermassive black hole (Sgr A*) in the Galactic center, powerful galactic microquasars, e.g. Cygnus X-3 and SS 433, or even pulsars, are all sensible options because of their ability to accelerate protons to hundreds of PeV — the segment in the cosmic-ray spectrum that remains poorly explored.

Identifying the main contributors to locally measured cosmic-ray fluxes is one of the highest-priority tasks of gamma-ray astronomy, but the objectives of the field cannot be reduced only to this aspect of cosmic-ray studies. The term "cosmic rays" has a broader meaning, particularly in the context of their interpretation as the fourth substance ("relativistic plasma") of the observable Universe, after matter, magnetic field and radiation. The study of the processes of acceleration, transport and radiation of particles in these cosmic-ray factories, regardless of the fraction of the latter's contributions to the local cosmic-ray "fog", is highly important for studying fundamental physical laws at extreme conditions. Gamma rays offer unique probes of physical conditions in the vicinity of black holes close to the event horizon, the formation of relativistic outflows (pulsar winds and AGN jets), the hydrodynamics of catastrophic events like SN explosions and GRB phenomenon, and the acceleration of particles by the accretion shocks in the Galaxy clusters.

In terms of observational cosmology, gamma rays can be used for sensitive probes of the Extragalactic Background Light (EBL) from UV to far-infrared wavelengths, and intergalactic magnetic fields. Gamma rays as annihilation or decay products of dark-matter particles have often been invoked to explain various types of cosmic-ray and gamma-ray "anomalies". Although no traces of dark matter have been conclusively demonstrated so far in the gamma-ray data, these "exotic" but not unrealistic interpretations give a unique cosmological and particle-physics flavor to gamma-ray astronomy.

Advances in very-high-energy gamma-ray astronomy can be described as a success story. This success came as a surprise, especially given the controversial path to the current status of gamma-ray astronomy. The question arises, therefore, as to why this success was made possible. The answer to this question can be formulated as a fortunate combination of two factors:

(1) The Universe appeared to be full of perfectly "designed accelerators": electron and proton TeVatrons where almost ideal conditions are universally formed regardless of their scale and nature. The effective acceleration in these objects is accompanied by effective radiation processes, both inside the accelerator and in the surrounding environments, resulting in the formation of extended structures — gamma-ray "halos" around pulsars, SNRs and AGNs. If the accelerator operates inside or close to the compact object, variable radiation is expected, either periodic as in pulsars or stochastic as in AGNs. Therefore, temporal studies (timing) are crucial for the interpretation of gamma-ray data. For haloes, the highest priority is the spatial distribution of radiation (morphology). And, of course, spectrometric measurements are critical for both compact objects and extended sources.
(2) The implementation of the imaging atmospheric Cherenkov technique: The stereoscopic observations of EASs with a handful of Imaging Atmospheric Cherenkov Telescopes (IACTs) can provide an energy threshold down to 30–50 GeV, an effective rejection of hadronic showers,

and good angular (a few arcminutes) and energy (approximately 15%) resolutions. Around 1 TeV, the sensitivity approaches 10^{-12} erg/cm^{2}s; this is an impressive achievement even by the standards of advanced astronomical disciplines. The vast collection areas exceeding 0.1 km^2 provide adequate gamma-ray photon statistics, which, combined with good energy and angular resolutions, allows adequate morphological, spectral and temporal studies. This makes the arrays of IACTs perfect multi-functional and multipurpose astronomical tools for exploring a broad range of non-thermal sources and phenomena.

The arrays of IACTs are designed to observe point-like or moderately extended objects with known celestial coordinates. Although their high sensitivity and their 5° field of view allow for effective all-sky surveys, their potential is limited to the study of very-extended structures, as well as to the detection of solitary or transient gamma-ray phenomena, such as GRBs. In this regard, the detection technique based on the direct registration of particles that comprise the Extensive Air Showers (EASs) is a complementary approach to the imaging Cherenkov technique. More importantly, these particle array detectors allow the extension of spectral studies beyond 100 TeV. The recent discovery of a large number of cosmic PeVatrons by detectors of the giant LHAASO facility has demonstrated the great potential of the particle array detectors.

The future of the VHE gamma ray astronomy is very bright with the advent of the next generation of ground-based gamma-ray facilities of both kinds: two large arrays consisting of tens of IACTs operated as a unique astronomical observatory, the Cherenkov Telescope Array Observatory (CTAO), and, on the other hand, two particle detector array experiments, i.e., LHAASO located in Tibet, and the Southern Wide-Field Gamma-Ray Observatory (SWGO), meant to be developed in the Southern hemisphere but still in its design phase. The complementarity between these two kinds of instruments is certainly a key ingredient to guarantee the giant leap that is expected from the field. Whereas large arrays of IACTs are highly sensitive, pointing instruments, but with a limited duty cycle and a narrow FoV, the wide-FoV particle detector arrays with their 24/7 duty cycle are ideal instruments to study extended sources and identify emission regions at tens of TeV as well as transient phenomena. However, it is remarkable to note that the imaging Cherenkov technique has reached such a maturity with the last generation of instruments of this kind that it has successfully succeeded in moving from a discovery-machine proprietary experimental approach to a more astronomical way of operating. As its name suggests, CTAO is going to be the first proposal-driven observatory at VHEs, providing accessible data products and support services to a wide community.

CTAO is a European Reasearch Insfrastructure Consortium (ERIC) with its statutory seat in Bologna (Italy), which will build and operate the Northern array located at the Observatorio Roque de los Muchachos in the Canary island of La Palma (Spain), the Southern array at the Paranal Observatory in the Atacama desert (Chile), with the Science Data Management Center in Zeuthen (Germany) where data will be off-line processed and reduced for further dissemination to the scientific community. The arrays, hosting tens of IACT of three different configurations, will provide unprecedented sensitivity of transient phenomena and flares on timescales down to hours, minutes or even seconds. For steady emission, the sensitivity will be improved by a factor 5 to 10 depending on the energy, and spanning over more than four decades in energy (from 20 GeV till hundreds of TeV).

To conclude, with the next generation of ground-based gamma-ray facilities that are either in their design or construction phase, we can anticipate groundbreaking discoveries at very-high and ultra-high energies that could dramatically change the current concepts and paradigms of the most energetic and extreme phenomena in the non-thermal Universe.

https://doi.org/10.1142/9789813275720_bmatter

Glossary

ADAF Advection Dominated Accretion Flow.
ADC Analog-to-Digital Converter.
AGILE Astro-rivelatore Gamma a Immagini LEggero.
AGN Active Galactic Nucleus.
aLIGO Advanced LIGO.
ALP Axion-like Particle.
ANN Artificial Neural Network.
ASHRA aaa.
AU Astronomical Unit.

BAT Burst Alert Telescope.
BBH Binary Black Hole.
BDT Boosted Decision Trees.
BH Black Hole.
BLR Broad-Line Region.
BLRG Broad-Line Radio Galaxy.

CALET CALorimetric Electron Telescope.
CANGAROO Collaboration of Australia and Nippon (Japan) for a Gamma-Ray Observatory in the Outback.
CASA-MIA Chicago Air Shower Array and Michigan Muon Array.
CCD Charged-Coupled Device.
CDM Cold Dark Matter.
CGRO Compton Gamma-Ray Observatory.
CIB Cosmic Infrared Background.
CMB Cosmic Microwave Background.
CMZ Central Molecular Zone.
CNN Convolutional Neural Network.
COB Cosmic Optical Background.
CP Charge-Parity.

CR Cosmic Ray.
CSFH Cosmic Star-Formation History.
CTA Cherenkov Telescope Array.
CTAO Cherenkov Telescope Array Observatory.

DAMPE DArk Matter Particle Explorer.
DESI Dark Energy Spectroscopic Instrument.
DM Dark Matter.
DRS Domino Ring Sampler.
DSA Diffuse Shock Acceleration.
dSph Dwarf Spheroidal Galaxy.

EAS Extensive Air Shower.
EBL Extragalactic Background Light.
ECPL Exponential Cutoff Power Law.
EGRET Energetic Gamma-Ray Experiment Telescope.
ERIC European Reasearch Insfrastructure Consortium.
EVPA Electron Vector Polarization Angle.

FACT First G-APD Cherenkov Telescope.
FADC Flash Analog-to-Digital Converter.
Fermi Fermi Gamma-Ray Space Telescope.
Fermi-LAT Fermi-Large Area Telescope.
FIR Far-Infrared.
FoV Field of View.
FPGA Field Programmable Gate Array.
FR Fanaroff–Riley.
FRB Fast Radio Burst.
FSRQ Flat-Spectrum Radio Quasar.

G-APD Geiger-mode avalanche photodiode.
GBM Gamma-Ray Burst Monitor.
GC Galactic Center.
GMC Giant Molecular Cloud.
GPS Galactic Plane Survey.
GRAND Giant Radio Array for Neutrino Detection.
GRB Gamma-Ray Burst.
GRID Gamma-Ray Imaging Detector.
GUT Grand Unified Theory.
GW Gravitational Wave.

H.E.S.S. High-Energy Stereoscopic System.
HAWC High-Altitude Water Cherenkov.
HBL High-frequency BL Lac.
HE High-Energy.
HEGRA High-Energy Gamma-Ray.

IACT Imaging Atmospheric Cherenkov Telescope.
IBL Intermediate BL Lac.
IC Inverse Compton.
IGMF Inter-Galactic Magnetic Field.
IGRB Isotropic Gamma-Ray Background.
IMBH Intermediate Mass Black Holes.
IR InfraRed.
ISM InterStellar Medium.

KN Klein-Nishina.

LAT Large Area Telescope.
LBL Low Bl Lac.
LC Light Cylinder.
LHAASO The Large High-Altitude Air-Shower Observatory.
LIDAR Light Detection And Ranging.
LIV Lorentz Invariance Violation.
LMC Large Magellanic Cloud.
LSP Light SUSY Particle.
LST Large-Size Telescope.

MACE Major Atmospheric Cherenkov Experiment Telescope.
MAGIC Major Atmospheric Gamma-Ray Imaging Cherenkov.
MC Monte Carlo.
MHD Magneto Hydro Dynamic.
MIR Mid-Infrared.
MM Magnetic Monopole.
MSP MilliSecond Pulsar.

NFW Domanda a Michele.
NIR Near-Infrared.
NLR Narrow Line Region.
NLRG Narrow Line Radio Galaxy.

NS Neutron Star.
NSB Night Sky Background.

OSS Optical Support System.

PAO Pierre Auger Observatory.
PBH Primordial Black Hole.
PIC Particle-In-Cell.
PL Power Law.
PMT Photo-Multiplier Tube.
POEMMA Probe of Extreme Multi-Messenger Astrophysics.
PQ Peccei-Quinn.
PSF Point Spread Function.
PWN Pulsar Wind Nebula.
PWNe Pulsar Wind Nebulae.

QCD Quantum Chromodynamics.
QLT Quasi-Linear Theory.
QSO Quasi-Stellar Object.

RCP Resistive Plate Chambers.
RF Random Forest.
RHS Right-Hand Side.
RMS Root Mean Square.

SBH Supermassive Black Hole.
SCA Switched Capacitor Array.
SED Spectral Energy Distribution.
SFR Star-Forming Region.
SM Standard Model.
SN SuperNova.
SNe SuperNovae.
SNR SuperNova Remnant.
SSC Synchrotron Self Compton.
SST Small-Sized Telescope.
SUSY Supersymmetry.
SWGO Southern Wide-Field Gamma-Ray Observatory.

TACTIC TeV Atmospheric Cherenkov Telescope with Imaging Camera.
TDE Tidal Disruption Event.

UHE Ultra High Energy.
UHECR Ultra-High-Energy Cosmic Ray.
UV UltraViolet.

VERITAS Very-Energetic Radiation Imaging Telescope Array System.
VHE Very-High-Energy.
VLA Very Large Array.
VLBI Very-Long Baseline Interferometer.

WIMP Weakly Interacting Massive Particle.
WISP Weakly Interacting Slim Particle.
WR Wolf Rayet.

YSC Young Stellar Cluster.

https://doi.org/10.1142/9789813275720_bmatter

References

Aab, A. *et al.* (2015). Improved limit to the diffuse flux of ultrahigh energy neutrinos from the Pierre Auger Observatory, *Phys. Rev. D* **D91**, 9 p. 092008, doi:10.1103/PhysRevD.91.092008.

Aab, A. *et al.* (2016). Search for ultra-relativistic magnetic monopoles with the Pierre Auger Observatory, *Phys. Rev. D* **94**, 8, p. 082002, doi:10.1103/PhysRevD.94.082002.

Aab, A. *et al.* (2017). Observation of a large-scale anisotropy in the arrival directions of cosmic rays above $8{\times}10^{18}$ eV, *Science* **357**, pp. 1266–1270 doi:10.1126/science.aan4338.

Aartsen, M. G. *et al.* (2013). Observation of cosmic-ray anisotropy with the icetop air shower array, *ApJ* **765**, 1, p. 55.

Aartsen, M. G. *et al.* (2014a). IceCube-Gen2: a vision for the future of neutrino astronomy in Antarctica, *ArXiv e-prints*.

Aartsen, M. G. *et al.* (2014b). Observation of high-energy astrophysical neutrinos in three years of IceCube data, *Phys. Rev. Lett.* **113**, p. 101101, doi:10.1103/PhysRevLett.113.101101.

Aartsen, M. G. *et al.* (2014). Search for non-relativistic magnetic monopoles with IceCube, *Eur. Phys. J. C* **C74**, 7, p. 2938, doi:10.1140/epjc/s10052-014-2938-8.

Aartsen, M. G. *et al.* (2015). Searches for time-dependent neutrino sources with IceCube data from 2008 to 2012, *ApJ* **807**, p. 46, doi:10.1088/0004-637X/807/1/46.

Aartsen, M. G. *et al.* (2016). Observation and characterization of a cosmic muon neutrino flux from the Northern hemisphere using six years of IceCube data, *ApJ* **833**, p. 3, doi:10.3847/0004-637X/833/1/3.

Aartsen, M. G. *et al.* (2016a). Search for astrophysical tau neutrinos in three years of IceCube data, *Phys. Rev. D* **93**, 2, p. 022001, doi:10.1103/PhysRevD.93.022001.

Aartsen, M. G. *et al.* (2016b). Searches for relativistic magnetic monopoles in IceCube, *Eur. Phys. J. C* **C76**, 3, p. 133, doi:10.1140/epjc/s10052-016-3953-8.

Aartsen, M. G. *et al.* (2017a). All-sky search for time-integrated neutrino emission from astrophysical sources with 7 yr of IceCube data, *ApJ* **835**, p. 151, doi:10.3847/1538-4357/835/2/151.

Aartsen, M. G. *et al.* (2017b). Constraints on galactic neutrino emission with seven years of IceCube data, *ApJ* **849**, 67, doi:10.3847/1538-4357/aa8dfb.

Aartsen, M. G. *et al.* (2017c). Extending the search for muon neutrinos coincident with gamma-ray bursts in IceCube data, *ApJ* **843**, p. 112, doi:10.3847/1538-4357/aa7569.

Aartsen, M. G. *et al.* (2017d). Multiwavelength follow-up of a rare IceCube neutrino multiplet, *A&A* **607**, p. A115, doi:10.1051/0004-6361/201730620.

Aartsen, M. G. *et al.* (2017e). The contribution of Fermi-2LAC blazars to diffuse TeV-PeV neutrino flux, *ApJ* **835**, p. 45, doi:10.3847/1538-4357/835/1/45.

Aartsen, M. G. *et al.* (2017f). The IceCube neutrino observatory: instrumentation and online systems, *J. Instrum.* **12**, p. P03012, doi:10.1088/1748-0221/12/03/P03012.

Aartsen, M. G. *et al.* (2017). The IceCube realtime alert system, *Astrop. Phys.* **92**, pp. 30–41, doi:10.1016/j.astropartphys.2017.05.002.

Aartsen, M. G. *et al.* (2018). Multimessenger observations of a flaring blazar coincident with high-energy neutrino IceCube-170922A, *Science* **361**, eaat1378, doi:10.1126/science.aat1378.

Aartsen, M. G. *et al.* (2020a). Characteristics of the diffuse astrophysical electron and tau neutrino flux with six years of icecube high energy cascade data, *Phys. Rev. Lett.* **125**, p. 121104, doi:10.1103/PhysRevLett.125.121104, URL https://link.aps.org/doi/10.1103/PhysRevLett.125.121104.

Aartsen, M. G. *et al.* (2020b). Time-integrated neutrino source searches with 10 years of IceCube data, *Phys. Rev. Lett.* **124**, 5, p. 051103, doi:10.1103/PhysRevLett.124.051103.

Abbasi, R. *et al.* (2012). Observation of anisotropy in the galactic cosmic-ray arrival directions at 400 tev with IceCube, *ApJ* **746**, 1, p. 33.

Abbasi, R. *et al.* (2013). Evidence for high-energy extraterrestrial neutrinos at the IceCube detector, *Science* **342**, p. 1242856, 10.1126/science.1242856.

Abbasi, R. *et al.* (2013). Search for relativistic magnetic monopoles with IceCube, *Phys. Rev. D* **D87**, 2, p. 022001, doi:10.1103/PhysRevD.87.022001.

Abbasi, R. *et al.* (2021). IceCube high-energy starting event sample: Description and flux characterization with 7.5 years of data, *Phys. Rev. D* **104**, 2, 022002, doi:10.1103/PhysRevD.104.022002.

Abbott, B. P. *et al.* (2009). LIGO: the Laser Interferometer Gravitational-Wave Observatory, *Reports on Progress in Physics* **72**, 7, p. 076901, doi:10.1088/0034-4885/72/7/076901.

Abbott, B. P. *et al.* (2016a). Binary black hole mergers in the first Advanced Ligo observing run, *Physical Review X* **6**, p. 041015, doi:10.1103/PhysRevX.6.041015.

Abbott, B. P. *et al.* (2016b). GW151226: Observation of Gravitational Waves from a 22-Solar-Mass Binary Black Hole Coalescence, *Phys. Rev. Lett.* **116**, 24, p. 241103, doi:10.1103/PhysRevLett.116.241103.

Abbott, B. P. *et al.* (2016c). Observation of gravitational waves from a binary black hole merger, *Phys. Rev. Lett.* **116**, p. 061102, doi:10.1103/PhysRevLett.116.061102.

Abbott, B. P. *et al.* (2017a). Gravitational waves and gamma-ray from a binary neutron star merger: GW170817 and GRB 170817A, *ApJL* **848**, p. L13, doi:10.3847/2041-8213/aa920c.

Abbott, B. P. *et al.* (2017b). GW170104: observation of a 50-solar-mass binary black hole coalescence at redshift 0.2, *Phys. Rev. Lett.* **118**, 22, p. 221101, doi:10.1103/PhysRevLett.118.221101.

Abbott, B. P. *et al.* (2017c). Gw 170814: a three-detector observation of gravitational waves from a binary black hole coalescence, *Phys. Rev. Lett.* **119**, p. 141101, doi:10.1103/PhysRevLett.119.141101.

Abbott, B. P. *et al.* (2017d). GW 170817: observation of gravitational waves from a binary neutron star inspiral, *Phys. Rev. Lett.* **119**, 16, p. 161101, doi:10.1103/PhysRevLett.119.161101.

Abbott, T. M. C. *et al.* (2018). The Dark Energy Survey data release 1, *ApJS* **239**, 2, p. 18, doi:10.3847/1538-4365/aae9f0.

Abdalla, H., Adam, R., Aharonian, F., Ait Benkhali, F. and *et al.* (2019a). A very-high-energy component deep in the γ-ray burst afterglow, *Nature* **575**, 7783, pp. 464–467, doi:10.1038/s41586-019-1743-9.

Abdalla, H. *et al.* (2016a). H.E.S.S. limits on line-like dark matter signatures in the 100 GeV to 2 TeV energy range close to the Galactic Center, *Phys. Rev. Lett.* **117**, 15, p. 151302, doi:10.1103/PhysRevLett.117.151302.

Abdalla, H. *et al.* (2016b). Search for dark matter annihilations towards the inner Galactic halo from 10 years of observations with H.E.S.S., *Phys. Rev. Lett.* **117**, 11, p. 111301, doi:10.1103/PhysRevLett.117.111301.

Abdalla, H. *et al.* (2017a). Characterizing the γ-ray long-term variability of PKS 2155-304 with H.E.S.S. and Fermi-LAT, *A&A* **598**, p. A39, doi:10.1051/0004-6361/201629419.

Abdalla, H. *et al.* (2017b). Measurement of the EBL spectral energy distribution using the VHE γ-ray spectra of H.E.S.S. blazars, *A&A* **606**, p. A59, doi:10.1051/0004-6361/201731200.

Abdalla, H. *et al.* (2017c). TeV gamma-ray observations of the binary neutron star merger GW170817 with H.E.S.S., *ApJL* **850**, p. L22, doi:10.3847/2041-8213/aa97d2.

Abdalla, H. *et al.* (2018a). Characterising the VHE diffuse emission in the central 200 parsecs of our Galaxy with H.E.S.S., *A&A* **612**, p. A9, doi:10.1051/0004-6361/201730824.

Abdalla, H. *et al.* (2018b). Deeper H.E.S.S. observations of Vela Junior (RX J0852.0-4622): morphology studies and resolved spectroscopy, *A&A* **612**, p. A7, doi:10.1051/0004-6361/201630002.

Abdalla, H. *et al.* (2018c). Detection of variable VHE γ-ray emission from the extra-galactic γ-ray binary LMC P3, *A&A* **610**, p. L17, doi:10.1051/0004-6361/201732426.

Abdalla, H. *et al.* (2018d). First ground-based measurement of Sub-20 GeV to 100 GeV γ-rays from the Vela Pulsar with H.E.S.S. II, *A&A* .

Abdalla, H. *et al.* (2018e). H.E.S.S. observations of RX J1713.7-3946 with improved angular and spectral resolution: evidence for gamma-ray emission extending beyond the X-ray emitting shell, *A&A* **612**, p. A6, doi:10.1051/0004-6361/201629790.

Abdalla, H. *et al.* (2018f). Population study of Galactic supernova remnants at very highγ-ray energies with H.E.S.S., *A&A* **612**, p. A3, doi:10.1051/0004-6361/201732125.

Abdalla, H. *et al.* (2018g). Search for γ-ray line signals from dark matter annihilations in the inner Galactic halo from 10 years of observations with H.E.S.S., *Phys. Rev. Lett.* **120**, 20, p. 201101, doi:10.1103/PhysRevLett.120.201101.

Abdalla, H. *et al.* (2018h). Searches for gamma-ray lines and 'pure WIMP' spectra from dark matter annihilations in dwarf galaxies with H.E.S.S., *J. Cosmol. Astropart. Phys.* **11**, p. 037, doi:10.1088/1475-7516/2018/11/037.

Abdalla, H. *et al.* (2018i). The γ-ray spectrum of the core of Centaurus A as observed with H.E.S.S. and Fermi-LAT, *A&A* **619**, p. A71, doi:10.1051/0004-6361/201832640.

Abdalla, H. *et al.* (2018j). The H.E.S.S. Galactic plane survey, *A&A* **612**, p. A1, doi:10.1051/0004-6361/201732098.

Abdalla, H. *et al.* (2018k). The population of TeV pulsar wind nebulae in the H.E.S.S. Galactic Plane Survey, *A&A* **612**, p. A2, doi:10.1051/0004-6361/201629377.

Abdalla, H. *et al.* (2018l). The starburst galaxy NGC 253 revisited by H.E.S.S. and Fermi-LAT, *A&A* **617**, p. A73, doi:10.1051/0004-6361/201833202.

Abdalla, H. *et al.* (2019b). A very-high-energy component deep in the gamma-ray burst afterglow, *Nature* **575**, 7783, pp. 464–467, doi:10.1038/s41586-019-1743-9.

Abdalla, H. *et al.* (2020a). Detection of very-high-energy gamma-ray emission from the colliding wind binary η Car with H.E.S.S., *A&A* **635**, p. A167, doi:10.1051/0004-6361/201936761.

Abdalla, H. *et al.* (2020b). Probing the magnetic field in the GW170817 outflow using H.E.S.S. observations, *ApJL* **894**, 2, p. L16, doi:10.3847/2041-8213/ab8b59.

Abdalla, H. *et al.* (2020c). Resolving acceleration to very high energies along the jet of centaurus a, *Nature* **582**, p. 356–359, doi:10.1038/s41586-020-2354-1.

Abdalla, H. *et al.* (2020d). Search for dark matter signals towards a selection of recently detected DES dwarf galaxy satellites of the Milky Way with H.E.S.S., *Phys. Rev. D* **102**, 6, p. 062001, doi:10.1103/PhysRevD.102.062001.

Abdalla, H. *et al.* (2021a). Evidence of 100 TeV gamma-ray emission from HESS J1702-420: a new PeVatron candidate, *A&A* **653**, p. A152, doi:10.1051/0004-6361/202140962.

Abdalla, H. *et al.* (2021b). Revealing x-ray and gamma ray temporal and spectral similarities in the GRB 190829A afterglow, *Science* **372**, 6546, pp. 1081–1085, doi:10.1126/science.abe8560.

Abdalla, H. *et al.* (2021). Search for dark matter annihilation in the wolf-lundmark-melotte dwarf irregular galaxy with h.e.s.s. *Phys. Rev. D* **103**, p. 102002, doi:10.1103/PhysRevD.103.102002.

Abdalla, H. *et al.* (2021a). Search for dark matter annihilation signals from unidentified Fermi-LAT objects with H.E.S.S., *ApJ* **918**, 1, p. 17, doi:10.3847/1538-4357/abff59.

Abdalla, H. *et al.* (2021b). Sensitivity of the Cherenkov Telescope Array for probing cosmology and fundamental physics with gamma-ray propagation, *J. Cosmol. Astropart. Phys.* **2021**, 2, p. 048, doi:10.1088/1475-7516/2021/02/048.

Abdalla, H. *et al.* (2022). Time-resolved hadronic particle acceleration in the recurrent nova rs ophiuchi, **376**, pp. 70–80, doi:10.1126/science.abn0567.

Abdo, A. A. *et al.* (2007). TeV gamma-ray sources from a survey of the Galactic Plane with Milagro, *ApJL* **664**, pp. L91–L94, doi:10.1086/520717.

Abdo, A. A. *et al.* (2009a). A limit on the variation of the speed of light arising from quantum gravity effects, *Nature* **462**, pp. 331–334, doi:10.1038/nature08574.

Abdo, A. A. *et al.* (2009b). A population of gamma-ray millisecond pulsars seen with the Fermi Large Area Telescope, *Science* **325**, 5942, p. 848, doi:10.1126/science.1176113.

Abdo, A. A. *et al.* (2009c). The large-scale cosmic-ray anisotropy as observed with milagro, *ApJ* **698**, 2, p. 2121.

Abdo, A. A. *et al.* (2009d). Milagro Observations of Multi-TeV Emission from Galactic Sources in the Fermi Bright Source List, *ApJL* **700**, pp. L127–L131, doi:10.1088/0004-637X/700/2/L127.

Abdo, A. A. *et al.* (2010). Fermi Large Area Telescope first source catalog, *ApJS* **188**, pp. 405–436, doi:10.1088/0067-0049/188/2/405.

Abdo, A. A. *et al.* (2011a). Fermi Large Area Telescope observations of Markarian 421: the missing piece of its spectral energy distribution, *ApJ* **736**, 2, p. 131, doi:10.1088/0004-637X/736/2/131.

Abdo, A. A. *et al.* (2011b). Insights into the high-energy gamma-ray emission of Markarian 501 from extensive multifrequency observations in the Fermi era *ApJ* **727**, p. 129, doi:10.1088/0004-637X/727/2/129.

Abdo, A. A. *et al.* (2011c). Observations of the young supernova remnant RX J1713.7-3946 with the Fermi Large Area Telescope, *ApJ* **734**, p. 28, doi:10.1088/0004-637X/734/1/28.

Abdo, A. A. *et al.* (2013). The second *Fermi*-lat catalog of gamma-ray pulsars, *ApJ* **208**, 2, p. 17, doi:10.1088/0067-0049/208/2/17.

Abdo, A. A. *et al.* (2015). Milagro limits and HAWC sensitivity for the rate-density of evaporating Primordial Black Holes, *Astrop. Phys.* **64**, pp. 4–12, doi:10.1016/j.astropartphys.2014.10.007.

Abdollahi, S. *et al.* (2017). Cosmic-ray electron-positron spectrum from 7 GeV to 2 TeV with the Fermi Large Area Telescope, *Phys. Rev. D* **D95**, 8, p. 082007, doi:10.1103/PhysRevD.95.082007.

Abdollahi, S. *et al.* (2018). A gamma-ray determination of the Universe's star formation history, *Science* **362**, 6418, pp. 1031–1034, doi:10.1126/science.aat8123.

Abdollahi, S. *et al.* (2020). Fermi large area telescope fourth source catalog, *ApJS* **247**, 1, p. 33, doi:10.3847/1538-4365/ab6bcb, URL http://dx.doi.org/10.3847/1538-4365/ab6bcb.

Abeysekara, A. *et al.* (2019). Searching for dark matter sub-structure with HAWC, **2019**, 07, pp. 022–022, doi:10.1088/1475-7516/2019/07/022.

Abeysekara, A. U. *et al.* (2015). gamma rays from the quasar pks 1441+25: story of an escape, *ApJL* **815**, p. L22, doi:10.1088/2041-8205/815/2/L22.

Abeysekara, A. U. *et al.* (2017a). Extended gamma-ray sources around pulsars constrain the origin of the positron flux at Earth, *Science* **358**, pp. 911–914, doi:10.1126/science.aan4880.

Abeysekara, A. U. *et al.* (2017b). Observation of the Crab nebula with the HAWC gamma-ray observatory, *ApJ* **843**, p. 39, doi:10.3847/1538-4357/aa7555.

Abeysekara, A. U. *et al.* (2017c). The 2HWC HAWC observatory gamma-ray catalog, *ApJ* **843**, p. 40, doi:10.3847/1538-4357/aa7556.

Abeysekara, A. U. *et al.* (2018). A very-high energy gamma-ray survey toward the Cygnus region of the Galaxy, *ApJ* **861**, p. 134, doi:10.3847/1538-4357/aac4a2.

Abeysekara, A. U. *et al.* (2018). VERITAS observations of the BL Lac object TXS 0506+056, *ApJ* **861**, 2, p. L20, doi:10.3847/2041-8213/aad053.

Abeysekara, A. U. *et al.* (2018). Very-high-energy particle acceleration powered by the jets of the microquasar SS 433, *Nature* **562**, 7725, pp. 82–85, doi:10.1038/s41586-018-0565-5.

Abeysekara, A. U. *et al.* (2019). Measurement of the extragalactic background light spectral energy distribution with VERITAS, *ApJ* **885**, 2, p. 150, doi:10.3847/1538-4357/ab4817.

Abeysekara, A. U. *et al.* (2020). Multiple galactic sources with emission above 56 tev detected by HAWC, *Phys. Rev. Lett.* **124**, p. 021102, doi:10.1103/PhysRevLett.124.021102.

Abeysekara, A. U. *et al.* (2021). HAWC observations of the acceleration of very-high-energy cosmic rays in the Cygnus Cocoon, *Nature Astronomy* **5**, pp. 465–471, doi:10.1038/s41550-021-01318-y.

Abraham, J. *et al.* (2007). Correlation of the highest-energy cosmic rays with nearby extragalactic objects, *Science* **318**, p. 938, doi:10.1126/science.1151124.

Abramowski, A. *et al.* (2011). HESS J1943+213: a candidate extreme BL Lacertae object, *A&A* **529**, p. A49.

Abramowski, A. *et al.* (2011). H.E.S.S. observations of the globular clusters NGC 6388 and M 15 and search for a dark matter signal, *ApJ* **735**, p. 12, doi:10.1088/0004-637X/735/1/12.

Abramowski, A. *et al.* (2011a). Revisiting the Westerlund 2 field with the H.E.S.S. telescope array, *A&A* **525**, p. A46, doi:10.1051/0004-6361/201015290.

Abramowski, A. *et al.* (2011b). Search for a dark matter annihilation signal from the Galactic Center Halo with H.E.S.S., *Phys. Rev. Lett.* **106**, 16, pp. 161301–+, doi:10.1103/PhysRevLett.106.161301.

Abramowski, A. *et al.* (2011c). Search for Lorentz invariance breaking with a likelihood fit of the PKS 2155-304 flare data taken on MJD 53944, *Astrop. Phys.* **34**, pp. 738–747, doi:10.1016/j.astropartphys.2011.01.007.

Abramowski, A. *et al.* (2012a). Constraints on the gamma-ray emission from the cluster-scale AGN outburst in the Hydra A galaxy cluster, *A&A* **545**, A103, doi:10.1051/0004-6361/201219655.

Abramowski, A. *et al.* (2012b). Discovery of extended VHE γ-ray emission from the vicinity of the young massive stellar cluster Westerlund 1, *A&A* **537**, p. A114, doi:10.1051/0004-6361/201117928.

Abramowski, A. *et al.* (2012c). Identification of HESS J1303-631 as a pulsar wind nebula through γ-ray, X-ray, and radio observations, *A&A* **548**, p. A46, doi 10.1051/0004-6361/201219814.

Abramowski, A. *et al.* (2012d). H.E.S.S. observations of the Carina nebula and its enigmatic colliding wind binary Eta Carinae, *MNRAS* **424**, pp. 128–135, doi:10.1111/j.1365-2966.2012.21180.x.

Abramowski, A. *et al.* (2012). Search for dark matter annihilation signals from the Fornax galaxy Ccuster with H.E.S.S., *ApJ* **750**, p. 123, doi:10.1088/0004-637X/750/2/123.

Abramowski, A. *et al.* (2012). Spectral analysis and interpretation of the γ-ray emission from the starburst galaxy NGC 253, *ApJ* **757**, 158, doi:10 1088/0004-637X/757/2/158.

Abramowski, A. *et al.* (2013a). Constraints on axion-like particles with H.E.S.S. from the irregularity of the PKS 2155-304 energy spectrum, *Phys. Rev. D* **88**, 10, 102003, doi:10.1103/PhysRevD.88.102003.

Abramowski, A. *et al.* (2013b). Measurement of the extragalactic background light imprint on the spectra of the brightest blazars observed with H.E.S.S., *A&A* **550**, A4, doi:10.1051/0004-6361/201220355.

Abramowski, A. *et al.* (2013c). Search for photon line-like signatures from dark matter annihilations with H.E.S.S., *Phys. Rev. Lett.* **110**, p. 041301, doi:10.1103/PhysRevLett.110.041301.

Abramowski, A. *et al.* (2014a). Discovery of the hard spectrum VHE γ-ray source HESS J1641-463, *ApJL* **794**, p. L1, doi:10.1088/2041-8205/794/1/L1.

Abramowski, A. *et al.* (2014b). Long-term monitoring of PKS 2155-304 with ATOM and H.E.S.S.: investigation of optical/γ-ray correlations in different spectral states, *A&A* **571**, p. A39, doi:10.1051/0004-6361/201424142.

Abramowski, A. *et al.* (2014c). Search for dark matter annihilation signatures in H.E.S.S. observations of dwarf spheroidal Galaxies, *Phys. Rev. D* **D90**, p. 112012, doi:10.1103/PhysRevD.90.112012.

Abramowski, A. *et al.* (2014d). Search for extended γ-ray emission around AGN with H.E.S.S. and Fermi-LAT, *A&A* **562**, p. A145, doi:10.1051/0004-6361/201322510.

Abramowski, A. *et al.* (2015). Constraints on an annihilation signal from a core of constant dark matter density around the Milky Way center with H.E.S.S , *Phys. Rev. Lett.* **114**, 8, p. 081301, doi:10.1103/PhysRevLett.114.081301.

Abramowski, A. *et al.* (2015a). Discovery of variable VHE γ-ray emission from the binary system 1FGL J1018.6-5856, *A&A* **577**, p. A131, doi:10.1051/0004-6361/201525699.

Abramowski, A. *et al.* (2015b). The exceptionally powerful TeV γ-ray emitters in the Large Magellanic Cloud, *Science* **347**, pp. 406–412, doi:10.1126/science.1261313.

Abramowski, A. *et al.* (2016). Acceleration of petaelectronvolt protons in the Galactic Centre, *Nature* **531**, pp. 476–479, doi:10.1038/nature17147.

Abramowski, A. and others (2011). H.E.S.S. constraints on dark matter annihilations towards the Sculptor and Carina dwarf galaxies, *Astrop. Phys.* **34**, pp. 608–616, doi:10.1016/j.astropartphys.2010.12.006.

Acciari, V. A., Nava, L. *et al.* (2020). Bounds on Lorentz invariance violation from MAGIC observation of GRB 190114C, *Phys. Rev. Lett.* **125**, p. 021301.

Acciari, V. A. *et al.* (2008). Observation of gamma-ray emission from the galaxy M87 above 250 GeV with VERITAS, *ApJ* **679**, pp. 397–403, doi:10.1086/587458.

Acciari, V. A. *et al.* (2009a). A connection between star formation activity and cosmic rays in the starburst galaxy M82, *Nature* **462**, pp. 770–772, doi:10.1038/nature08557.

Acciari, V. A. *et al.* (2009b). Observation of extended very high-energy emission from the supernova remnant IC 443 with VERITAS, *ApJL* **698**, pp. L133–L137, doi:10.1088/0004-637X/698/2/L133.

Acciari, V. A. *et al.* (2009c). Radio imaging of the very high-energy gamma-ray emission region in the central engine of a radio galaxy, *Science* **325**, p. 444, doi:10.1126/science.1175406.

Acciari, V. A. *et al.* (2009d). VERITAS upper limit on the VHE emission from the radio galaxy NGC 1275, *ApJ* **706**, pp. L275–L280, doi:10.1088/0004-637X/706/2/L275.

Acciari, V. A. *et al.* (2010). VERITAS search for VHE gamma-ray emission from dwarf spheroidal galaxies, *ApJ* **720**, pp. 1174–1180, doi:10.1088/0004-637X/720/2/1174.

Acciari, V. A. *et al.* (2011a). Discovery of TeV gamma-ray emission from Tycho's supernova remnant, *ApJL* **730**, p. L20, doi:10.1088/2041-8205/730/2/L20.

Acciari, V. A. *et al.* (2011b). Spectral energy distribution of Markarian 501: quiescent state versus extreme outburst, *ApJ* **729**, p. 2, doi:10.1088/0004-637X/729/1/2.

Acciari, V. A. *et al.* (2014). Observation of Markarian 421 in TeV gamma rays over a 14-year time span, *Astrop. Phys.* **54**, pp. 1–10, doi:10.1016/j.astropartphys.2013.10.004.

Acciari, V. A. *et al.* (2018a). Constraining dark matter lifetime with a deep gamma-ray survey of the Perseus galaxy cluster with MAGIC, *Phys. Dark Univ.* **22**, pp. 38–47, doi:10.1016/j.dark.2018.08.002.

Acciari, V. A. *et al.* (2018b). Constraining very-high-energy and optical emission from FRB 121102 with the MAGIC telescopes, *MNRAS* **481**, 2, pp. 2479–2486, doi:10.1093/mnras/sty2422.

Acciari, V. A. *et al.* (2018). Detection of persistent VHE gamma-ray emission from pks 1510–089 by the MAGIC telescopes during low states between 2012 and 2017, *A&A* **619**, p. A159.

Acciari, V. A. *et al.* (2019). Constraints on gamma-ray and neutrino emission from NGC 1068 with the MAGIC telescopes, *ApJ* **883**, 2, 135, doi:10.3847/1538-4357/ab3a51.

Acciari, V. A. *et al.* (2019). Deep observations of the globular cluster M15 with the MAGIC telescopes, *MNRAS* **484**, 2, pp. 2876–2885, doi:10.1093/mnras/stz179.

Acciari, V. A. *et al.* (2019). Measurement of the extragalactic background light using MAGIC and *Fermi*-LAT gamma-ray observations of blazars up to z = 1, *MNRAS* **486**, 3, pp. 4233–4251, doi:10.1093/mnras/stz943.

Acciari, V. A. *et al.* (2019). Observation of inverse Compton emission from a long gamma-ray burst, *Nature* **575**, 7783, pp. 459–463, doi:10.1038/s41586-019-1754-6.

Acciari, V. A. *et al.* (2019a). Teraelectronvolt emission from the γ-ray burst GRB 190114C, *Nature* **575**, 7783, pp. 455–458, doi:10.1038/s41586-019-1750-x.

Acciari, V. A. *et al.* (2019b). Teraelectronvolt emission from the gamma-ray burst GRB 190114C, *Nature* **575**, 7783, pp. 455–458, doi:10.1038/s41586-019-1750-x.

Acciari, V. A. *et al.* (2020). A search for dark matter in Triangulum II with the MAGIC telescopes, *Phys. Dark Univ.* **28**, p. 100529, doi:10.1016/j.dark.2020.100529.

Acciari, V. A. *et al.* (2020a). MAGIC observations of the diffuse gamma-ray emission in the vicinity of the Galactic center, *A&A* **642**, p. A190, doi:10.1051/0004-6361/201936896.

Acciari, V. A. *et al.* (2020b). Unraveling the complex behavior of mrk 421 with simultaneous x-ray and vhe observations during an extreme flaring activity in 2013 april, *ApJS* **248**, 2, p. 29, doi:10.3847/1538-4365/ab89b5.

Acciari, V. A. *et al.* (2022). Combined searches for dark matter in dwarf spheroidal galaxies observed with the MAGICtelescopes, including new data from Coma Berenices and Draco, *Phys. Dark Univ.* **35**, p. 100912, doi:10.1016/j.dark.2021.100912.

Acciari, V. A. *et al.* (2022). Proton acceleration in thermonuclear nova explosions revealed by gamma rays, *Nature Astronomy* **6**, pp. 689–697, doi:10.1038/s41550-022-01640-z.

Acernese, F. *et al.* (2015). Advanced Virgo: a second-generation interferometric gravitational wave detector, *Classical and Quantum Gravity* **32**, 2, p. 024001, doi:10.1088/0264-9381/32/2/024001.

Acero, F. *et al.* (2009a). Detection of gamma rays from a starburst galaxy, *Science* **326**, p. 1080, doi:10.1126/science.1178826.

Acero, F. *et al.* (2009b). H.E.S.S. upper limits on very high energy gamma-ray emission from the microquasar GRS 1915+105, *A&A* **508**, 3, pp. 1135–1140, doi:10.1051/0004-6361/200913389.

Acero, F. *et al.* (2010a). First detection of VHE γ-rays from SN 1006 by H.E.S.S., *A&A* **516**, p. A62, doi:10.1051/0004-6361/200913916.

Acero, F. *et al.* (2010b). Localizing the VHE γ-ray source at the Galactic Centre, *MNRAS* **402**, 3, pp. 1877–1882, doi:10.1111/j.1365-2966.2009.16014.x.

Acero, F. *et al.* (2015). Fermi Large Area Telescope third source catalog, *ApJS* **218**, p. 23, doi:10.1088/0067-0049/218/2/23.

Acero, F. *et al.* (2016). Development of the model of Galactic Interstellar emission for standard point-source analysis of Fermi Large Area Telescope data, *ApJS* **223**, 2, p. 26, doi:10.3847/0067-0049/223/2/26.

Acharya, A. *et al.* (2021). Sensitivity of the Cherenkov Telescope Array to a dark matter signal from the Galactic centre, *JCAP* **01**, p. 057, doi:10.1088/1475-7516/2021/01/057.

Acharya, B. S., Actis, M. and CTA Consortium (2013). Introducing the CTA concept, *Astrop. Phys.* **43**, pp. 3–18, doi:10.1016/j.astropartphys.2013.01.007.

Acharya, B. S. *et al.* (2019). *Science with the Cherenkov Telescope Array*, doi: 10.1142/10986.

Achterberg, A. *et al.* (2010). Search for relativistic magnetic monopoles with the amanda-ii neutrino telescope, *Eur. Phys. J. C* **69**, pp. 361–378.

Ackermann, M. *et al.* (2007). Neutrino triggered target of opportunity (NToO) test run with AMANDA and MAGIC, in *Proceedings of the 30th International Cosmic Ray Conference — PoS(ICRC2007), Merida (Mexico)*, Universidad Nacional Autonoma de Mexico City, pp. 1257–1260.

Ackermann, M. *et al.* (2011). A cocoon of freshly accelerated cosmic rays detected by fermi in the cygnus superbubble, *Science* **334**, 6059, pp. 1103–1107, doi: 10.1126/science.1210311.

Ackermann, M. *et al.* (2012). Constraints on the Galactic halo dark matter from *Fermi*-LAT diffuse measurements, *ApJ* **761**, 2, p. 91, doi:10.1088/0004-637x/761/2/91.

Ackermann, M. *et al.* (2012a). Fermi Large Area Telescope study of cosmic rays and the interstellar medium in nearby molecular clouds, *ApJ* **755**, 1, p. 22, doi:10.1088/0004-637X/755/1/22.

Ackermann, M. *et al.* (2012b). GeV observations of star-forming Galaxies with the Fermi Large Area Telescope, *ApJ* **755**, p. 164, doi:10.1088/0004-637X/755/2/164.

Ackermann, M. *et al.* (2012c). In-flight measurement of the absolute energy scale of the Fermi Large Area Telescope, *Astrop. Phys.* **35**, pp. 346–353, doi:10.1016/j.astropartphys.2011.10.007.

Ackermann, M. *et al.* (2012d). Measurement of separate cosmic-ray electron and positron spectra with the Fermi Large Area Telescope, *Phys. Rev. Lett.* **108**, p. 011103, doi:10.1103/PhysRevLett.108.011103.

Ackermann, M. *et al.* (2012e). *Fermi*-LAT observations of the diffuse gamma-ray emission: implications for cosmic rays and the interstellar medium, *ApJ* **750**, p. 3, doi:10.1088/0004-637X/750/1/3.

Ackermann, M. *et al.* (2012f). gamma-ray observations of the Orion molecular clouds with the Fermi Large Area Telescope, *ApJ* **756**, 1, p. 4, doi:10.1088/0004-637X/756/1/4.

Ackermann, M. *et al.* (2012g). Search for dark matter satellites using *Fermi*-LAT, *ApJ* **747**, p. 121, doi:10.1088/0004-637X/747/2/121.

Ackermann, M. *et al.* (2012h). The imprint of the extragalactic background light in the gamma-ray spectra of blazars, *Science* **338**, p. 1190, doi: 10.1126/science.1227160.

Ackermann, M. *et al.* (2013). The first *Fermi*-LAT gamma-ray burst catalog, *ApJS* **209**, 11, doi:10.1088/0067-0049/209/1/11.

Ackermann, M. *et al.* (2014a). Fermi establishes classical novae as a distinct class of gamma-ray sources, *Science* **345**, pp. 554–558, doi:10.1126/science.1253947.

Ackermann, M. *et al.* (2014b). Search for cosmic-ray-induced gamma-ray emission in Galaxy Clusters, *ApJ* **787**, p. 18, doi:10.1088/0004-637X/787/1/18.

Ackermann, M. *et al.* (2015a). The spectrum of isotropic diffuse gamma-ray emission between 100 MeV and 820 GeV, *ApJ* **799**, p. 86, doi:10.1088/0004-637X/799/1/86.

Ackermann, M. *et al.* (2015b). The third catalog of active Galactic nuclei detected by the Fermi Large Area Telescope, *ApJ* **810**, p. 14, doi:10.1088/0004-637X/810/1/14.

Ackermann, M. *et al.* (2016a). 2FHL: the second catalog of hard *Fermi*-LAT sources, *ApJS* **222**, 1, p. 5, doi:10.3847/0067-0049/222/1/5.

Ackermann, M. *et al.* (2016b). Minute-timescales >100 mev γ-ray variability during the giant outburst of quasar 3c 279 obeserved by fermi-lat in 2015 june, *ApJL* **824**, 2, p. L20, doi:10.3847/2041-8205/824/2/l20, URL http://dx.doi.org/10.3847/2041-8205/824/2/L20.

Ackermann, M. *et al.* (2016a). Resolving the extragalactic γ-Ray background above 50 GeV with the Fermi Large Area Telescope, *Phys. Rev. Lett.* **116**, p. 15, 151105, doi:10.1103/PhysRevLett.116.151105.

Ackermann, M. *et al.* (2016b). Search for gamma-ray emission from the Coma Cluster with six years of *Fermi*-LAT data, *ApJ* **819**, p. 149, doi:10.3847/0004-637X/819/2/149.

Ackermann, M. *et al.* (2017). The Fermi Galactic Center GeV excess and implications for dark matter, *ApJ* **840**, 1, p. 43, doi:10.3847/1538-4357/aa6cab.

Ackermann, M. *et al.* (2018). The Search for spatial extension in high-latitude sources detected by the Fermi Large Area Telescope, *ApJS* **237**, 2, p. 32, doi:10.3847/1538-4365/aacdf7.

Ackermann, M. *et al.* (2019). Fundamental physics with high-energy cosmic neutrinos, *Bull. Am. Astron. Soc.* **51**, p. 215.

Actis, M. *et al.* (2011). Design concepts for the Cherenkov Telescope Array CTA: an advanced facility for ground-based high-energy gamma-ray astronomy, *Experimental Astronomy* **32**, pp. 193–316, doi:10.1007/s10686-011-9247-0.

Ade, P. A. R. *et al.* (2015). POLARBEAR constraints on cosmic birefringence and primordial magnetic fields, *Phys. Rev. D* **92**, p. 123509, doi:10.1103/PhysRevD.92.123509.

Ade, P. A. R. *et al.* (2016). Planck 2015 results. XIX. Constraints on primordial magnetic fields, *A&A* **594**, p. A19, doi:10.1051/0004-6361/201525821.

Adrian-Martinez, S. *et al.* (2011). First search for point sources of high energy cosmic neutrinos with the ANTARES neutrino telescope, *ApJL* **743**.

Adrián-Martínez, S. *et al.* (2016). Letter of intent for KM3net 2.0, *J. Phys. G* **43**, 8, p. 084001, doi:10.1088/0954-3899/43/8/084001, URL https://doi.org/10.1088/0954-3899/43/8/084001.

Adrián-Martínez, S. *et al.* (2016). Letter of intent for KM3Net 2.0, *J. Phys. G* **43**, 8, p. 084001, doi:10.1088/0954-3899/43/8/084001.

Adriani, O. *et al.* (2009). An anomalous positron abundance in cosmic rays with energies 1.5-100 GeV, *Nature* **458**, pp. 607–609, doi:10.1038/nature07942.

Adriani, O. *et al.* (2011). Pamela measurements of cosmic-ray proton and helium spectra, *Science* **332**, 6025, pp. 69–72, doi:10.1126/science.1199172.

Adriani, O. *et al.* (2018). Extended measurement of the cosmic-ray electron and positron spectrum from 11 gev to 4.8 tev with the calorimetric electron telescope on the international space station, *Phys. Rev. Lett.* **120**, p. 261102, doi:10.1103/PhysRevLett.120.261102.

Adrián-Martínez, S. *et al.* (2016). Optical and X-ray early follow-up of ANTARES neutrino alerts, *J. Cosmol. Astropart. Phys.* **1602**, 02, p. 062, doi:10.1088/1475-7516/2016/02/062.

Ageron, M. *et al.* (2011). ANTARES: The first undersea neutrino telescope, *Nucl. Instrum. Meth. Phys. Res. A* **656**, pp. 11–38, doi:10.1016/j.nima.2011.06.103.

Aghamousa, A. *et al.* (2016). The DESI experiment part I: science, targeting, and survey design, arXiv e-prints, arXiv:1611.00036.

Aghanim, N. *et al.* (2020). Planck 2018 results. vi. cosmological parameters. *A&A* **641**, p. A6, doi:10.1051/0004-6361/201833910, URL http://dx.doi.org/10.1051/0004-6361/201833910.

Aglietta, M. *et al.* (2009). Evolution of the cosmic-ray anisotropy above 1014 ev, *ApJL* **692**, 2, p. L130.

Aguilar, M. *et al.* (2013). First result from the alpha magnetic spectrometer on the international space station: precision measurement of the positron fraction in primary cosmic rays of 0.5–350 gev, *Phys. Rev. Lett.* **110**, p. 141102, doi:10.1103/PhysRevLett.110.141102.

Aguilar, M. *et al.* (2014). Electron and positron fluxes in primary cosmic rays measured with the Alpha Magnetic Spectrometer on the International Space Station, *Phys. Rev. Lett.* **113**, p. 121102, doi:10.1103/PhysRevLett.113.121102.

Aguilar, M. *et al.* (2015a). Precision measurement of the helium flux in primary cosmic rays of rigidities 1.9 gv to 3 tv with the alpha magnetic spectrometer on the international space station, *Phys. Rev. Lett.* **115**, p. 211101, doi:10.1103/PhysRevLett.115.211101.

Aguilar, M. *et al.* (2015b). Precision measurement of the proton flux in primary cosmic rays from rigidity 1 gv to 1.8 tv with the alpha magnetic spectrometer on the international space station, *Phys. Rev. Lett.* **114**, p. 171103, doi:10.1103/PhysRevLett.114.171103.

Aharonian, F. *et al.* (2003). Detection of TeV gamma-rays from the BL Lac 1ES 1959+650 in its low states and during a major outburst in 2002, *A&A* **406**, pp. L9–L13, doi:10.1051/0004-6361:20030838.

Aharonian, F. *et al.* (2008). Discovery of very high energy gamma-ray emission coincident with molecular clouds in the W 28 (G6.4-0.1) field, *A&A* **481**, pp. 401–410, doi:10.1051/0004-6361:20077765.

Aharonian, F. *et al.* (2008). Search for gamma-rays from dark matter annihilations around intermediate mass black holes with the H.E.S.S. experiment, *Phys. Rev. D* **D78**, p. 072008, doi:10.1103/PhysRevD.78.072008.

Aharonian, F. *et al.* (2009a). Constraints on the multi-TeV particle population in the Coma galaxy cluster with H.E.S.S. observations, *A&A* **502**, pp. 437–443, doi:10.1051/0004-6361/200912086.

Aharonian, F. *et al.* (2009b). Discovery of very high energy γ-Ray emission from Centaurus a with H.E.S.S., *ApJL* **695**, pp. L40–L44, doi:10.1088/0004-637X/695/1/L40.

Aharonian, F. *et al.* (2009c). Probing the ATIC peak in the cosmic-ray electron spectrum with H.E.S.S., *A&A* **508**, p. 561, doi:10.1051/0004-6361/200913323.

Aharonian, F. *et al.* (2009d). Very high energy gamma-ray observations of the galaxy clusters Abell 496 and Abell 85 with H.E.S.S., *A&A* **495**, pp. 27–35, doi:10.1051/0004-6361:200811372.

Aharonian, F. A. (2001). gamma rays from molecular clouds, *Space Sci. Rev.* **99**, pp. 187–196, doi:10.1023/A:1013845015364.

Aharonian, F. A. (2004). *Very high energy cosmic gamma radiation : a crucial window on the extreme Universe*, World Scientific Publishing Co, doi 10.1142/4657.

Aharonian, F. A. and Atoyan, A. M. (1996). On the emissivity of π^0-decay gamma radiation in the vicinity of accelerators of galactic cosmic rays. *A&A* **309**, pp. 917–928.

Aharonian, F. A., Atoyan, A. M. and Volk, H. J. (1995). High energy electrons and positrons in cosmic rays as an indicator of the existence of a nearby cosmic tevatron, *A&A* **294**, pp. L41–L44.

Aharonian, F. A., Khangulyan, D. and Costamante, L. (2008a). Formation of hard very high energy gamma-ray spectra of blazars due to internal photon-photon absorption, *MNRAS* **387**, pp. 1206–1214, doi:10.1111/j.1365-2966.2008.13315.x.

Aharonian, F. A., Yang, R. and Wilhelmi, E. d. O. (2019). Massive stars as major factories of Galactic cosmic rays, *Nature Astronomy* **3**, pp. 561–567.

Aharonian, F. A. *et al.* (1997). The potential of ground based arrays of imaging atmospheric Cherenkov telescopes. I. Determination of shower parameters, *Astrop. Phys.* **6**, pp. 343–368, doi:10.1016/S0927-6505(96)00069-2.

Aharonian, F. A. *et al.* (2000). The energy spectrum of TeV gamma rays from the Crab Nebula as measured by the HEGRA system of imaging air Cherenkov telescopes, *ApJ* **539**, pp. 317–324, doi:10.1086/309225.

Aharonian, F. A. *et al.* (2001). Evidence for TeV gamma-ray emission from Cassiopeia A, *A&A* **370**, pp. 112–120, doi:10.1051/0004-6361:20010243.

Aharonian, F. A. *et al.* (2002). An unidentified TeV source in the vicinity of Cygnus OB2, *A&A* **393**, pp. L37–L40, doi:10.1051/0004-6361:20021171.

Aharonian, F. A. *et al.* (2004a). High-energy particle acceleration in the shell of a supernova remnant, *Nature* **432**, pp. 75–77, doi:10.1038/nature02960.

Aharonian, F. A. *et al.* (2004b). The Crab nebula and pulsar between 500 GeV and 80 TeV: observations with the HEGRA stereoscopic air Cherenkov telescopes, *ApJ* **614**, pp. 897–913, doi:10.1086/423931.

Aharonian, F. A. *et al.* (2004c). Very high energy gamma rays from the direction of Sagittarius A*, *A&A* **425**, pp. L13–L17, doi:10.1051/0004-6361:200400055.

Aharonian, F. A. *et al.* (2005a). Discovery of the binary pulsar PSR B1259-63 in very-high-energy gamma rays around periastron with H.E.S.S., *A&A* **442**, pp. 1–10, doi:10.1051/0004-6361:20052983.

Aharonian, F. A. *et al.* (2005b). Discovery of very high energy gamma rays associated with an X-ray binary, *Science* **309**, pp. 746–749, doi:10.1126/science.1113764.

Aharonian, F. A. *et al.* (2006a). A low level of extragalactic background light as revealed by γ-rays from blazars, *Nature* **440**, pp. 1018–1021, doi:10.1038/nature04680.

Aharonian, F. A. *et al.* (2006b). Energy dependent γ-ray morphology in the pulsar wind nebula HESS J1825-137, *A&A* **460**, pp. 365–374, doi:10.1051/0004-6361:20065546.

Aharonian, F. A. *et al.* (2006c). Fast variability of tera-electronvolt γ rays from the radio galaxy M87, *Science* **314**, pp. 1424–1427, doi:10.1126/science.1134408.

Aharonian, F. A. *et al.* (2006d). Observations of the Crab nebula with H.E.S.S., *A&A* **457**, pp. 899–915.

Aharonian, F. A. *et al.* (2006e). H.E.S.S. observations of the Galactic Center region and their possible dark matter interpretation, *Phys. Rev. Lett.* **97**, p. 221102, doi:10.1103/PhysRevLett.97.221102.

Aharonian, F. A. *et al.* (2007). An exceptional very high energy gamma-ray flare of PKS 2155-304, *ApJL* **664**, pp. L71–L74, doi:10.1086/520635.

Aharonian, F. A. *et al.* (2008b). Observations of the Sagittarius dwarf galaxy by the H.E.S.S. experiment and search for a dark matter signal, *Astrop. Phys.* **29**, pp. 55–62, doi:10.1016/j.astropartphys.2007.11.007.

Aharonian, F. A. *et al.* (2008c). The energy spectrum of cosmic-ray electrons at TeV energies, *Phys. Rev. Lett.* **101**, p. 261104, doi:10.1103/PhysRevLett.101.261104.

Aharonian, F. A. *et al.* (2009e). A search for a dark matter annihilation signal toward the Canis Major overdensity with H.E.S.S., *ApJ* **691**, pp. 175–181, doi:10.1088/0004-637X/691/1/175.

Aharonian, F. A. *et al.* (2009f). Simultaneous multiwavelength observations of the second exceptional γ-ray flare of PKS 2155-304 in July 2006, *A&A* **502**, pp. 749–770, doi:10.1051/0004-6361/200912128.

Aharonian, F. A. *et al.* (2013). TeV gamma rays from blazars beyond z=1? *Phys. Rev. D* **87**, 6, p. 063002, doi:10.1103/PhysRevD.87.063002.

Aharonian, F. A. *et al.* (2020). Probing the sea of galactic cosmic rays with *Fermi*-LAT, *Phys. Rev. D* **101**, 8, 083018, doi:10.1103/PhysRevD.101.083018.

Aharonian, F. A. and others (2009). Spectrum and variability of the Galactic Center VHE gamma-ray source HESS J1745-290, *A&A* **503**, p. 817, doi:10.1051/0004-6361/200811569.

Aharonov, Y. and Bohm, D. (1959). Significance of electromagnetic potentials in the quantum theory, *Phys. Rev.* **115**, pp. 485–491, doi:10.1103/PhysRev.115.485.

Ahlers, M. and Halzen, F. (2018). Opening a new window onto the Universe with IceCube, ArXiv e-prints.

Ahnen, M. L. *et al.* (2016a). Deep observation of the NGC 1275 region with MAGIC: search of diffuse gamma-ray emission from cosmic rays in the Perseus cluster, *A&A* **589**, p. A33, doi:10.1051/0004-6361/201527846.

Ahnen, M. L. *et al.* (2016b). Detection of very high energy gamma-ray emission from the gravitationally lensed blazar QSO B0218+357 with the MAGIC telescopes, *A&A* **595**, p. A98, doi:10.1051/0004-6361/201629461.

Ahnen, M. L. *et al.* (2016c). Limits to dark matter annihilation cross-section from a combined analysis of MAGIC and *Fermi*-LAT observations of dwarf satellite galaxies, *J. Cosmol. Astropart. Phys.* **1602**, 02, p. 039, doi:10.1088/1475-7516/2016/02/039.

Ahnen, M. L. *et al.* (2016d). Super-orbital variability of ls i +61°303 at tev energies, *A&A* **591**, p. A76, doi:10.1051/0004-6361/201527964.

Ahnen, M. L. *et al.* (2017a). A cut-off in the TeV gamma-ray spectrum of the SNR Cassiopeia A, *MNRAS* **472**, pp. 2956–2962, doi:10.1093/mnras/stx2079.

Ahnen, M. L. *et al.* (2017b). Constraining Lorentz invariance violation using the Crab pulsar emission observed up to TeV energies by MAGIC, *ApJS* **232**, p. 9, doi:10.3847/1538-4365/aa8404.

Ahnen, M. L. *et al.* (2017c). Multiwavelength observations of a VHE gamma-ray flare from PKS 1510-089 in 2015, *A&A* **603**, p. A29, doi:10.1051/0004-6361/201629960.

Ahnen, M. L. *et al.* (2017d). Performance of the MAGIC telescopes under moonlight, *Astrop. Phys.* **94**, pp. 29–41, doi:10.1016/j.astropartphys.2017.08.001.

Ahnen, M. L. *et al.* (2018a). Indirect dark matter searches in the dwarf satellite galaxy Ursa Major II with the MAGIC telescopes, *J. Cosmol. Astropart. Phys.* **1803**, 03, p. 009, doi:10.1088/1475-7516/2018/03/009.

Ahnen, M. L. *et al.* (2018b). Limits on the flux of tau neutrinos from 1 eV to 3 EeV with the MAGIC telescopes, *Astrop. Phys.* **102**, pp. 77–88, doi:10.1016/j.astropartphys.2018.05.002.

Ajello, M. *et al.* (2017). 3FHL: The third catalog of hard *Fermi*-LAT sources, *ApJS* **232**, 2, p. 18, doi:10.3847/1538-4365/aa8221.

Ajello, M. *et al.* (2020). The Fourth Catalog of Active Galactic Nuclei Detected by the Fermi Large Area Telescope, *ApJ* **892**, p. 2, 105, doi:10.3847/1538-4357/ab791e.

Ajello, M. o. (2016). Search for spectral irregularities due to photon-axion-like-Particle oscillations with the Fermi Large Area Telescope, *Phys. Rev. Lett.* **116**, 16, p. 161101, doi:10.1103/PhysRevLett.116.161101.

Akutsu, T. *et al.* (2018). Construction of KAGRA: an underground gravitational-wave observatory, *Progress of theoretical and experimental physics* **2018**, 1, p. 013F01, doi:10.1093/ptep/ptx180.

Albert, A., *et al.* (2020a). Constraining the local burst rate density of primordial black holes with HAWC, *J. Cosmol. Astropart. Phys.* **2020**, 4, p. 026, doi: 10.1088/1475-7516/2020/04/026.

Albert, A. *et al.* (2017). Dark Matter limits from dwarf spheroidal galaxies with the HAWC gamma-ray observatory, *ApJ* **853**, 2, p. 154, doi:10.3847/1538-4357/aaa6d8.

Albert, A. *et al.* (2019). Science Case for a Wide Field-of-View Very-High-Energy Gamma-Ray Observatory in the Southern Hemisphere, arXiv e-prints, arXiv:1902.08429.

Albert, A. *et al.* (2020b). 3HWC: The Third HAWC Catalog of Very-high-energy Gamma-Ray Sources, *ApJ* **905**, 1, p. 76, doi:10.3847/1538-4357/abc2d8.

Albert, A. *et al.* (2020a). 3HWC: The third hawc catalog of very-high-energy gamma-ray sources, *The Astrophys. J.* **905**, 1, p. 76, doi:10.3847/1538-4357/abc2d8, URL https://dx.doi.org/10.3847/1538-4357/abc2d8.

Albert, A. *et al.* (2020b). Constraints on Lorentz invariance violation from HAWC observations of gamma rays above 100 TeV, *Phys. Rev. Lett.* **124**, p. 131101, doi:10.1103/PhysRevLett.124.131101.

Albert, A. *et al.* (2020). HAWC J2227+610 and its association with G106.3+2.7, a new potential Galactic PeVatron, *ApJL* **896**, 2, p. L29, doi:10.3847/2041-8213/ab96cc.

Albert, A. *et al.* (2021). Evidence of 200 TeV photons from HAWC J1825-134, *ApJL* **907**, 2, p. L30, doi:10.3847/2041-8213/abd77b.

Albert, A. *et al.* (2021). Probing the sea of cosmic rays by measuring gamma-ray emission from passive giant molecular clouds with HAWC, *ApJ* **914**, 2, p. 106, doi:10.3847/1538-4357/abfc47.

Albert, J. *et al.* (2006). Observation of gamma rays from the Galactic center with the MAGIC telescope, *ApJ* **638**, pp. L101–L104, doi:10.1086/501164.

Albert, J. *et al.* (2007a). Discovery of very-high energy gamma radiation from IC 443 with the MAGIC telescope, *ApJL* **664**, pp. L87–L90, doi:10.1086/520957.

Albert, J. *et al.* (2007b). Observation of VHE gamma-ray from Cassiopeia A with the MAGIC telescope, *A&A* **474**, pp. 937–940, doi:10.1051/0004-6361:20078168.

Albert, J. *et al.* (2007c). Unfolding of differential energy spectra in the MAGIC experiment, *Nucl. Instrum. Meth. Phys. Res. A* **583**, pp. 494–506, doi:10.1016/j.nima.2007.09.048.

Albert, J. *et al.* (2007d). Variable very-high energy gamma-ray emission from Markarian 501, *ApJ* **669**, 2, pp. 862–883, doi:10.1086/521382.

Albert, J. *et al.* (2008a). Implementation of the random forest method for the imaging atmospheric Cherenkov telescope MAGIC, *Nucl. Instrum. Meth. Phys. Res. A* **A588**, pp. 424–432, doi:10.1016/j.nima.2007.11.068.

Albert, J. *et al.* (2008b). MAGIC observations of the unidentified gamma-ray source TeV J2032+4130, *ApJL* **675**, p. L25, doi:10.1086/529520.

Albert, J. *et al.* (2008c). Probing quantum gravity using photons from a flare of the active galactic nucleus Markarian 501 observed by the MAGIC telescope, *Phys. Let. B* **668**, pp. 253–257, doi:10.1016/j.physletb.2008.08.053.

Albert, J. *et al.* (2008d). Upper limit for gamma-ray emission above 140 GeV from the dwarf spheroidal galaxy Draco, *ApJ* **679**, pp. 428–431, doi:10.1086/529135.

Albert, J. *et al.* (2008e). Very-high energy gamma-ray observations of strong flaring activity in M87 in 2008 February, *ApJL* **685**, p. L23, doi:10.1086/592348.

Albert, J. *et al.* (2008f). Very-high energy gamma rays from a distant quasar: how transparent is the Universe? *Science* **320**, p. 1752, doi:10.1126/science.1157087.

Aleksić, J., others and Ghisellini, G. (2012). MAGIC observations of the giant radio galaxy M87 in a low-emission state between 2005 and 2007, *A&A* **544**, A96, doi:10.1051/0004-6361/201117827.

Aleksic, J., Rico, J. and Martinez, M. (2012). Optimized analysis method for indirect dark matter searches with Imaging Air Cherenkov Telescopes, *J. Cosmol. Astropart. Phys.* **1210**, p. 032, doi:10.1088/1475-7516/2012/10/032.

Aleksić, J. *et al.* (2010a). MAGIC constraints on gamma-ray emission from Cygnus X-3, *ApJ* **721**, 1, pp. 843–855, doi:10.1088/0004-637X/721/1/843.

Aleksić, J. *et al.* (2010b). Detection of very high-energy gamma-ray emission from the Perseus cluster head-tail galaxy IC 310 by the MAGIC telescopes, *ApJL* **723**, pp. L207–L212, doi:10.1088/2041-8205/723/2/L207.

Aleksić, J. *et al.* (2010c). Search for an extended VHE gamma-ray emission from Mrk 421 and Mrk 501 with the MAGIC telescope, *A&A* **524**, p. A77, doi:10.1051/0004-6361/201014747.

Aleksić, J. *et al.* (2010d). MAGIC gamma-ray telescope observation of the Perseus cluster of galaxies: implications for cosmic rays, dark matter and NGC 1275, *ApJ* **710**, pp. 634–647, doi:10.1088/0004-637X/710/1/634.

Aleksić, J. *et al.* (2011a). A search for very high-energy gamma-ray emission from Scorpius X-1 with the MAGIC telescopes, *ApJL* **735**, 1, p. L5, doi:10.1088/2041-8205/735/1/L5.

Aleksić, J. *et al.* (2011b). MAGIC discovery of very high energy emission from the FSRQ PKS 1222+21, *ApJL* **730**, L8, doi:10.1088/2041-8205/730/1/L8.

Aleksić, J. *et al.* (2011c). Observations of the Crab pulsar between 25 and 100 GeV with the MAGIC I telescope, *ApJ* **742**, p. 43, doi:10.1088/0004-637X/742/1/43.

Aleksić, J. *et al.* (2011d). Searches for dark matter annihilation signatures in the Segue 1 satellite galaxy with the MAGIC-I telescope, *J. Cosmol. Astropart. Phys.* **1106**, p. 035, doi:10.1088/1475-7516/2011/06/035.

Aleksić, J. *et al.* (2012a). Constraining cosmic rays and magnetic fields in the Perseus galaxy cluster with TeV observations by the MAGIC telescopes, *A&A* **541**, p. A99, doi:10.1051/0004-6361/201118502.

Aleksić, J. *et al.* (2012b). Detection of very high-energy gamma-ray emission from NGC 1275 by the MAGIC telescopes, *A&A* **539**, L2, doi:10.1051/0004-6361/201118668.

Aleksić, J. *et al.* (2012c). Discovery of VHE gamma-ray from the blazar 1ES 1215+303 with the MAGIC telescopes and simultaneous multi-wavelength observations, *A&A* **544**, A142, doi:10.1051/0004-6361/201219133.

Aleksić, J. *et al.* (2012d). Performance of the MAGIC stereo system obtained with Crab nebula data, *Astrop. Phys.* **35**, pp. 435–448, doi:10.1016/j.astropartphys.2011.11.007.

Aleksić, J. *et al.* (2013). Very high energy gamma-ray observation of the peculiar transient event Swift J1644+57 with the MAGIC telescopes and *Agile*, *A&A* **552**, A112, doi:10.1051/0004-6361/201321197.

Aleksić, J. *et al.* (2014a). Black hole lightning due to particle acceleration at subhorizon scales, *Science* **346**, pp. 1080–1084, doi:10.1126/science.1256183.

Aleksić, J. *et al.* (2014b). Optimized dark matter searches in deep observations of Segue 1 with MAGIC, *J. Cosmol. Astropart. Phys.* **1402**, p. 008, doi:10.1088/1475-7516/2014/02/008.

Aleksić, J. *et al.* (2015a). Measurement of the Crab nebula spectrum over three decades in energy with the MAGIC telescopes, *J. High-Energy Astroph.* **5**, pp. 30–38, doi:10.1016/j.jheap.2015.01.002.

Aleksić, J. *et al.* (2015b). Very high-energy gamma-ray from the Universe's middle age: detection of the z = 0.940 blazar PKS 1441+25 with MAGIC, *ApJL* **815**, p. L23, doi:10.1088/2041-8205/815/2/L23.

Aleksić, J. *et al.* (2016a). The major upgrade of the MAGIC telescopes, part I: the hardware improvements and the commissioning of the system, *Astrop. Phys.* **72**, pp. 61–75, doi:10.1016/j.astropartphys.2015.04.004.

Aleksić, J. *et al.* (2016b). The major upgrade of the MAGIC telescopes, part II: a performance study using observations of the Crab nebula, *Astrop. Phys.* **72**, pp. 76–94, doi:10.1016/j.astropartphys.2015.02.005.

Aliu, E. *et al.* (2008). Observation of pulsed gamma rays above 25 gev from the crab pulsar with MAGIC, *Science* **322**, p. 1221, doi:10.1126/science.1164718.

Aliu, E. *et al.* (2009a). Improving the performance of the single-dish Cherenkov telescope MAGIC through the use of signal timing, *Astrop. Phys.* **30**, pp. 293–305, doi:10.1016/j.astropartphys.2008.10.003.

Aliu, E. *et al.* (2009b). MAGIC upper limits on the VHE gamma-ray emission from the satellite galaxy Willman 1, *ApJ* **697**, pp. 1299–1304, doi:10.1088/0004-637X/697/2/1299.

Aliu, E. *et al.* (2011a). Detection of pulsed gamma rays above 100 GeV from the Crab Pulsar, *Science* **334**, p. 69, doi:10.1126/science.1208192.

Aliu, E. *et al.* (2011b). VERITAS observations of the unusual extragalactic transient Swift J164449.3+573451, *ApJL* **738**, 2, p. L30, doi:10.1088/2041-8205/738/2/L30.

Aliu, E. *et al.* (2012). VERITAS deep observations of the dwarf spheroidal galaxy Segue 1, *Phys. Rev. D* **D85**, p. 062001, doi:10.1103/PhysRevD.85.062001.

Aliu, E. *et al.* (2014a). A Search for enhanced very-high energy gamma-ray emission from the 2013 March Crab nebula flare, *ApJL* **781**, p. L11, doi:10.1088/2041-8205/781/1/L11.

Aliu, E. *et al.* (2014b). Observations of the unidentified gamma-ray source tev j2032+4130 by VERITAS, *ApJ* **783**, p. 16, doi:10.1088/0004-637X/783/1/16.

Aliu, E. *et al.* (2015). A search for pulsations from Geminga above 100 GeV with VERITAS, *ApJ* **800**, p. 61, doi:10.1088/0004-637X/800/1/61.

Aloisio, R., Berezinsky, V. and Gazizov, A. (2012). Transition from galactic to extragalactic cosmic rays, *Astrop. Phys.* **39**, pp. 129–143, doi:10.1016/j.astropartphys.2012.09.007.

Aloisio, R., Blasi, P. and Serpico, P. D. (2015). Nonlinear cosmic ray Galactic transport in the light of AMS-02 and Voyager data, *A&A* **583**, p. A95, doi:10.1051/0004-6361/201526877.

Aloisio, R. *et al.* (2018). *Selected topics in cosmic ray physics*, doi:10.1007/978-3-319-65425-6_1.

Alvarez-Muñiz, J. *et al.* (2018). Comprehensive approach to tau-lepton production by high-energy tau neutrinos propagating through the Earth, *Phys. Rev. D* **97**, 2, p. 023021, doi:10.1103/PhysRevD.97.023021, [Erratum: Phys.Rev.D 99, 069902 (2019)].

Alvarez-Muñiz, J. *et al.* (2018). The giant radio array for neutrino detection (GRAND): science and design, .

Alves Batista, R. and Saveliev, A. (2021). The Gamma-Ray Window to Intergalactic Magnetism, *Universe* **7**, 7, p. 223, doi:10.3390/universe7070223.

Alves Batista, R. *et al.* (2015). Effects of uncertainties in simulations of extragalactic UHECR propagation, using CRPropa and SimProp, *J. Cosmol. Astropart. Phys.* **10**, 063, doi:10.1088/1475-7516/2015/10/063.

Alves Batista, R. *et al.* (2016). CRPropa 3-a public astrophysical simulation framework for propagating extraterrestrial ultra-high energy particles, *J. Cosmol. Astropart. Phys.* **2016**, 05, p. 038.

Amato, E. and Arons, J. (2006). Heating and nonthermal particle acceleration in relativistic, transverse magnetosonic shock waves in proton-electron-positron plasmas, *ApJ* **653**, 1, pp. 325–338, doi:10.1086/508050.

Ambrogi, L., Celli, S. and Aharonian, F. A. (2018). On the potential of Cherenkov Telescope Arrays and KM3 Neutrino Telescopes for the detection of extended sources, *Astrop. Phys.* **100**, pp. 69–79, doi:10.1016/j.astropartphys.2018.03.001.

Ambrosi, G. *et al.* (2017). Direct detection of a break in the teraelectronvolt cosmic-ray spectrum of electrons and positrons, *Nature* **552**, pp. 63–66, doi:10.1038/nature24475.

Ambrosio, M. *et al.* (2002). Final results of magnetic monopole searches with the MACRO experiment, *Eur. Phys. J. C* **C25**, pp. 511–522, doi:10.1140/epjc/s2002-01046-9.

Amelino-Camelia, G. (2013). Quantum-Spacetime Phenomenology, *Living Rev. Relativ.* **16**, p. 5, doi:10.12942/lrr-2013-5.

Amelino-Camelia, G. and Smolin, L. (2009). Prospects for constraining quantum gravity dispersion with near term observations, *Phys. Rev. D* **80**, p. 084017, doi:10.1103/PhysRevD.80.084017.

Amenomori, M. *et al.* (2015). Search for Gamma Rays above 100 TeV from the Crab Nebula with the Tibet Air Shower Array and the 100 m^2 muon Detector, *ApJ* **813**, 2, p. 98, doi:10.1088/0004-637X/813/2/98.

Amenomori, M. *et al.* (2017). Northern sky galactic cosmic ray anisotropy between 10 and 1000 tev with the tibet air shower array, *ApJ* **836**, 2, p. 153.

Amenomori, M. *et al.* (2021). First detection of sub-PeV Diffuse gamma rays from the Galactic disk: evidence for ubiquitous Galactic cosmic rays beyond PeV energies, *Phys. Rev. Lett.* **126**, 14, p. 141101, doi:10.1103/PhysRevLett.126. 141101.

Amiri, M. *et al.* (2019). A second source of repeating fast radio bursts, *Nature* **566**, pp. 235–238, doi:10.1038/s41586-018-0864-x.

Anastassopoulos, V. *et al.* (2017). New CAST limit on the axion-photon interaction, *Nature Physics* **13**, pp. 584–590, doi:10.1038/nphys4109.

Anderhub, H. *et al.* (2013). Design and operation of FACT - the first G-APD Cherenkov telescope, *Journal of Instrumentation* **8**, P06008, doi:10.1088/1748-0221/8/06/P06008.

Ando, S. *et al.* (2020). Structure formation models weaken limits on WIMP dark matter from dwarf spheroidal galaxies, *Phys. Rev. D* **102**, 6, p. 061302, doi:10.1103/PhysRevD.102.061302.

Andrews, S. K. *et al.* (2018). Modelling the cosmic spectral energy distribution and extragalactic background light over all time, *MNRAS* **474**, pp. 898–916, doi:10.1093/mnras/stx2843.

Andriamonje, S. *et al.* (2007). An improved limit on the axion photon coupling from the CAST experiment, *J. Cosmol. Astropart. Phys.* **4**, 010, doi:10.1088/1475-7516/2007/04/010.

Ansoldi, S. *et al.* (2016). Teraelectronvolt pulsed emission from the Crab Pulsar detected by MAGIC, *A&A* **585**, A133, doi:10.1051/0004-6361/201526853.

Antoni, T. *et al.* (2004). Large-scale cosmic-ray anisotropy with kascade, *ApJ* **604**, 2, p. 687.

Antoni, T. *et al.* (2005). Kascade measurements of energy spectra for elemental groups of cosmic rays: Results and open problems, *Astrop. Phys.* **24**, 1-2, pp. 1–25, doi:10.1016/j.astropartphys.2005.04.001, cited By 389.

Apel, W. *et al.* (2013). *KASCADE*-grande measurements of energy spectra for elemental groups of cosmic rays, *Astrop. Phys.* **47**, pp. 54–66, doi:https://doi.org/10.1016/j.astropartphys.2013.06.004.

Apel, W. D. *et al.* (2009). Energy spectra of elemental groups of cosmic rays: Update on the KASCADE unfolding analysis, *Astrop. Phys.* **31**, pp. 86–91, doi:10.1016/j.astropartphys.2008.11.008.

Archambault, S. *et al.* (2016). Exceptionally bright TeV flares from the binary LS I +61 303, *ApJL* **817**, L7, doi:10.3847/2041-8205/817/1/L7.

Archambault, S. *et al.* (2017). Dark matter constraints from a joint analysis of dwarf spheroidal galaxy observations with VERITAS, *Phys. Rev. D* **95**, 8, p. 082001, doi:10.1103/PhysRevD.95.082001.

Archambault, S. *et al.* (2017a). gamma-ray observations of Tycho supernova remnant with VERITAS and *Fermi*-LAT, *ApJ* **836**, p. 23, doi:10.3847/1538-4357/836/1/23.

Archambault, S. *et al.* (2017b). Search for magnetically broadened cascade emission from blazars with VERITAS, *ApJ* **835**, p. 288, doi:10.3847/1538-4357/835/2/288.

Archer, A. *et al.* (2014). Very-high energy observations of the galactic center region by VERITAS in 2010-2012, *ApJ* **790**, 2, p. 149, doi:10.1088/0004-637X/790/2/149.

Archer, A. *et al.* (2015). Search for pulsed emission in archival VERITAS data, ArXiv e-prints, arXiv:1509.01201.

Archer, A. *et al.* (2016). TeV gamma-ray observations of the Galactic center ridge by VERITAS, *ApJ* **821**, p. 129, doi:10.3847/0004-637X/821/2/129.

Archer, A. *et al.* (2018). HESS J1943+213: an extreme blazar shining through the galactic plane, *ApJ* **862**, p. 41, doi:10.3847/1538-4357/aacbd0.

Archer, A. *et al.* (2018). Measurement of cosmic-ray electrons at TeV energies by VERITAS, *Phys. Rev. D* **D98**, 6, p. 062004, doi:10.1103/PhysRevD.98.062004.

Archidiacono, M. *et al.* (2015). Future cosmological sensitivity for hot dark matter axions, *J. Cosmol. Astropart. Phys.* **5**, p. 050, doi:10.1088/1475-7516/2015/05/050.

Ardi, E. and Baumgardt, H. (2020). Depletion of dark matter within globular clusters, *J. Phys. Conf. Ser.* **1503**, 1, p. 012023, doi:10.1088/1742-6596/1503/1/012023.

Arias, P. *et al.* (2012). WISPy cold dark matter, *J. Cosmol. Astropart. Phys.* **2012**, 06, pp. 013–013, doi:10.1088/1475-7516/2012/06/013.

Arina, C. *et al.* (2010). Intense gamma-ray lines from hidden vector dark matter decay, *J. Cosmol. Astropart. Phys.* **03**, p. 024, doi:10.1088/1475-7516/2010/03/024.

Arlen, T. A. *et al.* (2012). Constraints on cosmic rays, magnetic fields, and dark matter from gamma-ray observations of the Coma cluster of galaxies with VERITAS and *Fermi*-LAT, *ApJ* **757**, p. 123, doi:10.1088/0004-637X/757/2/123.

Arlen, T. C. *et al.* (2014). Intergalactic Magnetic Fields and Gamma-Ray Observations of Extreme TeV Blazars, *ApJ* **796**, 18, doi:10.1088/0004-637X/796/1/18.

Armand, C. *et al.* (2021). Combined dark matter searches towards dwarf spheroidal galaxies with *Fermi*-LAT, HAWC, H.E.S.S., MAGIC, and

VERITAS, in *Proceedings of the 37th International Cosmic Ray Conference — PoS(ICRC2021), Berlin (Germany)*, Vol. 358, p. 12, doi:10.22323/1.358.0012.

Armstrong, J. W., Rickett, B. J. and Spangler, S. R. (1995). Electron density power spectrum in the local interstellar medium, *ApJ* **443**, pp. 209–221, doi:10.1086/175515.

Arons, J. and Scharlemann, E. T. (1979). Pair formation above pulsar polar caps: structure of the low altitude acceleration zone. *ApJ* **231**, pp. 854–879, doi:10.1086/157250.

Arvanitaki, A. *et al.* (2010). String axiverse, *Phys. Rev. D* **81**, 12, p. 123530, doi:10.1103/PhysRevD.81.123530.

Asano, K. *et al.* (2014). Time-dependent models for blazar emission with the second-order Fermi acceleration, *ApJ* **780**, 1, p. 64, doi:10.1088/0004-637X/780/1/64.

Asaoka, Y. and Sasaki, M. (2013). Cherenkov τ shower earth-skimming method for pev–eev ν_τ observation with ashra, *Astrop. Phys.* **41**, pp. 7–16, doi:http://dx.doi.org/10.1016/j.astropartphys.2012.10.001.

Ashkar, H. *et al.* (2021). The H.E.S.S. gravitational wave rapid follow-up program, *J. Cosmol. Astropart. Phys.* **2021**, 3, 045, doi:10.1088/1475-7516/2021/03/045.

Atoyan, A., Buckley, J. and Krawczynski, H. (2006). A gamma-ray burst remnant in Our Galaxy: HESS J1303-631, *ApJL* **642**, pp. L153–L156, doi:10.1086/504712.

Atoyan, A. and Dermer, C. D. (2001). High-energy neutrinos from photomeson processes in blazars, *Phys. Rev. Lett.* **87**, p. 221102, doi:10.1103/PhysRevLett.87.221102.

Atoyan, A. M., Aharonian, F. A. and Völk, H. J. (1995). Electrons and positrons in the galactic cosmic rays, *Phys. Rev. D* **52**, pp. 3265–3275, doi:10.1103/PhysRevD.52.3265.

Axford, W. I. (1981). The acceleration of cosmic rays by shock waves, *Annals of the New York Academy of Sciences* **375**, pp. 297–313, doi:10.1111/j.1749-6632.1981.tb33702.x.

Aynutdinov, V. *et al.* (2005). The Baikal neutrino experiment: from NT200 to NT200+, *Int. Cosmic Ray Conf.* **5**, p. 75.

Baade, W. and Zwicky, F. (1934). Remarks on Super-Novae and Cosmic Rays, *Physical Review* **46**, pp. 76–77, doi:10.1103/PhysRev.46.76.2.

Babcock, H. W. (1939). The rotation of the Andromeda Nebula, *Lick observatory bulletin* **498**, pp. 41–51, doi:10.5479/ADS/bib/1939LicOB.19.41B.

Baghmanyan, V. *et al.* (2020). Evidence of cosmic-ray excess from local giant molecular clouds, *ApJL* **901**, 1, p. L4, doi:10.3847/2041-8213/abb5f8.

Bahcall, J. and Waxman, E. (2001). High-energy astrophysical neutrinos: the upper bound is robust, *Phys. Rev. D* **64**, 2, p. 023002, doi:10.1103/PhysRevD.64.023002.

Bahcall, J. N., Bahcall, N. A. and Shaviv, G. (1968). Present status of the theoretical predictions for the ^{37}Cl solar-neutrino experiment, *Phys. Rev. Lett.* **20**, pp. 1209–1212, doi:10.1103/PhysRevLett.20.1209.

Bähre, R. *et al.* (2013). Any light particle search II —Technical Design Report, *JINST* **8**, p. T09001, doi:10.1088/1748-0221/8/09/T09001.

Bai, X. *et al.* (2019). The Large High Altitude Air Shower Observatory (LHAASO) Science White Paper, .

Bailes, M. *et al.* (2021). Gravitational-wave physics and astronomy in the 2020s and 2030s, *Nat. Rev. Phys.* **3**, 5, pp. 344–366, doi:10.1038/s42254-021-00303-8, URL https://doi.org/10.1038/s42254-021-00303-8.

Baillon, P. *et al.* (1994). Detection of MultiTeV gamma rays in the THEMISTOCLE experiment: new results, in *International Symposium on Cosmic Ray Physics in Tibet*, Tibet University, Lhasa, pp. 121–134.

Baixeras, C. *et al.* (2004). Commissioning and first tests of the MAGIC telescope, *Nucl. Instrum. Meth. Phys. Res. A* **518**, pp. 188–192, doi:10.1016/j.nima.2003.10.057.

Band, D. *et al.* (1993). BATSE Observations of Gamma-Ray Burst Spectra. I. Spectral Diversity, *ApJ* **413**, p. 281, doi:10.1086/172995.

Barack, L. *et al.* (2019). Black holes, gravitational waves and fundamental physics: a roadmap, *Classi. Quant. Grav.* **36**, 14, p. 143001, doi:10.1088/1361-6382/ab0587, URL http://dx.doi.org/10.1088/1361-6382/ab0587.

Barbiellini, G. *et al.* (2001). The next generation of high-energy gamma-ray detectors for satellites: the AGILE silicon tracker, in S. Ritz, N. Gehrels and C. R. Shrader. eds., *Gamma 2001: Gamma-ray Astrophysics*, *American Institute of Phys. Conference Series*, Vol. 587, pp. 754–758, doi:10.1063/1.1419494.

Baret, B. and Van Elewyck, V. (2011). High-energy neutrino astronomy: Detection methods and first achievements, *Rept. Prog. Phys.* **74**, p. 046902, doi:10.1088/0034-4885/74/4/046902.

Barkov, M. V. *et al.* (2012). Rapid TeV variability in blazars as a result of jet-star interaction, *ApJ* **749**, 2, p. 119, doi:10.1088/0004-637X/749/2/119.

Barrau, A. (1997). Cherenkov imaging camera for the gamma-ray astrophysics experiment CAT, *Nucl. Instrum. Meth. Phys. Res. A* **387**, 1, pp. 69–73, doi:10.1016/S0168-9002(96)00963-1.

Barrau, A., Gorecki, A. and Grain, J. (2008). An original constraint on the Hubble constant: $h > 0.74$, *MNRAS* **389**, 2, pp. 919–924, doi:10.1111/j.1365-2966.2008.13607.x.

Barrau, A. *et al.* (1998). The CAT imaging telescope for very-high-energy gamma-ray astronomy. *Nuclear Instruments and Methods in Physics Research A* **416** 2, pp. 278–292, doi:10.1016/S0168-9002(98)00749-9.

Barres de Almeida, U. (2021). The southern wide-field gamma-ray observatory, *Astron. Nachr.* **342**, 1-2, pp. 431–437, doi:10.1002/asna.202113946.

Barrio, J. A. *et al.* (1998). The MAGIC telescope Design study for the construction of a 17 m Cerenkov telescope for gamma-astronomy above 10 GeV, Letter of intent.

Bartels, R. and Edwards, T. (2019). Comment on "Understanding the γ-ray emission from the globular cluster 47 Tuc: evidence for dark matter?", *Phys. Rev. D* **100**, p. 6, doi:10.1103/physrevd.100.068301, URL http://dx.doi.org/10.1103/PhysRevD.100.068301.

Bartels, R., Krishnamurthy, S. and Weniger, C. (2016). Strong support for the millisecond pulsar origin of the Galactic center GeV excess, *Phys. Rev. Lett.* **116**, 5, p. 051102, doi:10.1103/PhysRevLett.116.051102.

Bartoli, B. *et al.* (2012). Observation of TeV gamma rays from the Cygnus Rrgion with the ARGO-YB experiment, *ApJL* **745**, L22, doi:10.1088/2041-8205/745/2/L22.

Bartoli, B. *et al.* (2014). Identification of the tev gamma-ray source argo j2031+4157 with the cygnus cocoon, *ApJ* **790**, 2, p. 152, doi:10.1088/0004-637x/790/2/152.

Bartoli, B. *et al.* (2015). Knee of the cosmic hydrogen and helium spectrum below 1 PeV measured by ARGO-YBJ and a Cherenkov telescope of LHAASO, *Phys. Rev. D* **92**, 9, p. 092005, doi:10.1103/PhysRevD.92.092005.

Bartos, I. (2017). Multimessenger prospects with gravitational waves and neutrinos after ligo's first discovery, *J. Phys. Conf. Ser.* **888**, 1, p. 012001.

Batković, I. *et al.* (2021). Axion-Like particle searches with IACTs, *Universe* **7**, p. 185, doi:10.3390/universe7060185.

Baumgardt, H. (2017). N-body modelling of globular clusters: masses, mass-to-light ratios and intermediate-mass black holes, *MNRAS* **464**, 2, pp. 2174–2202, doi:10.1093/mnras/stw2488.

Becherini, Y. *et al.* (2011). A new analysis strategy for detection of faint γ-ray sources with Imaging Atmospheric Cherenkov Telescopes, *Astrop. Phys.* **34**, pp. 858–870, doi:10.1016/j.astropartphys.2011.03.005.

Bechtol, K. *et al.* (2017). Evidence against star-forming galaxies as the dominant source of IceCube neutrinos, *ApJ* **836**, p. 47, doi:10.3847/1538-4357/836/1/47.

Becker, J. K. (2008). High-energy neutrinos in the context of multimessenger physics, *Phys. Rept.* **458**, pp. 173–246, doi:10.1016/j.physrep.2007.10.006.

Beckmann, V. and Shrader, C. R. (2012). *Active Galactic Nuclei.*

Bednarek, W. (2012). On the origin of sub-TeV gamma-ray pulsed emission from rotating neutron stars, *MNRAS* **424**, 3, pp. 2079–2085, doi:10.1111/j.1365-2966.2012.21354.x.

Bednarz, J. and Ostrowski, M. (1998). Energy spectra of cosmic rays accelerated at ultrarelativistic shock waves, *Phys. Rev. Lett.* **80**, 18, pp. 3911–3914, doi:10.1103/PhysRevLett.80.3911.

Begelman, M. C., Fabian, A. C. and Rees, M. J. (2008). Implications of very rapid TeV variability in blazars, *MNRAS* **384**, 1, pp. L19–L23, doi:10.1111/j.1745-3933.2007.00413.x.

Beilicke, M. *et al.* (2012). The Galactic Center Region Imaged by VERITAS from 2010-2012, *AIP Conf. Proc.* **1505**, pp. 462–465, doi:10.1063/1.4772297.

Belikov, A. V., Buckley, M. R. and Hooper, D. (2012). Searching for dark matter subhalos in the Fermi-LAT second source catalog, *Phys. Rev. D* **86**, 4, p. 043504, doi:10.1103/PhysRevD.86.043504.

Belikov, A. V., Goodenough, L. and Hooper, D. (2011). No indications of axionlike particles from Fermi, *Phys. Rev. D* **83**, 6, p. 063005, doi:10.1103/PhysRevD.83.063005.

Bell, A. R. (1978a). The acceleration of cosmic rays in shock fronts. I, *MNRAS* **182**, pp. 147–156, doi:10.1093/mnras/182.2.147.

Bell, A. R. (1978b). The acceleration of cosmic rays in shock fronts. II, *MNRAS* **182**, pp. 443–455, doi:10.1093/mnras/182.3.443.

Beneke, M. *et al.* (2018). Energetic γ-rays from TeV scale dark matter annihilation resummed, *Phys. Let. B* **786**, pp. 347–354, doi:10.1016/j.physletb.2018.10.008, [Erratum: *Phys.Lett.B 810*, 135831 (2020)].

Benito, M., Cuoco, A. and Iocco, F. (2019). Handling the uncertainties in the galactic dark matter distribution for particle dark matter searches, *J. Cosmol. Astropart. Phys.* **03**, p. 033, doi:10.1088/1475-7516/2019/03/033.

Benito, M., Iocco, F. and Cuoco, A. (2021). Uncertainties in the Galactic dark matter distribution: an update, *Phys. Dark Univ.* **32**, p. 100826, doi:10.1016/j.dark.2021.100826.

Berezinskii, V. S. *et al.* (1990). Astrophysics of cosmic rays.

Berezinsky, V. S. and Zatsepin, G. T. (1969). Cosmic rays at ultrahigh-energies (neutrino?), *Phys. Lett.* **28B**, pp. 423–424, doi:10.1016/0370-2693(69)90341-4.

Berg, M. *et al.* (2017). Constraints on axion-like particles from X-ray observations of NGC 1275, *ApJ* **847**, p. 101, doi:10.3847/1538-4357/aa8b16.

Berger, E. (2014). Short-duration gamma-ray bursts, *Ann. Rev. Astron. Astrophys.* **52**, pp. 43–105, doi:10.1146/annurev-astro-081913-035926.

Bergstrom, L. and Hooper, D. (2006). Dark matter and gamma rays from Draco: MAGIC, GLAST and CACTUS, *Phys. Rev. D* **D73**, p. 063510, doi:10.1103/PhysRevD.73.063510.

Bergstrom, L. and Kaplan, J. (1994). Gamma-ray lines from TeV dark matter. *Astrop. Phys.*. **2**, pp. 261–268, doi:10.1016/0927-6505(94)90005-1.

Bergstrom, L. and Ullio, P. (1997). Full one loop calculation of neutralino annihilation into two photons, *Nucl. Phys. B* **504**, pp. 27–44, doi:10.1016/S0550-3213(97)00530-0.

Bergstrom, L., Ullio, P. and Buckley, J. H. (1998). Observability of gamma-rays from dark matter neutralino annihilations in the Milky Way halo, *Astrop. Phys.* **9**, pp. 137–162, doi:10.1016/S0927-6505(98)00015-2.

Berlin, A. and Hooper, D. (2014). Stringent constraints on the dark matter annihilation cross section from subhalo searches with the Fermi Gamma-Ray Space Telescope, *Phys. Rev. D* **89**, 1, p. 016014, doi:10.1103/PhysRevD.89.016014.

Bernardini, E. (2005). Multi-messenger studies with AMANDA/IceCube: observations and strategies, .

Bernlöhr, K. (2008). Simulation of imaging atmospheric Cherenkov telescopes with CORSIKA and sim_ telarray, *Astrop. Phys.* **30**, pp. 149–158, doi:10.1016/j.astropartphys.2008.07.009.

Bertone, G. and Hooper, D. (2016). A history of dark matter, *Rev. Mod. Phys.*

Bertone, G. *et al.* (2005a). A new signature of dark matter annihilations: gamma-rays from intermediate-mass black holes, *Phys. Rev. D* **D72**, p. 103517, doi:10.1103/PhysRevD.72.103517.

Bertone, G. *et al.* (2005b). Particle dark matter: evidence, candidates and constraints, *Phys. Rep.* **405**, pp. 279–390, doi:10.1016/j.physrep.2004.08.031.

Bertone, G. *et al.* (2009). Dark matter annihilation around intermediate mass black holes: an update, *New. J. Phys* **11**, p. 105016, doi:10.1088/1367-2630/11/10/105016.

Bertoni, B., Hooper, D. and Linden, T. (2015). Examining the *Fermi*-LAT third source catalog in search of dark matter subhalos, *J. Cosmol. Astropart. Phys.* **12**, 035, doi:10.1088/1475-7516/2015/12/035.

Bertoni, B., Hooper, D. and Linden, T. (2016). Is the gamma-ray source 3FGL J2212.5+0703 a dark matter subhalo? *J. Cosmol. Astropart. Phys.* **5**, 049, doi:10.1088/1475-7516/2016/05/049.

Bhatt, N. *et al.* (2021). Status update of MACE gamma-ray telescope, arXiv e-prints, arXiv:2107.04297.

Bhattacharjee, P. (2000). Origin and propagation of extremely high energy cosmic rays, *Phys. Rep.* **327**, pp. 109–247, doi:10.1016/S0370-1573(99)00101-5.

Bignami, G. F., Caraveo, P. A. and Lamb, R. C. (1983). An identification for "GEMINGA" (2CG 195+04) 1E 0630+178 : a unique object in the error box of the high-energy gamma-ray source. *ApJL* **272**, pp. L9–L13, doi:10.1086/184107.

Bignami, G. F. *et al.* (1975). The COS-B experiment for gamma-ray astronomy. *Space Sci. Instrument.* **1**, pp. 245–268.

Biland, A. *et al.* (2007). The active mirror control of the MAGIC telescopes, in *Proceedings of the 30th International Cosmic Ray Conference — PoS(ICRC2007), Merida (Mexico)*, Vol. 3, Universidad Nacional Autonoma de Mexico City, pp. 1353–1356.

Bionta, R. M. *et al.* (1987). Observation of a neutrino burst in coincidence with supernova 1987a in the large magellanic cloud, *Phys. Rev. Lett.* **58**, pp. 1494–1496, doi:10.1103/PhysRevLett.58.1494.

Bird, R. *et al.* (2016a). Observing the cosmic ray moon shadow with VERITAS, *PoS* **ICRC2015**, p. 852, doi:10.22323/1.236.0852.

Bird, R. *et al.* (2017). Observing FRB 121102 with VERITAS; searching for associated TeV emission, in *Proceedings of the 35th International Cosmic Ray Conference — PoS(ICRC2017), Busan (South Korea)*, Vol. 301, p. 621, doi:10.22323/1.301.0621.

Bird, S. *et al.* (2016b). Did LIGO detect dark matter? *Phys. Rev. Lett.* **116**, 20, p. 201301, doi:10.1103/PhysRevLett.116.201301.

Bishop, C. (1995). *Neural Networks for Pattern Recognition*, Oxford, Clarendon Press, New York.

Biteau, J. and Williams, D. A. (2015). The extragalactic background light, the Hubble constant, and anomalies: conclusions from 20 years of TeV gamma-ray observations, *ApJ* **812**, 1, 60, doi:10.1088/0004-637X/812/1/60.

Biteau, J. *et al.* (2020). Progress in unveiling extreme particle acceleration in persistent astrophysical jets, *Nat. Astron.* **4**, pp. 124–131, doi:10.1038/s41550-019-0988-4.

Blanch, O. and Martinez, M. (2005a). Exploring the gamma ray horizon with the next generation of gamma ray telescopes. part 1: Theoretical predictions, *Astrop. Phys.* **23**, 6, pp. 588 – 597, doi:https://doi.org/10.1016/j.astropartphys.2005.03.008, URL http://www.sciencedirect.com/science/article/pii/S0927650505000642.

Blanch, O. and Martinez, M. (2005b). Exploring the gamma-ray horizon with the next generation of gamma-ray telescopes. part 2: Extracting cosmological parameters from the observation of gamma-ray sources, *Astrop. Phys.* **23**, 6, pp. 598 – 607, doi:https://doi.org/10.1016/j.astropartphys.2005.03.009, URL http://www.sciencedirect.com/science/article/pii/S0927650505000630

Blanch, O. and Martinez, M. (2005c). Exploring the gamma ray horizon with the next generation of gamma ray telescopes. part 3: Optimizing the observation schedule of gamma ray sources for the extraction of cosmological parameters *ap* **23**, 6, pp. 608–615, doi:https://doi.org/10.1016/j.astropartphys.2005.03.010, URL http://www.sciencedirect.com/science/article/pii/S0927650505000629.

Blanch, O. *et al.* (2020). GRB 201216C: MAGIC detection in very high energy gamma rays, *GRB Coordinates Network* **29075**, p. 1.

Blandford, R. and Eichler, D. (1987). Particle acceleration at astrophysical shocks: a theory of cosmic ray origin, *Phys. Rep.* **154**, pp. 1–75, doi:10.1016/0370-1573(87)90134-7.

Blandford, R. D. and Ostriker, J. P. (1978). Particle acceleration by astrophysical shocks, *ApJL* **221**, pp. L29–L32, doi:10.1086/182658.

Blasi, P. (2013). The origin of galactic cosmic rays, *A&A Rev.* **21**, 70, doi:10.1007/s00159-013-0070-7.

Blasi, P., Burles, S. and Olinto, A. V. (1999). Cosmological Magnetic Field Limits in an Inhomogeneous Universe, *ApJL* **514**, pp. L79–L82, doi:10.1086/311958.

Blasi, P. and Colafrancesco, S. (1999). Cosmic rays, radio halos and non-thermal x-ray emission in clusters of galaxies, *Astrop. Phys.* **12**, pp. 169–183, doi:10.1016/S0927-6505(99)00079-1,10.1016/S0927-6505(99)00079-1.

Bock, R. K. *et al.* (2004). Methods for multidimensional event classification: a case study using images from a Cherenkov gamma-ray telescope, *Nucl. Instrum. Meth. Phys. Res. A* **516**, pp. 511–528, doi:10.1016/j.nima.2003.08.157.

Boella, G. *et al.* (1997). BeppoSAX, the wide band mission for X-ray astronomy, *Astronom. Astrophys. Suppl.* **122**, pp. 299–307, doi:10.1051/aas:1997136.

Bogovalov, S. V. (1999). On the physics of cold MHD winds from oblique rotators, *A&A* **349**, pp. 1017–1026.

Bogovalov, S. V. (2014). Magnetocentrifugal acceleration of bulk motion of plasma in pulsar magnetosphere, *MNRAS* **443**, 3, pp. 2197–2203, doi:10.1093/mnras/stu1283.

Bolmont, J. *et al.* (2014). The camera of the fifth H.E.S.S. telescope. Part I: System description, *Nucl. Instrum. Meth. Phys. Res. A* **761**, pp. 46–57, doi:10.1016/j.nima.2014.05.093.

Bonnivard, V., Maurin, D. and Walker, M. G. (2016). Contamination of stellar-kinematic samples and uncertainty about dark matter annihilation profiles in ultrafaint dwarf galaxies: the example of Segue I, *MNRAS* **462**, 1, pp. 223–234, doi:10.1093/mnras/stw1691.

Bonnivard, V. *et al.* (2015a). Dark matter annihilation and decay in dwarf spheroidal galaxies: the classical and ultrafaint dSphs, *MNRAS* **453**, 1, pp. 849–867, doi:10.1093/mnras/stv1601.

Bonnivard, V. *et al.* (2015b). Spherical Jeans analysis for dark matter indirect detection in dwarf spheroidal galaxies - Impact of physical parameters and triaxiality, *MNRAS* **446**, pp. 3002–3021, doi:10.1093/mnras/stu2296.

Borione, A. *et al.* (1993). CASA-MIA: A "precision" EAS detector, in L. Jones. ed., *The Seventh International Symposium on very High energy cosmic-ray interactions*, *American Institute of Physics Conference Series*, Vol. 276, pp. 207–211, doi:10.1063/1.43832.

Borkowski, K. J. *et al.* (2017). Asymmetric expansion of the youngest galactic supernova remnant g1.9+0.3, **837**, 1, p. L7, doi:10.3847/2041-8213/aa618c.

Borla Tridon, D. *et al.* (2011). Measurement of the cosmic electron spectrum with the MAGIC telescopes, in *Proceedings of the 32th International Cosmic Ray Conference — PoS(ICRC2011), Beijing (China)*, Vol. 6, pp. 47–50, doi: 10.7529/ICRC2011/V06/0680.

Borwankar, C. *et al.* (2016). Simulation studies of MACE-I: trigger rates and energy thresholds, *Astrop. Phys.* **84**, pp. 97–106, doi:10.1016/j.astropartphys. 2016.06.009.

Bosma, A. (1978). *The Distribution and Kinematics of Neutral Hydrogen in Spiral Galaxies of Various Morphological Types*, Groningen Univ.

Boylan-Kolchin, M. *et al.* (2011). Too big to fail? The puzzling darkness of massive Milky Way subhaloes, *MNRAS* **415**, p. L40.

Bradford, J. D. *et al.* (2011). Structure and dynamics of the Globular Cluster Palomar 13, *ApJ* **743**, p. 167, doi:10.1088/0004-637X/743/2/167, [Erratum: Astrophys.J. 778, 85 (2013)].

Breiman, L. (2001). Random forests, *Machine learning* **45**, 1, pp. 5–32, doi:10. 1023/A:1010933404324.

Breiman, L. *et al.* (1984). *Classification and regression trees*, Chapman and Hall/CRC, ISBN 0412048418, 9780412048418.

Bringmann, T. (2009). Particle Models and the small-scale structure of dark matter, *New J.Phys.* **11**, p. 105027, doi:10.1088/1367-2630/11/10/105027.

Bringmann, T., Bergstrom, L. and Edsjo, J. (2008). New gamma-ray contributions to supersymmetric dark matter annihilation, *J. High Energy Phys.* **01**, p. 049, doi:10.1088/1126-6708/2008/01/049.

Bringmann, T. and Calore, F. (2014). Significant Enhancement of Neutralino Dark Matter Annihilation from Electroweak Bremsstrahlung, *Phys. Rev. Lett.* **112**, p. 071301, doi:10.1103/PhysRevLett.112.071301.

Brock, M. N. *et al.* (1992). Gamma-Ray Burst Locations from the Burst and Transient Source Experiment, in W. S. Paciesas and G. J. Fishman. eds., *American Institute of Physics Conference Series*, *American Institute of Physics Conference Series*, Vol. 265, p. 383.

Brockway, J. W., Carlson, E. D. and Raffelt, G. G. (1996). SN 1987A gamma-ray limits on the conversion of pseudoscalars, *Phys. Let. B* **383**, pp. 439–443, doi:10.1016/0370-2693(96)00778-2.

Broderick, A. E., Chang, P. and Pfrommer, C. (2012). The cosmological impact of luminous TeV blazars. I. Implications of plasma instabilities for the intergalactic magnetic field and extragalactic gamma-ray background, *ApJ* **752**, p. 22, doi:10.1088/0004-637X/752/1/22.

Brooks, A. M. and Zolotov, A. (2014). Why baryons matter: the kinematics of dwarf spheroidal satellites, *ApJ* **786**, p. 87, doi:10.1088/0004-637X/786/2/87.

Brose, R. *et al.* (2020). Cosmic-ray acceleration and escape from post-adiabatic supernova remnants, *A&A* **634**, p. A59, doi:10.1051/0004-6361/201936567.

Brown, A. M. *et al.* (2018). Understanding the gamma-ray emission from the globular cluster 47 Tuc: evidence for dark matter? *Phys. Rev. D* **98**, p. 4, doi:10.1103/physrevd.98.041301, URL http://dx.doi.org/10.1103/PhysRevD.98.041301.

Brown, A. M. *et al.* (2019a). Reply to "Comment on 'Understanding the γ-ray emission from the globular cluster 47 Tuc: Evidence for dark matter?"', *Phys. Rev. D* **100**, 6, p. 068302, doi:10.1103/PhysRevD.100.068302.

Brown, A. M. *et al.* (2019b). The glow of annihilating dark matter in Omega Centauri, archive eprints.

Brown, A. M. *et al.* (2021). Trinity: an imaging air Cherenkov telescope to search for ultra-high energy neutrinos, in *Proceedings of the 37th International Cosmic Ray Conference — PoS(ICRC2021), Berlin (Germany)*, Vol. 395, p. 1179, doi:10.22323/1.395.1179.

Brun, P. *et al.* (2011). Searches for dark matter subhaloes with wide-field Cherenkov telescope surveys, *Phys. Rev. D* **83**, 1, 015003, doi:10.1103/PhysRevD.83.015003.

Brunetti, G. and Jones, T. W. (2014). Cosmic rays in galaxy clusters and their non-thermal emission, *Int. J. Mod. Phys. D* **23**, p. 1430007, doi:10.1142/S0218271814300079.

Brunetti, G., Zimmer, S. and Zandanel, F. (2017). Relativistic protons in the Coma galaxy cluster: first gamma-ray constraints ever on turbulent reacceleration, *MNRAS* **472**, pp. 1506–1525, doi:10.1093/mnras/stx2092.

Brunetti, G. *et al.* (2012). Probing the origin of giant radio haloes through radio and gamma-ray data: the case of the Coma cluster, *MNRAS* **426**, pp. 956–968, doi:10.1111/j.1365-2966.2012.21785.x.

Buchmuller, W. *et al.* (2007). Gravitino dark matter in R-parity breaking vacua. *J. High Energy Phys.* **03**, p. 037, doi:10.1088/1126-6708/2007/03/037.

Buckley, M. R. and Hooper, D. (2010). Dark matter subhalos in the Fermi first source catalog, *Phys. Rev. D* **82**, 6, 063501, doi:10.1103/PhysRevD.82.063501.

Bühler, R. and Blandford, R. (2014). The surprising Crab pulsar and its nebula: a review, *Rep. Prog. Phys.* **77**, 6, 066901, doi:10.1088/0034-4885/77/6/066901.

Bulanov, S. V. and Dogel, V. A. (1974). The Influence of the energy dependence of the diffusion coefficient on the spectrum of the electron component of cosmic rays and the radio background radiation of the Galaxy, *Ap&SS* **29**, pp. 305–318, doi:10.1007/BF02639066.

Bulgarelli, A. *et al.* (2019). Second AGILE catalogue of gamma-ray sources, *A&A* **627**, A13, doi:10.1051/0004-6361/201834143.

Bullock, J. S. and Boylan-Kolchin, M. (2017). small-scale challenges to the ΛCDM paradigm, *Ann. Rev. Astron. Astrophys.* **55**, pp. 343–387, doi: 10.1146/annurev-astro-091916-055313.

Burbidge, E. M. *et al.* (1957). Synthesis of the elements in stars, *Rev. Mod. Phys.* **29**, pp. 547–650, doi:10.1103/RevModPhys.29.547, URL https://link.aps.org/doi/10.1103/RevModPhys.29.547.

Burkert, A. (1996). The Structure of dark matter halos in dwarf galaxies, *IAU Symp.* **171**, p. 175, doi:10.1086/309560, [Astrophys. J.447,L25(1995)].

Burrows, D. N. *et al.* (2005). The swift X-ray telescope, *Space Sci. Rev.* **120**, 3-4, pp. 165–195, doi:10.1007/s11214-005-5097-2.

Buson, S. and Cheung, C. C. (2014). Fermi LAT detection of a hard spectrum gamma-ray flare from gravitationally lensed blazar S3 0218+357, *The Astronomer's Telegram* **6316**, p. 1.

Bykov, A. M. (2001). Particle acceleration and nonthermal phenomena in superbubbles, *Space Sci. Rev.* **99**, pp. 317–326.

Bykov, A. M. and Toptygin, I. N. (2001). A model of particle acceleration to high energies by multiple supernova explosions in OB associations, *Astronomy Letters* **27**, pp. 625–633, doi:10.1134/1.1404456.

Cahill-Rowley, M. *et al.* (2015). Complementarity of dark matter searches in the phenomenological MSSM, *Phys. Rev. D* **D91**, 5, p. 055011, doi:10.1103/PhysRevD.91.055011.

Callan, C. G. (1982). Disappearing dyons, *Phys. Rev. D* **25**, pp. 2141–2146, doi: 10.1103/PhysRevD.25.2141.

Calore, F., Cholis, I. and Weniger, C. (2015). Background model systematics for the Fermi GeV excess, *J. Cosmol. Astropart. Phys.* **1503**, p. 038, doi: 10.1088/1475-7516/2015/03/038.

Calore, F., Hütten, M. and Stref, M. (2019). Gamma-ray sensitivity to dark matter subhalo modelling at high latitudes, *Galaxies* **7**, 4, p. 90, doi: 10.3390/galaxies7040090.

Calore, F. *et al.* (2017). Realistic estimation for the detectability of dark matter subhalos using Fermi-LAT catalogs, *Phys. Rev. D* **96**, 6, 063009, doi:10.1103/PhysRevD.96.063009.

Cao, Z. *et al.* (2021). Ultrahigh-energy photons up to 1.4 petaelectronvolts from 12 gamma-ray Galactic sources, *Nature* **594**, 7861, pp. 33–36, doi:10.1038/s41586-021-03498-z.

Caputo, R. *et al.* (2018). Exploring the particle nature of dark matter with the All-sky Medium Energy Gamma-ray Observatory (AMEGO), in *American Astronomical Society Meeting Abstracts #231*, *American Astronomical Society Meeting Abstracts*, Vol. 231, p. 355.49.

Carlson, E., Linden, T. and Profumo, S. (2016). Cosmic-Ray injection from star-forming regions, *Phys. Rev. Lett.* **117**, 11, p. 111101, doi:10.1103/PhysRevLett.117.111101.

Carr, B., Kühnel, F. and Sandstad, M. (2016). Primordial black holes as dark matter, *Phys. Rev. D* **94**, p. 8, doi:10.1103/PhysRevD.94.083504.

Carr, B. J. (2005). Primordial black holes: Do they exist and are they useful? in *59th Yamada Conference on Inflating Horizon of Particle Astrophysics and Cosmology Tokyo, Japan, June 20-24, 2005.*

Carr, B. J. *et al.* (2010). New cosmological constraints on primordial black holes, *Phys. Rev. D* **D81**, p. 104019, doi:10.1103/PhysRevD.81.104019.

Casanova, S. *et al.* (2010a). Modeling the gamma-ray emission produced by runaway cosmic rays in the rnvironment of RX J1713.7-3946, *Pub. of the Astron. Soc. Japan* **62**, pp. 1127–1134, doi:10.1093/pasj/62.5.1127.

Casanova, S. *et al.* (2010b). Molecular clouds as cosmic-ray barometers, *Pub. of the Astron. Soc. Japan* **62**, pp. 769–777, doi:10.1093/pasj/62.3.769.

Cassam-Chenaï, G. *et al.* (2004). XMM-Newton observations of the supernova remnant RX J1713.7-3946 and its central source, *A&A* **427**, pp. 199–216, doi:10.1051/0004-6361:20041154.

Cassanyes, M. (2015). MAGIC sensitivity to primordial black hole bursts and modelization of BH chromospheres and gamma-ray emission spectra, .

Casse, M. and Paul, J. A. (1980). Local gamma rays and cosmic-ray acceleration by supersonic stellar winds, *ApJ* **237**, pp. 236–243, doi:10.1086/157863.

Catanese, M. and Weekes, T. C. (1999). Very-high energy gamma-ray astronomy, *Pub. Astron. Soc. Pacific* **111**, 764, pp. 1193–1222, doi:10.1086/316435.

Cecchini, S. *et al.* (2016). Energy losses of magnetic monopoles in aluminum, iron and copper.

Celli, S. *et al.* (2019). Supernova remnants in clumpy media: particle propagation and gamma-ray emission, *MNRAS* **487**, 3, pp. 3199–3213, doi:10.1093/mnras/stz1425.

Cerri, S. S. *et al.* (2017). A signature of anisotropic cosmic-ray transport in the gamma-ray sky, *J. Cosmol. Astropart. Phys.* **1710**, 10, p. 019, doi:10.1088/1475-7516/2017/10/019.

Cerutti, B., Philippov, A. and Spitkovsky, A. (2016). Modelling high-energy pulsar light curves from first principles, *MNRAS* **457**, 3, pp. 2401–2414, doi:10.1093/mnras/stw124.

Cerutti, B. and Philippov, A. A. (2017). Dissipation of the striped pulsar wind, *A&A* **607**, p. A134, doi:10.1051/0004-6361/201731680.

Cerutti, B. *et al.* (2015). Particle acceleration in axisymmetric pulsar current sheets, *MNRAS* **448**, p. 606–619, doi:10.1093/mnras/stv042.

Cesarsky, C. J. (1980). Cosmic-ray confinement in the galaxy, *Ann. Rev. Astron. Astrophys.* **18**, pp. 289–319, doi:10.1146/annurev.aa.18.090180.001445.

Cesarsky, C. J. and Montmerle, T. (1983). Gamma rays from active regions in the galaxy - he possible contribution of stellar winds, *Space Sci. Rev.* **36**, pp. 173–193, doi:10.1007/BF00167503.

Chakrabarty, I. (1999). Gravitational waves: an introduction, arxiv Physics e-prints.

Chandran, B. D. G. (2000). Scattering of energetic particles by anisotropic magnetohydrodynamic turbulence with a Goldreich-Sridhar power spectrum, *Phys. Rev. Lett.* **85**, pp. 4656–4659, doi:10.1103/PhysRevLett.85.4656.

Chang, J. *et al.* (2008). An excess of cosmic ray electrons at energies of 300-800 GeV, *Nature* **456**, pp. 362–365, doi:10.1038/nature07477.

Chatterjee, S. *et al.* (2017). A direct localization of a fast radio burst and its host, *Nature* **541**, pp. 58–61, doi:10.1038/nature20797.

Cheng, K. S., Ho, C. and Ruderman, M. (1986a). Energetic radiation from rapidly spinning pulsars. I. Outer magnetosphere gaps, *ApJ* **300**, p. 500, doi:10.1086/163829.

Cheng, K. S., Ho, C. and Ruderman, M. (1986b). Energetic radiation from rapidly spinning pulsars. II. VELA and Crab, *ApJ* **300**, p. 522, doi:10.1086/163830.

Chernyakova, M. *et al.* (2012). Superorbital modulation of X-ray emission from gamma-ray binary LSI +61 303, *ApJL* **747**, 2, L29, doi:10.1088/2041-8205/747/2/L29.

Chiaberge, M. *et al.* (2000). Does the unification of BL Lac and FR I radio galaxies require jet velocity structures? *A&A* **358**, pp. 104–112.

Chiappo, A. *et al.* (2017). Dwarf spheroidal J-factors without priors: a likelihood-based analysis for indirect dark matter searches, *MNRAS* **466**, 1, pp. 669–676, doi:10.1093/mnras/stw3079.

Chikashige, Y., Fujii, Y. and Mima, K. (1978). Lorentz invariant string regularized dimensionally by means of hyperfunctions. II, *Prog. of Theoret. Phys.* **59**, pp. 274–288, doi:10.1143/PTP.59.274.

Chitnis, V. R. and Bhat, P. N. (1998). Čerenkov photon density fluctuations in extensive air showers, *Astrop. Phys.* **9**, pp. 45–63, doi:10.1016/S0927-6505(98)00007-3.

Cholis, I., Karwal, T. and Kamionkowski, M. (2018). Features in the spectrum of cosmic-ray positrons from pulsars, *Phys. Rev. D* **D97**, 12, p. 123011, doi:10.1103/PhysRevD.97.123011.

Cicoli, M., Goodsell, M. D. and Ringwald, A. (2012). The type IIB string axiverse and its low-energy phenomenology, *J. High Energy Phys.* **10**, 146, doi:10.1007/JHEP10(2012)146.

Cirelli, M. *et al.* (2011). PPPC 4 DM ID: a poor particle physicist cookbook for dark matter indirect detection, *J. Cosmol. Astropart. Phys.* **1103**, p. 051, doi:10.1088/1475-7516/2011/03/051.

Cocconi, G. (1954). Influence of the Earth's magnetic field on the extensive air showers, *Phys. Rev.* **93**, pp. 646–647, doi:10.1103/PhysRev.93.646.2.

Colafrancesco, S., Profumo, S. and Ullio, P. (2006). Multi-frequency analysis of neutralino dark matter annihilations in the Coma cluster, *A&A* **455**, p. 21, doi:10.1051/0004-6361:20053887.

Colafrancesco, S. *et al.* (2011). On the DM interpretation of the origin of non-thermal phenomena in galaxy clusters, *A&A* **527**, p. A80.

Colin, P. *et al.* (2011). Probing the CR positron/electron ratio at few hundreds GeV through Moon shadow observation with the MAGIC telescopes, doi:10.7529/ICRC2011/V06/1114.

Conlon, J. P. (2006). The QCD axion and moduli stabilisation, *J. High Energy Phys.* **5**, p. 078, doi:10.1088/1126-6708/2006/05/078.

Connaughton, V. *et al.* (1997). A search for TeV counterparts to BATSE gamma-ray bursts, *ApJ* **479**, 2, pp. 859–867, doi:10.1086/303918.

Connaughton, V. *et al.* (1998). A search for tev gamma-ray bursts on a 1-second time-scale, *Astrop. Phys.* **8**, 3, pp. 179–191, doi:https://doi.org/10.1016/S0927-6505(97)00055-8.

Contopoulos, I. (2016). Are there two types of pulsars? *MNRAS* **463**, p. L94–L98, doi:10.1093/mnrasl/slw166.

Contopoulos, I., Kazanas, D. and Fendt, C. (1999). The axisymmetric pulsar magnetosphere, *ApJ* **511**, 1, pp. 351–358, doi:10.1086/306652.

Coronado-Blázquez, J. and Sánchez-Conde, M. A. (2020). Constraints to dark matter annihilation from high-latitude HAWC unidentified sources, *Galaxies* **8**, 1, p. 5, doi:10.3390/galaxies8010005.

Coronado-Blázquez, J. *et al.* (2019a). Spectral and spatial analysis of the dark matter subhalo candidates among *Fermi* Large Area Telescope unidentified sources, *J. Cosmol. Astropart. Phys.* **11**, p. 045, doi:10.1088/1475-7516/2019/11/045.

Coronado-Blázquez, J. *et al.* (2019b). Unidentified gamma-ray sources as targets for indirect dark matter detection with the Fermi-Large Area Telescope, *J. Cosmol. Astropart. Phys.* **07**, p. 020, doi:10.1088/1475-7516/2019/07/020.

Coronado-Blázquez, J. *et al.* (2021). Sensitivity of the Cherenkov Telescope Array to dark subhalos, *Phys. Dark Univ.* **32**, p. 100845, doi:10.1016/j.dark.2021.100845.

Coroniti, F. V. (1990). Magnetically striped relativistic magnetohydrodynamic winds: the Crab nebula revisited, *ApJ* **349**, p. 538, doi:10.1086/168340.

Cortina, J. and González, J. C. (2001). The cosmic electron background in low energy imaging atmospheric Cherenkov telescopes: effect of the geomagnetic field, *Astrop. Phys.* **15**, pp. 203–210, doi:10.1016/S0927-6505(00)00147-X.

Creasey, P. *et al.* (2019). Globular clusters formed within dark haloes I: present-day abundance, distribution, and kinematics, *MNRAS* **482**, 1, pp. 219–230, doi:10.1093/mnras/sty2701.

Crick, F. (1989). The recent excitement about neural networks, *Nature* **337**, pp. 129–132, doi:10.1038/337129a0.

Csáki, C., Kaloper, N. and Terning, J. (2002). Dimming supernovae without cosmic acceleration, *Phys. Rev. Lett.* **88**, 16, 161302, doi:10.1103/PhysRevLett.88.161302.

Csáki, C. *et al.* (2003). Super-GZK photons from photon-axion mixing, *J. Cosmol. Astropart. Phys.* **5**, 005, doi:10.1088/1475-7516/2003/05/005.

Da Vela, P. *et al.* (2018). Study of the IACT angular acceptance and point spread function, *Astrop. Phys.* **98**, pp. 1–8, doi:10.1016/j.astropartphys.2018.01.002.

Daniel, M. K. *et al.* (2007). The VERITAS standard data analysis, in *Proceedings of the 30th International Cosmic Ray Conference — PoS(ICRC2007), Merida (Mexico)*, Vol. 3, Universidad Nacional Autonoma de Mexico City, pp. 1325–1328.

Daniel, M. K. *et al.* (2015). The camera calibration strategy of the Cherenkov Telescope Array, in A. S. Borisov, V. G. Denisova, Z. M. Guseva, E. A. Kanevskaya, M. G. Kogan, A. E. Morozov, V. S. Puchkov, S. E. Pyatovsky, G. P. Shoziyoev, M. D. Smirnova, A. V. Vargasov, V. I. Galkin, S. I. Nazarov and R. A. Mukhamedshin. eds., *Proceedings of the 34th International*

Cosmic Ray Conference — PoS(ICRC2015), The Hague (Netherlands), World Scientific, p. 941.

Daum, A. *et al.* (1997). First results on the performance of the HEGRA IACT array, *Astrop. Phys.* **8**, pp. 1–11, doi:10.1016/S0927-6505(97)00031-5.

D'Avanzo, P. (2015). Short gamma-ray bursts: A review, *J. High Energy Astrophys.* **7**, pp. 73–80, doi:https://doi.org/10.1016/j.jheap.2015.07.002, swift 10 years of discovery, a novel approach to time domain astronomy.

Davies, J. M. and Cotton, E. S. (1957). Design of the quartermaster solar furnace, *Solar Energy* **1**, 2, pp. 16–22, doi:10.1016/0038-092X(57)90116-0.

Davis, R., Harmer, D. S. and Hoffman, K. C. (1968). Search for neutrinos from the sun, *Phys. Rev. Lett.* **20**, pp. 1205–1209, doi:10.1103/PhysRevLett.20.1205.

Dazzi, F. *et al.* (2021). The stereoscopic analog trigger of the magic telescopes, *IEEE Trans. Nucl. Sci.* **68**, 7, pp. 1473–1486, doi:10.1109/TNS.2021.3079262.

De Angelis, A., Galanti, G. and Roncadelli, M. (2011). Relevance of axionlike particles for very-high-energy astrophysics, *Phys. Rev. D* **84**, 10, 105030, doi:10.1103/PhysRevD.84.105030.

De Angelis, A., Mansutti, O. and Roncadelli, M. (2008). Axion-like particles, cosmic magnetic fields and gamma-ray astrophysics, *Phys. Let. B* **659**, pp. 847–855, doi:10.1016/j.physletb.2007.12.012.

De Angelis, A., Roncadelli, M. and Mansutti, O. (2007). Evidence for a new light spin-zero boson from cosmological gamma-ray propagation? *Phys. Rev. D* **76**, 12, p. 121301, doi:10.1103/PhysRevD.76.121301.

De Angelis, A. *et al.* (2009). Photon propagation and the very high energy gamma-ray spectra of blazars: how transparent is the Universe? *MNRAS* **394**, pp. L21–L25, doi:10.1111/j.1745-3933.2008.00602.x.

De Angelis, A. *et al.* (2017). Science with e-ASTROGAM (A space mission for MeV-GeV gamma-ray astrophysics), ArXiv e-prints.

de Boer, W. *et al.* (2017). Molecular clouds as origin of the Fermi gamma-ray GeV excess, *Phys. Rev. D* **D96**, 4, p. 043012, doi:10.1103/PhysRevD.96.043012.

De Jager, O. C. and Harding, A. K. (1992). The Expected high-Energy to ultra-high-energy gamma-ray spectrum of the Crab Nebula, *ApJ* **396**, p. 161, doi:10.1086/171706.

de la Calle Pérez, I. and Biller, S. D. (2006). Extending the sensitivity of air Cherenkov telescopes, *Astrop. Phys.* **26**, pp. 69–90, doi:10.1016/j.astropartphys.2006.05.002.

De Lotto, B. *et al.* (2016). MAGIC electromagnetic follow-up of gravitational wave alerts, *IAU Symp.* **324**, pp. 287–290, doi:10.1017/S1743921317001430.

De Menezes, R. *et al.* (2021). Multi-Messenger observations of the *Fermi*-LAT blazar 4FGL J0658.6+0636 consistent with an IceCube high-energy neutrino, in *Proceedings of the 37th International Cosmic Ray Conference — PoS(ICRC2021), Berlin (Germany)*, Vol. **395**, p. 955, doi:10.22323/1.395.0955.

de Naurois, M. and Rolland, L. (2009). A high performance likelihood reconstruction of gamma rays for imaging atmospheric cherenkov telescopes, *Astrop. Phys.* **32**, 5, pp. 231–252, doi:10.1016/j.astropartphys.2009.09.001.

Dean, A. J. *et al.* (2008). Polarized gamma-ray emission from the crab, *Science* **321**, 5893, pp. 1183–1185, doi:10.1126/science.1149056.

Del Zanna, L. *et al.* (2006). Simulated synchrotron emission from pulsar wind nebulae, *A&A* **453**, 2, pp. 621–633, doi:10.1051/0004-6361:20064858.

Delagnes, E. *et al.* (2006). Sam: A new ghz sampling ASIC for the H.E.S.S.-ii front-end electronics, *Nucl. Instrum. Meth. Phys. Res. A* **567**, 1, pp. 21–26, doi:10.1016/j.nima.2006.05.052.

DeMarco, D., Blasi, P. and Stanev, T. (2007). Numerical propagation of high energy cosmic rays in the Galaxy: I. Technical issues, *J. Cosmol. Astropart. Phys.* **6**, 027, doi:10.1088/1475-7516/2007/06/027.

Den Brok, M. *et al.* (2014). The central mass and mass-to-light profile of the Galactic globular cluster M15, *MNRAS* **438**, 1, pp. 487–493, doi:10.1093/mnras/stt2221.

Derishev, E. and Piran, T. (2021). GRB Afterglow Parameters in the Era of TeV Observations: The Case of GRB 190114C, *ApJ* **923**, 2, 135, doi:10.3847/1538-4357/ac2dec.

Dermer, C. D. and Giebels, B. (2016). Active galactic nuclei at gamma-ray energies, *Comptes Rendus Physique* **17**, 6, pp. 594–616, doi:10.1016/j.crhy.2016.04.004.

Dermer, C. D. *et al.* (2011). Time delay of cascade radiation for TeV blazars and the measurement of the intergalactic magnetic field, *ApJL* **733**, L21, doi:10.1088/2041-8205/733/2/L21.

Desai, A. *et al.* (2019). A GeV-TeV Measurement of the extragalactic background light, *ApJL* **874**, 1, L7, doi:10.3847/2041-8213/ab0c10.

Di Bernardo, G. *et al.* (2013). Cosmic ray electrons, positrons and the synchrotron emission of the Galaxy: consistent analysis and implications, *J. Cosmol. Astropart. Phys.* **1303**, p. 036, doi:10.1088/1475-7516/2013/03/036.

Di Cintio, A. *et al.* (2013). The dependence of dark matter profiles on the stellar to halo mass ratio: a prediction for cusps vs cores, doi:10.1093/mnras/stt1891, URL http://dx.doi.org/10.1093/mnras/stt1891.

Di Mauro, M., Manconi, S. and Donato, F. (2020). Evidences of low-diffusion bubbles around Galactic pulsars, *Phys. Rev. D* **101**, 10, p. 103035, doi:10.1103/PhysRevD.101.103035.

Di Mauro, M. and Winkler, M. W. (2021). Multimessenger constraints on the dark matter interpretation of the Fermi-LAT Galactic center excess, *Phys. Rev. D* **103**, 12, p. 123005, doi:10.1103/PhysRevD.103.123005.

Di Mauro, M. *et al.* (2014). Interpretation of AMS-02 electrons and positrons data, *J. Cosmol. Astropart. Phys.* **1404**, p. 006, doi:10.1088/1475-7516/2014/04/006.

Di Sciascio, G. and LHAASO Collaboration (2016). The LHAASO experiment: From gamma-ray astronomy to cosmic rays, *Nucl. Particle Phys. Proc.* **279**, pp. 166–173, doi:10.1016/j.nuclphysbps.2016.10.024.

Di Valentino, E. *et al.* (2021). In the realm of the Hubble tension-a review of solutions, *Class. Quant. Grav.* **38**, 15, 153001, doi:10.1088/1361-6382/ac086d.

Dicus, D. A. *et al.* (1978). Astrophysical bounds on the masses of axions and Higgs particles, *Phys. Rev. D* **18**, pp. 1829–1834, doi:10.1103/PhysRevD.18.1829.

Diemand, J. *et al.* (2008). Clumps and streams in the local dark matter distribution, *Nature* **454**, pp. 735–738, doi:10.1038/nature07153.

Dirac, P. A. M. (1931). Quantised singularities in the electromagnetic field, *Proc. Roy. Soc. Lond. Seri. A* **133**, pp. 60–72, doi:10.1098/rspa.1931.0130.

Dodelson, S. (2003). *Modern Cosmology*, Academic Press, Amsterdam, ISBN 9780122191411, URL http://www.slac.stanford.edu/spires/find/books/www?cl=QB981:D62:2003.

Doeleman, S. S. *et al.* (2012). Jet-launching structure resolved near the supermassive black hole in M87, *Science* **338**, p. 355, doi:10.1126/science.1224768.

Domainko, W. *et al.* (2009). New results from H.E.S.S. observations of galaxy clusters, *Int. J. Mod. Phys. D* **18**, pp. 1627–1631, doi:10.1142/S021827180901545X.

Domingo-Santamaria, E. *et al.* (2005). The DISP analysis method for point-like or extended gamma source searches/studies with the MAGIC Telescope, in *Proceedings, 29th International Cosmic Ray Conference (ICRC 2005): Pune, India, August 3-11, 2005*, Vol. 5, Tata Institute of Fundamental Research, p. 363.

Domínguez, A. and Ajello, M. (2015). Spectral analysis of Fermi-LAT blazars above 50 GeV, *ApJL* **813**, L34, doi:10.1088/2041-8205/813/2/L34.

Domínguez, A. and Prada, F. (2013). Measurement of the expansion rate of the universe from γ-ray attenuation, *ApJL* **771**, L34, doi:10.1088/2041-8205/771/2/L34.

Domínguez, A., Sánchez-Conde, M. A. and Prada, F. (2011a). Axion-like particle imprint in cosmological very-high-energy sources, *J. Cosmol. Astropart. Phys.* **11**, 020, doi:10.1088/1475-7516/2011/11/020.

Domínguez, A. *et al.* (2011b). Extragalactic background light inferred from AEGIS galaxy-SED-type fractions, *MNRAS* **410**, pp. 2556–2578, doi:10.1111/j.1365-2966.2010.17631.x.

Domínguez, A. *et al.* (2013). Detection of the cosmic γ-ray horizon from multiwavelength observations of blazars, *ApJ* **770**, 77, doi:10.1088/0004-637X/770/1/77.

Domínguez, A. *et al.* (2019). A new measurement of the hubble constant and matter content of the universe using extragalactic background light γ-ray attenuation, *ApJ* **885**, 2, 137, doi:10.3847/1538-4357/ab4a0e.

Donato, F. *et al.* (2004). Antiprotons in cosmic rays from neutralino annihilation, *Phys. Rev. D* **D69**, p. 063501, doi:10.1103/PhysRevD.69.063501.

Doro, M. (2014). A decade of dark matter searches with ground-based Cherenkov telescopes, *Nucl. Instrum. Meth. Phys. Res. A* **A742**, pp. 99–106, doi:10.1016/j.nima.2013.12.010.

Doro, M. *et al.* (2007). Dark Matter searches at intermediate mass black hole with the MAGIC telescope, in *Proceedings of the 30th International Cosmic Ray Conference — PoS(ICRC2007), Merida (Mexico)*, Universidad Nacional Autonoma de Mexico City, p. 550.

Doro, M. *et al.* (2013). Dark matter and fundamental physics with the Cherenkov Telescope Array, *Astrop. Phys.* **43**, pp. 189–214, doi:10.1016/j.astropartphys.2012.08.002.

Doro, M. *et al.* (2017). Rare events searches with Cherenkov Telescopes, in *Eur. Phys. J. Web Conf.*, *Eur. Phys. J. Web Conf.*, Vol. 136, p. 01003, doi:10.1051/epjconf/201713601003.

Driver, S. P. (2021). Measuring energy production in the Universe over all wavelengths and all time, arXiv e-prints, arXiv:2102.12089.

Driver, S. P. *et al.* (2016). Measurements of extragalactic background light from the far UV to the far IR from deep ground- and space-based galaxy counts, *ApJ* **827**, 108, doi:10.3847/0004-637X/827/2/108.

Drlica-Wagner, A. *et al.* (2015). Eight ultra-faint galaxy candidates discovered in year two of the dark energy survey, *ApJ* **813**, 2, p. 109, doi:10.1088/0004-637X/813/2/109.

Drlica-Wagner, A. *et al.* (2018). Dark energy survey year 1 results: Photometric data set for cosmology, *ApJS* **235**, 2, p. 33, doi:10.3847/1538-4365/aab4f5.

Drlica-Wagner, A. *et al.* (2019). Probing the fundamental nature of dark matter with the large synoptic survey telescope, arXiv e-prints.

Druillole, F. *et al.* (2001). The analogue ring sampler: an asic front-end electronics of the antares neutrino telescope, in *Fuel Cells Bulletin*, Vol. 1, ISBN 978-0-7803-7324-2, pp. 138–142, doi:10.1109/NSSMIC.2001.1008427.

Drury, L. O. . (2012). Origin of cosmic rays, *Astrop. Phys.* **39**, pp. 52–60, doi:10.1016/j.astropartphys.2012.02.006.

Drury, L. O., Aharonian, F. A. and Voelk, H. J. (1994). The gamma-ray visibility of supernova remnants. A test of cosmic ray origin, *A&A* **287** pp. 959–971.

Dubus, G. (2013). Gamma-ray binaries and related systems, *A&A Rev.* **21**, 64, doi:10.1007/s00159-013-0064-5.

Dubus, G. (2015). Gamma-ray emission from binaries in context, *Comptes Rendus Physique* **16**, 6-7, pp. 661–673, doi:10.1016/j.crhy.2015.08.014.

Dubus, G., Cerutti, B. and Henri, G. (2008). The modulation of the gamma-ray emission from the binary LS 5039, *A&A* **477**, 3, pp. 691–700, doi:10.1051/0004-6361:20078261.

Dubus, G. *et al.* (2013). Surveys with the Cherenkov telescope array, *Astrop. Phys.* **43**, pp. 317–330, doi:10.1016/j.astropartphys.2012.05.020.

Dubus, G. *et al.* (2017). Sizing up the population of gamma-ray binaries, *A&A* **608**, A59, doi:10.1051/0004-6361/201731084.

Duffy, L. D. *et al.* (2006). High resolution search for dark-matter axions, *Phys. Rev. D* **74**, 1, 012006, doi:10.1103/PhysRevD.74.012006.

Duivenvoorden, S. *et al.* (2020). Have we seen all the galaxies that comprise the cosmic infrared background at 250 μm $\leq \lambda \leq$ 500 μm? *MNRAS* **491**, 1, pp. 1355–1368, doi:10.1093/mnras/stz3110.

Dumora, D. *et al.* (2001). Artificial neural networks for the shower reconstruction of gamma-showers in the energy range [20-300] GeV, *Nucl. Phys. B Proc. Suppl.* **97**, pp. 255–258, doi:10.1016/S0920-5632(01)01277-4.

Durrer, R. and Neronov, A. (2013). Cosmological magnetic fields: their generation, evolution and observation, *A&A Rev.* **21**, 62, doi:10.1007/s00159-013-0062-7.

Dwek, E., Arendt, R. G. and Krennrich, F. (2005). The near-infrared background: Interplanetary dust or primordial stars? *ApJ* **635**, 2, p. 784.

Dwek, E. and Krennrich, F. (2013). The extragalactic background light and the gamma-ray opacity of the universe, *Astropar. Phys.* **43**, pp. 112–133, doi:https://doi.org/10.1016/j.astropartphys.2012.09.003, URL https://www.sciencedirect.com/science/article/pii/S0927650512001752, seeing the High-Energy Universe with the Cherenkov Telescope Array - The Science Explored with the CTA.

Dwek, E. and Krennrich, F. (2013). The extragalactic background light and the gamma-ray opacity of the universe, *Astrop. Phys.* **43**, pp. 112–133, doi:10.1016/j.astropartphys.2012.09.003.

Eichmann, B. *et al.* (2018). Ultra-high-energy cosmic rays from radio galaxies, *J. Cosmol. Astropart. Phys.* **2**, p. 036, doi:10.1088/1475-7516/2018/02/036.

Einasto, J., Kaasik, A. and Saar, E. N. N. (1974). Dynamic evidence on massive coronas of galaxies, *Nature* **250**, p. 309.

Ellis, J., Mavromatos, N. and Nanopoulos, D. (2008). Derivation of a vacuum refractive index in a stringy space-time foam model, *Phys. Lett. B* **665**, 5, pp. 412–417, doi:https://doi.org/10.1016/j.physletb.2008.06.029.

Ellison, D. C., Drury, L. O. and Meyer, J.-P. (1997). Galactic cosmic rays from supernova remnants. II. Shock acceleration of gas and dust, *ApJ* **487**, pp. 197–217, doi:10.1086/304580.

Ellison, D. C. *et al.* (2010). Efficient cosmic ray acceleration, hydrodynamics, and self-consistent thermal X-ray emission applied to supernova remnant RX J1713.7-3946, *ApJ* **712**, pp. 287–293, doi:10.1088/0004-637X/712/1/287.

Elmegreen, B. G. and Scalo, J. (2004). Interstellar turbulence I: observations and processes, *Ann. Rev. Astron. Astrophys.* **42**, pp. 211–273, doi:10.1146/annurev.astro.41.011802.094859.

Enomoto, R. *et al.* (2002). Design study of CANGAROO-III, stereoscopic imaging atmospheric Cherenkov telescopes for sub-TeV $/\gamma$-ray detection, *Astrop. Phys.* **16**, pp. 235–244, doi:10.1016/S0927-6505(01)00112-8.

Essey, W. and Kusenko, A. (2010). A new interpretation of the gamma-ray observations of distant active galactic nuclei, *Astrop. Phys.* **33**, pp. 81–85, doi:10.1016/j.astropartphys.2009.11.007.

Essey, W. and Kusenko, A. (2012). On weak redshift dependence of gamma-ray spectra of distant blazars, *ApJL* **751**, p. L11, doi:10.1088/2041-8205/751/1/L11.

Essey, W. *et al.* (2011). Role of line-of-sight cosmic-ray interactions in forming the spectra of distant blazars in TeV gamma rays and high-energy neutrinos, *ApJ* **731**, p. 51, doi:10.1088/0004-637X/731/1/51.

Eungwanichayapant, A. and Aharonian, F. A. (2009). Very High Energy Gamma Rays from $e^{\pm}$ Pair Halos, *Int. J. Mod. Phys. D* **18**, pp. 911–927, doi:10.1142/S0218271809014832.

Evoli, C., Gaggero, D. and Grasso, D. (2015). Secondary antiprotons as a Galactic dark matter probe, *J. Cosmol. Astropart. Phys.* **1512**, 12, p. 039, doi:10.1088/1475-7516/2015/12/039.

Evoli, C., Linden, T. and Morlino, G. (2018a). Self-generated cosmic-ray confinement in TeV halos: implications for TeV gamma-ray emission and the positron excess, *Phys. Rev. D* **98**, 6, p. 063017, doi:10.1103/PhysRevD.98.063017.

Evoli, C. *et al.* (2008). Cosmic-Ray Nuclei, antiprotons and gamma rays in the Galaxy: a new diffusion model, *J. Cosmol. Astropart. Phys.* **0810**, p. 018, doi:10.1088/1475-7516/2008/10/018.

Evoli, C. *et al.* (2012). Common solution to the cosmic ray anisotropy and gradient problems, *Phys. Rev. Lett.* **108**, 21, 211102, doi:10.1103/PhysRevLett.108.211102.

Evoli, C. *et al.* (2017). Cosmic-ray propagation with DRAGON2: I. numerical solver and astrophysical ingredients, *J. Cosmol. Astropart. Phys.* **2**, 015, doi:10.1088/1475-7516/2017/02/015.

Evoli, C. *et al.* (2018b). Cosmic-ray propagation with dragon2: Ii. nuclear interactions with the interstellar gas, *J. Cosmol. Astropart. Phys.* **2018**, 07, pp. 006–006, doi:10.1088/1475-7516/2018/07/006.

Fairbairn, M. *et al.* (2014). The CTA sensitivity to Lorentz-violating effects on the gamma-ray horizon, *J. Cosmol. Astropart. Phys.* **6**, 005, doi:10.1088/1475-7516/2014/06/005.

Fang, J. and Zhang, L. (2010). Multiband emission from pulsar wind nebulae: a possible injection spectrum, *A&A* **515**, A20, doi:10.1051/0004-6361/200913615.

Fargion, D. (1999). Horizontal Tau air showers from mountains in deep vally: Traces of Ultrahigh neutrino tau, in *International Cosmic Ray Conference*, *International Cosmic Ray Conference*, Vol. 2, pp. 396–+.

Fargion, D. (2002). Discovering ultra high energy neutrinos by horizontal and upward tau air-showers: evidences in terrestrial gamma flashes? *ApJ* **570**, pp. 909–925, doi:10.1086/339772.

Fargion, D., De Sanctis Lucentini, P. G. and De Santis, M. (2004). Tau air showers from earth, *ApJ* **613**, pp. 1285–1301, doi:10.1086/423124.

Farina, M. *et al.* (2017). The photo-philic QCD axion, *J. High Energy Phys.* **2017**, 95, doi:10.1007/JHEP01(2017)095.

Farmer, A. J. and Goldreich, P. (2004). Wave damping by MHD turbulence and its effect upon cosmic ray propagation in the ISM, *ApJ* **604**, pp. 671–674, doi:10.1086/382040.

Fazio, G. G., Helmken, H. F., Rieke, G. H. and Weekes, T. C. (1968). An experiment to search for discrete sources of cosmic gamma rays in the 10^{11} to 10^{12} eV region, *Canad. J. Phys.* **46**, 10, pp. S451–S455, doi:10.1139/p68-268.

Fegan, D. J. (1997). TOPICAL REVIEW: ?/hadron separation at TeV energies, *J. Phys. G Nuclear Phys.* **23**, pp. 1013–1060, doi:10.1088/0954-3899/23/9/004.

Feng, J., Tomassetti, N. and Oliva, A. (2016). Bayesian analysis of spatial-dependent cosmic-ray propagation: Astrophysical background of antiprotons and positrons, *Phys. Rev. D* **94**, 12, p. 123007, doi:10.1103/PhysRevD.94.123007.

Feng, J. L. (2013). Naturalness and the status of supersymmetry, *Ann. Rev. Nucl. Part. Sci.* **63**, p. 351, uCI-TR-2013-01.

Feng, J. L. and Kumar, J. (2008). The WIMPless miracle: Dark-matter particles without weak-scale masses or weak interactions, *Phys. Rev. Lett.* **101**, p. 231301, doi:10.1103/PhysRevLett.101.231301.

Feng, Q. and Lin, T. T. Y. (2017). The analysis of VERITAS muon images using convolutional neural networks, in M. Brescia, S. G. Djorgovski, E. D. Feigelson, G. Longo and S. Cavuoti. eds., *Astroinformatics*, *IAU Symposium*, Vol. 325, Empty, pp. 173–179, doi:10.1017/S1743921316012734.

Feretti, L. *et al.* (2012). Clusters of galaxies: Observational properties of the diffuse radio emission, *A&A* **20**, p. 54, doi:10.1007/s00159-012-0054-z.

Ferrand, G. and Marcowith, A. (2010). On the shape of the spectrum of cosmic rays accelerated inside superbubbles, *A&A* **510**, p. A101, doi:10.1051/0004-6361/200913520.

Ferrand, G. and Safi-Harb, S. (2012). A census of high-energy observations of Galactic supernova remnants, *Adv. Space Res.* **49**, pp. 1313–1319, doi:10.1016/j.asr.2012.02.004.

Fichtel, C. E. *et al.* (1975). High-energy gamma-ray results from the second Small Astronomy Satellite. *ApJ* **198**, pp. 163–182, doi:10.1086/153590.

Finkbeiner, D. P. (2004). Microwave ism emission observed by wmap, *ApJ* **614**, pp. 186–193, doi:10.1086/423482.

Finke, J. D., Razzaque, S. and Dermer, C. D. (2010). Modeling the extragalactic background light from stars and dust, *ApJ* **712**, pp. 238–249, doi:10.1088/0004-637X/712/1/238.

Finke, J. D. *et al.* (2015). Constraints on the intergalactic magnetic field with gamma-ray observations of blazars, *ApJ* **814**, 20, doi:10.1088/0004-637X/814/1/20.

Fishman, G. J. *et al.* (1994). The first BATSE gamma-ray burst catalog, *ApJS* **92**, p. 229, doi:10.1086/191968.

Fitoussi, T. *et al.* (2017). Physics of cosmological cascades and observable properties, *MNRAS* **466**, 3, pp. 3472–3487, doi:10.1093/mnras/stw3365, URL http://dx.doi.org/10.1093/mnras/stw3365.

Fomin, V. P. *et al.* (1994). New methods of atmospheric Cherenkov imaging for gamma-ray astronomy. I. The false source method, *Astrop. Phys.* **2**, pp. 137–150, doi:10.1016/0927-6505(94)90036-1.

Fornasa, M. and Sánchez-Conde, M. A. (2015). The nature of the diffuse gamma-ray background, *Phys. Rep.* **598**, pp. 1–58, doi:10.1016/j.physrep.2015.09.002.

Fornieri, O. *et al.* (2021). Phenomenology of CR-scattering on pre-existing MHD modes, in *Proceedings of the 37th International Cosmic Ray Conference — PoS(ICRC2021), Berlin (Germany)*, Vol. 395, p. 426, doi:10.22323/1.395.0426.

Fossati, G. *et al.* (2008). Multiwavelength observations of Markarian 421 in 2001 March: An unprecedented view on the X-Ray/TeV correlated variability, *ApJ* **677**, 2, pp. 906–925, doi:10.1086/527311.

Franceschini, A. (2021). Photon-photon interactions and the opacity of the universe in gamma rays, *Universe* **7**, 5, p. 146, doi:10.3390/universe7050146.

Franceschini, A. and Rodighiero, G. (2017). The extragalactic background light revisited and the cosmic photon-photon opacity, *A&A* **603**, p. A34, doi: 10.1051/0004-6361/201629684.

Franceschini, A., Rodighiero, G. and Vaccari, M. (2008). Extragalactic optical-infrared background radiation, its time evolution and the cosmic photon-photon opacity, *A&A* **487**, pp. 837–852, doi:10.1051/0004-6361: 200809691.

Frieman, J. A. *et al.* (1995). Cosmology with ultralight pseudo Nambu-Goldstone bosons, *Phys. Rev. Lett.* **75**, pp. 2077–2080, doi:10.1103/PhysRevLett.75. 2077.

Fritz, K. D. (1989). Synchrotron emission spectra from shockwaves in active galactic nuclei: An energy and space dependent diffusion coefficient, *A&A* **214**, pp. 14–28.

Fruck, C. and Gaug, M. (2015). Atmospheric monitoring in MAGIC and data corrections, in *European Physics Journal of Web Conference*, Vol. 89, Empty, p. 02003, doi:10.1051/epjconf/20158902003.

Fukami, S. *et al.* (2022). Very-high-energy gamma-ray emission from GRB 201216C detected by MAGIC, in *37th International Cosmic Ray Conference*, p. 788, doi:10.22323/1.395.0788.

Fukui, Y. *et al.* (2012). A detailed study of the molecular and atomic gas toward the gamma-ray supernova remnant RX J1713.7-3946: Spatial TeV gamma-ray and interstellar medium gas correspondence, *ApJ* **746**, p. 82, doi:10. 1088/0004-637X/746/1/82.

Funk, S. (2015). Ground- and space-based gamma-ray astronomy, **65**, pp. 245–277, doi:10.1146/annurev-nucl-102014-022036.

Funk, S. and Hinton, J. A. (2013). Comparison of *Fermi*-LAT and CTA in the region between 10–100 GeV, *Astrop. Phys.* **43**, pp. 348–355, doi:10.1016/j. astropartphys.2012.05.018.

Funk, S. *et al.* (2004). The trigger system of the H.E.S.S. telescope array, *Astrop. Phys.* **22**, 3–4, pp. 285–296, doi:10.1016/j.astropartphys.2004.08.001.

Funk, S. *et al.* (2007). Energy dependent morphology in the PWN candidate HESS J1825-137, arXiv e-prints, arXiv:0709.3125.

Gabici, S. (2013). Cosmic rays and molecular clouds, in D. F. Torres and O. Reimer. eds., *Cosmic Rays in Star-Forming Environments, Astrophysics and Space Science Proceedings*, Vol. 34, p. 221, doi:10.1007/ 978-3-642-35410-6_16.

Gabici, S. and Aharonian, F. A. (2014). Hadronic gamma rays from RX J1713.7-3946? *MNRAS* **445**, pp. L70–L73, doi:10.1093/mnrasl/slu132.

Gabici, S., Aharonian, F. A. and Blasi, P. (2007). Gamma rays from molecular clouds, *Ap&SS* **309**, pp. 365–371, doi:10.1007/s10509-007-9427-6.

Gabici, S., Aharonian, F. A. and Casanova, S. (2009). Broad-band non-thermal emission from molecular clouds illuminated by cosmic rays from nearby supernova remnants, *MNRAS* **396**, pp. 1629–1639, doi:10.1111/j.1365-2966.2009.14832.x.

Gabici, S. and Blasi, P. (2003). The gamma-ray background from large scale structure formation, *Astrop. Phys.* **19**, pp. 679–689, doi:10.1016/S0927-6505(03)00106-3.

Gabici, S. *et al.* (2010). Constraints on the cosmic ray diffusion coefficient in the W28 region from gamma-ray observations.

Gabici, S. *et al.* (2019). The origin of Galactic cosmic rays: Challenges to the standard paradigm, *Int. J. Mod. Phys. D* **28**, 15, p. 1930022, doi:10.1142/S0218271819300222.

Gaensler, B. M. and Slane, P. O. (2006). The evolution and structure of pulsar wind nebulae, *Ann. Rev. Astron. Astrophys.* **44**, 1, pp. 17–47, doi:10.1146/annurev.astro.44.051905.092528.

Gaggero, D. and Valli, M. (2018). Impact of cosmic-ray physics on dark matter indirect searches, *Adv. in High Energy Phys.*

Gaggero, D. *et al.* (2013). Three-dimensional model of cosmic-ray lepton propagation reproduces data from the Alpha Magnetic Spectrometer on the International Space Station, *Phys. Rev. Lett.* **111**, p. 021102, doi:10.1103/PhysRevLett.111.021102.

Gaggero, D. *et al.* (2015a). gamma-ray sky points to radial gradients in cosmic-ray transport, *Phys. Rev. D* **D91**, 8, p. 083012, doi:10.1103/PhysRevD.91.083012.

Gaggero, D. *et al.* (2015b). Towards a realistic astrophysical interpretation of the gamma-ray Galactic center excess, *J. Cosmol. Astropart. Phys.* **1512**, 12, p. 056, doi:10.1088/1475-7516-2015-12-056,10.1088/1475-7516/2015/12/056.

Gaggero, D. *et al.* (2017). Diffuse cosmic rays shining in the Galactic Center: A Novel Interpretation of H.E.S.S. and *Fermi*-LAT gamma-ray data, *Phys. Rev. Lett.* **119**, 3, 031101, doi:10.1103/PhysRevLett.119.031101.

Gaisser, T. K. (1990). Cosmic rays and particle physics.

Gajdus, M. *et al.* (2015). Pulsations from the Vela pulsar down to 20 GeV with H.E.S.S. II, in A. S. Borisov, V. G. Denisova, Z. M. Guseva, E. A. Kanevskaya, M. G. Kogan, A. E. Morozov, V. S. Puchkov, S. E. Pyatovsky, G. P. Shoziyoev, M. D. Smirnova, A. V. Vargasov, V. I. Galkin, S. I. Nazarov and R. A. Mukhamedshin. eds., *Proceedings of the 34th International Cosmic Ray Conference — PoS(ICRC2015), The Hague (Netherlands)*, Vol. 34, World Scientific, p. 841.

Galante, N. and for the VERITAS Collaboration (2009). Observation of radio galaxies and clusters of galaxies with VERITAS, ArXiv e-prints.

Galanti, G. and Roncadelli, M. (2018). Extragalactic photon-axion-like particle oscillations up to 1000 TeV, *J. High-Energy Astroph.* **20**, pp. 1–17, doi:10.1016/j.jheap.2018.07.002.

Galanti, G. *et al.* (2020). Hint at an axion-like particle from the redshift dependence of blazar spectra, *MNRAS* **493**, 2, pp. 1553–1564, doi:10.1093/mnras/stz3410.

Galbraith, W. and Jelley, J. V. (1953). Light pulses from the night sky associated with cosmic rays, *Nature* **171**, 4347, pp. 349–350, doi:10.1038/171349a0.

Gao, L. *et al.* (2012). Where will supersymmetric dark matter first be seen? *MNRAS* **419**, p. 1721, doi:10.1111/j.1365-2966.2011.19836.x.

Garrison-Kimmel, S. *et al.* (2019). The local group on FIRE: Dwarf galaxy populations across a suite of hydrodynamic simulations, *MNRAS* **487**, 1, p. 1380–1399, doi:10.1093/mnras/stz1317, URL http://dx.doi.org/10.1093/mnras/stz1317.

Gaug, M. *et al.* (2007). Tau neutrino search with the MAGIC telescope, in *Proceedings of the 30th International Cosmic Ray Conference — PoS(ICRC2007), Merida (Mexico)*, Vol. 3, pp. 1273–1278.

Gazizov, A. and Kowalski, M. P. (2005). ANIS: High energy neutrino generator for neutrino telescopes, *Comput. Phys. Commun.* **172**, pp. 203–213, doi:10.1016/j.cpc.2005.03.113.

Gehrels, N. and Shrader, C. R. (1996). Status of the Compton gamma ray observatory, *Astronom. Astrophys. Suppl.* **120**, pp. 1–4.

Gelfand, J. D., Slane, P. O. and Zhang, W. (2009). A dynamical model for the evolution of a pulsar wind nebula inside a non-radiative supernova remnant, *ApJ* **703**, 2, pp. 2051–2067, doi:10.1088/0004-637X/703/2/2051.

Gelmini, G. B., Kalashev, O. E. and Semikoz, D. V. (2008). GZK photons as ultra-high-energy cosmic rays, *Sov. J. Exper. Theoret. Phys.* **106**, pp. 1061–1082, doi:10.1134/S106377610806006X.

Genolini, Y. (2018). Theoretical interpretations of *DAMPE* first results: A critical review,.

Geringer-Sameth, A., Koushiappas, S. M. and Walker, M. (2015). Dwarf galaxy annihilation and decay emission profiles for dark matter experiments, *ApJ* **801**, 2, p. 74, doi:10.1088/0004-637X/801/2/74.

Geringer-Sameth, A. *et al.* (2013). The VERITAS dark matter program, arXiv e-prints.

Gershtein, S. and Zeldovich, Y. A. (1966). Rest mass of muonic neutrino and cosmology, *JETP Lett.* **4**, p. 120.

Ghisellini, G., Righi, C., Costamante, L. and Tavecchio, F. (2017). The Fermi blazar sequence, *MNRAS* **469**, 1, pp. 255–266, doi:10.1093/mnras/stx806.

Ghisellini, G., Tavecchio, F. and Chiaberge, M. (2005). Structured jets in TeV BL Lac objects and radiogalaxies. Implications for the observed properties, *A&A* **432**, 2, pp. 401–410, doi:10.1051/0004-6361:20041404.

Ghisellini, G. *et al.* (1993). Relativistic bulk motion in active galactic nuclei, *ApJ* **407**, p. 65, doi:10.1086/172493.

Ghosh, D., Sachdeva, D. and Singh, P. (2021). Future constraints on primordial black Holes from XGIS-THESEUS.

Giacinti, G., Kachelrieß, M. and Semikoz, D. V. (2013). Anisotropic cosmic ray diffusion and its implications for gamma-ray astronomy, *Phys. Rev. D* **88**, 2, p. 023010, doi:10.1103/PhysRevD.88.023010.

Giacinti, G., Mitchell, A. M. W., López-Coto, R., Joshi, V., Parsons, R. D. and Hinton, J. A. (2020a). Halo fraction in TeV-bright pulsar wind nebulae, *A&A* **636**, p. A113, doi:10.1051/0004-6361/201936505.

Giacinti, G. *et al.* (2020b). Halo fraction in tev-bright pulsar wind nebulae, *A&A* **636**, p. A113, doi:10.1051/0004-6361/201936505.

Giacomelli, G. and Patrizii, L. (2001). Magnetic monopoles, in *NATO Advanced Research Workshop on Cosmic Radation: From Astronomy to Particle Physics Oujda, Morocco, March 21-23, 2001.*

Giannios, D., Uzdensky, D. A. and Begelman, M. C. (2010). Fast TeV variability from misaligned minijets in the jet of M87, *MNRAS* **402**, 3, pp. 1649–1656, doi:10.1111/j.1365-2966.2009.16045.x.

Gilmore, R. C. (2012). Constraining the near-infrared background light from Population III stars using high-redshift gamma-ray sources, *MNRAS* **420**, pp. 800–809, doi:10.1111/j.1365-2966.2011.20092.x.

Gilmore, R. C., Somerville, R. S., Primack, J. R. and Domínguez, A. (2012). Semi-analytic modelling of the extragalactic background light and consequences for extragalactic gamma-ray spectra, *MNRAS* **422**, pp. 3189–3207, doi:10.1111/j.1365-2966.2012.20841.x.

Ginzburg, V. L. and Syrovatskii, S. I. (1964). *The Origin of Cosmic Rays.*

Giroletti, M. *et al.* (2004). Parsec-scale properties of Markarian 501, *ApJ* **600**, 1, pp. 127–140, doi:10.1086/379663.

Giroletti, M. *et al.* (2008). The jet of Markarian 501 from millions of Schwarzschild radii down to a few hundreds, *A&A* **488**, 3, pp. 905–914, doi:10.1051/0004-6361:200809784.

Goldreich, P. and Julian, W. H. (1969). Pulsar electrodynamics, *ApJ* **157**, p. 869, doi:10.1086/150119.

Goldreich, P. and Sridhar, S. (1995). Toward a theory of interstellar turbulence. 2: Strong alfvenic turbulence, *ApJ* **438**, pp. 763–775, doi:10.1086/175121.

Goldstein, A. *et al.* (2017). An ordinary short gamma-ray burst with extraordinary implications: *Fermi*-GBM Detection of GRB 170817A, *ApJL* **848**, 2, L14, doi:10.3847/2041-8213/aa8f41.

Goodenough, L. and Hooper, D. (2009). Possible evidence for dark matter annihilation in the inner milky way from the fermi gamma ray space telescope.

Goodfellow, I., Bengio, Y. and Courville, A. (2016). *Deep Learning*, MIT Press, http://www.deeplearningbook.org.

Gora, D. and Bernardini, E. (2016). Detection of tau neutrinos by imaging air Cherenkov telescopes, *Astrop. Phys.* **82**, pp. 77–85, doi:10.1016/j.astropartphys.2016.06.002.

Gora, D., Roth, M. and Tamburro, A. (2007). A MC approach to simulate up- and down-going neutrino showers including local topographic conditions, *Astrop. Phys.* **26**, pp. 402–413, doi:10.1016/j.astropartphys.2006.08.008.

Gora, D. *et al.* (2017a). MAGIC gamma-ray telescopes hunting for neutrinos and their sources, in *Proceedings, 27th International Conference on Neutrino Physics and Astrophysics (Neutrino 2016): London, United Kingdom, July 4-9, 2016*, Vol. 888, p. 012147, doi:10.1088/1742-6596/888/1/012147.

Gora, D. *et al.* (2017b). Sensitivity for Tau neutrinos at PeV energies and beyond with the MAGIC telescopes, in *Proceedings of the 35th International Cosmic Ray Conference — PoS(ICRC2017), Busan (South Korea)*, Vol. 301, p. 992, doi:10.22323/1.301.0992.

Goret, P. *et al.* (1993). Observations of TeV gamma rays from the Crab Nebula, *A&A* **270**, pp. 401–406.

Gould, R. J. and Schréder, G. P. (1967a). Opacity of the Universe to high-energy photons, *Physical Review* **155**, pp. 1408–1411, doi:10.1103/PhysRev.155.1408.

Gould, R. J. and Schréder, G. P. (1967b). Pair production in photon-photon collisions, *Phys. Rev.* **155**, pp. 1404–1407, doi:10.1103/PhysRev.155.1404.

Granot, J. and Sari, R. (2002). The shape of spectral breaks in gamma-ray burst afterglows, *ApJ* **568**, 2, pp. 820–829, doi:10.1086/338966.

Grasso, D. and Rubinstein, H. R. (2001). Magnetic fields in the early Universe. *Phys. Rep.* **348**, pp. 163–266, doi:10.1016/S0370-1573(00)00110-1.

Green, A. M. and Kavanagh, B. J. (2021). Primordial Black Holes as a dark matter candidate, *J. Phys. G* **48**, 4, p. 043001, doi:10.1088/1361-6471/abc534.

Griest, K. and Kamionkowski, M. (1990). Unitarity limits on the mass and radius of dark matter particles, *Phys. Rev. Lett.* **64**, p. 615, doi:10.1103/PhysRevLett.64.615.

Grifols, J. A., Massó, E. and Toldrà, R. (1996). Gamma rays from SN 1987A due to pseudoscalar conversion, *Phys. Rev. Lett.* **77**, pp. 2372–2375, doi:10.1103/PhysRevLett.77.2372.

Guedes Lang, R., Martínez-Huerta, H. and de Souza, V. (2018). Limits on the Lorentz invariance violation from UHECR astrophysics, *ApJ* **853**, p. 23, doi:10.3847/1538-4357/aa9f2c.

Guo, J. *et al.* (2021). Implications of axion-like particles from the Fermi-LAT and H.E.S.S. observations of PG 1553+113 and PKS 2155−304, *Chin. Phys. C* **45**, 2, p. 025105, doi:10.1088/1674-1137/abcd2e.

Gupta, S. *et al.* (2020). Realistic modelling of wind and supernovae shocks in star clusters: Addressing ^{22}Ne/^{20}Ne and other problems in Galactic cosmic rays, *MNRAS* **493**, 3, pp. 3159–3177, doi:10.1093/mnras/staa286.

Hagedorn, R. (1968). Hadronic matter near the boiling point, *Nuovo Cim.* **A56**, pp. 1027–1057, doi:10.1007/BF02751614.

Hahn, J. *et al.* (2014). Impact of aerosols and adverse atmospheric conditions on the data quality for spectral analysis of the H.E.S.S. telescopes, *Astrop. Phys.* **54**, pp. 25–32, doi:10.1016/j.astropartphys.2013.10.003.

Hall, J. *et al.* (2003). Veritas cfds, *Inter. Cosmic Ray Conf.* **5**, p. 2851.

Halzen, F., Vazques, R. and Zas, E. (1991). Seperating gamma-ray signals By Cherenkov Imaging: Neural network optimization, *Inter. Cosmic Ray Conf.* **1**, p. 504.

Halzen, F. *et al.* (1991). Gamma-rays and energetic particles from primordial black holes, *Nature* **353**, pp. 807–815, doi:10.1038/353807a0.

Hanna, D. S. *et al.* (2002). The STACEE-32 ground based gamma-ray detector, *Nucl. Instrum. Meth. Phys. Res. A* **491**, 1–2, pp. 126–151, doi:10.1016/S0168-9002(02)01126-9.

Harding, A. K., Tademaru, E. and Esposito, L. W. (1978). A curvature — radiation — pair-production model for gamma-ray pulsars. *ApJ* **225**, pp. 226–236, doi:10.1086/156486.

Hargis, J. R., Willman, B. and Peter, A. H. G. (2014). Too many, too few, or just right? The predicted number and distribution of Milky Way dwarf galaxies, *ApJ* **795**, 1, p. L13, doi:10.1088/2041-8205/795/1/L13.

Harrison, E. R. (1970). Generation of magnetic fields in the radiation ERA, *MNRAS* **147**, 3, pp. 279–286, doi:10.1093/mnras/147.3.279, URL http://dx.doi.org/10.1093/mnras/147.3.279.

Hartmann, D. (2018). Advancing the MeV frontier with AMEGO.

Hassan, T. *et al.* (2017). Monte Carlo performance studies for the site selection of the Cherenkov Telescope Array, *Astrop. Phys.* **93**, pp. 76–85, doi:10.1016/j.astropartphys.2017.05.001.

Hawking, S. (1971). Gravitationally collapsed objects of very low mass, *MNRAS* **152**, p. 75.

Hawking, S. W. (1974). Black hole explosions, *Nature* **248**, pp. 30–31, doi:10.1038/248030a0.

Hayashi, K., Chiba, M. and Ishiyama, T. (2020). Diversity of dark matter density profiles in the Galactic dwarf spheroidal satellites, *ApJ* **904**, 1, p. 45, doi:10.3847/1538-4357/abbe0a.

He, D.-Z. *et al.* (2019). Prospect for dark matter annihilation signatures from gamma-ray observation of dwarf galaxies by LHAASO, *Phys. Rev. D* **100**, 8, p. 083003, doi:10.1103/PhysRevD.100.083003.

Heck, D. *et al.* (1998). *CORSIKA: A Monte Carlo code to simulate extensive air showers.*, Forschungszentrum Karlsruhe GmbH, Karlsruhe (Germany).

Heck, D. *et al.* (1998). CORSIKA: A Monte Carlo code to simulate extensive air showers.

Heger, A. *et al.* (2003). How massive single stars end their life, *ApJ* **591**, pp. 288–300, doi:10.1086/375341.

Heiles, C. and Troland, T. H. (2004). The Millennium Arecibo 21 Centimeter Absorption-Line Survey. III. Techniques for spectral polarization and results for stokes V, *ApJS* **151**, pp. 271–297, doi:10.1086/381753.

Heiter, C. *et al.* (2018). Production and propagation of ultra-high energy photons using CRPropa 3, *Astrop. Phys.* **102**, pp. 39–50, doi:10.1016/j.astropartphys.2018.05.003.

Helder, E. A. *et al.* (2012). Observational signatures of particle acceleration in supernova remnants, *Space Sci. Rev.* **173**, 1–4, pp. 369–431, doi:10.1007/s11214-012-9919-8.

Helgason, K. and Kashlinsky, A. (2012). Reconstructing the gamma-ray photon optical depth of the Universe to z $\sim$ 4 from multiwavelength Galaxy survey data, *ApJL* **758**, p. L13, doi:10.1088/2041-8205/758/1/L13.

Helmi, A. *et al.* (2012). Not too big, not too small: The dark haloes of the dwarf spheroidals in the Milky Way, *MNRAS* **428**, 2, pp. 1696–1703.

Helmi, A. *et al.* (2018). Gaia data release 2: Kinematics of globular clusters and dwarf galaxies around the Milky Way, *A&A* **616**, p. A12, doi:10.1051/0004-6361/201832698, [Erratum: *Astron. Astrophys.* **637**, C3 (2020), Erratum: Astron.Astrophys. 642, C1 (2020)].

Hernquist, L. (1990). An analytical model for spherical galaxies and bulges, *ApJ* **356**, p. 359, doi:10.1086/168845.

Heß, M. *et al.* (1999). The time structure of Cherenkov images generated by TeV γ rays and by cosmic rays, *Astrop. Phys.* **11**, pp. 363–377, doi:10.1016/S0927-6505(99)00003-1.

Higdon, J. C. and Lingenfelter, R. E. (2005). OB associations, supernova-generated superbubbles, and the source of cosmic rays, *ApJ* **628**, pp. 738–749, doi:10.1086/430814.

Hillas, A. M. (1982). Angular and energy distributions of charged particles in electron-photon cascades in air, *J. Phys. G* **8**, pp. 1461–1473, doi:10.1088/0305-4616/8/10/016.

Hillas, A. M. (1985a). Cerenkov light images of EAS produced by primary gamma, in F. C. Jones. ed., *International Cosmic Ray Conference*, Vol. 3, Empty, pp. 445–448.

Hillas, A. M. (1985b). *The Origin of Ultra-High-Energy Cosmic Rays*. p. 277.

Hillas, A. M. (1996). Differences between gamma-ray and hadronic showers, *Space Sci. Rev.* **75**, pp. 17–30, doi:10.1007/BF00195021.

Hillas, A. M. (2006). Cosmic rays: Recent progress and some current questions, arXiv Astrophysics e-prints.

Hinton, J. A. and Hofmann, W. (2009). Teraelectronvolt astronomy, *Ann. Rev. Astron. Astrophys.* **47**, pp. 523–565, 10.1146/annurev-astro-082708-101816.

Hirata, K. *et al.* (1987). Observation of a neutrino burst from the supernova sn 1987a, *Phys. Rev. Lett.* **58**, pp. 1490–1493, doi:10.1103/PhysRevLett.58.1490.

Hiroshima, N., Ando, S. and Ishiyama, T. (2018). Modeling evolution of dark matter substructure and annihilation boost, *Phys. Rev. D* **D97**, 12, p. 123002, doi:10.1103/PhysRevD.97.123002.

Hochmuth, K. A. and Sigl, G. (2007). Effects of axion-photon mixing on gamma-ray spectra from magnetized astrophysical sources, *Phys. Rev. D* **76**, 12, 123011, doi:10.1103/PhysRevD.76.123011.

Hofmann, W. (2001). Status of the high energy stereoscopic system (H.E.S.S.) project, in *Proceedings, X International Cosmic Ray Conference (ICRC 2001): Hamburg, Germany 07-15 August 2001*, Copernicus Gesellschaft 2001, pp. 2785–2788.

Hofmann, W. (2003). Intercalibration of Cherenkov telescopes in telescope arrays. *Astrop. Phys.* **20**, pp. 1–3, doi:10.1016/S0927-6505(03)00137-3.

Hofmann, W. (2006). Performance Limits for Cherenkov Instruments, ArXiv Astrophysics e-prints.

Hofmann, W. (2020). On angular resolution limits for air shower arrays, *Astropart. Phys.* **123**, 102479, doi:10.1016/j.astropartphys.2020.102479.

Hofmann, W. *et al.* (1999). Comparison of techniques to reconstruct VHE gamma-ray showers from multiple stereoscopic Cherenkov images, *Astrop. Phys.* **12**, pp. 135–143, doi:10.1016/S0927-6505(99)00084-5.

Holder, J. *et al.* (2005). Exploiting VERITAS timing information, in *Proceedings, 29th International Cosmic Ray Conference (ICRC 2005): Pune, India, August 3-11, 2005*, Vol. 5, Tata Institute of Fundamental Research, p. 383.

Holder, J. *et al.* (2006). The first VERITAS telescope, *Astrop. Phys.* **25**, 6, pp. 391–401, doi:10.1016/j.astropartphys.2006.04.002.

Holder, J. *et al.* (2008). Status of the VERITAS Observatory, in F. A. Aharonian, W. Hofmann and F. Rieger. eds., *American Institute of Physics Conference Series, American Institute of Physics Conference Series*, Vol. 1085, pp. 657–660, doi:10.1063/1.3076760.

Holler, M. *et al.* (2015). Observations of the Crab Nebula with H.E.S.S. phase II, in A. S. Borisov, V. G. Denisova, Z. M. Guseva, E. A. Kanevskaya, M. G. Kogan, A. E. Morozov, V. S. Puchkov, S. E. Pyatovsky, G. P. Shoziyoev, M. D. Smirnova, A. V. Vargasov, V. I. Galkin, S. I. Nazarov and R. A. Mukhamedshin. eds., *Proceedings of the 34th International Cosmic Ray Conference — PoS(ICRC2015), The Hague (Netherlands)*, World Scientific, p. 847.

Hooper, D. (2006). *Dark Cosmos: In Search of Our Universe's Missing Mass and Energy*.

Hooper, D. and Goodenough, L. (2011). Dark matter annihilation in The Galactic center as seen by the Fermi Gamma Ray Space Telescope, *Phys. Let. B* **B697**, pp. 412–428, doi:10.1016/j.physletb.2011.02.029.

Hooper, D. and Linden, T. (2018). Millisecond pulsars, TeV halos, and implications for the Galactic Center gamma-ray excess, *Phys. Rev. D* **98**, 4, p. 043005, doi:10.1103/PhysRevD.98.043005.

Hooper, D. and Serpico, P. D. (2007). Detecting axion-like particles with gamma ray telescopes, *Phys. Rev. Lett.*, pp. 8–11, doi:10.1103/PhysRevLett.99.231102.

Hooper, D. and Witte, S. J. (2017). Gamma rays from dark matter subhalos revisited: Refining the predictions and constraints, *J. Cosmol. Astropart. Phys.* **4**, 018, doi:10.1088/1475-7516/2017/04/018.

Hooper, D. *et al.* (2017). HAWC observations strongly favor pulsar interpretations of the cosmic-ray positron excess, *Phys. Rev. D* **96**, 10, p. 103013, doi:10.1103/PhysRevD.96.103013.

Hörandel, J. R. (2003). On the knee in the energy spectrum of cosmic rays, *Astrop. Phys.* **19**, pp. 193–220, doi:10.1016/S0927-6505(02)00198-6.

Horiuchi, S., Beacom, J. F. and Dwek, E. (2009). Diffuse supernova neutrino background is detectable in Super-Kamiokande, *Phys. Rev. D* **79**, 8, p. 083013, doi:10.1103/PhysRevD.79.083013.

Horns, D. and Meyer, M. (2012). Indications for a pair-production anomaly from the propagation of VHE gamma-rays, *J. Cosmol. Astropart. Phys.* **2**, 033, doi:10.1088/1475-7516/2012/02/033.

Horns, D. *et al.* (2007). Nucleonic gamma-ray production in pulsar wind nebulae, *Ap& SS* **309**, 1–4, pp. 189–195, doi:10.1007/s10509-007-9444-5.

Horns, D. *et al.* (2012). Hardening of TeV gamma spectrum of active galactic nuclei in galaxy clusters by conversions of photons into axionlike particles, *Phys. Rev. D* **86**, 7, 075024, doi:10.1103/PhysRevD.86.075024.

Hovatta, T. *et al.* (2016). Optical polarization of high-energy BL Lacertae objects, *A&A* **596**, p. A78, doi:10.1051/0004-6361/201628974.

Huang, Y. *et al.* (2022). LHAASO observed GRB 221009A with more than 5000 VHE photons up to around 18 TeV, *GRB Coordinates Network* **32677**, p. 1.

Hubble, E. and Humason, M. L. (1931). The velocity-distance relation among extra-galactic nebulae, *ApJ* **74**, p. 43, doi:10.1086/143323.

Hui, C. M. (2008). VERITAS distant laser calibration and atmospheric monitoring, in F. A. Aharonian, W. Hofmann and F. Rieger. eds., *American Instit. of Physics Conference Series*, Vol. 1085, pp. 735–737, doi:10.1063/1.3076783.

Hütten, M. *et al.* (2016). Dark matter substructure modelling and sensitivity of the Cherenkov Telescope Array to Galactic dark halos, *J. Cosmol. Astropart. Phys.* **9**, 047, doi:10.1088/1475-7516/2016/09/047.

Ibarra, A., Lopez Gehler, S. and Pato, M. (2012). Dark matter constraints from box-shaped gamma-ray features, *J. Cosmol. Astropart. Phys.* **1207**, p. 043, doi:10.1088/1475-7516/2012/07/043.

Ibata, R. *et al.* (2013). Do globular clusters possess Dark Matter halos? A case study in NGC 2419, *MNRAS* **428**, p. 3648, doi:10.1093/mnras/sts302.

Ichikawa, K. *et al.* (2017). Foreground effect on the *J*-factor estimation of classical dwarf spheroidal galaxies, *MNRAS* **468**, 3, pp. 2884–2896, doi: 10.1093/mnras/stx682.

Ichikawa, K. *et al.* (2018). Foreground effect on the *J*-factor estimation of ultrafaint dwarf spheroidal galaxies, *MNRAS* **479**, 1, pp. 64–74, doi:10.1093/mnras/sty1387.

Inada, T. *et al.* (2021). Upper limits on the WIMP annihilation cross section from a joint analysis of dwarf spheroidal satellite galaxy observations with the MAGIC telescopes, in *Proceedings of the 37th International Cosmic Ray Conference — PoS(ICRC2021), Berlin (Germany)*, Vol. 395, p. 512, doi: 10.22323/1.395.0512.

Inoue, S., Aharonian, F. A. and Sugiyama, N. (2005). Hard X-Ray and gamma-ray emission induced by ultra-high-energy protons in cluster accretion shocks *ApJL* **628**, pp. L9–L12, doi:10.1086/432602.

Inoue, T. *et al.* (2012). Toward understanding the cosmic-ray acceleration at young supernova remnants interacting with interstellar clouds: Possible applications to RX J1713.7-3946, *ApJ* **744**, p. 71, doi:10.1088/0004-637X/744/1/71.

Irastorza, I. G. and Redondo, J. (2018). New experimental approaches in the search for axion-like particles, *Progress in Particle and Nuclear Physics* **102**, pp. 89–159, doi:10.1016/j.ppnp.2018.05.003.

Irastorza, I. G. *et al.* (2013). Future axion searches with the International Axion Observatory (IAXO), in *J. Phys. Conf. Ser., J. Phys. Conf. Ser.*, Vol. 460, p. 012002, doi:10.1088/1742-6596/460/1/012002.

Issa, M. R. and Wolfendale, A. W. (1981). gamma rays from the cosmic ray irradiation of local molecular clouds, *Nature* **292**, pp. 430–433, doi:10.1038/292430a0.

Jacob, U. and Piran, T. (2008). Inspecting absorption in the spectra of extra-galactic gamma-ray sources for insight into Lorentz invariance violation, *Phys. Rev. D* **78**, 12, p. 124010, doi:10.1103/PhysRevD.78.124010.

Jansson, R. and Farrar, G. R. (2012). The Galactic magnetic field, *ApJL* **761**, p. L11, doi:10.1088/2041-8205/761/1/L11.

Jardin-Blicq, A. and Joshi, V. (2017). HAWC upgrade for multi-TeV γ-ray detection, in *6th International Symposium on High Energy Gamma-Ray Astronomy*, *American Institute of Physics Conference Series*, Vol. 1792, p. 070010, doi:10.1063/1.4969007.

Jedamzik, K. and Saveliev, A. (2019). Stringent limit on primordial magnetic fields from the cosmic microwave background radiation, *Phys. Rev. Lett.* **123**, p. 021301, doi:10.1103/PhysRevLett.123.021301, URL https://link.aps.org/doi/10.1103/PhysRevLett.123.021301.

Jeltema, T. E., Kehayias, J. and Profumo, S. (2009). Gamma rays from clusters and groups of galaxies: Cosmic rays versus dark matter, *Phys. Rev. D* **D80**, p. 023005, doi:10.1103/PhysRevD.80.023005.

Johannesson, G. *et al.* (2016). Bayesian analysis of cosmic-ray propagation: Evidence against homogeneous diffusion, *ApJ* **824**, 1, p. 16, doi:10.3847/0004-637X/824/1/16.

Johnson, W. N. *et al.* (1993). The oriented scintillation spectrometer experiment: Instrument description, *ApJS* **86**, p. 693, doi:10.1086/191795.

Johnston, S. *et al.* (1992). A high-frequency survey of the southern galactic plane for pulsars, **255**, p. 401, doi:10.1093/mnras/255.3.401.

Jokipii, J. R. (1966). Cosmic-ray propagation. I. Charged particles in a random magnetic field, *ApJ* **146**, p. 480, doi:10.1086/148912.

Jokipii, J. R. and Parker, E. N. (1968). Random walk of magnetic lines of force in astrophysics, *Phys. Rev. Lett.* **21**, pp. 44–47, doi:10.1103/PhysRevLett.21.44.

Jones, F. C. (1994). A theoretical review of diffusive shock acceleration, *ApJS* **90**, p. 561, doi:10.1086/191875.

Jones, T. W. (2004). Ultra-high-energy cosmic rays and clusters, *J. Korean Astron. Soc.* **37**, pp. 421–426, doi:10.5303/JKAS.2004.37.5.421.

Joshi, J. C. *et al.* (2018). Very high-energy gamma-ray signature of ultrahigh-energy cosmic ray acceleration in Centaurus A, *MNRAS* **478**, pp. L1–L6, doi:10.1093/mnrasl/sly064.

Jungman, G. and Kamionkowski, M. (1995). Gamma rays from neutralino annihilation, *Phys. Rev. D* **51**, pp. 3121–3124, doi:10.1103/PhysRevD.51.3121.

Kachelriess, M., Ostapchenko, S. and Tomas, R. (2012). ELMAG: A Monte Carlo simulation of electromagnetic cascades on the extragalactic background light and in magnetic fields, *Comput. Phys. Commun.* **183**, 4, pp. 1036–1043, doi: https://doi.org/10.1016/j.cpc.2011.12.025.

Kahlhoefer, F. *et al.* (2019). Diversity in density profiles of self-interacting dark matter satellite halos, *J. Cosmol. Astropart. Phys.* **12**, p. 010, doi:10.1088/1475-7516/2019/12/010.

Kahniashvili, T., Brandenburg, A. and Tevzadze, A. G. (2016). The evolution of primordial magnetic fields since their generation, *Physica Scripta* **91**, 10, p. 104008, doi:10.1088/0031-8949/91/10/104008.

Kalapotharakos, C. and Contopoulos, I. (2009). Three-dimensional numerical simulations of the pulsar magnetosphere: Preliminary results, *A&A* **496**, 2, pp. 495–502, doi:10.1051/0004-6361:200810281.

Kalashev, O. E. *et al.* (2002). Ultra-high-energy neutrino fluxes and their constraints, *Phys. Rev. D* **66**, 6, p. 063004, doi:10.1103/PhysRevD.66.063004.

Kanbach, G. *et al.* (1989). The project EGRET (energetic gamma-ray experiment telescope) on NASA's Gamma-Ray Observatory GRO, *Space Sci. Rev.* **49** 1–2, pp. 69–84, doi:10.1007/BF00173744.

Kang, H., Rachen, J. P. and Biermann, P. L. (1997). Contributions to the cosmic ray flux above the ankle: Clusters of galaxies, *MNRAS* **286**, pp. 257–267. doi:10.1093/mnras/286.2.257.

Kang, H. and Ryu, D. (2013). Diffusive shock acceleration at cosmological shock waves, *ApJ* **764**, p. 95, doi:10.1088/0004-637X/764/1/95.

Kang, H., Ryu, D. and Jones, T. W. (1996). Cluster accretion shocks as possible acceleration sites for ultra-high-energy protons below the Greisen cutoff, *ApJ* **456**, p. 422, doi:10.1086/176666.

Kaplan, D. L. *et al.* (2008). A precise proper motion for the Crab Pulsar, and the difficulty of Testing spin-kick alignment for young neutron Stars, *ApJ* **677**, pp. 1201–1215, doi:10.1086/529026.

Kartavtsev, A., Raffelt, G. and Vogel, H. (2017). Extragalactic photon-ALP conversion at CTA energies, *J. Cosmol. Astropart. Phys.* **1**, p. 024, doi: 10.1088/1475-7516/2017/01/024.

Kasen, D. *et al.* (2017). Origin of the heavy elements in binary neutron-star mergers from a gravitational-wave event, *Nature* **551**, pp. 80–84, doi:10.1038/nature24453.

Kelley-Hoskins, N. (2018). The VERITAS dark matter and astroparticle physics program (tevpa 2018 conference).

Kelner, S. R. and Aharonian, F. A. (2008). Energy spectra of gamma rays, electrons, and neutrinos produced at interactions of relativistic protons with low energy radiation, *Phys. Rev. D* **78**, 3, p. 034013, doi:10.1103/PhysRevD.78.034013.

Kelner, S. R., Aharonian, F. A. and Bugayov, V. V. (2006). Energy spectra of gamma rays, electrons, and neutrinos produced at proton-proton interactions in the very high energy regime, *Phys. Rev. D* **74**, p. 034018, doi:10.1103/PhysRevD.74.034018.

Kelner, S. R., Aharonian, F. A. and Bugayov, V. V. (2006). Energy spectra of gamma rays, electrons, and neutrinos produced at proton-proton interactions in the very high energy regime, *Phys. Rev. D* **74**, 3, p. 034018, doi:10.1103/PhysRevD.74.034018.

Kennel, C. F. and Coroniti, F. V. (1984). Magnetohydrodynamic model of Crab nebula radiation, *ApJ* **283**, pp. 710–730, doi:10.1086/162357.

Kephart, T. W. and Weiler, T. J. (1996). Magnetic monopoles as the highest energy cosmic ray primaries, *Astrop. Phys.* **4**, pp. 271–279, doi:10.1016/0927-6505(95)00043-7.

Kerszberg, D. *et al.* (2017). The cosmic-ray electron spectrum measured up to 20 tev with h.e.s.s. .

Keshet, U. *et al.* (2003). Gamma rays from intergalactic shocks, *ApJ* **585**, pp. 128–150, doi:10.1086/345946.

Keshet, U. *et al.* (2017). Preliminary evidence for a virial shock around the Coma galaxy cluster, *ApJ* **845**, p. 24, doi:10.3847/1538-4357/aa794b.

Khaire, V. and Srianand, R. (2019). New synthesis models of consistent extragalactic background light over cosmic time, *MNRAS* **484**, 3, pp. 4174–4199, doi:10.1093/mnras/stz174.

Khangulyan, D. *et al.* (2007). TeV light curve of PSR B1259-63/SS2883, *MNRAS* **380**, 1, pp. 320–330, doi:10.1111/j.1365-2966.2007.12075.x.

Khangulyan, D. *et al.* (2018). On the anomalously large extension of the pulsar wind nebula HESS J1825-137, *ApJ* **860**, 1, p. 59, doi:10.3847/1538-4357/aac20f.

Khlopov, M. Yu. (2010). Primordial black holes, *Res. Astron. Astrophys.* **10**, pp. 495–528, doi:10.1088/1674-4527/10/6/001.

Khoze, V. V. and Ro, G. (2014). Dark matter monopoles, vectors and photons, *J. High-Energy Astroph.* **10**, p. 061, doi:10.1007/JHEP10(2014)061.

Kifune, T. (1999). Invariance violation extends the cosmic-ray horizon? *ApJL* **518**, pp. L21–L24, doi:10.1086/312057.

Kimura, S. S. *et al.* (2017). High-energy neutrino emission from short gamma-ray bursts: Prospects for coincident detection with gravitational waves, *ApJL* **848**, p. L4, doi:10.3847/2041-8213/aa8d14.

Kirk, J., Skjaraasen, O. and Gallant, Y. (2002). Pulsed radiation from neutron star winds, *å* **388**, 2, pp. L29–L32, doi:10.1051/0004-6361:20020599.

Kislat, F. and Krawczynski, H. (2015). Search for anisotropic Lorentz invariance violation with γ-rays, *Phys. Rev. D* **92**, p. 045016, doi:10.1103/PhysRevD.92.045016.

Kissmann, R. (2014). PICARD: A novel code for the Galactic cosmic ray propagation problem, *Astrop. Phys.* **55**, pp. 37–50, doi:10.1016/j.astropartphys.2014.02.002.

Kiuchi, R. *et al.* (2009). CANGAROO-III search for TeV gamma rays from two clusters of galaxies, *ApJ* **704**, pp. 240–246, doi:10.1088/0004-637X/704/1/240.

Klebesadel, R. W., Strong, I. B. and Olson, R. A. (1973). Observations of gamma-ray bursts of cosmic origin, *ApJL* **182**, p. L85, doi:10.1086/181225.

Klepach, E. G., Ptuskin, V. S. and Zirakashvili, V. N. (2000). Cosmic ray acceleration by multiple spherical shocks, *Astrop. Phys.* **13**, pp. 161–172, doi:10.1016/S0927-6505(99)00108-5.

Klop, N. *et al.* (2017). Impact of axisymmetric mass models for dwarf spheroidal galaxies on indirect dark matter searches, *Phys. Rev. D* **95**, p. 12, doi:10.1103/physrevd.95.123012.

Klypin, A. A. *et al.* (1999). Where are the missing galactic satellites? *ApJ* **522**, pp. 82–92, doi:10.1086/307643.

Kohnle, A. *et al.* (1996). Stereoscopic imaging of air showers with the first two HEGRA Cherenkov telescopes, *Astrop. Phys.* **5**, pp. 119–131, doi:10.1016/0927-6505(96)00011-4.

Komissarov, S. S. and Lyubarsky, Y. E. (2003). The origin of peculiar jet-torus structure in the Crab nebula, *MNRAS* **344**, 4, pp. L93–L96, doi:10.1046/j.1365-8711.2003.07097.x.

Konopelko, A. *et al.* (2007). Observations of the unidentified TeV gamma-ray source TeV J2032+4130 with the Whipple Observatory 10 m Telescope, *ApJ* **658**, pp. 1062–1068, doi:10.1086/511262.

Koul, R. *et al.* (2007). The TACTIC atmospheric Cherenkov imaging telescope, *Nucl. Instrum. Meth. Phys. Res. A* **578**, pp. 548–564, doi:10.1016/j.nima.2007.06.011.

Koushan, S. *et al.* (2021). GAMA/DEVILS: Constraining the cosmic star formation history from improved measurements of the 0.3–2.2 μm extragalactic background light, *MNRAS* **503**, 2, pp. 2033–2052, doi:10.1093/mnras/stab540.

Koushiappas, S. M., Bullock, J. S. and Dekel, A. (2004). Massive black hole seeds from low angular momentum material, *MNRAS* **354**, p. 292, doi:10.1111/j.1365-2966.2004.08190.x.

Kouveliotou, C. *et al.* (1993). Identification of two classes of gamma-ray bursts, *ApJL* **413**, p. L101, doi:10.1086/186969.

Krause, M., Pueschel, E. and Maier, G. (2017). Improved γ/hadron separation for the detection of faint γ-ray sources using boosted decision trees, *Astrop. Phys.* **89**, pp. 1–9, doi:10.1016/j.astropartphys.2017.01.004.

Kraushaar, W. L. *et al.* (1972). High-energy cosmic gamma-ray observations from the OSO-3 Satellite, *ApJ* **177**, p. 341, doi:10.1086/151713.

Kraushaar, W. L. *et al.* (1973). High-energy cosmic gamma-ray observations from the OSO-3 Satellite, *ApJ* **186**, pp. 401–402, doi:10.1086/152507.

Krawczynski, H., Coppi, P. S. and Aharonian, F. A. (2002). Time-dependent modelling of the Markarian 501 X-ray and TeV gamma-ray data taken during 1997 March and April, *MNRAS* **336**, 3, pp. 721–735, doi:10.1046/j.1365-8711.2002.05750.x.

Krawczynski, H. *et al.* (2006). Gamma hadron separation methods for the VERITAS array of four imaging atmospheric Cherenkov telescopes, *Astrop. Phys.* **25**, pp. 380–390, doi:10.1016/j.astropartphys.2006.03.011.

Krennrich, F. (2009). Gamma ray astronomy with atmospheric Cherenkov telescopes: The future, *New J. Phys.* **11**, 11, p. 115008, doi:10.1088/1367-2630/11/11/115008.

Krennrich, F., Bohec, S. L. and Weekes, T. C. (2000). Detection techniques of microsecond gamma-ray bursts using ground-based telescopes, *The Astrophys. J.* **529**, 1, p. 506, URL http://stacks.iop.org/0004-637X/529/i=1/a=506.

Krymskii, G. F. (1977). A regular mechanism for the acceleration of charged particles on the front of a shock wave, *Akademiia Nauk SSSR Doklady* **234**, pp. 1306–1308.

Kuhlen, M., Vogelsberger, M. and Angulo, R. (2012). Numerical simulations of the dark Universe: State of the art and the next decade, *Phys. Dark Univ.* **1**, pp. 50–93, doi:10.1016/j.dark.2012.10.002.

Kushnir, D. and Waxman, E. (2009). Nonthermal emission from clusters of galaxies, *J. Cosmol. Astropart. Phys.* **8**, p. 002, doi:10.1088/1475-7516/2009/08/002.

Laha, R., Muñoz, J. B. and Slatyer, T. R. (2020). Integral constraints on primordial black holes and particle dark matter, *Phys. Rev. D* **101**, p. 12, doi:10.1103/physrevd.101.123514, URL http://dx.doi.org/10.1103/PhysRevD.101.123514.

Lane, R. R. *et al.* (2010). Halo globular clusters observed with AAOmega: Dark matter content, metallicity and tidal heating, *MNRAS* **406**, pp. 2732–2742, doi:10.1111/j.1365-2966.2010.16874.x.

Lang, R. G., Martínez-Huerta, H. and de Souza, V. (2019). Improved limits on Lorentz invariance violation from astrophysical gamma-ray sources, *Phys. Rev. D* **99**, p. 043015, doi:10.1103/PhysRevD.99.043015, URL https://link.aps.org/doi/10.1103/PhysRevD.99.043015.

Langacker, P., Peccei, R. D. and Yanagida, T. (1986). Invisible axions and light neutrinos. Are they connected? *Modern Phys. Lett. A* **1**, pp. 541–552, doi:10.1142/S0217732386000683.

Lauer, T. R. *et al.* (2021). New Horizons observations of the cosmic optical background, *ApJ* **906**, 2, p. 77, doi:10.3847/1538-4357/abc881.

Lauer, T. R. *et al.* (2022). Anomalous flux in the cosmic optical background detected with New Horizons observations, *ApJL* **927**, 1, p. L8, doi:10.3847/2041-8213/ac573d.

Lazar, A. *et al.* (2020). A dark matter profile to model diverse feedback-induced core sizes of ΛCDM haloes, *MNRAS* **497**, 2, pp. 2393–2417, doi:10.1093/mnras/staa2101.

Le Bohec, S. *et al.* (1998). A new analysis method for very high definition imaging atmospheric Cherenkov telescopes as applied to the cat telescope, *Nucl. Instrum. Meth. Phys. Res. A* **416**, 2, pp. 425–437, doi:10.1016/S0168-9002(98)00750-5.

Lebohec, S. and Holder, J. (2003). The cosmic ray background as a tool for relative calibration of atmospheric Cherenkov telescopes, *Astrop. Phys.* **19**, pp. 221–233, doi:10.1016/S0927-6505(02)00201-3.

LeBohec, S., Krennrich, F. and Sleege, G. (2005). SGARFACE: A Novel detector for microsecond gamma ray bursts, *Astrop. Phys.* **23**, pp. 235–248, doi:10.1016/j.astropartphys.2005.01.003.

Lecun, Y., Bengio, Y. and Hinton, G. (2015). Deep learning, *Nature* **521**, pp. 436–444, doi:10.1038/nature14539.

Lee, S. K. *et al.* (2016). Evidence for unresolved gamma-ray point sources in the inner galaxy, *Phys. Rev. Lett.* **116**, 5, p. 051103, doi:10.1103/PhysRevLett.116.051103.

Lefa, E., Kelner, S. R. and Aharonian, F. A. (2012). On the spectral shape of radiation due to inverse Compton scattering close to the maximum cutoff, *ApJ* **753**, 2, p. 176, doi:10.1088/0004-637X/753/2/176.

Levenson, L. R. and Wright, E. L. (2008). Probing the 3.6 μm CIRB with Spitzer in Three DIRBE Dark Spots, *ApJ* **683**, pp. 585–596, 10.1086/589808.

LeyVa, J. *et al.* (2021). Exploring mev gamma rays from dark matter annihilation and evaporating primordial black holes in the grams experiment, in *Proceedings of the 37th International Cosmic Ray Conference — PoS(ICRC2021), Berlin (Germany)*, Vol. 395, p. 552, doi:10.22323/1.395.0552.

Li, H.-J., Bi, X.-J. and Yin, P.-F. (2021a). Searching for axion-like particles with the blazar observations of MAGIC and Fermi-LAT, arXiv e-prints, arXiv:2110.13636.

Li, H.-J. *et al.* (2021b). Limits on axionlike particles from Mrk 421 with 4.5-year period observations by ARGO-YBJ and Fermi-LAT, *Phys. Rev. D* **103**, 8, p. 083003, doi:10.1103/PhysRevD.103.083003.

Li, J. *et al.* (2012). Unveiling the super-orbital modulation of LS I +61°303 in X-rays, *ApJL* **744**, 1, p. L13, doi:10.1088/2041-8205/744/1/L13.

Li, T.-P. and Ma, Y.-Q. (1983). Analysis methods for results in gamma-ray astronomy, *ApJ* **272**, pp. 317–324, doi:10.1086/161295.

Liang, Y.-F. *et al.* (2019). Constraints on axion-like particle properties with TeV gamma-ray observations of Galactic sources, *J. Cosmol. Astropart. Phys.* **06**, p. 042, doi:10.1088/1475-7516/2019/06/042.

Limyansky, B. (2019). The third fermi pulsar catalog, in *AAS/High Energy Astrophysics Division, AAS/High Energy Astrophysics Division*, Vol. 17, p. 109.32.

Linton, E. T. *et al.* (2006). A new search for primordial black hole evaporations using the Whipple gamma-ray telescope, *J. Cosmol. Astropart. Phys.* **0601**, p. 013, doi:10.1088/1475-7516/2006/01/013.

Lipari, P. (2018). Spectral features in the cosmic ray fluxes, *Astrop. Phys.* **97**, pp. 197–204, doi:10.1016/j.astropartphys.2017.11.008.

Lisanti, M. (2017). Lectures on Dark Matter Physics, in *New Front. Fields Strings*, WORLD SCIENTIFIC, ISBN 978-981-314-943-4, pp. 399–446, doi:10.1142/9789813149441_0007.

Long, G. *et al.* (2021). Probing μeV ALPs with future LHAASO observation of AGN γ-ray spectra.

López-Coto, R. *et al.* (2021). Prospects for the observation of primordial black hsole evaporation with the southern wide field of view gamma-ray observatory, *J. Cosmol. Astropart. Phys.* p. 40, doi:10.1088/1475-7516/2021/08/040.

Lorenz, E. and Martinez, M. (2005). High energy astrophysics: The MAGIC telescope, *Astron. Geophys.* **46**, 6, pp. 6.21–6.25, doi:10.1111/j.1468-4004.2005.46621.x.

Lucek, S. G. and Bell, A. R. (2000). Non-linear amplification of a magnetic field driven by cosmic ray streaming, *MNRAS* **314**, pp. 65–74, doi:10.1046/j.1365-8711.2000.03363.x.

Lunardini, C. (2016). Diffuse supernova neutrinos at underground laboratories, *Astrop. Phys.* **79**, pp. 49–77, doi:10.1016/j.astropartphys.2016.02.005.

Lyne, A. G. *et al.* (2015). The binary nature of PSR J2032+4127, *MNRAS* **451**, 1, pp. 581–587, doi:10.1093/mnras/stv236.

Lyutikov, M. (2013). Pulsar high energy emission due to inverse Compton scattering, *Nucl. Phys. B Proc. Suppl.* **239**, pp. 61–63, doi:10.1016/j.nuclphysbps.2013.05.009.

MacGibbon, J. H. *et al.* (2015). *Primordial Black Holes,* in *5th International Fermi Symposium Nagoya, Japan, October 20–24, 2014.*

Madau, P. and Pozzetti, L. (2000). Deep galaxy counts, extragalactic background light and the stellar baryon budget, *MNRAS* **312**, pp. L9–L15, doi:10.1046/j.1365-8711.2000.03268.x.

Madau, P. and Rees, M. J. (2001). Massive black holes as population III remnants, *ApJL* **551**, pp. L27–L30, doi:10.1086/319848.

Madejski, G. G. and Sikora, M. (2016). Gamma-ray observations of active galactic nuclei, *Ann. Rev. Astron. Astrophys.* **54**, pp. 725–760, doi:10.1146/annurev-astro-081913-040044.

Maier, G. and Knapp, J. (2007). Cosmic-ray events as background in imaging atmospheric Cherenkov telescopes, *Astrop. Phys.* **28**, pp. 72–81, doi:10.1016/j.astropartphys.2007.04.009.

Majumdar, J., Calore, F. and Horns, D. (2018). Search for gamma-ray spectral modulations in Galactic pulsars, *J. Cosmol. Astropart. Phys.* **4**, 048, doi:10.1088/1475-7516/2018/04/048.

Malkov, M. A. and Drury, L. O. (2001). Nonlinear theory of diffusive acceleration of particles by shock waves, *Rep. Prog. Phys.* **64**, pp. 429–481, doi:10.1088/0034-4885/64/4/201.

Malkov, M. A. *et al.* (2013). Analytic solution for self-regulated collective escape of cosmic rays from their acceleration sites, *ApJ* **768**, 1, p. 73, doi:10.1088/0004-637X/768/1/73.

Malone, K. *et al.* (2017). The gamma-ray sky above 50 TeV with the HAWC Observatory, in *APS April Meeting Abstracts*, p. X4.008.

Malyshev, D. *et al.* (2018). Improved limit on axion-like particles from gamma-ray data on Perseus cluster, ArXiv e-prints.

Manconi, S., Calore, F. and Donato, F. (2021). Dissecting the inner Galaxy with gamma-ray pixel count statistics, in *Proceedings of the 37th International Cosmic Ray Conference — PoS(ICRC2021), Berlin (Germany)*, Vol. 395, p. 677, doi:10.22323/1.395.0677.

Margutti, R. *et al.* (2017). The electromagnetic counterpart of the binary neutron star merger LIGO/Virgo GW170817. V. rising X-ray emission from an off-axis jet, *ApJL* **848**, 2, p. L20, doi:10.3847/2041-8213/aa9057.

Markov, M. (1960). On high-energy neutrinos, in *Proceedings of Annual Int. Conf on High Energy Physics*, Vol. 1, Rochester, UK, p. 578.

Martí-Devesa, G. and Reimer, O. (2021). η Carinae with Fermi-LAT: Two full orbits and the third periastron, *A&A* **654**, p. A44, doi:10.1051/0004-6361/202140451.

Martín, J., Torres, D. F. and Rea, N. (2012). Time-dependent modelling of pulsar wind nebulae: Study on the impact of the diffusion-loss approximations, *MNRAS* **427**, 1, pp. 415–427, doi:10.1111/j.1365-2966.2012.22014.x.

Martínez, M. and Errando, M. (2009). A new approach to study energy-dependent arrival delays on photons from astrophysical sources, *Astrop. Phys.* **31**, pp. 226–232, doi:10.1016/j.astropartphys.2009.01.005.

Martínez-Huerta, H., Lang, R. G. and de Souza, V. (2020). Lorentz invariance violation tests in astroparticle physics, *Symmetry* **12**, p. 8, doi:10.3390/sym12081232, URL https://www.mdpi.com/2073-8994/12/8/1232.

Mashchenko, S. and Sills, A. (2005a). Globular clusters with dark matter halos. 1. Initial relaxation, *ApJ* **619**, p. 243, doi:10.1086/426132.

Mashchenko, S. and Sills, A. (2005b). Globular clusters with dark matter halos 2. Evolution in tidal field, *ApJ* **619**, p. 258, doi:10.1086/426133.

Massaro, F., Thompson, D. J. and Ferrara, E. C. (2015). The extragalactic gamma-ray sky in the Fermi era, *A&A Rev.* **24**, p. 2, doi:10.1007/s00159-015-0090-6.

Matsuura, S. *et al.* (2017). New spectral evidence of an unaccounted component of the near-infrared extragalactic background light from the CIBER, *ApJ* **839**, p. 7, doi:10.3847/1538-4357/aa6843.

Mattila, K. and Väisänen, P. (2019). Extragalactic background light: Inventory of light throughout the cosmic history, *Contemp. Phys.* **60**, 1, pp. 23–44, doi:10.1080/00107514.2019.1586130.

Mattila, K. *et al.* (2017a). Extragalactic background light: A measurement at 400 nm using dark cloud shadow — II. Spectroscopic separation of the dark cloud's light, and results, *MNRAS* **470**, pp. 2152–2169, doi:10.1093/mnras/stx1296.

Mattila, K. *et al.* (2017b). Extragalactic background light: A measurement at 400 nm using dark cloud shadow*^{y} — I. Low surface brightness spectrophotometry in the area of Lynds 1642, *MNRAS* **470**, pp. 2133–2151, doi:10.1093/mnras/stx1295.

Mattox, J. R. *et al.* (1996). The likelihood analysis of EGRET data, *ApJ* **461**, p. 396, doi:10.1086/177068.

Maurin, D. *et al.* (2001). Cosmic rays below z = 30 in a diffusion model: New constraints on propagation parameters, *ApJ* **555**, pp. 585–596, doi:10.1086/321496.

Mayer, M. *et al.* (2013). Rapid gamma-ray flux variability during the 2013 march crab nebula flare, *ApJ* **775**, 2, p. L37, doi:10.1088/2041-8205/775/2/l37.

Mazin, D. and Raue, M. (2007). New limits on the density of the extragalactic background light in the optical to the far infrared from the spectra of all known TeV blazars, *A&A* **471**, 2, pp. 439–452, doi:10.1051/0004-6361:20077158.

Meegan, C. *et al.* (2009). The Fermi gamma-ray burst monitor, *ApJ* **702**, 1, pp. 791–804, doi:10.1088/0004-637X/702/1/791.

Meegan, C. A. *et al.* (1996). The hird BATSE gamma-ray burst Catalog, *ApJS* **106**, p. 65, doi:10.1086/192329.

Meekins, J. F. *et al.* (1971). Physical sciences: X-rays from the Coma cluster of galaxies, *Nature* **231**, p. 107.

Mena, O. and Razzaque, S. (2013). Hints of an axion-like particle mixing in the GeV gamma-ray blazar data? *J. Cosmol. Astropart. Phys.* **11**, p. 023, doi: 10.1088/1475-7516/2013/11/023.

Mestre, E. *et al.* (2021). Probing the hadronic nature of the gamma-ray emission associated with Westerlund 2, *MNRAS* **505**, 2, pp. 2731–2740, doi:10.1093/mnras/stab1455.

Metzger, B. D. (2017). Kilonovae, *Liv. Rev. Relativ.* **20**, p. 3, doi:10.1007/s41114-017-0006-z.

Meyer, M. and Conrad, J. (2014). Sensitivity of the Cherenkov Telescope Array to the detection of axion-like particles at high gamma-ray opacities, *J. Cosmol. Astropart. Phys.* **12**, p. 016, doi:10.1088/1475-7516/2014/12/016.

Meyer, M., Horns, D. and Raue, M. (2013). First lower limits on the photon-axion-like particle coupling from very high energy gamma-ray observations, *Phys. Rev. D* **87**, 3, p. 035027, doi:10.1103/PhysRevD.87.035027.

Meyer, M., Horns, D. and Zechlin, H.-S. (2010). The Crab Nebula as a standard candle in very high-energy astrophysics, *A&A* **523**, p. A2, doi:10.1051/0004-6361/201014108.

Michel, F. (1969). Hydromagnetic plasma acceleration by rapidly rotating astrophysical objects, *Phys. Rev. Lett.* **23**, 5, pp. 247–249, doi:10.1103/PhysRevLett.23.247.

Michel, F. C. (1973). Rotating magnetosphere: A simple relativistic model, *ApJ* **180**, pp. 207–226, doi:10.1086/151956.

Michelson, P. F., Atwood, W. B. and Ritz, S. (2010). Fermi gamma-ray space Telescope: High-energy results from the first year, *Rep. Prog. Phys.* **73**, 7, p. 074901, doi:10.1088/0034-4885/73/7/074901.

Milgrom, M. (1983). A modification of the Newtonian dynamics as a possible alternative to the hidden mass hypothesis, *ApJ* **270**, pp. 365–370, doi:10.1086/161130.

Mirabal, N. *et al.* (2012). Fermi's SIBYL: Mining the gamma-ray sky for dark matter subhaloes, *MNRAS* **424**, pp. L64–L68, doi:10.1111/j.1745-3933.2012.01287.x.

Mirabal, N. *et al.* (2016). 3FGL Demographics outside the galactic plane using supervised machine learning: Pulsar and dark matter subhalo interpretations, *ApJ* **825**, p. 69, doi:10.3847/0004-637X/825/1/69.

Mirabel, I. F. (2006). Very energetic gamma-ray from microquasars and binary pulsars, *Science* **312**, p. 1759, doi:10.1126/science.1129815.

Mirizzi, A. and Montanino, D. (2009). Stochastic conversions of TeV photons into axion-like particles in extragalactic magnetic fields, *J. Cosmol. Astropart. Phys.* **12**, p. 004, doi:10.1088/1475-7516/2009/12/004.

Mirizzi, A., Raffelt, G. G. and Serpico, P. D. (2005). Photon-axion conversion as a mechanism for supernova dimming: Limits from CMB spectral distortion, *Phys. Rev. D* **72**, p. 2, 023501, doi:10.1103/PhysRevD.72.023501.

Mirizzi, A., Raffelt1, G. G. and Serpico, P. D. (2008). Photon-axion conversion in intergalactic magnetic fields and cosmological consequences, in M. Kuster, G. Raffelt and B. Beltrán. eds., *Axions, Lecture Notes in Physics*, Vol. 741, Springer Berlin Heidelberg, p. 115.

Mirzoian, R., Laatiaoui, M. and Teshima, M. (2006). Very high quantum efficiency pmts with bialkali photo-cathode, *Nucl. Instrum. Meth. Phys. Res. A* **A567**, pp. 230–232, doi:10.1016/j.nima.2006.05.094.

Mirzoyan, R. (2017).

Mirzoyan, R. and Mukherjee, R. (2017). TeV gamma-ray emission from PSR J2032+4127/MT91 213 at periastron,.

Mochol, I. and Petri, J. (2015). Very high energy emission as a probe of relativistic magnetic reconnection in pulsar winds, *MNRAS* **449**, 1, pp. L51–L55, doi: 10.1093/mnrasl/slv018.

Mohrmann, L. (2015). *Characterizing Cosmic Neutrino Sources*, Ph.D. thesis, Humboldt-Universität zu Berlin, Mathematisch-Naturwissenschaftliche Fakultät, doi:http://dx.doi.org/10.18452/17377.

Moliné, A. *et al.* (2017). Characterization of subhalo structural properties and implications for dark matter annihilation signals, *MNRAS* **466**, 4, pp. 4974–4990, doi:10.1093/mnras/stx026.

Montanari, A., Moulin, E. and Malyshev, D. (2021). Search for dark matter annihilation signals from the Galactic center with the H.E.S.S. inner Galaxy survey, in *Proceedings of the 37th International Cosmic Ray Conference — PoS(ICRC2021), Berlin (Germany)*, Vol. 395, p. 511, doi:10.22323/1.395.0511.

Montmerle, T. and Cesarsky, C. (1979). Gamma-ray emission from snobs, *Inter. Cosmic Ray Conf.* **1**, p. 191.

Moore, B. *et al.* (1999). Dark matter substructure within galactic halos, *ApJ* **524**, pp. L19–L22, doi:10.1086/312287.

Moriarty, P. *et al.* (1997). Strategies for selection of small γ-ray showers using the imaging atmospheric Cherenkov technique, *Astrop. Phys.* **7**, pp. 315–327, doi:10.1016/S0927-6505(97)00033-9.

Morrison, P. (1957). On the origins of cosmic rays, *Rev. Mod. Phys.* **29**, pp. 235–243, doi:10.1103/RevModPhys.29.235.

Moskalenko, I. V. and Strong, A. W. (1998). Production and propagation of cosmic ray positrons and electrons, *ApJ* **493**, pp. 694–707, doi:10.1086/305152.

Moskalenko, I. V. *et al.* (2009). On the possible association of ultra-high-energy cosmic rays with nearby active galaxies, *ApJ* **693**, pp. 1261–1274, doi:10.1088/0004-637X/693/2/1261.

Mücke, A. *et al.* (2000). Photomeson production in astrophysical sources, *Nucl. Phys. B Proc. Suppl.* **80**, p. 08/10.

Mukherjee, R. (2001). EGRET (GeV) Blazars, in F. A. Aharonian and H. J. Völk. eds., *High Energy Gamma-Ray Astronomy: International Symposium, American Institute of Physics Conference Series*, Vol. 558, pp. 324–337, doi: 10.1063/1.1370801.

Mukherjee, R. *et al.* (1997). EGRET Observations of high-energy gamma-ray emission from blazars: An update, *ApJ* **490**, 1, pp. 116–135, doi:10.1086/304851.

Muraishi, H. *et al.* (2000). Evidence for TeV gamma-ray emission from the shell type SNR RX J1713.7-3946, *A&A* **354**, pp. L57–L61.

Murase, K. (2009). Ultrahigh-energy photons as a probe of nearby transient ultrahigh-energy cosmic-ray sources and possible Lorentz-invariance violation, *Phys. Rev. Lett.* **103**, 8, p. 081102, doi:10.1103/PhysRevLett.103.081102.

Murase, K., Guetta, D. and Ahlers, M. (2016). Hidden cosmic-ray accelerators as an origin of TeV-PeV cosmic neutrinos, *Phys. Rev. Lett.* **116**, 7, p. 071101, doi:10.1103/PhysRevLett.116.071101.

Murase, K., Inoue, S. and Nagataki, S. (2008). Cosmic rays above the second knee from clusters of galaxies and associated high-energy neutrino emission, *ApJL* **689**, p. L105, doi:10.1086/595882.

Muslimov, A. and Harding, A. (2003). Extended acceleration in slot gaps and pulsar high-energy emission, *ApJ* **588**, pp. 430–440, doi:10.1086/368162.

Nagayama, M. *et al.* (2021). Sensitivity of indirect detection of neutralino dark matter by Sommerfeld enhancement mechanism, *archive eprints*.

Naumann, C. *et al.* (2013). Nectar: New electronics for the Cherenkov telescope array, in *Proceedings of International Workshop on New Photon-Detectors — PoS(PhotoDet2012)*, Vol. 158, SISSA Medialab, p. 030, doi:10.22323/1.158.0030.

Navarro, J. F., Frenk, C. S. and White, S. D. M. (1996). The structure of cold dark matter halos, *ApJ* **462**, pp. 563–575, doi:10.1086/177173.

Navarro, J. F. *et al.* (2004). The inner structure of ΛCDM haloes — III. Universality and asymptotic slopes, *MNRAS* **349**, 3, pp. 1039–1051, doi: 10.1111/j.1365-2966.2004.07586.x.

Nayerhoda, A. *et al.* (2019). Gamma-ray diffuse emission from the Galactic plane with HAWC data, in *Proceedings of the 36th International Cosmic Ray Conference — PoS(ICRC2019), Madison (USA)*, Vol. 358, p. 750, doi: 10.22323/1.358.0750.

Neronov, A., Malyshev, D. and Semikoz, D. V. (2017). Cosmic-ray spectrum in the local Galaxy, *A&A* **606**, p. A22, doi:10.1051/0004-6361/201731149.

Neronov, A. and Semikoz, D. V. (2009). Sensitivity of γ-ray telescopes for detection of magnetic fields in the intergalactic medium, *Phys. Rev. D* **80**, 12, p. 123012, doi:10.1103/PhysRevD.80.123012.

Neronov, D. V. and Semikoz, A. (2007). A method of measurement of extra-galactic magnetic fields by TeV gamma ray telescopes, *JETP Lett.* **85**, 10, pp. 473–477, doi:10.1134/S0021364007100013.

Nieto, D. *et al.* (2011a). A search for possible dark matter subhalos as IACT targets in the first *Fermi*-LAT source catalog.

Nieto, D. *et al.* (2011b). The search for galactic dark matter clump candidates with Fermi and MAGIC, in *Proceedings of the 32th International Cosmic Ray Conference — PoS(ICRC2011), Beijing (China)*, p. xx.

Nieto, D. *et al.* (2015). Hunting for dark matter subhalos among the Fermi-LAT sources with VERITAS, arXiv e-prints.

Nieto, D. *et al.* (2017). Exploring deep learning as an event classification method for the Cherenkov Telescope Array, in *Proceedings of the 35th International Cosmic Ray Conference — PoS(ICRC2017), Busan (South Korea)*, Vol. 301, p. 809, doi:10.22323/1.301.0809.

Nikishov, A. I. (1962). Absorption of high-enegy photons in the universe, *Sov. Phys. JETP* **14**, 2, pp. 393–394.

Nilsson, K. *et al.* (2018). Long-term optical monitoring of TeV emitting blazars. I. Data analysis, *A&A* **620**, p. A185, doi:10.1051/0004-6361/201833621.

Ninci, D. *et al.* (2019). Searching for dark matter decay signals in the Galactic Halo with the MAGIC telescopes, **358**, p. 538, doi:10.22323/1.358.0538.

Nolan, P. *et al.* (2012). Fermi large area telescope second source catalog, *ApJS* **199**, p. 31, doi:10.1088/0067-0049/199/2/31.

Obertacke Pollmann, A. (2017). Luminescence of water or ice as a new detection method for magnetic monopoles, *Eur. Phys. J. Web Conf.* **164**, p. 07019, doi:10.1051/epjconf/201716407019.

Ogawa, S. (2019). Observation of optical transients and search for PeV-EeV Tau neutrinos with Ashra-1, in *Proceedings of the 36th International Cosmic Ray Conference — PoS(ICRC2019), Madison (USA)*, Vol. 358, p. 970, doi:10.22323/1.358.0970.

Ohm, S., van Eldik, C. and Egberts, K. (2009). γ/hadron separation in very-high-energy γ-ray astronomy using a multivariate analysis method, *Astrop. Phys.* **31**, pp. 383–391, doi:10.1016/j.astropartphys.2009.04.001.

Ohtani, Y. *et al.* (2021). Cross-calibration and combined analysis of the CTA-LST prototype and the MAGIC telescopes, in *Proceedings of the 37th International Cosmic Ray Conference — PoS(ICRC2021), Berlin (Germany)*, Vol. 395, p. 724, doi:10.22323/1.395.0724.

Olinto, A. V. *et al.* (2020). The POEMMA (Probe of Extreme Multi-Messenger Astrophysics) Observatory.

Onishi, T., Mizuno, N. and Fukui, Y. (2004). New view of molecular gas distribution of the southern sky: Co surveys with nanten, in S. Pfalzner, C. Kramer, C. Straubmeier and A. Heithausen. eds., *The Dense Interstellar Medium in Galaxies*, Springer Berlin Heidelberg, Berlin, Heidelberg, pp. 203–209.

Orlando, E. and Strong, A. (2013). Galactic synchrotron emission with cosmic ray propagation models, *MNRAS* **436**, p. 2127, doi:10.1093/mnras/stt1718.

Orr, M. R., Krennrich, F. and Dwek, E. (2011). Strong new constraints on the extragalactic background light in the near- to mid-infrared, *ApJ* **733**, p. 77, doi:10.1088/0004-637X/733/2/77.

Otte, A. N. (2019). Studies of an air-shower imaging system for the detection of ultrahigh-energy neutrinos, *Phys. Rev. D* **99**, 8, p. 083012, doi:10.1103/PhysRevD.99.083012.

Pagels, H. and Primack, J. R. (1982). Supersymmetry, cosmology and new TeV physics, *Phys. Rev. Lett.* **48**, p. 223, doi:10.1103/PhysRevLett.48.223.

Paiano, S. *et al.* (2018). The redshift of the BL Lac object TXS 0506+056, *ApJL* **854**, p. L32, doi:10.3847/2041-8213/aaad5e.

Paoletti, R. *et al.* (2007). The trigger system of the magic telescope, *IEEE Trans. Nucl. Sci.* **54**, pp. 404–409, doi:10.1109/TNS.2007.892649.

Paoletti, R. *et al.* (2008). The global trigger system of the magic telescope array, *IEEE Trans. Nucl. Sci.*, pp. 2781–2783, doi:10.1109/NSSMIC.2008.4774948.

Paré, E. *et al.* (2002). CELESTE: An atmospheric Cherenkov telescope for high energy gamma astrophysics, *Nucl. Instrum. Meth. Phys. Res. A* **490**, 1-2, pp. 71–89, doi:10.1016/S0168-9002(02)01005-7.

Paredes, J. M. (1987). Ph.D. thesis.

Parizot, E. *et al.* (2004). Superbubbles and energetic particles in the Galaxy. I. Collective effects of particle acceleration, *A&A* **424**, pp. 747–760, doi:10.1051/0004-6361:20041269.

Park, N. *et al.* (2015). Performance of the VERITAS experiment, in *Proceedings of the 34th International Cosmic Ray Conference — PoS(ICRC2015), The Hague (Netherlands)*, Vol. 34, World Scientific, p. 771.

Parker, E. N. (1970). The origin of magnetic fields, *ApJ* **160**, p. 383, doi:10.1086/150442.

Parsons, R. D. and Hinton, J. A. (2014). A monte carlo template based analysis for air-cherenkov arrays, *Astrop. Phys.* **56**, pp. 26–34, doi:10.1016/j.astropartphys.2014.03.002.

Patrizii, L. and Spurio, M. (2015). Status of searches for magnetic monopoles, *Ann. Rev. Nucl. Part. Sci.* **65**, pp. 279–302, doi:10.1146/annurev-nucl-102014-022137.

Peccei, R. D. and Quinn, H. R. R. D. (1977). CP conservation in the presence of pseudoparticles, *Phys. Rev. Lett.* **38**, pp. 1440–1443, doi:10.1103/PhysRevLett.38.1440.

Perkins, J. S. (2008). VERITAS Observations of the coma cluster of galaxies, in F. A. Aharonian, W. Hofmann and F. Rieger. eds., *American Institute of Physics Conference Series*, Vol. 1085, pp. 569–572, doi:10.1063/1.3076736.

Perkins, J. S. *et al.* (2006). TeV gamma-ray observations of the Perseus and Abell 2029 galaxy clusters, *ApJ* **644**, pp. 148–154, doi:10.1086/503321.

Perkins, J. S. *et al.* (2006). TeV gamma-ray observations of the Perseus and Abell 2029 galaxy clusters, *ApJ* **644**, pp. 148–154, doi:10.1086/503321.

Peron, G. *et al.* (2020). On the gamma-ray emission of W44 and its surroundings, *ApJL* **896**, 2, p. L23, doi:10.3847/2041-8213/ab93d1.

Peron, G. *et al.* (2021). Probing the cosmic-ray density in the inner Galaxy, *ApJL* **907**, 1, p. L11, doi:10.3847/2041-8213/abcaa9.

Petri, J. (2012). High-energy emission from the pulsar striped wind: A synchrotron model for gamma-ray pulsars, *MNRAS* **424**, p. 2023, doi:10.1111/j.1365-2966.2012.21350.x.

Pfrang, K., Hassan, T. and Pueschel, E. (2021a). Optical microlensing by primordial black holes with IACTs, in *Proceedings of the 37th International Cosmic Ray Conference — PoS(ICRC2021), Berlin (Germany)*, Vol. 395, p. 495, doi:10.22323/1.395.0495.

Pfrang, K. *et al.* (2021b). Deep learning transient detection with VERITAS, in *Proceedings of the 37th International Cosmic Ray Conference — PoS(ICRC2021), Berlin (Germany)*, Vol. 395, p. 522, doi:10.22323/1.395.0822.

Pfrommer, C. and Enßlin, T. A. (2004). Constraining the population of cosmic ray protons in cooling flow clusters with gamma-ray and radio observations: Are radio mini-halos of hadronic origin? *A&A* **413**, pp. 17–36.

Pfrommer, C., Enßlin, T. A. and Springel, V. (2008). Simulating cosmic rays in clusters of galaxies — II. A unified scheme for radio haloes and relics with predictions of the γ-ray emission, *MNRAS* **385**, pp. 1211–1241, doi:10.1111/j.1365-2966.2008.12956.x.

Piner, B. G. and Edwards, P. G. (2004). The parsec-scale structure and jet motions of the TeV Blazars 1ES 1959+650, PKS 2155-304, and 1ES 2344+514, *ApJ* **600**, 1, pp. 115–126, doi:10.1086/379769.

Pinzke, A. and Pfrommer, C. (2010). Simulating the gamma-ray emission from galaxy clusters: A universal cosmic ray spectrum and spatial distribution. *MNRAS* **409**, pp. 449–480, doi:10.1111/j.1365-2966.2010.17328.x.

Pinzke, A. *et al.* (2011). Prospects of detecting gamma-ray emission from galaxy clusters: Cosmic rays and dark matter annihilations, *Phys. Rev. D* **84** p. 123509.

Piro, L. *et al.* (2021). Multi-messenger-athena synergy white paper, arXiv e-prints, arXiv:2110.15677.

Poincare, H. (1906). *LÁstronomie* **158**.

Pollack, J. B. and Fazio, G. G. (1963). Production of π mesons and gamma radiation in the galaxy by cosmic rays, *Phys. Rev.* **131**, pp. 2684–2691, doi:10.1103/PhysRev.131.2684.

Polyakov, A. (1974). Particle spectrum in the quantum field theory, *JETP Lett.* **20**.

Porter, N. A. and Weekes, T. C. (1978). A search for high energy gamma-ray bursts from primordial black holes or other astronomical objects, *MNRAS* **183**, 2, pp. 205–210, doi:10.1093/mnras/183.2.205, URL http://dx.doi.org/10.1093/mnras/183.2.205.

Porter, T. and Weekes, N. A. (1979). Upper limits for gamma-ray bursts from primordial black holes, *Nature* **277**, p. 199.

Porth, O., Komissarov, S. S. and Keppens, R. (2013). Solution to the sigma problem of pulsar wind nebulae. *MNRAS* **431**, pp. L48–L52, doi:10.1093/mnrasl/slt006.

Porth, O. *et al.* (2017). Modelling jets, tori and flares in pulsar wind nebulae, *Space Sci. Rev.* **207**, 1-4, pp. 137–174, doi:10.1007/s11214-017-0344-x.

Pothast, M. *et al.* (2018). On the progressive hardening of the cosmic-ray proton spectrum in the inner galaxy, *J. Cosmol. Astropart. Phys.* **2018**, 10, pp. 045–045, doi:10.1088/1475-7516/2018/10/045.

Preskill, J. (1984). Magnetic monopoles, *Annu. Rev. Nucl. Particle Sci.* **34**, pp. 461–530, doi:10.1146/annurev.ns.34.120184.002333.

Primack, J. R. and Gross, M. A. K. (2000). *Hot Dark Matter in Cosmology*, Caldwell, David O.

Profumo, S. (2012). Dissecting cosmic-ray electron-positron data with Occam's razor: The role of known pulsars, *Central Eur. J. Phys.* **10**, pp. 1–31, doi: 10.2478/s11534-011-0099-z.

Profumo, S. (2013). TASI 2012 Lectures on Astrophysical Probes of Dark Matter, in *Search. New Phys. Small Large Scales*, WORLD SCIENTIFIC, ISBN 978-981-4525-21-3, pp. 143–189, doi:10.1142/9789814525220_0004.

Profumo, S. (2017). *An Introduction to Particle Dark Matter*, doi:10.1142/Q0001.

Profumo, S. *et al.* (2018). Lessons from HAWC pulsar wind nebulae observations: The diffusion constant is not a constant; pulsars remain the likeliest sources of the anomalous positron fraction; cosmic rays are trapped for long periods of time in pockets of inefficient diffusion, *Phys. Rev. D* **97**, 12, p. 123008, doi:10.1103/PhysRevD.97.123008.

Profumo, S. *et al.* (2018). Lessons from HAWC pulsar wind nebulae observations: The diffusion constant is not a constant; pulsars remain the likeliest sources of the anomalous positron fraction; cosmic rays are trapped for long periods of time in pockets of inefficient diffusion, *Phys. Rev. D* **97**, 12, p. 123008, doi:10.1103/PhysRevD.97.123008.

Protheroe, R. J. and Biermann, P. L. (1996). A new estimate of the extragalactic radio background and implications for ultra-high-energy gamma-ray propagation, *Astrop. Phys.* **6**, 1, pp. 45–54, doi:10.1016/S0927-6505(96)00041-2.

Puerto-Giménez, I. *et al.* (2013). Characterization of the candidate site for the Cherenkov Telescope Array at the Observatorio del Teide, in *Proceedings of the 33th International Cosmic Ray Conference — PoS(ICRC2013), Rio de Janeiro (Brazil)*, Ronald Cintra Shellard, p. 11.

Punch, M. *et al.* (1991). Supercuts: An improved method of selecting gamma-rays, *Inter. Cosmic Ray Conf.* **1**, p. 464.

Raccanelli, A., Vidotto, F. and Verde, L. (2018). Effects of primordial black holes quantum gravity decay on galaxy clustering, *J. Cosmol. Astropart. Phys.* , 08, pp. 003–003, doi:10.1088/1475-7516/2018/08/003, URL http://dx.doi.org/10.1088/1475-7516/2018/08/003.

Rafighi, I. *et al.* (2017). Plasma effects on relativistic pair beams from TeV blazars: PIC simulations and analytical predictions, *A&A* **607**, p. A112, doi:10.1051/0004-6361/201731127.

Ray, A. *et al.* (2021). Near future MeV telescopes can discover asteroid-mass primordial black hole dark matter, *Phys. Rev. D* **104**, 2, p. 023516, doi: 10.1103/PhysRevD.104.023516.

Read, J. I., Walker, M. G. and Steger, P. (2019). Dark matter heats up in dwarf galaxies, *MNRAS* **484**, 1, pp. 1401–1420, doi:10.1093/mnras/sty3404.

Read, J. I. *et al.* (2006). The importance of tides for the Local Group dwarf spheroidals, *MNRAS* **367**, pp. 387–399, doi:10.1111/j.1365-2966.2005.\break09959.x.

Recchia, S., Blasi, P. and Morlino, G. (2016). On the radial distribution of Galactic cosmic rays, *MNRAS* **462**, pp. L88–L92, doi:10.1093/mnrasl/slw136.

Recchia, S. *et al.* (2021). Do the Geminga, Monogem and PSR J0622+3749 gamma-ray halos imply slow diffusion around pulsars? *Phys. Rev. D* **104**, 12, p. 123017, doi:10.1103/PhysRevD.104.123017.

Regis, M. *et al.* (2021). The EMU view of the large magellanic cloud: Troubles for sub-TeV WIMPs, *J. Cosmol. Astropart. Phys.* **2021**, 11, p. 046, doi: 10.1088/1475-7516/2021/11/046, URL http://dx.doi.org/10.1088/1475-7516/2021/11/046.

Reimer, A. (2007). The redshift dependence of gamma-ray absorption in the environments of strong-line AGNs, *ApJ* **665**, pp. 1023–1029, doi:10.1086/519766.

Reimer, A., Pohl, M. and Reimer, O. (2006). Nonthermal high-energy emission from colliding winds of massive stars, *ApJ* **644**, pp. 1118–1144, doi:10.1086/503598.

Reimer, O. *et al.* (2003). EGRET upper limits on the high-energy gamma-ray emission of galaxy clusters, *ApJ* **588**, pp. 155–164.

Remy, Q. *et al.* (2021). Survey of the galactic plane with the Cherenkov telescope array, in *Proceedings of the 37th International Cosmic Ray Conference — PoS(ICRC2021), Berlin (Germany)*, Vol. 886, doi:10.22323/1.395.0886.

Reynolds, P. T. and Fegan, D. J. (1995). Neural network classification of TeV gamma-ray images, *Astrop. Phys.* **3**, pp. 137–150, doi:10.1016/0927-6505(94)00037-4.

Reynolds, S. P. (2008). Supernova remnants at high energy. *Ann. Rev. Astron. Astrophys.* **46**, pp. 89–126, doi:10.1146/annurev.astro.46.060407.145237.

Reynolds, S. P. and Chevalier, R. A. (1984). Evolution of pulsar-driven supernova remnants. *ApJ* **278**, pp. 630–648, doi:10.1086/161831.

Rico, J. (2020). Gamma-Ray dark matter searches in Milky Way satellites — A comparative review of data analysis methods and current results, *Galaxies* **8**, 1, p. 25, doi:10.3390/galaxies8010025.

Rieger, F. M. (2017). Gamma-rays from non-blazar AGN, in *6th International Symposium on High Energy Gamma-Ray Astronomy, American Institute of Physics Conference Series*, Vol. 1792, p. 020008, doi:10.1063/1.4968893.

Rieger, F. M. and Mannheim, K. (2000). Particle acceleration by rotating magnetospheres in active galactic nuclei, *A&A* **353**, pp. 473–478.

Riess, A. G. *et al.* (2018). New parallaxes of galactic cepheids from spatially scanning the hubble space telescope: Implications for the Hubble constant, *ApJ* **855**, p. 136, doi:10.3847/1538-4357/aaadb7.

Riess, A. G. *et al.* (2019). Large magellanic cloud Cepheid standards provide a 1% foundation for the determination of the Hubble constant and stronger evidence for physics beyond ΛCDM, *ApJ* **876**, 1, p. 85, doi:10.3847/1538-4357/ab1422.

Rinchiuso, L. *et al.* (2018). Hunting for heavy winos in the galactic center, *Phys. Rev. D* **98**, 12, p. 123014, doi:10.1103/PhysRevD.98.123014.

Rissi, M. *et al.* (2009). A new sum trigger to provide a lower energy threshold for the magic telescope, *IEEE Trans. Nucl. Sci.* **56**, 6, pp. 3840–3843, doi: 10.1109/TNS.2009.2030802.

Ritt, S. (2004). The drs chip: Cheap waveform digitizing in the GHz range, *Nucl. Instrum. Meth. Phys. Res. A* **518**, 1–2, pp. 470–471, doi:10.1016/j.nima.2003. 11.059.

Ritt, S. (2008). Design and performance of the 6 ghz waveform digitizing chip drs4, *IEEE Trans. Nucl. Sci.* pp. 1512–1515, doi:10.1109/NSSMIC.2008. 4774700.

Rocha, M. *et al.* (2013). Cosmological simulations with self-interacting dark matter I: Constant density cores and substructure, *MNRAS* **430**, pp. 81–104, doi:10.1093/mnras/sts514.

Rodrigues, X. *et al.* (2021). Multiwavelength and neutrino emission from blazar pks 1502+106, *ApJ* **912**, 1, p. 54, doi:10.3847/1538-4357/abe87b.

Romani, R. and Watters, K. (2010). Constraining pulsar magnetosphere geometry with gamma-ray light curves, *ApJ* **714**, 1, pp. 810–824, doi:10.1088/ 0004-637x/714/1/810.

Romani, R. W. and Yadigaroglu, I. A. (1995). Gamma-ray pulsars: Emission zones and viewing geometries, *ApJ* **438**, p. 314, doi:10.1086/175076.

Romoli, C., Taylor, A. M. and Aharonian, F. A. (2017). Cut-off characterisation of energy spectra of bright fermi sources: Current instrument limits and future possibilities, *Astrop. Phys.* **88**, pp. 38–45, doi:10.1016/j.astropartphys.2016. 12.007.

Rovelli, C. (2008). Loop quantum gravity, *Living Rev. in Relativ.* **11**, p. 5, doi: 10.12942/lrr-2008-5.

Rowell, G. P. (2003). A new template background estimate for source searching in TeV gamma-ray astronomy, *A&A* **410**, pp. 389–396, doi:10.1051/0004-6361: 20031194.

Rubakov, V. (1982). Adler-bell-jackiw anomaly and fermion-number breaking in the presence of a magnetic monopole, *Nucl. Phys. B* **203**, 2, pp. 311–348, doi:https://doi.org/10.1016/0550-3213(82)90034-7.

Rubin, V. C., Ford, W. K. and Thonnard, N. (1978). Extended rotation curves of high-luminosity spiral galaxies. IV — Systematic dynamical properties, SA through SC, *ApJL* **225**, pp. L107–L111, doi:10.1086/182804.

Rubin, V. C., Thonnard, N. and Ford, W. K., Jr. (1980). Rotational properties of 21 SC galaxies with a large range of luminosities and radii, from NGC 4605 /R = 4kpc/ to UGC 2885 /R = 122 kpc/, *Astrophys. J.* **238**, p. 471, doi:10.1086/158003.

Rubtsov, G. I. and Troitsky, S. V. (2014). Breaks in gamma-ray spectra of distant blazars and transparency of the Universe, *Soviet Journal of Experimental and Theoretical Physics Letters* **100**, pp. 355–359, doi:10.1134/ S0021364014180088.

Salamon, M. H., Stecker, F. W. and de Jager, O. C. (1994). A new method for determining the Hubble constant from sub-TeV gamma-ray observations, *ApJL* **423**, pp. L1–L4, doi:10.1086/187221.

Saldana-Lopez, A. *et al.* (2021). An observational determination of the evolving extragalactic background light from the multiwavelength HST/CANDELS survey in the Fermi and CTA era, *MNRAS* **507**, 4, pp. 5144–5160, doi: 10.1093/mnras/stab2393.

Sanchez, D. A., Fegan, S. and Giebels, B. (2013). Evidence for a cosmological effect in γ-ray spectra of BL Lacertae, *A&A* **554**, A75, doi:10.1051/0004-6361/201220631.

Sánchez-Conde, M. A. and Prada, F. (2014). The flattening of the concentration — mass relation towards low halo masses and its implications for the annihilation signal boost, *MNRAS* **442**, 3, pp. 2271–2277, doi: 10.1093/mnras/stu1014.

Sánchez-Conde, M. A. *et al.* (2009). Hints of the existence of axionlike particles from the gamma-ray spectra of cosmological sources, *Phys. Rev. D* **79**, 12, p. 123511, doi:10.1103/PhysRevD.79.123511.

Sanchez-Conde, M.-A. *et al.* (2011). Dark matter searches with Cherenkov telescopes: Nearby dwarf galaxies or local galaxy clusters? *J. Cosmol. Astropart. Phys.* **1112**, p. 011, doi:10.1088/1475-7516/2011/12/011.

Sanchez-Conde, M. A. *et al.* (2013). The hunt for axionlike particles with the Cherenkov Telescope Array, arXiv e-prints.

Sanders, R. H. (2014). *The Dark Matter Problem.*

Santander, M. *et al.* (2017). Searching for VHE gamma-ray emission associated with IceCube astrophysical neutrinos using FACT, H.E.S.S., MAGIC, and VERITAS, **301**, p. 618, doi:10.22323/1.301.0618.

Sasaki, M. *et al.* (2018). Primordial black holes — perspectives in gravitational wave astronomy, *Class. Quant. Grav.* **35**, 6, p. 063001, doi:10.1088/1361-6382/aaa7b4.

Satalecka, K. *et al.* (2021). Searching for VHE gamma-ray emission associated with IceCube neutrino alerts using FACT, H.E.S.S., MAGIC, and VERITAS, in *Proceedings of the 37th International Cosmic Ray Conference — PoS(ICRC2021), Berlin (Germany)*, Vol. 395, p. 960, doi:10.22323/1.395.0960.

Sawala, T. *et al.* (2015). Bent by baryons: The low-mass galaxy-halo relation, *MNRAS* **448**, 3, pp. 2941–2947, doi:10.1093/mnras/stu2753.

Schaeffer, R., Declais, Y. and Jullian, S. (1987). The neutrino emission of sn1987a, *Nature* **330**, pp. 142–144, doi:10.1038/330142a0.

Schlickeiser, R. (1989). Cosmic-Ray transport and acceleration. I. Derivation of the kinetic equation and application to cosmic rays in static cold media, *ApJ* **336**, p. 243, doi:10.1086/167009.

Schoenfelder, V. (1991). The imaging gamma-ray telescope COMPTEL aboard GRO, *Adv. Space Res.* **11**, 8, pp. 313–322, doi:10.1016/0273-1177(91)90183-K.

Schoenfelder, V. *et al.* (1993). Instrument description and performance of the imaging gamma-ray telescope COMPTEL aboard the Compton gamma-ray observatory, *ApJS* **86**, p. 657, doi:10.1086/191794.

Schoonenberg, D. *et al.* (2016). Dark matter subhalos and unidentified sources in the Fermi 3FGL source catalog, *J. Cosmol. Astropart. Phys.* **5**, p. 028, doi:10.1088/1475-7516/2016/05/028.

Schroedter, M. *et al.* (2009). Search for Short Bursts of Gamma Rays Above 100 MeV from the Crab using VERITAS and SGARFACE, *ArXiv e-prints*.

Schumann, M. (2019). Direct detection of WIMP dark matter: Concepts and status, *J. Phys. G* **46**, 10, p. 103003, doi:10.1088/1361-6471/ab2ea5.

Schure, K. M. *et al.* (2010). Time-dependent particle acceleration in supernova remnants in different environments, *MNRAS* **406**, 4, pp. 2633–2649, doi: 10.1111/j.1365-2966.2010.16857.x.

Schwartz, D. A. (2004). The development and scientific impact of the Chandra X-ray observatory, *Inter. J. Modern Phys. D* **13**, 7, pp. 1239–1247, doi:10. 1142/S0218271804005377.

Schwinger, J. (1951). On gauge invariance and vacuum polarization, *Phys. Rev.* **82**, 5, pp. 664–679, doi:10.1103/PhysRev.82.664.

Schwinger, J. (1969). A magnetic model of matter, *Science* **165**, pp. 757–761, doi:10.1126/science.165.3895.757.

Sembroski, G. and Kertzman, M. P. (1991). The use of neural nets to determine the source directions of VHE gamma ray air showers detected by an atmospheric Cherenkov air Shower Array, *Inter. Cosmic Ray Conf.* **1**, p. 500.

Sharma, M. *et al.* (2017). Sensitivity estimate of the MACE gamma-ray telescope, *Nucl. Instrum. Meth. Phys. Res. A* **851**, pp. 125–131, doi:10.1016/j.nima. 2017.01.005.

Shayduk, M. *et al.* (2005). A new image cleaning method for the MAGIC Telescope, *Inter. Cosmic Ray Conf.* **5**, p. 223.

Shilon, I. *et al.* (2019). Application of Deep Learning methods to analysis of Imaging Atmospheric Cherenkov Telescopes data, *Astrop. Phys.* **105**, pp. 44–53, doi:10.1016/j.astropartphys.2018.10.003.

Shnir, Y. M. (2005). *Magnetic monopoles*, Text and Monographs in Physics, Springer, Berlin/Heidelberg, ISBN 9783540252771, 9783540290827, doi:10. 1007/3-540-29082-6.

Sikivie, P. (1983). Experimental tests of the 'invisible' axion, *Phys. Rev. Lett.* **51**, pp. 1415–1417, doi:10.1103/PhysRevLett.51.1415.

Sikivie, P. (2008). Axion Cosmology, in M. Kuster, G. Raffelt and B. Beltrán. eds., *Axions, Lecture Notes in Physics, Berlin Springer Verlag*, Vol. 741, p. 19.

Simet, M., Hooper, D. and Serpico, P. D. (2008). Milky Way as a kiloparsec-scale axionscope, *Phys. Rev. D* **77**, 6, p. 063001, doi:10.1103/PhysRevD.77. 063001.

Sironi, L. and Giannios, D. (2014). Relativistic pair beams from TeV blazars: A source of reprocessed GeV emission rather than intergalactic heating, *ApJ* **787**, p. 49, doi:10.1088/0004-637X/787/1/49.

Sitarek, J. *et al.* (2013). Analysis techniques and performance of the Domino Ring Sampler version 4 based readout for the MAGIC telescopes, *Nucl. Instrum. Meth. Phys. Res. A* **723**, pp. 109–120, doi:10.1016/j.nima.2013.05.014.

Sitarek, J. *et al.* (2018). Nature of the low-energy, γ-like background for the Cherenkov Telescope Array, *Astrop. Phys.* **97**, pp. 1–9, doi:10.1016/j.astropartphys.2017.10.005.

Sjostrand, T., Mrenna, S. and Skands, P. Z. (2006). PYTHIA 6.4 physics and manual, *J. High Energy Phys.* **05**, p. 026, doi:10.1088/1126-6708/2006/05/026.

Skilling, J. (1975). Cosmic ray streaming. I — effect of alfven waves on particles, *MNRAS* **172**, pp. 557–566.

Slane, P. (2017). *Pulsar Wind Nebulae*, Springer International Publishing AG, doi:10.1007/978-3-319-21846-5_95.

Smith, D. A. (2006). Review of the solar array telescopes, arXiv e-prints, astro-ph/0608251.

Snodin, A. P. *et al.* (2016). Global diffusion of cosmic rays in random magnetic fields, *MNRAS* **457**, pp. 3975–3987, doi:10.1093/mnras/stw217.

Sobczynska, D. (2007). Natural limit on the γ/hadron separation for a stand alone air Cherenkov telescope, *J. Phys. G Nucl. Phys.* **34**, pp. 2279–2288, doi:10.1088/0954-3899/34/11/005.

Sobczyńska, D. (2009a). Limits to the energy resolution of a single air Cherenkov telescope at low energies, *J. Phys. G* **36**, 4, p. 045201, doi:10.1088/0954-3899/36/4/045201.

Sobczyńska, D. (2009b). The background from single electromagnetic subcascades for a stereo system of air Cherenkov telescopes, *J. Phys. G* **36**, 12, p. 125201, doi:10.1088/0954-3899/36/12/125201.

Spengler, G. (2009). *Signatures of Ultrarelativistic Magnetic Monopoles in Imaging Cherenkov Telescopes*, Bsc, Humboldt-University of Berlin.

Spitkovsky, A. (2006). Time-dependent force-free pulsar magnetospheres: Axisymmetric and oblique rotators, *ApJ* **648**, p. L51, doi:10.1086/507518.

Spitler, L. G. *et al.* (2016). A repeating fast radio burst, *Nature* **531**, 7593 pp. 202–205, doi:10.1038/nature17168.

Springel, V. *et al.* (2008a). The aquarius project: The subhaloes of galactic haloes, *MNRAS* **391**, pp. 1685–1711, doi:10.1111/j.1365-2966.2008.14066.x.

Springel, V. *et al.* (2008b). The Aquarius project: The subhalos of galactic halos, *MNRAS* **391**, pp. 1685–1711, doi:10.1111/j.1365-2966.2008.14066.x.

Sridhar, S. and Goldreich, P. (1994). Toward a theory of interstellar turbulence. 1: Weak Alfvenic turbulence, *ApJ* **432**, pp. 612–621, doi:10.1086/174600.

Stadel, J. *et al.* (2009). Quantifying the heart of darkness with GHALO — a multibillion particle simulation of a galactic halo, *MNRAS* **398**, pp. L21–L25, doi:10.1111/j.1745-3933.2009.00699.x.

Stamatescu, V. *et al.* (2011). Timing analysis techniques at large core distances for multi-TeV gamma ray astronomy, *Astrop. Phys.* **34**, pp. 886–896, doi: 10.1016/j.astropartphys.2011.03.008.

Staszak, D. (2015). A cosmic-ray electron spectrum with VERITAS, in A. S. Borisov, V. G. Denisova, Z. M. Guseva, E. A. Kanevskaya, M. G. Kogan, A. E. Morozov, V. S. Puchkov, S. E. Pyatovsky, G. P. Shoziyoev, M. D. Smirnova, A. V. Vargasov, V. I. Galkin, S. I. Nazarov and R. A. Mukhamedshin. eds., *Proceedings of the 34th International Cosmic Ray Conference — PoS(ICRC2015), The Hague (Netherlands)*, Vol. 34, World Scientific, p. 411, doi:10.22323/1.236.0411.

Stawarz, Ł. and Petrosian, V. (2008). On the momentum diffusion of radiating ultrarelativistic electrons in a turbulent magnetic field, *ApJ* **681**, 2, pp. 1725–1744, doi:10.1086/588813.

Stecker, F. W., Baring, M. G. and Summerlin, E. J. (2007). Blazar gamma-rays, shock acceleration, and the extragalactic background light, *ApJL* **667**, pp. L29–L32, doi:10.1086/522005.

Stecker, F. W. and Scully, S. T. (2009). Is the Universe more transparent to gamma rays than previously thought? *ApJL* **691**, pp. L91–L94, doi:10.1088/0004-637X/691/2/L91.

Steigman, G., Dasgupta, B. and Beacom, J. F. (2012). Precise relic WIMP abundance and its impact on searches for dark matter annihilation, *Phys. Rev. D* **D86**, p. 023506, doi:10.1103/PhysRevD.86.023506.

Stein, R. *et al.* (2021). A tidal disruption event coincident with a high-energy neutrino, *Nature Astronomy* **5**, pp. 510–518, doi:10.1038/s41550-020-01295-8.

Strigari, L. E. (2013a). Galactic searches for dark matter, *Phys. Rep.* , p. 1–88.

Strigari,
L. E. (2013b). Galactic searches for dark matter, *Phys. Rep.* **531**, pp. 1–88, doi:10.1016/j.physrep.2013.05.004.

Strigari, L. E. (2018). Dark matter in dwarf spheroidal galaxies and indirect detection: A review, *Rept. Prog. Phys.* **81**, 5, p. 056901, doi:10.1088/1361-6633/aaae16.

Strigari, L. E. *et al.* (2007). Precise constraints on the dark matter content of MilkyWay dwarf galaxies for gamma-ray experiments, *Phys. Rev. D* **75**, 8, p. 083526, doi:10.1103/PhysRevD.75.083526.

Strigari, L. E. *et al.* (2008). A common mass scale for satellite galaxies of the Milky Way, *Nature* **454**, pp. 1096–1097.

Strong, A. W. and Moskalenko, I. V. (1998). Propagation of cosmic-ray nucleons in the galaxy, *ApJ* **509**, pp. 212–228, doi:10.1086/306470.

Strong, A. W., Moskalenko, I. V. and Ptuskin, V. S. (2007). Cosmic-ray propagation and interactions in the galaxy, *Ann. Rev. Nucl. Part. Sci.* **57**, pp. 285–327, doi:10.1146/annurev.nucl.57.090506.123011.

Strong, A. W. *et al.* (2004). The distribution of cosmic-ray sources in the galaxy, gamma-rays, and the gradient in the co-to-h2 relation, *A&A* **422**, pp. L47–L50,
doi:10.1051/0004-6361:20040172.

Su, M., Slatyer, T. R. and Finkbeiner, D. P. (2010). Giant gamma-ray bubbles from *Fermi*-LAT: AGN activity or bipolar Galactic wind? *ApJ* **724**, pp. 1044–1082, doi:10.1088/0004-637X/724/2/1044.

Sudoh, T., Linden, T. and Beacom, J. F. (2019). TeV halos are everywhere: Prospects for new discoveries, *Phys. Rev. D* **100**, 4, p. 043016, doi:10.1103/PhysRevD.100.043016.

Swanenburg, B. N. *et al.* (1981). Second COS-B catalogue of high-energy gamma-ray sources. *ApJL* **243**, pp. L69–L73, doi:10.1086/183445.

't Hooft, G. (1974). A planar diagram theory for strong interactions, *Nucl. Phys. B* **79**, doi:10.1016/0550-3213(74)90154-0.

Takata, J. *et al.* (2006). A two-dimensional electrodynamical outer gap model for γ-ray pulsars: γ-ray spectrum, *MNRAS* **366**, 4, pp. 1310–1328, doi:10.1111/j.1365-2966.2006.09904.x.

Tamborra, I., Ando, S. and Murase, K. (2014). Star-forming galaxies as the origin of diffuse high-energy backgrounds: Gamma-ray and neutrino connections, and implications for starburst history, *J. Cosmol. Astropart. Phys.* **9**, p. 043, doi:10.1088/1475-7516/2014/09/043.

Tamm, E. and Frank, M. (1937). *Dokl. Akad. Nauk SSSR (Akad. of Science of the USSR)* **14**, p. 107.

Tamura, N. *et al.* (2018). Prime Focus Spectrograph (PFS) for the Subaru telescope: Ongoing integration and future plans, in C. J. Evans, L. Simard and H. Takami. eds., *Ground-based and Airborne Instrumentation for Astronomy VII, Society of Photo-Optical Instrumentation Engineers (SPIE) Conference Series*, Vol. 10702, p. 107021C, doi:10.1117/12.2311871.

Tanabashi, M. *et al.* (2018). Review of particle physics, *Phys. Rev. D* **98**, p. 030001.

Tanabashi, M. *et al.* (2018). Review of particle physics, *Phys. Rev. D* **98**, p. 030001, doi:10.1103/PhysRevD.98.030001.

Tanaka, S. J. and Takahara, F. (2013). Properties of young pulsar wind nebulae: TeV detectability and pulsar properties, *MNRAS* **429**, 4, pp. 2945–2952, doi:10.1093/mnras/sts528.

Tanaka, T. *et al.* (2008). Study of nonthermal emission from SNR RX J1713.7-3946 with Suzaku, *ApJ* **685**, pp. 988–1004, doi:10.1086/591020.

Tavecchio, F. and Bonnoli, G. (2016). On the detectability of Lorentz invariance violation through anomalies in the multi-TeV γ-ray spectra of blazars, *A&A* **585**, p. A25, doi:10.1051/0004-6361/201526071.

Tavecchio, F. and Ghisellini, G. (2008). Spine-sheath layer radiative interplay in subparsec-scale jets and the TeV emission from M87, *MNRAS* **385**, 1, pp. L98–L102, doi:10.1111/j.1745-3933.2008.00441.x.

Tavecchio, F. and Ghisellini, G. (2012). "Flat" broad line region and gamma-ray absorption in blazars, arXiv e-prints, arXiv:1209.2291.

Tavecchio, F. and Ghisellini, G. (2014). On the spine-layer scenario for the very high-energy emission of NGC 1275, *MNRAS* **443**, 2, pp. 1224–1230, doi:10.1093/mnras/stu1196.

Tavecchio, F. *et al.* (2012). Evidence for an axion-like particle from PKS 1222+216? *Phys. Rev. D* **86**, p. 085036, doi:10.1103/PhysRevD.86.085036.

Tavernier, T. *et al.* (2019). Search for primordial black hole evaporation with H.E.S.S., in *Proceedings of the 36th International Cosmic Ray Conference — PoS(ICRC2019), Madison (USA)*, Vol. 358, p. 804, 10.22323/1.358.0804.

Tavernier, T. *et al.* (2021). Limits on primordial black hole evaporation from H.E.S.S. observations, in *Proceedings of the 37th International Cosmic Ray Conference — PoS(ICRC2021), Berlin (Germany)*, Vol. 395, p. 518, doi: 10.22323/1.395.0518.

Taylor, A. M., Vovk, I. and Neronov, A. (2011). Extragalactic magnetic fields constraints from simultaneous GeV-TeV observations of blazars, *A&A* **529**, p. A144, doi:10.1051/0004-6361/201116441.

Tchekhovskoy, A., Spitkovsky, A. and Li, J. G. (2013). Time-dependent 3d magnetohydrodynamic pulsar magnetospheres: Oblique rotators, *MNRAS* **435**, p. 1, doi:10.1093/mnrasl/slt076.

Tchekhovskoy, A., Spitkovsky, A. and Li, J. G. (2013). Time-dependent 3D magnetohydrodynamic pulsar magnetospheres: Oblique rotators. *MNRAS* **435**, pp. L1–L5, doi:10.1093/mnrasl/slt076.

Tepe, A. *et al.* (2012). HAWC — The high altitude water Cherenkov detector, doi:10.1088/1742-6596/375/1/052026.

Ter Haar, D. (1950). Cosmogonical problems and stellar energy, *Rev. Mod. Phys.* **22**, pp. 119–152, doi:10.1103/RevModPhys.22.119.

Terral, P. and Ferrière, K. (2017). Constraints from Faraday rotation on the magnetic field structure in the Galactic halo, *A&A* **600**, A29, doi:10.1051/0004-6361/201629572.

Tluczykont, M. and Fontaine, G. (2005). A second level trigger for H.E.S.S. phase 2, in *Proceedings of the Conference Towards a Network of Atmospheric Cherenkov Detectors VII*, Palaiseau, France, pp. 509–518.

Tluczykont, M. *et al.* (2014). The HiSCORE concept for gamma-ray and cosmic-ray astrophysics beyond 10 TeV, *Astrop. Phys.* **56**, pp. 42–53, doi:10.1016/j.astropartphys.2014.03.004.

Tollerud, E. J. *et al.* (2008). Hundreds of Milky Way satellites? Luminosity bias in the satellite luminosity function, *ApJ* **688**, pp. 277–289, doi:10.1086/592102.

Tollet, E. *et al.* (2016). Nihao — iv: Core creation and destruction in dark matter density profiles across cosmic time, *MNRAS* **456**, 4, pp. 3542–3552, doi: 10.1093/mnras/stv2856, URL http://dx.doi.org/10.1093/mnras/stv2856.

Tomassetti, N. (2015). Cosmic-ray protons, nuclei, electrons, and antiparticles under a two-halo scenario of diffusive propagation, *Phys. Rev. D* **92**, 8, p. 081301, doi:10.1103/PhysRevD.92.081301.

Tomozeiu, M., Mayer, L. and Quinn, T. (2016). Tidal stirring of satellites with shallow density profiles prevents them from being too big to fail, *ApJ* **827**, 1, p. L15, doi:10.3847/2041-8205/827/1/l15.

Tompkins, D. R. (1965). Total energy loss and Čerenkov emission from monopoles, *Phys. Rep.* **138**, pp. B248–B250.

Torres, D. F. (2018). Order parameters for the high-energy spectra of pulsars, *Nature Astronomy* **2**, pp. 247–256, doi:10.1038/s41550-018-0384-5.

Torres, D. F., Domingo-Santamaría, E. and Romero, G. E. (2004). High-energy gamma rays from stellar associations, *ApJL* **601**, pp. L75–L78, doi:10.1086/381803.

Torres, D. F. *et al.* (2017). A search for transitions between states in redbacks and black widows using seven years of *Fermi*-LAT observations, *ApJ* **836**, 1, p. 68, doi:10.3847/1538-4357/836/1/68.

Tramacere, A. *et al.* (2009). Swift observations of the very intense flaring activity of Mrk 421 during 2006. I. Phenomenological picture of electron acceleration and predictions for MeV/GeV emission, *A&A* **501**, 3, pp. 879–898, doi:10.1051/0004-6361/200810865.

Trotta, R. *et al.* (2011). Constraints on cosmic-ray propagation models from a global Bayesian analysis, *ApJ* **729**, p. 106, doi:10.1088/0004-637X/729/2/106.

Tsuboi, M., Handa, T. and Ukita, N. (1999). Dense molecular clouds in the Galactic Center region. I. observations and data, *ApJS* **120**, pp. 1–39, doi:10.1086/313165.

Tsuchiya, K. *et al.* (2004). Detection of sub-TeV gamma rays from the Galactic Center direction by CANGAROO-II, *ApJL* **606**, pp. L115–L118, doi:10.1086/421292.

Tulin, S. and Yu, H.-B. (2018). Dark Matter self-interactions and small scale structure, *Phys. Rep.* **730**, pp. 1–57, doi:10.1016/j.physrep.2017.11.004.

Uchiyama, Y. *et al.* (2007). Extremely fast acceleration of cosmic rays in a supernova remnant, *Nature* **449**, pp. 576–578, doi:10.1038/nature06210.

Ueno, K. *et al.* (2012). Search for GUT monopoles at Super–Kamiokande, *A&A* **36**, pp. 131–136, doi:10.1016/j.astropartphys.2012.05.008.

Ulrich, M. (1996). *Eine neue Bildanalysemethode für Abbildende Atmosphärische Cherenkov Teleskop e und die zweidimensionale Richtungsrekonstruktion von TeV-Photonen des Krebs-Nebels*, Ph.D. Thesis, Ruprecht-Karls-Universität, Heidelberg, Germany.

Urry, C. M. and Padovani, P. (1995). Unified schemes for radio-loud active galactic nuclei, *Pub. Astron. Soc. Pacific* **107**, p. 803, doi:10.1086/133630.

Vacanti, G. *et al.* (1994). Muon ring images with an atmospheric Čerenkov telescope, *Astrop. Phys.* **2**, pp. 1–11, doi:10.1016/0927-6505(94)90012-4.

Vachaspati, T. (2021). Progress on Cosmological Magnetic Fields, *Rep. Prog. Phys.* **84**, p. 7, doi:10.1088/1361-6633/ac03a9.

Vafin, S., Rafighi, I., Pohl, M. and Niemiec, J. (2018). The electrostatic instability for realistic pair distributions in blazar/EBL cascades, *ApJ* **857**, p. 43, doi:10.3847/1538-4357/aab552.

van der Swaluw, E. *et al.* (2001). Pulsar wind nebulae in supernova remnants. Spherically symmetric hydrodynamical simulations, *A&A* **380**, pp. 309–317, doi:10.1051/0004-6361:20011437.

Vannoni, G. *et al.* (2011). Acceleration and radiation of ultra-high energy protons in galaxy clusters, *A&A* **536**, p. A56, doi:10.1051/0004-6361/200913568.

Vasileiadis, G. *et al.* (2017). Raman LIDARs and atmospheric calibration for the Cherenkov telescope array, in *Proceedings of the 35th International Cosmic*

Ray Conference — PoS(ICRC2017), Busan (South Korea), Vol. 358, p. 814, doi:10.22323/1.358.0814.

Vasileiou, V. *et al.* (2013). Constraints on Lorentz invariance violation from Fermi-Large Area Telescope observations of gamma-ray bursts, *Phys. Rev. D* **87**, p. 122001, doi:10.1103/PhysRevD.87.122001, URL https://link.aps.org/doi/10.1103/PhysRevD.87.122001.

Vazza, F. *et al.* (2017). Simulations of extragalactic magnetic fields and of their observables, *Class. Quant. Gravi.* **34**, 23, p. 234001, doi:10.1088/1361-6382/aa8e60.

Vieyro, F. L. *et al.* (2017). Collective non-thermal emission from an extragalactic jet interacting with stars, *A&A* **604**, p. A57, doi:10.1051/0004-6361/201630333.

Vink, J. (2012). Supernova remnants: The X-ray perspective, *A&A Rev.* **20**, p. 49, doi:10.1007/s00159-011-0049-1.

Visinelli, L. and Gondolo, P. (2010). Axion cold dark matter in nonstandard cosmologies, *Phys. Rev. D* **81**, 6, p. 063508, doi:10.1103/PhysRevD.81.063508.

Vitagliano, E., Tamborra, I. and Raffelt, G. (2019). Grand unified neutrino spectrum at Earth.

Vitale, V., Morselli, A. and for the Fermi/LAT Collaboration (2009). Indirect Search for Dark Matter from the center of the Milky Way with the Fermi-Large Area Telescope, arXiv e-prints.

Voelk, H. J. and Forman, M. (1982). Cosmic rays and gamma-rays from OB stars, *ApJ* **253**, pp. 188–198, doi:10.1086/159623.

Voit, G. (2005). Tracing cosmic evolution with clusters of galaxies, *Rev. Mod. Phys.* **77**, pp. 207–258, doi:10.1103/RevModPhys.77.207.

Voit, G. M. (2005). Tracing cosmic evolution with clusters of galaxies, *Rev. Mod. Phys.* **77**, pp. 207–258, doi:10.1103/RevModPhys.77.207.

Volpi, D. *et al.* (2008). Non-thermal emission from relativistic MHD simulations of pulsar wind nebulae: From synchrotron to inverse Compton, *A&A* **485**, 2, pp. 337–349, doi:10.1051/0004-6361:200809424.

Vovk, I., Strzys, M. and Fruck, C. (2018). A spatial likelihood analysis for MAGIC telescope data, *A&A* **619**, p. A7, doi:10.1051/0004-6361/201833139.

Wagner, A. *et al.* (2010). Search for hidden sector photons with the ADMX detector, *Phys. Rev. Lett.* **105**, 17, p. 171801, doi:10.1103/PhysRevLett.105.171801.

Wakely, S. P. and Horan, D. (2007). TeVCat: An online catalog for very high energy gamma-ray astronomy, in *Proceedings of the 30th International Cosmic Ray Conference — PoS(ICRC2007), Merida (Mexico)*, Vol. 3, Universidad Nacional Autonoma de Mexico City, pp. 1341–1344.

Wang, J. *et al.* (2020). Universal structure of dark matter haloes over a mass range of 20 orders of magnitude, *Nature* **585**, 7823, pp. 39–42, doi:10.1038/s41586-020-2642-9.

Wantz, O. and Shellard, E. P. S. (2010). Axion cosmology revisited, *Phys. Rev. D* **82**, 12, p. 123508, doi:10.1103/PhysRevD.82.123508.

Waxman, E. and Bahcall, J. (1999). High-energy neutrinos from astrophysical sources: An upper bound, *Phys. Rev. D* **59**, 2, p. 023002, doi:10.1103/PhysRevD.59.023002.

Weekes, T. C. *et al.* (1989). Observation of TeV gamma rays from the Crab nebula using the atmospheric Cerenkov imaging technique, *ApJ* **342**, pp. 379–395, doi:10.1086/167599.

Weinberg, S. (1978). A new light boson? *Phys. Rev. Lett.* **40**, pp. 223–226, doi:10.1103/PhysRevLett.40.223.

Weinstein, A. (2007). The VERITAS trigger system, in *Proceedings of the 30th International Cosmic Ray Conference — PoS(ICRC2007), Merida (Mexico)*, Vol. 3, Universidad Nacional Autonoma de Mexico City, pp. 1539–1542.

Wentzel, D. G. (1974). Cosmic-ray propagation in the galaxy: Collective effects, *Ann. Rev. Astron. Astrophys.* **12**, pp. 71–96, doi:10.1146/annurev.aa.12.090174.000443.

Werner, M. *et al.* (2015). Spiral arms as cosmic ray source distributions, *Astrop. Phys.* **64**, pp. 18–33, doi:10.1016/j.astropartphys.2014.10.005.

Wesson, P. S. (1991). Olbers's paradox and the spectral intensity of the extragalactic background light, *ApJ* **367**, pp. 399–406, doi:10.1086/169638.

White, S. D. M., Frenk, C. S. and Davis, M. (1983). Clustering in a Neutrino Dominated Universe, *ApJL* **274**, pp. L1–L5, doi:10.1086/161425.

Wick, S. *et al.* (2003). Signatures for a cosmic flux of magnetic monopoles, *A&A* **18**, p. 663.

Wilczek, F. (1978). Problem of strong p and t invariance in the presence of instantons, *Phys. Rev. Lett.* **40**, pp. 279–282, doi:10.1103/PhysRevLett.40.279.

Wilczek, F. (1982). Axions and family symmetry breaking, *Phys. Rev. Lett.* **49**, pp. 1549–1552, doi:10.1103/PhysRevLett.49.1549.

Will, M. (2017). Atmospheric monitoring at the site of the MAGIC telescopes, in *Eur. Phys. J. Web Conf.*, Vol. 144, p. 01002, doi:10.1051/epjconf/201714401002.

Winston, R. (1970). Light collection within the framework of geometrical optics, *J. Opt. Soc. Am. (1917–1983)* **60**, pp. 245–247.

Witten, E. (1984). Some properties of O(32) superstrings, *Phys. Let. B* **149**, pp. 351–356, doi:10.1016/0370-2693(84)90422-2.

Wood, M. *et al.* (2008). A search for dark matter annihilation with the whipple 10m telescope, *ApJ* **678**, p. 594.

Wouters, D. and Brun, P. (2013). Constraints on axion-like particles from X-ray observations of the Hydra galaxy cluster, *ApJ* **772**, p. 44, doi:10.1088/0004-637X/772/1/44.

Xia, Z.-Q. *et al.* (2018). Searching for spectral oscillations due to photon-axionlike particle conversion using the Fermi-LAT observations of bright supernova remnants, *Phys. Rev. D* **97**, 6, p. 063003, doi:10.1103/PhysRevD.97.063003.

Xia, Z.-Q. *et al.* (2019). Searching for the possible signal of the photon-axionlike particle oscillation in the combined GeV and TeV spectra of supernova remnants, *Phys. Rev. D* **100**, 12, p. 123004, doi:10.1103/PhysRevD.100.123004.

Tanaka, Y. T., Buson, S. and K, D. (2017).

Yamaguchi, H., Sawada, M. and Bamba, A. (2010). Searching for diffuse non-thermal x-rays from the superbubbles n11 and n51d in the large magellanic cloud, *ApJ* **715**, 1, p. 412.

Yan, H. and Lazarian, A. (2002). Scattering of cosmic rays by magnetohydrodynamic interstellar turbulence, *Phys. Rev. Lett.* **89**, p. 281102, doi: 10.1103/PhysRevLett.89.281102.

Yan, H. and Lazarian, A. (2008). Cosmic-ray propagation: Nonlinear diffusion parallel and perpendicular to mean magnetic field, *ApJ* **673**, pp. 942–953, doi:10.1086/524771.

Yang, R., Aharonian, F. A. and Evoli, C. (2016). Radial distribution of the diffuse gamma-ray emissivity in the galactic disk, *Phys. Rev. D* **93**, p. 12, doi:10.1103/physrevd.93.123007.

Yang, R.-Z., de Oña Wilhelmi, E. and Aharonian, F. A. (2018). Diffuse gamma-ray emission in the vicinity of young star cluster Westerlund 2, *A&A* **611**, p. A77, doi:10.1051/0004-6361/201732045.

York, D. G. *et al.* (2000). The sloan digital sky survey: Technical summary, *ApJ* **120**, pp. 1579–1587, doi:10.1086/301513.

Young, S. and Byrnes, C. T. (2020). Initial clustering and the primordial black hole merger rate, *J. Cosmol. Astropart. Phys.* **2020**, 03, pp. 004–004, doi: 10.1088/1475-7516/2020/03/004.

Yuan, Q. and Feng, L. (2018). Dark Matter particle Explorer observations of high-energy cosmic ray electrons plus positrons and their physical implications, *Sci. China Phys. Mech. Astron.* **61**, 10, p. 101002, doi:10.1007/s11433-018-9226-y.

Yuan, Q. *et al.* (2017). Propagation of cosmic rays in the AMS-02 era, *Phys. Rev. D* **D95**, 8, p. 083007, doi:10.1103/PhysRevD.95.083007.

Zabalza, V. (2015). Naima: A python package for inference of particle distribution properties from nonthermal spectra, in *Proceedings of the 34th International Cosmic Ray Conference — PoS(ICRC2015), The Hague (Netherlands)*, Vol. 34, p. 922.

Zabalza, V. *et al.* (2013). Unraveling the high-energy emission components of gamma-ray binaries, *A&A* **551**, A17, doi:10.1051/0004-6361/201220589.

Zandanel, F. and Ando, S. (2014). Constraints on diffuse gamma-ray emission from structure formation processes in the Coma cluster, *MNRAS* **440**, pp. 663–671, doi:10.1093/mnras/stu324.

Zandanel, F., Pfrommer, C. and Prada, F. (2014). On the physics of radio haloes in galaxy clusters: Scaling relations and luminosity functions, *MNRAS* **438**, pp. 124–144, doi:10.1093/mnras/stt2250.

Zanin, R. *et al.* (2016). Gamma rays detected from Cygnus X-1 with likely jet origin, *A&A* **596**, A55, doi:10.1051/0004-6361/201628917.

Zavala, J. and Afshordi, N. (2016). Universal clustering of dark matter in phase space, *MNRAS* **457**, 1, pp. 986–992, doi:10.1093/mnras/stw048.

Zavattini, E. *et al.* (2006). Experimental observation of optical rotation generated in vacuum by a magnetic field, *Phys. Rev. Lett.* **96**, 11, p. 110406, doi:10.1103/PhysRevLett.96.110406.

Zavattini, E. *et al.* (2008). New PVLAS results and limits on magnetically induced optical rotation and ellipticity in vacuum, *Phys. Rev. D* **77**, 3, p. 032006, doi:10.1103/PhysRevD.77.032006.

Zechlin, H.-S. and Horns, D. (2012). Unidentified sources in the Fermi-LAT second source catalog: The case for DM subhalos, *J. Cosmol. Astropart. Phys.* **11**, p. 050, doi:10.1088/1475-7516/2012/11/050.

Zechlin, H.-S. *et al.* (2012). Dark matter subhaloes as gamma-ray sources and candidates in the first Fermi-LAT catalogue, *A&A* **538**, p. A93, doi:10.1051/0004-6361/201117655.

Zel'dovich, Y. B. and Novikov, I. (1967). The hypothesis of cores retarded during expansion and the hot cosmological model, *Sov. Astron.* **10**, p. 602.

Zemcov, M. *et al.* (2017). Measurement of the cosmic optical background using the long range reconnaissance imager on New Horizons, *Nat. Communi.* **8**, p. 15003, doi:10.1038/ncomms15003.

Zhang, C. *et al.* (2018). New bounds on axionlike particles from the Fermi Large Area Telescope observation of PKS 2155-304, *Phys. Rev. D* **97**, 6, p. 063009, doi:10.1103/PhysRevD.97.063009.

Zhao, H. (1996). Analytical models for galactic nuclei, *MNRAS* **278**, pp. 488–496, doi:10.1093/mnras/278.2.488.

Zioutas, K. *et al.* (2005). First results from the CERN axion solar telescope, *Phys. Rev. Lett.* **94**, 12, p. 121301, doi:10.1103/PhysRevLett.94.121301.

Zirakashvili, V. N. and Aharonian, F. A. (2010). Nonthermal radiation of young supernova remnants: The case of RX J1713.7-3946, *ApJ* **708**, pp. 965–980, doi:10.1088/0004-637X/708/2/965.

Zitzer, B. *et al.* (2013). The VERITAS upgraded telescope-level trigger systems: Technical details and performance characterization, arXiv e-prints, **1307**. URL http://adsabs.harvard.edu/abs/2013arXiv1307.8360Z.

Zwicky, F. (1933). *Helvetica Physica Acta* **6**, p. 110.

https://doi.org/10.1142/9789813275720_bmatter

Index

A

B

H

I

J

K

L

M

N

www.ingramcontent.com/pod-product-compliance
Lightning Source LLC
LaVergne TN
LVHW010326220926
848484LV00004BB/847

9789813275713